MATRIX THEORY AND FINITE MATHEMATICS

INTERNATIONAL SERIES IN
PURE AND APPLIED MATHEMATICS

William Ted Martin, E. H. Spanier,
G. Springer, and P. J. Davis
CONSULTING EDITORS

AHLFORS: Complex Analysis
BUCK: Advanced Calculus
BUSACKER AND SAATY: Finite Graphs and Networks
CHENEY: Introduction to Approximation Theory
CHESTER: Techniques in Partial Differential Equations
CODDINGTON AND LEVINSON: Theory of Ordinary Differential Equations
COHN: Conformal Mapping on Riemann Surfaces
CONTE AND DE BOOR: Elementary Numerical Analysis: An Algorithmic Approach
DENNEMEYER: Introduction to Partial Differential Equations and Boundary Value Problems
DETTMAN: Mathematical Methods in Physics and Engineering
EPSTEIN: Partial Differential Equations
GOLOMB AND SHANKS: Elements of Ordinary Differential Equations
GRAVES: The Theory of Functions of Real Variables
GREENSPAN: Introduction to Partial Differential Equations
GRIFFIN: Elementary Theory of Numbers
HAMMING: Numerical Methods for Scientists and Engineers
HILDEBRAND: Introduction to Numerical Analysis
HOUSEHOLDER: The Numerical Treatment of a Single Nonlinear Equation

KALMAN, FALB, AND ARBIB: Topics in Mathematical Systems Theory
LASS: Vector and Tensor Analysis
LEPAGE: Complex Variables and the Laplace Transform for Engineers
MCCARTY: Topology: An Introduction with Applications to Topological Groups
MONK: Introduction to Set Theory
MOORE: Elements of Linear Algebra and Matrix Theory
MOSTOW AND SAMPSON: Linear Algebra
MOURSUND AND DURIS: Elementary Theory and Application of Numerical Analysis
NEF: Linear Algebra
PEARL: Matrix Theory and Finite Mathematics
PIPES AND HARVILL: Applied Mathematics for Engineers and Physicists
RALSTON: A First Course in Numerical Analysis
RITGER AND ROSE: Differential Equations with Applications
RITT: Fourier Series
ROSSER: Logic for Mathematicians
RUDIN: Principles of Mathematical Analysis
SAATY AND BRAM: Nonlinear Mathematics
SAGAN: Introduction to the Calculus of Variations
SIMMONS: Differential Equations with Applications and Historical Notes
SIMMONS: Introduction to Topology and Modern Analysis
SNEDDON: Elements of Partial Differential Equations
SNEDDON: Fourier Transforms
STRUBLE: Nonlinear Differential Equations
WEINSTOCK: Calculus of Variations
WEISS: Algebraic Number Theory
ZEMANIAN: Distribution Theory and Transform Analysis

McGraw-Hill
Book
Company
New York
St. Louis
San Francisco
Düsseldorf
Johannesburg
Kuala Lumpur
London
Mexico
Montreal
New Delhi
Panama
Rio de Janeiro
Singapore
Sydney
Toronto

MARTIN PEARL
Professor of Mathematics
University of Maryland

Mathematician, National Bureau of Standards

Matrix Theory and Finite Mathematics

The Library·
Colby College-New Hampshire
New London, New Hampshire

This book was set in Times New Roman.
The editors were Jack L. Farnsworth and Norma Frankel;
the designer was Rolando Morales;
and the production supervisor was Thomas J. Lo Pinto.
The drawings were done by Reproduction Drawings Ltd.
The printer and binder was R. R. Donnelley & Sons Company.

QA
263
P34

**MATRIX THEORY AND
FINITE MATHEMATICS**

Copyright © 1973 by McGraw-Hill, Inc. All rights reserved. Printed in the United States of America. No part of this publication may be reproduced, stored in a retrieval system, or transmitted, in any form or by any means, electronic, mechanical, photocopying, recording, or otherwise, without the prior written permission of the publisher.

2 3 4 5 6 7 8 9 0 DODO 7 9 8 7 6 5 4 3

Library of Congress Cataloging in Publication Data

Pearl, Martin, 1928—
    Matrix theory and finite mathematics.

    (International series in pure and applied mathematics)
    1. Matrices. I. Title.
QA263.P34   512.9'43   72-6823
ISBN 0-07-049027-9

CONTENTS

	Preface	ix
0	**FIELDS**	**1**
0.1	Fields	1
1	**MATRICES**	**5**
1.1	Introduction and Notation	5
1.2	The Arithmetic of Matrices	8
1.3	Vector Spaces	16
1.4	The Row Space and the Column Space of a Matrix	30
1.5	The Steinitz Replacement Theorem	35
1.6	Nonsingular Matrices	47
1.7	The Elementary Operations	54
1.8	Permutations	67
1.9	Determinants	74
1.10	Linear Transformations and Invariant Spaces	95
1.11	Polynomials	114
1.12	The Decomposition Theorems	130
1.13	The Canonical Forms	156

2 THE THEORY OF GAMES — 168

- 2.1 Examples of Games — 168
- 2.2 Maxmin and Minmax — 189
- 2.3 A Mathematical Introduction to the Theory of Games — 196
- 2.4 Linear Inequalities — 202
- 2.5 The Fundamental Theorem — 213
- 2.6 Relations among Games — 218
- 2.7 Extreme Optimal Strategies — 229
- 2.8 Linear Programming — 240

3 FINITE MARKOV CHAINS — 254

- 3.1 Examples of Finite Markov Chains — 254
- 3.2 Infinite Processes — 273
- 3.3 A Mathematical Introduction to the Theory of Finite Markov Chains — 283
- 3.4 Periodic Transition Matrices — 301
- 3.5 Recurrence Probabilities — 313

4 THE THEORY OF GRAPHS — 332

- 4.1 Some Examples — 332
- 4.2 Undirected Graphs — 345
- 4.3 The Vector Space of Sets — 355
- 4.4 Matrices and Undirected Graphs — 360
- 4.5 Directed Graphs — 387
- 4.6 Matrices and Trees — 404

Answers to Selected Exercises — 423

Indexes — 443

    Subject Index

    Index of Symbols

PREFACE

The term "applied mathematics" stands for a conglomeration of ordinary and partial differential equations, Fourier series, Laplace transforms, and other similar topics from the domain of classical analysis. Until recently applied mathematics (without quotation marks) consisted almost exclusively of "applied mathematics" and numerical analysis. However, since the end of World War II two factors have combined to change the situation radically. Foremost is the advent of the computer. Whereas it had long been practically possible to solve a great many continuous, and hence infinite, problems when only hand computation was available, it usually happened that the amount of computation necessary to solve a meaningful finite problem was prohibitively large. At about the same time that the computer became an indispensable tool in applied mathematics, enormous contributions were being made to our knowledge of finite mathematics. It is interesting to note that of the vast quantity of mathematics developed in the last quarter of a century, virtually all the material that can be presented at the undergraduate level comes from the realm of finite mathematics.

One consequence of finite mathematics becoming an integral part of applied mathematics is the recent resurgence of interest in matrices, as a subject both for research and for undergraduate courses. The basic course in matrix theory was greatly changed after World War II by shifting the emphasis to finite-dimensional linear transformations. Today there is a growing feeling that both points of view are necessary. The aim of this book is to provide a unified treatment of matrix algebra coupled with an introduction to three of the most important modern applications of matrices.

Chapter 1 contains most of the material usually encountered in a one-semester undergraduate course in vectors and matrices. After much soul-searching I finally decided not to add a section dealing with the orthogonal and unitary reduction of normal matrices. I was also greatly tempted to include some discussion of the generalized inverse of a matrix. The sole reason for excluding both these topics is that they are not used elsewhere in the book. Thus, as things now stand, virtually all the material covered in Chap. 1 is needed for at least one of the three later chapters. It is interesting to note just how much matrix theory is required in the later chapters.

The vectors and matrices of Chap. 1 take their entries from an arbitrary field. In the application of Chap. 2 (The Theory of Games) this field is specialized to be the field of real numbers. When discussing Markov chains in Chap. 3 it is more convenient to use the field of complex numbers. In Chap. 4 (The Theory of Graphs) the theorems of Chap. 1 are applied both to matrices with real entries and to matrices whose entries are from the field with two elements, $\mathbb{Z}_2$. This provides an unusually striking example, especially to the reader who is not a professional mathematician, of the power and beauty of abstract mathematics.

After Chap. 1 the remaining chapters may be covered in any order. The following dependency diagram should prove useful:

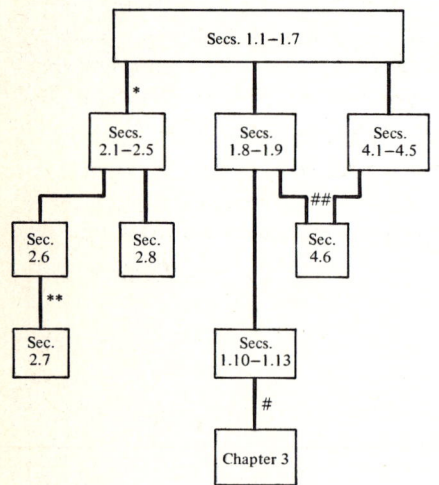

* Section 1.10 is used in the proof of Theorem 2.4.10 (Farkas' theorem) and in the preceding lemma. If Farkas' theorem is stated without proof, only Secs. 1.1 to 1.7 are needed for Chap. 2 (but see next note).

** Theorem 1.9.32 is required for Lemma 2.7.18 and Theorem 2.7.21. However, not much is lost by leaving them out.

\# The Jordan canonical form is used in Sec. 3.2 and is basic to all of Chap. 3.

\#\# Both Secs. 1.8 to 1.9 and Secs. 4.1 to 4.5 are needed for Sec. 4.6.

In order to achieve this flexibility it was occasionally necessary to repeat a definition or the discussion of a particular point. It seems a very small price to pay.

In the numerous times I have taught courses from the various preliminary versions of this book (at Imperial College of the University of London and at the University of Maryland) I have found that Chap. 1 requires slightly more than a semester's time. With the current trend of introducing more and more linear algebra into the standard calculus sequence it will undoubtedly soon be possible to cover the first chapter in a significantly shorter time. Another means for decreasing the time needed for Chap. 1 is to assign Secs. 1.8 and 1.11 as independent reading at the very beginning of the semester. Neither of these sections depends on any previous material in the course and the student can study them at his leisure.

Chapter 4 fits conveniently into a semester, but I have repeatedly found that somewhat more than a semester is needed to complete either Chap. 2 or Chap. 3. On the other hand, colleagues who have used my notes in their courses have been able to construct a very satisfactory semester course by taking roughly the first third of each of Chaps. 2, 3, and 4.

I was faced with a notational dilemma in Chap. 3. The standard matrix notation for a Markov chain calls for the probability vector to be written as a row vector so that the entry a_{ij} of the transition matrix is the probability of going from state S_i to state S_j. Unfortunately, this leads to the stationary vector being the characteristic vector of A^T. In addition, in Sec. 1.10, when a linear transformation acts on a vector, the corresponding matrix appears on the left and the vector is written as a column vector. For these reasons, I decided to express the probability vectors as column vectors and to have a_{ij} represent the probability of going from state S_j to state S_i.

Many of the results in Chaps. 2 to 4 are often obtained without relying on the machinery of matrix theory. My purpose here, however, is to demonstrate how matrices can be used and, therefore, whenever possible I have presented proofs employing matrices. On the other hand, I have also tried to provide the reader with a reference to an alternate approach.

I am deeply indebted to Professor George A. Barnard, who originally suggested that I teach a course at Imperial College along the lines of this book. At the same time I would like to thank Professor Walter Hayman of Imperial College, whose generous hospitality to me as a visiting lecturer in 1964–1965 provided me with the facilities for preparing a preliminary edition. Professors Richard Good, Howard Egan, and William Anderson taught several courses from subsequent preliminary editions and made a number of very valuable criticisms. Also, many ideas and suggestions came to me through numerous discussions with Dr. Alan Goldman, Jack Edmonds, Lambert Joel, and Joel Levy of the National Bureau of Standards. In addition, many kind students, both at Imperial College and at the University of Maryland, pointed out errors and commented on my lapses of various sorts. I would especially like to thank Betty Vanderslice for the care and patience she showed in typing endless versions of the manuscript. Finally, I wish to express my appreciation to the McGraw-Hill Book Company for several years of benign neglect.

MARTIN PEARL

MATRIX THEORY AND
FINITE MATHEMATICS

0
FIELDS

0.1 FIELDS

In Chap. 1 we will develop in great detail the basic facts concerning matrices whose entries are taken from a field. In the three subsequent chapters several quite different problems will be discussed that make heavy use of the theory of matrices. A striking feature of this treatment is that it will be appropriate to use a different field for each of the applications we study. Since we will not restrict ourselves to a specific field in Chap. 1, all the results we obtain there can be utilized equally well no matter which field is needed for a given application. Thus, by introducing the concept of an arbitrary field at this point, before launching into the study of matrices, we can develop at one time all the matrix tools which will be essential in later chapters.

Definition 0.1.1 A set $\mathbb{F}$, together with two operations denoted by $+$ and $\cdot$, called *addition* and *multiplication* respectively, is called a *field* provided the following 11 conditions hold:

   1 *Closure under addition:* $a + b$ belongs to $\mathbb{F}$ whenever a and b belong to $\mathbb{F}$.

2 *Associativity of addition:*
$$a + (b + c) = (a + b) + c$$
for all a, b, and c belonging to $\mathbb{F}$.

3 *Commutativity of addition:*
$$a + b = b + a$$
for all a and b belonging to $\mathbb{F}$.

4 *Additive identity:* there exists an element of $\mathbb{F}$, denoted by 0 and called *zero*, such that
$$a + 0 = 0 + a = a$$
for all a belonging to $\mathbb{F}$.

5 *Additive inverse:* For each element a of $\mathbb{F}$ there exists an element of $\mathbb{F}$, denoted by $-a$, such that
$$a + (-a) = (-a) + a = 0$$

6 *Closure under multiplication:* $a \cdot b$ belongs to $\mathbb{F}$ whenever a and b belong to $\mathbb{F}$.

7 *Associativity of multiplication:*
$$a \cdot (b \cdot c) = (a \cdot b) \cdot c$$
for all a, b, and c belonging to $\mathbb{F}$.

8 *Commutativity of multiplication:*
$$a \cdot b = b \cdot a$$
for all a and b belonging to $\mathbb{F}$.

9 *Multiplicative identity:* There exists an element of $\mathbb{F}$, distinct from 0, denoted by 1 and called *one*, such that
$$a \cdot 1 = 1 \cdot a = a$$
for all a belonging to $\mathbb{F}$.

10 *Multiplicative inverse:* For each element a of $\mathbb{F}$, other than 0, there exists an element of $\mathbb{F}$, denoted by a^{-1} and called the *inverse* of a, such that
$$a \cdot a^{-1} = a^{-1} \cdot a = 1$$

11 *The distributive laws:*
$$a \cdot (b + c) = a \cdot b + a \cdot c$$
$$(a + b) \cdot c = a \cdot c + b \cdot c$$
for all a, b, and c belonging to $\mathbb{F}$.

It is customary to write ab instead of $a \cdot b$. Also we occasionally write a/b instead of ab^{-1}.

Roughly speaking, a field is a set in which the four basic arithmetic operations (addition, subtraction, multiplication, and division by a nonzero element) can be performed in a "reasonable" way. Operations which have not been mentioned, e.g., finding a square root, may or may not be possible. Note that parts 4 and 9 of the definition require that a field contain at least two elements.

The most frequently encountered fields are:

1 The real numbers
2 The complex numbers
3 The rational numbers

Other examples of fields are given in the exercises below. We will see that one of them is used extensively in Chap. 4.

EXERCISES

1 Which of the following sets are fields (with the usual addition and multiplication)? For those sets which are not fields determine which of the 11 parts of Definition 0.1.1 are violated.
   (a) The numbers of the form $a + bi$, where a and b are rational numbers and $i = \sqrt{-1}$.
   (b) The numbers of the form $a + b\sqrt{2}$, where a and b are rational numbers.
   (c) The integers.
   (d) The set of all polynomials in x with real coefficients.
   (e) The set of all quotients of polynomials in x with real coefficients, $f(x)/g(x)$, $g(x) \neq 0$.
   (f) The subsets of the plane, with addition and multiplication defined to be union and intersection respectively.
2 Show that a field has exactly one additive identity and exactly one multiplicative identity.
3 Show that every element of a field has exactly one additive inverse.
4 Show that every element of a field other than 0 has exactly one multiplicative inverse. Show that 0 does not have a multiplicative inverse.
5 Prove that if a and b are elements of a field such that $ab = 0$ and $a \neq 0$, then $b = 0$.

6 Let $\mathbb{Z}$ denote the set of integers and let p be a prime in $\mathbb{Z}$. We partition $\mathbb{Z}$ into p subsets

$$0' = \{0, \quad p, \quad 2p, \ldots; \quad -p, \quad -2p, \ldots\}$$
$$1' = \{1, 1 + p, 1 + 2p, \ldots; \quad 1 - p, 1 - 2p, \ldots\}$$
$$2' = \{2, 2 + p, 2 + 2p, \ldots; \quad 2 - p, 2 - 2p, \ldots\}$$
$$\cdots\cdots\cdots\cdots\cdots\cdots\cdots\cdots\cdots\cdots\cdots\cdots\cdots$$
$$(p - 1)' = \{-1 + p, -1 + 2p, -1 + 3p, \ldots; \quad -1 - p, -1 - 2p, \ldots\}$$

that is, x' is the set of all integers which when divided by p leave a remainder of x. The set consisting of the p elements $0', 1', 2', \ldots, (p - 1)'$ is denoted by $\mathbb{Z}_p$. Show that if a and c are integers which belong to the same element of $\mathbb{Z}_p$ and b and d are integers which belong to the same element of $\mathbb{Z}_p$, then $a + b$ and $c + d$ belong to the same element of $\mathbb{Z}_p$ and ab and cd belong to the same element of $\mathbb{Z}_p$.

We define the operations of addition and multiplication in $\mathbb{Z}_p$ as follows: $a' + b'$ is defined to be the element of $\mathbb{Z}_p$ which contains $a + b$, and $a' \cdot b'$ is defined to be the element of $\mathbb{Z}_p$ which contains ab. Show that with these definitions $\mathbb{Z}_p$ is a field. [*Hint:* Show that if a is any integer which does not belong to $0'$, then no two of the integers $a, 2a, 3a, \ldots, (p - 1)a, pa$ belong to the same element of $\mathbb{Z}_p$. Where is the hypothesis that p is a prime used?] Show that if p is not a prime, then $\mathbb{Z}_p$ is not a field.

Write out the addition and multiplication tables for $\mathbb{Z}_2, \mathbb{Z}_3, \mathbb{Z}_5,$ and $\mathbb{Z}_6$.

1
MATRICES

1.1 INTRODUCTION AND NOTATION

Several quite different topics will be treated in these pages, but throughout there will be a single unifying concept—that of a matrix. This chapter is devoted to matrices themselves.[1] In the subsequent chapters we will introduce a few of the many subjects to which the material of this chapter can be applied, namely, zero-sum two-person games, linear programming, finite Markov chains, and directed and undirected graphs. Perhaps the most widely known and historically most important use for matrices is as a device for solving a system of simultaneous linear equations. A brief treatment of this problem is contained in Sec. 1.9.

[1] During the course of this chapter two digressions will be necessary. The first presents just enough information about finite permutations to enable us to define the determinant of a matrix. This material will be found in the first part of Sec. 1.8, ending with Definition 1.8.11. The second digression, devoted to the greatest common divisor and the least common multiple of polynomials in one variable, occupies all of Sec. 1.11. Since the discussion of each of these topics is self-contained and independent of the remainder of the text, a good deal of time can be saved if the reader is already familiar with the contents of these sections by the time they arise.

We will begin by defining a matrix with entries from a field $\mathbb{F}$ and continue with a study, in some depth, of the general theory of matrices with entries from $\mathbb{F}$. Although the treatment of the material throughout this chapter, including the proofs in the exercises, is in terms of an unspecified field $\mathbb{F}$ whose elements are referred to as *numbers* or *scalars*, readers who are unfamiliar with the concept of a field will lose very little (at least until Chap. 4) by always interpreting $\mathbb{F}$ as being either the set of all real numbers or the set of all complex numbers. However, in all exercises and examples of a numerical nature in this chapter, $\mathbb{F}$ is assumed to be the field of real numbers.

Definition 1.1.1 Let m and n be positive integers. A rectangular array of $m \cdot n$ numbers belonging to $\mathbb{F}$ arranged in m rows and n columns is called an $m \times n$ *matrix* with *entries* from $\mathbb{F}$. We call m and n the *row* and *column dimensions* of the matrix.

Matrices will be denoted by capital letters of the Latin alphabet and will be displayed in several ways. The simplest situation occurs when all the entries of the matrix are known. In such a case we write down the matrix explicitly. For example,

$$A = \begin{bmatrix} 1 & 3 & -4 \\ 0 & -6 & 2 \end{bmatrix}$$

is a 2×3 matrix all of whose entries are specified.

Often, especially when proving theorems about matrices, we do not wish (or are unable) to state exactly which number is in each position of the matrix. Indeed, we may even not wish to specify the dimensions. In such cases we will discuss an $m \times n$ matrix A and write

$$A = \begin{bmatrix} a_{11} & a_{12} & \cdots & a_{1n} \\ a_{21} & a_{22} & \cdots & a_{2n} \\ \cdots\cdots\cdots\cdots\cdots\cdots\cdots \\ a_{m1} & a_{m2} & \cdots & a_{mn} \end{bmatrix} \quad (1.1.2)$$

Elements of $\mathbb{F}$ (in particular, entries of a matrix) will be denoted either by lowercase Latin or lowercase Greek letters. Thus a_{21} stands for the number in the second row and the first column of A. When it is not necessary to illustrate all the entries of A, we will use the more condensed notation

$$A = [a_{ij}] \quad \begin{array}{l} i = 1, 2, \ldots, m \\ j = 1, 2, \ldots, n \end{array}$$

Either way we have used a *double-subscript notation*. The first subscript, i, of a_{ij} denotes the row in which the entry a_{ij} is found. The second subscript, j, denotes the column. Thus we say that a_{ij} is the *entry in the (i, j) position* of A. In order to characterize a matrix, it is necessary to give not only the numbers of which it is composed but also the positions within the matrix in which the numbers are located.

There is another important way of illustrating a matrix. A matrix can be *partitioned* into matrices of smaller dimensions. For example, the matrix A of (1.1.2) can be expressed as

$$A = \begin{bmatrix} A_{11} & A_{12} \\ A_{21} & A_{22} \end{bmatrix}$$

where

$$A_{11} = \begin{bmatrix} a_{11} & a_{12} & \cdots & a_{1t} \\ a_{21} & a_{22} & \cdots & a_{2t} \\ \cdots & \cdots & \cdots & \cdots \\ a_{s1} & a_{s2} & \cdots & a_{st} \end{bmatrix} \quad A_{12} = \begin{bmatrix} a_{1,t+1} & a_{1,t+2} & \cdots & a_{1n} \\ a_{2,t+1} & a_{2,t+2} & \cdots & a_{2n} \\ \cdots & \cdots & \cdots & \cdots \\ a_{s,t+1} & a_{s,t+2} & \cdots & a_{sn} \end{bmatrix}$$

$$A_{21} = \begin{bmatrix} a_{s+1,1} & a_{s+1,2} & \cdots & a_{s+1,t} \\ a_{s+2,1} & a_{s+2,2} & \cdots & a_{s+2,t} \\ \cdots & \cdots & \cdots & \cdots \\ a_{m1} & a_{m2} & \cdots & a_{mt} \end{bmatrix}$$

$$A_{22} = \begin{bmatrix} a_{s+1,t+1} & a_{s+1,t+2} & \cdots & a_{s+1,n} \\ a_{s+2,t+1} & a_{s+2,t+2} & \cdots & a_{s+2,n} \\ \cdots & \cdots & \cdots & \cdots \\ a_{m,t+1} & a_{m,t+2} & \cdots & a_{mn} \end{bmatrix}$$

Again we have used a double-subscript notation, this time for the *submatrices* into which the original matrix A was partitioned. We say that A_{ij} is the submatrix in the (i, j) *block position* of A (relative to this partition of A).

Definition 1.1.3 Two matrices are *equal* only if they are identical, i.e., if they have the same dimensions and if the entries in corresponding positions are equal. For

$$A = [a_{ij}] \quad \begin{matrix} i = 1, 2, \ldots, m \\ j = 1, 2, \ldots, n \end{matrix}$$

$$B = [b_{kl}] \quad \begin{matrix} k = 1, 2, \ldots, p \\ l = 1, 2, \ldots, q \end{matrix}$$

we understand $A = B$ to mean that:

1 $m = p$ and $n = q$

2 $a_{ij} = b_{ij}$ $\quad \begin{matrix} i = 1, 2, \ldots, m \\ j = 1, 2, \ldots, n \end{matrix}$

There are two kinds of matrices which have special roles in the general theory, and we use them often enough to warrant giving them special names.

Definition 1.1.4 Let A be an $m \times n$ matrix. If $m = n$, then A is said to be a *square matrix* of *order n*.

An $n \times 1$ matrix is called an *n-dimensional column vector*. A $1 \times n$ matrix is called an *n-dimensional row vector*.

Thus, in the context of this book, a vector is a particular type of matrix.

A matrix (and, in particular, a vector) all of whose entries are the zero element of $\mathbb{F}$ is called a *zero matrix* (or a *zero vector*) and is denoted by 0, irrespective of its dimensions. The dimensions of a zero matrix will usually be evident from the context. When this is not the case, we will specify the dimensions.

1.2 THE ARITHMETIC OF MATRICES

In the last section we defined a matrix and introduced a simple and efficient notation. Now we define the three basic arithmetic operations performed on matrices and present some of their elementary properties. These operations are:

1 Addition of matrices
2 Multiplication of a matrix by a scalar
3 Multiplication of matrices

Definition 1.2.1 Addition of matrices Let A and B be two $m \times n$ matrices,

$$A = [a_{ij}] \qquad B = [b_{ij}] \qquad \begin{array}{l} i = 1, 2, \ldots, m \\ j = 1, 2, \ldots, n \end{array}$$

We define $A + B$ to be the $m \times n$ matrix which has $a_{ij} + b_{ij}$ in the (i, j) position.

Corollary 1.2.2 Let A, B, and C be $m \times n$ matrices. Then

1 $A + (B + C) = (A + B) + C$

i.e., addition is defined and is *associative*,

2 $A + B = B + A$

i.e., addition is *commutative*.

PROOF (1) Since A, B, and C are $m \times n$ matrices, $B + C$ and $A + B$ are both defined by Definition 1.2.1 and are also $m \times n$ matrices. In the same way, $A + (B + C)$ and $(A + B) + C$ are both defined by Definition 1.2.1 and are $m \times n$ matrices.

In order to show that $A + (B + C) = (A + B) + C$ we must show that for any choice of i and j ($1 \leq i \leq m$, $1 \leq j \leq n$) the entry in the (i, j) position of $A + (B + C)$ is equal to the entry in the (i, j) position of $(A + B) + C$. For each choice of i and j, the entry in the (i, j) position of $A + (B + C)$ is $a_{ij} + (b_{ij} + c_{ij})$, where a_{ij}, b_{ij}, and c_{ij} are the entries in the (i, j) positions of A, B and C, respectively. Similarly, the entry in the (i, j) position of $(A + B) + C$ is $(a_{ij} + b_{ij}) + c_{ij}$. Since addition is associative in $\mathbb{F}$, both $A + (B + C)$ and $(A + B) + C$ have $a_{ij} + b_{ij} + c_{ij}$ in the (i, j) position. Hence $A + (B + C) = (A + B) + C$.

(2) Left as an exercise. ////

Since the addition of matrices is associative whenever it is defined, we may omit all the parentheses and write $A + B + C$ instead of $A + (B + C)$ or $(A + B) + C$. It follows from repeated applications of Definition 1.2.1 and the part 1 of Corollary 1.2.2 that the entry in the (i, j) position of any finite sum of $m \times n$ matrices is the sum of the entries in the (i, j) positions of the matrices. On the other hand, if not all matrices of the collection have the same dimensions, their sum is not defined.

Definition 1.2.3 Multiplication of a matrix by a scalar Let A be an $m \times n$ matrix, $A = [a_{ij}]$, and let α be an element of $\mathbb{F}$. Then αA and $A\alpha$ are both defined to be the $m \times n$ matrix which has αa_{ij} in the (i, j) position.

Thus multiplication of a matrix by a scalar is defined for all matrices A and all elements α of $\mathbb{F}$. Observe that multiplication of a matrix by a scalar does not change the dimensions of the matrix.

According to Definition 1.2.3, we have $1A = A$ for every matrix A and $0A = 0$, where the first zero is an element of $\mathbb{F}$ and the second zero is the zero matrix whose dimensions are the same as those of A.

Definition 1.2.4 Multiplication of matrices Let A be an $m \times n$ matrix and B be an $n \times p$ matrix

$$A = [a_{ij}] \quad B = [b_{jk}] \quad \begin{aligned} i &= 1, 2, \ldots, m \\ j &= 1, 2, \ldots, n \\ k &= 1, 2, \ldots, p \end{aligned}$$

AB is defined to be the $m \times p$ matrix which has

$$\sum_{j=1}^{n} a_{ij} b_{jk} = a_{i1} b_{1k} + a_{i2} b_{2k} + \cdots + a_{in} b_{nk}$$

in the (i, k) position.

Definition 1.2.4 requires that the number of columns of A be equal to the number of rows of B. If the number of columns of A is not equal to the number of rows of B, the product AB is not defined. Thus, for a pair of matrices A, B, it is possible that none, one, or both of the products AB and BA are defined. The product $A^2 = AA$ is defined only for those matrices A for which $m = n$, that is, only for square matrices. Then A^2 is also a square matrix and has the same order as A.

Considerable care must be exercised when operating with matrices. As we have seen (e.g., in Corollary 1.2.2), many of the common properties of arithmetic operations remain valid when applied to matrices of the proper dimensions. However, there are numerous exceptions, especially in dealing with the multiplication of matrices. Some of these are illustrated in the following example.

EXAMPLE 1.2.5 Let

$$A = \begin{bmatrix} 1 & 2 \\ 2 & 4 \end{bmatrix} \quad B = \begin{bmatrix} -2 & 2 \\ 1 & -1 \end{bmatrix} \quad C = \begin{bmatrix} 1 \\ 1 \end{bmatrix} \quad D = \begin{bmatrix} -1 \\ 2 \end{bmatrix} \quad \alpha = 4$$

Then

1. $A + B = \begin{bmatrix} -1 & 4 \\ 3 & 3 \end{bmatrix} \quad C + D = \begin{bmatrix} 0 \\ 3 \end{bmatrix}$

2. $\alpha A = \begin{bmatrix} 4 & 8 \\ 8 & 16 \end{bmatrix} \quad \alpha C = \begin{bmatrix} 4 \\ 4 \end{bmatrix}$

3. $AB = 0 \quad BA = \begin{bmatrix} 2 & 4 \\ -1 & -2 \end{bmatrix} \quad AC = \begin{bmatrix} 3 \\ 6 \end{bmatrix} = AD$

4. $A + C$ and CA are not defined   ////

Example 1.2.5 demonstrates several peculiarities of matrix multiplication. Although $A \neq 0$, $B \neq 0$, a simple calculation shows that $AB = 0$. Also, $AC = AD$, $A \neq 0$, and $C \neq D$. Thus the usual cancellation is generally not permitted. Later we will determine under what circumstances cancellation can be performed. Notice also that although AB and BA are both defined and have the same dimensions, they are not equal.

Let A, B, and C be $m \times n$, $p \times q$, and $s \times t$ matrices respectively. According to Definition 1.2.4, $A(BC)$ is defined if and only if both the conditions $n = p$ and $q = s$ are satisfied, and then $A(BC)$ is an $m \times t$ matrix. Exactly the same two conditions must be satisfied in order that $(AB)C$ be defined, and then $(AB)C$ is also an $m \times t$ matrix. Thus, if one of the products $A(BC)$ and $(AB)C$ is defined, both are defined and they have the same dimensions. In fact, we will show that the two products of A, B, and C, taken in that order, are equal.

Corollary 1.2.6 Let

$$A = [a_{ij}] \quad B = [b_{jk}] \quad C = [c_{kl}] \quad \begin{aligned} i &= 1, 1, \ldots, m \\ j &= 1, 2, \ldots, n \\ k &= 1, 2, \ldots, p \\ l &= 1, 2, \ldots, q \end{aligned}$$

be $m \times n$, $n \times p$, $p \times q$ matrices respectively. Then

$$A(BC) = (AB)C$$

i.e., multiplication of matrices is defined and *associative*.

PROOF To prove the corollary we must show that for each choice of i and l ($1 \leq i \leq m$, $1 \leq l \leq q$), the entry in the (i, l) position of $A(BC)$ is equal to the entry in the (i, l) position of $(AB)C$. Set

$$D = AB = [d_{ik}] \quad \begin{aligned} i &= 1, 2, \ldots, m \\ k &= 1, 2, \ldots, p \end{aligned}$$

$$F = BC = [f_{jl}] \quad \begin{aligned} j &= 1, 2, \ldots, n \\ l &= 1, 2, \ldots, q \end{aligned}$$

According to Definition 1.2.4,

$$d_{ik} = \sum_{j=1}^{n} a_{ij} b_{jk} \quad f_{jl} = \sum_{k=1}^{n} b_{jk} c_{kl}$$

The entry in the (i, l) position of $A(BC) = AF$ is

$$\sum_{j=1}^{n} a_{ij} f_{jl} = \sum_{j=1}^{n} a_{ij} \left(\sum_{k=1}^{p} b_{jk} c_{kl} \right) \quad (1.2.7)$$

and the entry in the (i, l) position of $(AB)C = DC$ is

$$\sum_{k=1}^{p} d_{ik} c_{kl} = \sum_{k=1}^{p} \left(\sum_{j=1}^{n} a_{ij} b_{jk} \right) c_{kl} \quad (1.2.8)$$

By expanding the right-hand sides of equations (1.2.7) and (1.2.8) and rearranging the terms it is easily seen that they are equal. Each expression consists of all terms of the form

$$a_{ij} b_{jk} c_{kl} \quad (1.2.9)$$

where i and l are fixed and are determined by the position we are considering, i.e., the (i, l) position, and where j takes on the values $1, 2, \ldots, n$ and k takes on the values $1, 2, \ldots, p$. There are the same np terms of the form (1.2.9) in each of the sums (1.2.7) and (1.2.8). Thus, each entry of $A(BC)$ is equal to the entry of $(AB)C$ in the same position, and consequently $A(BC) = (AB)C$. ////

Since matrix multiplication is always associative when it is defined, we may omit the parentheses and write ABC instead of $A(BC)$ or $(AB)C$.

At first glance Definition 1.2.4 may appear rather arbitrary and artificial. However, the following example shows one way in which this definition arises quite naturally. Let us suppose that $x_1, x_2, \ldots, x_p$ and $y_1, y_2, \ldots, y_n$ are related by the system of equations

$$\begin{aligned} y_1 &= b_{11} x_1 + b_{12} x_2 + \cdots + b_{1p} x_p \\ y_2 &= b_{21} x_1 + b_{22} x_2 + \cdots + b_{2p} x_p \\ &\cdots \cdots \cdots \cdots \cdots \cdots \cdots \cdots \cdots \cdots \\ y_n &= b_{n1} x_1 + b_{n2} x_2 + \cdots + b_{np} x_p \end{aligned} \quad (1.2.10)$$

Suppose further that $y_1, y_2, \ldots, y_n$ and $z_1, z_2, \ldots, z_m$ are related by the equations

$$\begin{aligned} z_1 &= a_{11} y_1 + a_{12} y_2 + \cdots + a_{1n} y_n \\ z_2 &= a_{21} y_1 + a_{22} y_2 + \cdots + a_{2n} y_n \\ &\cdots \cdots \cdots \cdots \cdots \cdots \cdots \cdots \cdots \cdots \\ z_m &= a_{m1} y_1 + a_{m2} y_2 + \cdots + a_{mn} y_n \end{aligned} \quad (1.2.11)$$

We can eliminate the y's completely by substituting equations (1.2.10) into equations (1.2.11). This expresses each of the z's in terms of the x's. Carrying out this computation, we get

$$\begin{aligned} z_1 &= c_{11} x_1 + c_{12} x_2 + \cdots + c_{1p} x_p \\ z_2 &= c_{21} x_1 + c_{22} x_2 + \cdots + c_{2p} x_p \\ &\cdots \cdots \cdots \cdots \cdots \cdots \cdots \cdots \cdots \cdots \\ z_m &= c_{m1} x_1 + c_{m2} x_2 + \cdots + c_{mp} x_p \end{aligned} \quad (1.2.12)$$

where

$$c_{ik} = a_{i1} b_{1k} + a_{i2} b_{2k} + \cdots + a_{in} b_{nk} = \sum_{j=1}^{n} a_{ij} a_{jk} \quad \begin{array}{l} i = 1, 2, \ldots, m \\ k = 1, 2, \ldots, p \end{array} \quad (1.2.13)$$

Let us form matrices from the coefficients in the systems of equations (1.2.10), (1.2.11), and (1.2.12). Set

$$A = [a_{ij}] \quad B = [b_{jk}] \quad C = [c_{ik}] \quad \begin{matrix} i = 1, 2, \ldots, m \\ j = 1, 2, \ldots, n \\ k = 1, 2, \ldots, p \end{matrix}$$

Then A, B, C are $m \times n$, $n \times p$, $m \times p$ matrices respectively. The adoption of Definition 1.2.4 allows us to write the systems of equations (1.2.10) to (1.2.12) in a very simple and compact way:

$$Y = BX \quad Z = AY \quad Z = CX \quad (1.2.14)$$

where X, Y, and Z are the column vectors

$$X = \begin{bmatrix} x_1 \\ x_2 \\ \vdots \\ x_p \end{bmatrix} \quad Y = \begin{bmatrix} y_1 \\ y_2 \\ \vdots \\ y_n \end{bmatrix} \quad Z = \begin{bmatrix} z_1 \\ z_2 \\ \vdots \\ z_m \end{bmatrix}$$

Substitution of the first equation of (1.2.14) into the second equation of (1.2.14) yields

$$Z = A(BX) \quad (1.2.15)$$

Then Definition 1.2.4 and equation (1.2.13) permit us to write $C = AB$, and hence the equations of (1.2.12) can be expressed as

$$Z = (AB)X$$

Throughout this section no use has been made of the fact that division is possible within $\mathbb{F}$. Later (specifically in Secs. 1.10, 1.12, and 1.13) we will need to consider matrices whose entries come from a set having all the properties of a field except those related to division.[1] It is clear that all the results of this section remain valid for such matrices.

EXERCISES

1 Find all 2×2 matrices which commute under the operation of multiplication with the matrix

(a) $\begin{bmatrix} 1 & 2 \\ 0 & 1 \end{bmatrix}$ \qquad (b) $\begin{bmatrix} 1 & 2 \\ 0 & -1 \end{bmatrix}$

---

[1] Such a system is called a *ring* in modern algebra. Division by a nonzero number in a field is equivalent to part 10 of Definition 0.1.1.

2 Find all 3 × 3 matrices which commute under the operation of multiplication with the matrix

$$\begin{bmatrix} 1 & 0 & 1 \\ 0 & 1 & -2 \\ 0 & 0 & 2 \end{bmatrix}$$

3 Let

$$A = \begin{bmatrix} 1 & 2 \\ -3 & 6 \end{bmatrix}$$

Find
(a) All 2 × 2 matrices X such that $AX = 0$.
(b) All 2 × 2 matrices X such that $XA = 0$.
(c) All 2 × 2 matrices X such that $AX = A$.
(d) All 2 × 2 matrices X such that $XA = A$.
(e) All 2 × 2 matrices X such that $XA = \begin{bmatrix} 1 & 0 \\ 0 & 1 \end{bmatrix}$.

4 Repeat Exercise 3 using the matrix $\begin{bmatrix} 1 & 2 \\ 3 & 6 \end{bmatrix}$ for A.
5 Find all 2 × 2 matrices X such that $X^2 = 0$.
6 Verify the computation leading to (1.2.13).
7 Prove part 2 of Corollary 1.2.2.
8 Let α, β be scalars, A, B be $m \times n$ matrices, and C be an $n \times p$ matrix. Prove each of the following:

$$(\alpha + \beta)A = \alpha A + \beta A \quad (1.2.16a)$$
$$\alpha(A + B) = \alpha A + \alpha B \quad (1.2.16b)$$
$$\alpha(AC) = (\alpha A)C = A(\alpha C) \quad (1.2.16c)$$
$$\alpha(\beta A) = (\alpha \beta)A \quad (1.2.16d)$$

9 Let A be an $m \times n$ matrix and B and C be $n \times p$ matrices. Prove that

$$A(B + C) = AB + AC \quad (1.2.17)$$

i.e., all the indicated operations are defined, and matrix multiplication is *left-distributive* with respect to matrix addition.

10 State and prove the right-distributive law of matrix multiplication with respect to matrix addition.
11 Let $A = [a_{ij}]$ and $B = [b_{jk}]$ be $m \times n$ and $n \times p$ matrices respectively and let $AB = C = [c_{ik}]$. We denote the *i*th *row* of A by $A_{(i)}$ and the *k*th *column* of B by $B^{(k)}$, that is,

$$A_{(i)} = [a_{i1} \quad a_{i2} \quad \cdots \quad a_{in}] \qquad B^{(k)} = \begin{bmatrix} b_{1k} \\ b_{2k} \\ \vdots \\ b_{nk} \end{bmatrix}$$

Prove:

$$c_{ik} = A_{(i)}B^{(k)} \qquad (1.2.18a)$$
$$C_{(i)} = a_{i1}B_{(1)} + a_{i2}B_{(2)} + \cdots + a_{in}B_{(n)} \qquad (1.2.18b)$$
$$C^{(k)} = A^{(1)}b_{1k} + A^{(2)}b_{2k} + \cdots + A^{(n)}b_{nk} \qquad (1.2.18c)$$

12   Show that the set of matrices of the form

$$\begin{bmatrix} a & b \\ -b & a \end{bmatrix}$$

where a and b are real numbers, form a field. Show that if a and b are permitted to be complex numbers, this set of matrices is not a field.

13   Let $A = [a_{ij}]$ be a square matrix of order n. A is called a *diagonal matrix* if $a_{ij} = 0$ whenever $i \neq j$. Show that multiplication of diagonal matrices of order n is commutative.

14   A diagonal matrix $A = [a_{ij}]$ of order n is called a *scalar matrix* if $a_{11} = a_{22} = \cdots = a_{nn}$. Show that a scalar matrix of order n commutes with every matrix of order n under multiplication.

15   The scalar matrix $A = [a_{ij}]$ of order n in which $a_{11} = a_{22} = \cdots = a_{nn} = 1$ is called the *identity matrix* of order n and is denoted by I_n (or simply by I when the order is clear from the context).[1] Show that for any $m \times n$ matrix A

$$I_m A = A I_n = A \qquad (1.2.19)$$

16   Prove that a matrix which commutes under multiplication with every square matrix of order n is a scalar matrix of order n.

17   Let $m_1, m_2, n_1, n_2, p_1, p_2$ be positive integers. For $i, j = 1, 2$, let A_{ij} be an $m_i \times n_j$ matrix and B_{ij} be an $n_i \times p_j$ matrix. Show that

$$\begin{bmatrix} A_{11} & A_{12} \\ A_{21} & A_{22} \end{bmatrix} \begin{bmatrix} B_{11} & B_{12} \\ B_{21} & B_{22} \end{bmatrix} = \begin{bmatrix} A_{11}B_{11} + A_{12}B_{21} & A_{11}B_{12} + A_{12}B_{22} \\ A_{21}B_{11} + A_{22}B_{21} & A_{21}B_{12} + A_{22}B_{22} \end{bmatrix}$$

$$(1.2.20)$$

---

[1] Note that the identity matrix of order n is denoted by I_n whereas $I_{(n)}$ stands for the nth row of some identity matrix I. According to these conventions, the nth row of the $m \times m$ identity matrix would be denoted by $I_{m(n)}$. However, this notation is quite cumbersome and will be avoided.

Generalize (1.2.20) to a product of two matrices A, B, where A is partitioned into r row blocks and s column blocks and B is partitioned into s row blocks and t column blocks.

1.3 VECTOR SPACES

As we will see, the language of this section has a strong geometric flavor and, in fact, had its beginnings in the analytic approach to the study of euclidean spaces. As every reader who has encountered plane or multidimensional analytic geometry knows, vectors in a plane can be identified with ordered pairs of real numbers, and vectors in a three-dimensional space can be identified with ordered triples of real numbers. Although it is possible to give a geometric interpretation to most of the theorems and definitions stated in this section in algebraic language, a detailed discussion of this material would carry us much too far afield. However, the reader can profit by keeping in mind that this interpretation exists. In fact, it will often happen that restating an algebraic definition or theorem in geometric terms and drawing a two- or three-dimensional diagram to illustrate it will help the reader see what is really going on.

Definition 1.3.1 Let $\mathbb{V}$ be a nonempty set of $m \times n$ matrices with entries from the field $\mathbb{F}$. $\mathbb{V}$ is called a *vector space* (over $\mathbb{F}$) provided the following two conditions are satisfied:

1. *Closure under addition:* $X + Y$ belongs to $\mathbb{V}$ whenever X and Y belong to $\mathbb{V}$.
2. *Closure under multiplication by a scalar:* αX belongs to $\mathbb{V}$ whenever X belongs to $\mathbb{V}$ and α belongs to $\mathbb{F}$.

If $\mathbb{U}$ and $\mathbb{V}$ are vector spaces over $\mathbb{F}$ and every element of $\mathbb{U}$ is also an element of $\mathbb{V}$, we say that $\mathbb{U}$ is a *subspace* of $\mathbb{V}$.

A vector space, then, is a set of matrices which is closed under the operations of addition and scalar multiplication.[1] Vector spaces arise in many

---

[1] In more abstract treatments than the one given here the general topic of vector spaces is sharply divided into two parts. A *vector space* over $\mathbb{F}$ is defined to be any set $\mathbb{V}$ of elements which are closed under addition and multiplication by a scalar and which satisfy certain combining axioms such as parts 1 and 2 of Corollary 1.2.2 and equations (1.2.16a), (1.2.16b), and (1.2.16d). If the vector spaces are permitted to be infinite-dimensional, the subject falls in the realm of analysis (see, for example, Taylor [9]). When the vector spaces are assumed to be finite-dimensional, many difficulties disappear and they can be studied algebraically. The vector spaces of this chapter all have finite dimension. For a more abstract treatment of finite-dimensional vector spaces see Halmos [6] or Greub [4]. The elements of the vector spaces we will discuss will always be matrices and usually be vectors (as defined in Definition 1.1.4). Note that the multiplication of matrices does not enter into the definition of a vector space.

different ways. In Examples 1.3.2 to 1.3.6 we describe some of the more important and more commonly encountered vector spaces. Several others will be found in the exercises.

EXAMPLE 1.3.2 For each choice of m and n the set of all $m \times n$ matrices with entries in $\mathbb{F}$ is a vector space which we denote by $\mathbb{F}_n^m$. Two important special cases of this example are:

1 The set of all m-dimensional column vectors with entries in $\mathbb{F}$ (denoted by $\mathbb{F}_1^m$ or by $\mathbb{F}^m$)
2 The set of all n-dimensional row vectors with entries from $\mathbb{F}$ (denoted by $\mathbb{F}_n^1$ or by $\mathbb{F}_n$) ////

EXAMPLE 1.3.3 The set consisting only of a zero matrix is a vector space. We call this vector space a *zero space* and denote it by 0. ////

EXAMPLE 1.3.4 For a fixed matrix A, the set of all multiples of A by elements of $\mathbb{F}$ is a vector space   For example, let

$$A = \begin{bmatrix} 1 & 2 & -1 \\ 0 & 1 & 1 \end{bmatrix}$$

Then $\mathbb{V} = \left\{ \alpha \begin{bmatrix} 1 & 2 & -1 \\ 0 & 1 & 1 \end{bmatrix} \middle| \alpha \in \mathbb{F} \right\} = \left\{ \begin{bmatrix} \alpha & 2\alpha & -\alpha \\ 0 & \alpha & \alpha \end{bmatrix} \middle| \alpha \in \mathbb{F} \right\}$
is a vector space. ////

EXAMPLE 1.3.5 For a fixed $m \times n$ matrix A and a fixed positive integer p, the set of all $n \times p$ matrices X satisfying the equation

$$AX = 0$$

is a vector space.   In the special case where X is a column vector (that is, $p = 1$) this vector space is called the *column null space* (or simply *null space*) of A and is denoted by $\mathbb{N}(A)$.

For example, suppose A is the matrix of Example 1.3.4 and we set $p = 1$. $\mathbb{N}(A)$ consists of all three-dimensional column vectors

$$X = \begin{bmatrix} x_1 \\ x_2 \\ x_3 \end{bmatrix}$$

such that
$$AX = \begin{bmatrix} 1 & 2 & -1 \\ 0 & 1 & 1 \end{bmatrix} \begin{bmatrix} x_1 \\ x_2 \\ x_3 \end{bmatrix} = \begin{bmatrix} 0 \\ 0 \end{bmatrix} = 0$$

The single matrix equation $AX = 0$ is equivalent to the pair of linear equations
$$x_1 + 2x_2 - x_3 = 0$$
$$x_2 + x_3 = 0$$

We can solve for two of these unknowns in terms of the third. For example,
$$x_2 = -x_3$$
$$x_1 = -2x_2 + x_3 = 3x_3$$

expresses x_1 and x_2 in terms of x_3. Substituting for x_1 and x_2, we have

$$\mathbb{N}(A) = \left\{ \begin{bmatrix} 3x_3 \\ -x_3 \\ x_3 \end{bmatrix} \middle| x_3 \in \mathbb{F} \right\} = \left\{ x_3 \begin{bmatrix} 3 \\ -1 \\ 1 \end{bmatrix} \middle| x_3 \in \mathbb{F} \right\} \qquad ////$$

EXAMPLE 1.3.6 For k fixed $m \times n$ matrices $A_1, A_2, \ldots, A_k$ the set of all matrices which can be expressed as
$$\alpha_1 A_1 + \alpha_2 A_2 + \cdots + \alpha_k A_k$$
where $\alpha_1, \alpha_2, \ldots, \alpha_k \in \mathbb{F}$, is a vector space of $m \times n$ matrices.[1]

For example, let
$$A_1 = \begin{bmatrix} 1 & 2 \\ 0 & 1 \end{bmatrix} \qquad A_2 = \begin{bmatrix} 0 & 0 \\ 1 & 1 \end{bmatrix}$$

Then
$$\mathbb{V} = \left\{ \alpha_1 \begin{bmatrix} 1 & 2 \\ 0 & 1 \end{bmatrix} + \alpha_2 \begin{bmatrix} 0 & 0 \\ 1 & 1 \end{bmatrix} \middle| \alpha_1, \alpha_2 \in \mathbb{F} \right\} = \left\{ \begin{bmatrix} \alpha_1 & 2\alpha_1 \\ \alpha_2 & \alpha_1 + \alpha_2 \end{bmatrix} \middle| \alpha_1, \alpha_2 \in \mathbb{F} \right\}$$
is a vector space. ////

The process used in Example 1.3.6 to obtain a vector space from a given finite set of matrices all of which have the same dimensions is sufficiently important to warrant giving it a name.

---

[1] What is the geometrical interpretation of this vector space when the matrices $A_1, A_2, \ldots, A_k$, are three-dimensional vectors considered to be points in three-dimensional euclidean space?

Definition 1.3.7 Let $A_1, A_2, \ldots, A_k$, B be $m \times n$ matrices. We say that B is a *linear combination* of $A_1, A_2, \ldots, A_k$ if there exist scalars $\alpha_1, \alpha_2, \ldots, \alpha_k$ such that

$$B = \alpha_1 A_1 + \alpha_2 A_2 + \cdots + \alpha_k A_k$$

Example 1.3.6 states that the set of all linear combinations of any finite set of matrices of the same dimensions is a vector space.

Definition 1.3.8 Let $\mathbb{V}$ be a vector space and let $A_1, A_2, \ldots, A_k$ be matrices. If:

1. $A_1, A_2, \ldots, A_k$ belong to $\mathbb{V}$
2. Every element of $\mathbb{V}$ can be expressed as a linear combination of $A_1, A_2, \ldots, A_k$

then the set

$$\mathbb{S} = \{A_1, A_2, \ldots, A_k\}$$

is said to *span* (or form a *spanning set* for) $\mathbb{V}$.

Every finite set of matrices of the same dimensions spans exactly one vector space, namely, the one constructed in Example 1.3.6. In Examples 1.3.4 and 1.3.5 we have vector spaces which are spanned by a single matrix. The vector space $\mathbb{V}$ of Example 1.3.6 is spanned by the pair of matrices

$$\mathbb{S}_1 = \left\{ \begin{bmatrix} 1 & 2 \\ 0 & 1 \end{bmatrix}, \begin{bmatrix} 0 & 0 \\ 1 & 1 \end{bmatrix} \right\} \quad (1.3.9)$$

A nonzero vector space has many spanning sets. For example, it is not difficult to show that each of the sets

$$\mathbb{S}_2 = \left\{ \begin{bmatrix} 2 & 4 \\ 0 & 2 \end{bmatrix}, \begin{bmatrix} 0 & 0 \\ 1 & 1 \end{bmatrix} \right\}$$

$$\mathbb{S}_3 = \left\{ \begin{bmatrix} -4 & -8 \\ 0 & -4 \end{bmatrix}, \begin{bmatrix} 0 & 0 \\ 2 & 2 \end{bmatrix}, \begin{bmatrix} 0 & 0 \\ 1 & 1 \end{bmatrix} \right\} \quad (1.3.10)$$

$$\mathbb{S}_4 = \left\{ \begin{bmatrix} 1 & 2 \\ 1 & 2 \end{bmatrix}, \begin{bmatrix} 2 & 4 \\ 3 & 5 \end{bmatrix}, \begin{bmatrix} -1 & -2 \\ 1 & 0 \end{bmatrix} \right\}$$

is a spanning set for the vector space $\mathbb{V}$ of Example 1.3.6.

Any finite set of matrices of a vector space $\mathbb{V}$ which contains a spanning set for $\mathbb{V}$ is itself a spanning set. Thus, spanning sets may contain more matrices than are necessary. It is left as an exercise to show that $\mathbb{S}_3$ and $\mathbb{S}_4$ above contain unnecessary matrices and that $\mathbb{S}_1$ and $\mathbb{S}_2$ do not. Note also that if $A_1, A_2, \ldots, A_k$ are matrices belonging to the vector space $\mathbb{U}$ and $\mathbb{V}$ is the vector space spanned by $A_1, A_2, \ldots, A_k$, then $\mathbb{V}$ is a subspace of $\mathbb{U}$.

Let us pause here to point out a danger. We have spoken about vector spaces which are obtained from a given spanning set. Conversely, should we be given a vector space, we would like to be able to find some spanning set for it. However, it is not until Sec. 1.5 that we are able to prove that every subspace of $\mathbb{F}_n^m$ has at least one spanning set. One vector space for which we can furnish a particularly simple spanning set is $\mathbb{F}^m$, the set of all m-dimensional column vectors. Clearly the m m-dimensional *unit column vectors*

$$I^{(1)} = \begin{bmatrix} 1 \\ 0 \\ \vdots \\ 0 \end{bmatrix}, \quad I^{(2)} = \begin{bmatrix} 0 \\ 1 \\ \vdots \\ 0 \end{bmatrix}, \quad \ldots, \quad I^{(m)} = \begin{bmatrix} 0 \\ 0 \\ \vdots \\ 1 \end{bmatrix} \quad (1.3.11)$$

belong to $\mathbb{F}^m$. In order to show that they span $\mathbb{F}^m$ we must prove that every element of $\mathbb{F}^m$ can be expressed as a linear combination of them. Let

$$X = \begin{bmatrix} x_1 \\ x_2 \\ \vdots \\ x_m \end{bmatrix}$$

be any element of $\mathbb{F}^m$. Then

$$X = x_1 I^{(1)} + x_2 I^{(2)} + \cdots + x_m I^{(m)}$$

as required.

There are other ways in which vector spaces arise. Specifically, new vector spaces can be formed by combining old ones.

Definition 1.3.12 Let $V_1, V_2, \ldots, V_k$ be vector spaces of $m \times n$ matrices. Then:

1 The set of matrices which are contained in all the vector spaces $\mathbb{V}_1, \mathbb{V}_2, \ldots, \mathbb{V}_k$ form a vector space called the *intersection* of $\mathbb{V}_1, \mathbb{V}_2, \ldots, \mathbb{V}_k$ and denoted by

$$\mathbb{V}_1 \cap \mathbb{V}_2 \cap \cdots \cap \mathbb{V}_k$$

2 The set of all $m \times n$ matrices which can be expressed in the form

$$V_1 + V_2 + \cdots + V_k$$

where $V_1 \in \mathbb{V}_1, V_2 \in \mathbb{V}_2, \ldots, V_k \in \mathbb{V}_k$, form a vector space called the *sum* of $\mathbb{V}_1, \mathbb{V}_2, \ldots, \mathbb{V}_k$ and denoted by

$$\mathbb{V}_1 + \mathbb{V}_2 + \cdots + \mathbb{V}_k$$

We have already used part 2 of Definition 1.3.12. If $\mathbb{V}_1, \mathbb{V}_2, \ldots, \mathbb{V}_k$ are vector spaces of the type of Example 1.3.4 and each $\mathbb{V}_i$ is spanned by a single matrix A_i, then Example 1.3.6 can be considered as applying part 2 of Definition 1.3.12 to $\mathbb{V}_1, \mathbb{V}_2, \ldots, \mathbb{V}_k$.

EXAMPLE 1.3.13 Let $\mathbb{V}$ be the vector space spanned by

$$\left\{ V_1 = \begin{bmatrix} 2 \\ -1 \\ 0 \\ 1 \end{bmatrix}, \quad V_2 = \begin{bmatrix} 1 \\ 0 \\ 0 \\ 1 \end{bmatrix}, \quad V_3 = \begin{bmatrix} 1 \\ 1 \\ 0 \\ 1 \end{bmatrix} \right\}$$

and let

$$A = \begin{bmatrix} 1 & 2 & 1 & 0 \\ -1 & -2 & -1 & 0 \end{bmatrix}$$

Find a spanning set for $\mathbb{V} \cap \mathbb{N}(A)$.

The vector

$$X = \begin{bmatrix} x_1 \\ x_2 \\ x_3 \\ x_4 \end{bmatrix}$$

is contained in $\mathbb{V}$ if and only if it can be represented as

$$X = \alpha_1 V_1 + \alpha_2 V_2 + \alpha_3 V_3 = \begin{bmatrix} 2\alpha_1 + \alpha_2 + \alpha_3 \\ -\alpha_1 + \alpha_3 \\ 0 \\ \alpha_1 + \alpha_2 + \alpha_3 \end{bmatrix} \quad (1.3.14)$$

for some $\alpha_1, \alpha_2, \alpha_3 \in \mathbb{F}$. Similarly X is contained in $\mathbb{N}(A)$ if and only if

$$AX = 0 \quad (1.3.15)$$

Substituting (1.3.14) into (1.3.15), we see that X belongs to $\mathbb{V} \cap \mathbb{N}(A)$ if and only if

$$1(2\alpha_1 + \alpha_2 + \alpha_3) + 2(-\alpha_1 + \alpha_3) = 0$$
$$-1(2\alpha_1 + \alpha_2 + \alpha_3) - 2(-\alpha_1 + \alpha_3) = 0$$

which reduces to the single condition

$$\alpha_2 = -3\alpha_3$$

Thus

$$V \cap N(A) = \{\alpha_1 V_1 - 3\alpha_3 V_2 + \alpha_3 V_3 \mid \alpha_1, \alpha_3 \in \mathbb{F}\}$$

$$= \left\{ \begin{bmatrix} 2\alpha_1 - 2\alpha_3 \\ -\alpha_1 + \alpha_3 \\ 0 \\ \alpha_1 - 2\alpha_3 \end{bmatrix} \middle| \alpha_1, \alpha_3 \in \mathbb{F} \right\} = \left\{ \alpha_1 \begin{bmatrix} 2 \\ -1 \\ 0 \\ 1 \end{bmatrix} + \alpha_3 \begin{bmatrix} -2 \\ 1 \\ 0 \\ -2 \end{bmatrix} \middle| \alpha_1, \alpha_3 \in \mathbb{F} \right\}$$

and

$$\left\{ \begin{bmatrix} 2 \\ -1 \\ 0 \\ 1 \end{bmatrix}, \begin{bmatrix} -2 \\ 1 \\ 0 \\ -2 \end{bmatrix} \right\}$$

is a spanning set for $V \cap N(A)$. ////

It seems intuitively desirable to get rid of the unnecessary matrices in a spanning set. A means of accomplishing this is contained in:

Lemma 1.3.16 Let V be a vector space spanned by $S = \{A_1, A_2, \ldots, A_k\}$ and let U be a vector space spanned by $k - 1$ of the elements of S. Then $U \subseteq V$. Moreover, $U = V$ if and only if the matrix removed from S can be expressed as a linear combination of the remaining elements of S.

PROOF It will be convenient for us to assume that the elements of S are ordered in such a way that U is spanned by $A_1, A_2, \ldots, A_{k-1}$. Each element of U is a linear combination of the matrices $A_1, A_2, \ldots, A_{k-1}$ and hence may be considered as a linear combination of $A_1, A_2, \ldots, A_{k-1}, A_k$, where the coefficient of A_k is zero. Thus every element of U belongs to V, and we have shown that $U \subseteq V$.

If $U = V$, then A_k belongs to U and hence can be expressed as a linear combination of $A_1, A_2, \ldots, A_{k-1}$. Conversely, suppose A_k can be expressed as

$$A_k = \alpha_1 A_1 + \alpha_2 A_2 + \cdots + \alpha_{k-1} A_{k-1} \quad (1.3.17)$$

for some $\alpha_1, \alpha_2, \ldots, \alpha_{k-1} \in \mathbb{F}$. We wish to show that $U = V$. By the preceding paragraph we have $U \subseteq V$, and it only remains to show that $V \subseteq U$. For each $V \in V$ there exist scalars $\beta_1, \beta_2, \ldots, \beta_{k-1}, \beta_k$ in $\mathbb{F}$ such that

$$V = \beta_1 A_1 + \beta_2 A_2 + \cdots + \beta_{k-1} A_{k-1} + \beta_k A_k \quad (1.3.18)$$

Substituting equation (1.3.17) into equation (1.3.18) yields

$$V = \beta_1 A_1 + \beta_2 A_2 + \cdots + \beta_{k-1} A_{k-1} + \beta_k(\alpha_1 A_1 + \alpha_2 A_2 + \cdots + \alpha_{k-1} A_{k-1})$$
$$= (\beta_1 + \beta_k \alpha_1) A_1 + (\beta_2 + \beta_k \alpha_2) A_2 + \cdots + (\beta_{k-1} + \beta_k \alpha_{k-1}) A_{k-1}$$

which shows that $V \in \mathbb{U}$. Hence $\mathbb{V} \subseteq \mathbb{U}$, and it follows that $\mathbb{U} = \mathbb{V}$ as required. ////

Definition 1.3.19 A finite set $A_1, A_2, \ldots, A_k$ of elements of a vector space $\mathbb{V}$ is *linearly dependent* if there exist scalars $\alpha_1, \alpha_2, \ldots, \alpha_k$ in $\mathbb{F}$, not all of which are zero, such that

$$\alpha_1 A_1 + \alpha_2 A_2 + \cdots + \alpha_k A_k = 0 \quad (1.3.20)$$

On the other hand, if the only instance in which equation (1.3.20) is satisfied is when $\alpha_1 = \alpha_2 = \cdots = \alpha_k = 0$, then $A_1, A_2, \ldots, A_k$ are *linearly independent*.

Definitions 1.3.7 and 1.3.19 and the process described in Lemma 1.3.16 are closely related. Both definitions are applicable to a set of matrices only when all the matrices concerned have the same dimensions. It is clear that if one of the matrices $A_1, A_2, \ldots, A_k$ can be expressed as a linear combination of the others, then a relation such as equation (1.3.20) exists in which not all of the scalars $\alpha_1, \alpha_2, \ldots, \alpha_k$ are zeros. Thus $A_1, A_2, \ldots, A_k$ are linearly dependent. A little care must be exercised in examining the converse, however. A finite set of matrices, all of the same dimensions and containing at least two matrices, in which it is not possible to express one of them as a linear combination of the others is easily shown to be linearly independent. On the other hand, a set consisting of a single matrix is a linearly independent set if the matrix is not a zero matrix, and it is a linearly dependent set if the matrix is a zero matrix.

EXAMPLE 1.3.21 Let

$$A_1 = \begin{bmatrix} 1 & 0 \\ 2 & -1 \end{bmatrix} \quad A_2 = \begin{bmatrix} 0 & 1 \\ -1 & 1 \end{bmatrix} \quad A_3 = \begin{bmatrix} 1 & 1 \\ 1 & 0 \end{bmatrix} \quad A_4 = \begin{bmatrix} 1 & 2 \\ 0 & 1 \end{bmatrix}$$

In order to determine the linear dependence or independence of A_1, A_2, A_3, A_4 we need to know whether the equation

$$\alpha_1 A_1 + \alpha_2 A_2 + \alpha_3 A_3 + \alpha_4 A_4 = 0$$

has any nonzero solutions. This matrix equation is equivalent to the system of linear equations

$$\alpha_1 + \alpha_3 + \alpha_4 = 0 \qquad \alpha_2 + \alpha_3 + 2\alpha_4 = 0$$
$$2\alpha_1 - \alpha_2 + \alpha_3 = 0 \qquad -\alpha_1 + \alpha_2 + \alpha_4 = 0$$

and it is easily seen that

$$\alpha_1 = 0 \quad \alpha_2 = 1 \quad \alpha_3 = 1 \quad \alpha_4 = -1$$

is a solution. Thus A_1, A_2, A_3, A_4 are linearly dependent. ////

According to Lemma 1.3.16, if a set of matrices span $\mathbb{V}$ and one of them can be expressed as a linear combination of the others, it is superfluous and can be discarded. The remaining matrices also span $\mathbb{V}$. This process can be repeated until we have a spanning set for $\mathbb{V}$ in which none of the matrices can be expressed as a linear combination of the others. If $\mathbb{V}$ is not the zero space, the matrices of this spanning set are linearly independent.

Definition 1.3.22 A set of matrices $\mathbb{S} = \{A_1, A_2, \ldots, A_k\}$ of a vector space $\mathbb{V}$ is a *basis* for $\mathbb{V}$ if:

1 $\mathbb{S}$ spans $\mathbb{V}$.
2 $\mathbb{S}$ is a linearly independent set.

EXAMPLE 1.3.23 Let $\mathbb{W}$ be the vector space spanned by the matrices A_1, A_2, A_3, A_4 of Example 1.3.21. We wish to find a basis for $\mathbb{W}$. According to Example 1.3.21, A_4 can be expressed as a linear combination of A_1, A_2 and A_3 (specifically, $A_4 = A_2 + A_3$) and, by Lemma 1.3.16, A_1, A_2, A_3 also span $\mathbb{W}$.

In order to determine whether A_1, A_2, A_3 form a basis for $\mathbb{W}$ we repeat the procedure of Example 1.3.21 and find that

$$A_1 + A_2 - A_3 = 0$$

Thus A_1, A_2, A_3 are linearly dependent. Since A_3 can be expressed as a linear combination of A_1 and A_2, it follows that A_1, A_2 also span $\mathbb{W}$.

In order to determine whether A_1, A_2 form a basis for $\mathbb{W}$ we need to know whether the equation

$$\beta_1 A_1 + \beta_2 A_2 = 0$$

has any solutions other than $\beta_1 = \beta_2 = 0$. The same procedure that we used before shows that there are no other solutions. Thus A_1, A_2 form a basis.

In exactly the same way it can be shown that any two of the matrices A_1, A_2, A_3, A_4 form a basis for $\mathbb{W}$. ////

In speaking of a basis for a vector space we have had to treat the zero space separately. It will simplify matters greatly if we now define an empty set of matrices to be linearly independent and to be a basis for a zero space.

It is not difficult to show that the unit column vectors are linearly independent and consequently form a basis for $\mathbb{F}^m$. We leave it as an exercise to show that in Example 1.3.6 the sets $\mathbb{S}_1$ and $\mathbb{S}_2$ of (1.3.9) and (1.3.10) are bases for $\mathbb{V}$ but $\mathbb{S}_3$ and $\mathbb{S}_4$ are not bases.

We stated earlier that it will not be until Sec. 1.5 that we can prove that every subspace of $\mathbb{F}_n^m$ has a spanning set. Since a basis is a particular type of spanning set, we cannot hope to show now that every subspace of $\mathbb{F}_n^m$ has a basis. However, the problem of finding a basis once a spanning set is obtained is relatively simple.

Theorem 1.3.24 If $\mathbb{S}$ is a spanning set for a vector space $\mathbb{V}$, then some subset of $\mathbb{S}$ is a basis for $\mathbb{V}$.

PROOF The proof consists of repeated use of Lemma 1.3.16 and makes use of the process described in Example 1.3.23. If $\mathbb{V} = 0$, then the empty set is a subset of $\mathbb{S}$ which is a basis for $\mathbb{V}$. Now suppose $\mathbb{V} \neq 0$. Then some of the matrices of $\mathbb{S}$ are nonzero. If none of the matrices of $\mathbb{S}$ can be expressed as a linear combination of the others, then $\mathbb{S}$ is a basis for $\mathbb{V}$. On the other hand, if one of the elements of $\mathbb{S}$ can be expressed as a linear combination of the others, then that element can be discarded and, by Lemma 1.3.16, the resulting set, say $\mathbb{S}_1$, again spans $\mathbb{V}$. If none of the matrices of $\mathbb{S}_1$ can be expressed as a linear combination of the others, then $\mathbb{S}_1$ is a basis for $\mathbb{V}$. Otherwise Lemma 1.3.16 can be applied again. Since $\mathbb{S}$ initially contains only a finite number of elements, after a finite number of steps the process ends and the resulting subset of $\mathbb{S}$ is a basis of $\mathbb{V}$. ////

There are many characterizations of a basis for a vector space. Probably the one most often used is:

Theorem 1.3.25 Let $\mathbb{S}$ span $\mathbb{V}$. Then $\mathbb{S}$ is a basis for $\mathbb{V}$ if and only if every element of $\mathbb{V}$ has a unique expression as a linear combination of the elements of $\mathbb{S}$.

PROOF Let $\mathbb{S} = \{A_1, A_2, \ldots, A_k\}$ span $\mathbb{V}$ and let each element of $\mathbb{V}$ have a unique representation as a linear combination of $A_1, A_2, \ldots, A_k$. We wish to show that $\mathbb{S}$ is a linearly independent set. Suppose the zero element of $\mathbb{V}$ can be expressed as

$$0 = \alpha_1 A_1 + \alpha_2 A_2 + \cdots + \alpha_k A_k$$

for some $\alpha_1, \alpha_2, \ldots, \alpha_k \in \mathbb{F}$. Clearly

$$0 = 0A_1 + 0A_2 + \cdots + 0A_k$$

Since the representation of each element of $\mathbb{V}$ is unique, it follows that

$$\alpha_1 = \alpha_2 = \cdots = \alpha_k = 0$$

and hence $\mathbb{S}$ is a linearly independent set.

Conversely let $\mathbb{S}$ be a basis for $\mathbb{V}$ and let

$$X = \alpha_1 A_1 + \alpha_2 A_2 + \cdots + \alpha_k A_k$$
$$X = \beta_1 A_1 + \beta_2 A_2 + \cdots + \beta_k A_k$$

be representations of an element X of $\mathbb{V}$. Subtracting one equation from the other, we have

$$0 = X - X = (\alpha_1 - \beta_1)A_1 + (\alpha_2 - \beta_2)A_2 + \cdots + (\alpha_k - \beta_k)A_k$$

Since $A_1, A_2, \ldots, A_k$ are linearly independent, it follows that

$$\alpha_1 - \beta_1 = \alpha_2 - \beta_2 = \cdots = \alpha_k - \beta_k = 0$$

Thus the representation for X is unique. ////

We close this section by pointing out a dilemma that our notation forces upon us. In Lemma 1.3.16, in Definitions 1.3.7, 1.3.8, and 1.3.19, and in numerous definitions and theorems to come the order implicit in writing down a set of elements of a vector space, for example, $\mathbb{S} = \{A_1, A_2, \ldots, A_k\}$, is of no importance. (Remember, however, that in defining a matrix it was stressed that the order in which the entries appear within the matrix is of extreme importance.) When speaking of spanning sets or of linearly independent sets, for example, the order in which the matrices are given is completely immaterial. The difficulty would disappear if only there were a simple way of exhibiting such sets without giving their elements some order. This problem becomes especially acute in the next section when the rows or columns of an $m \times n$ matrix are considered to be elements of some subspace of $\mathbb{F}^m$ or $\mathbb{F}_n$.

EXERCISES

1 Show that

$$\begin{bmatrix} 1 \\ 1 \\ 0 \end{bmatrix} \quad \begin{bmatrix} 0 \\ 1 \\ 1 \end{bmatrix} \quad \begin{bmatrix} 1 \\ 2 \\ -1 \end{bmatrix}$$

form a basis for $\mathbb{F}^3$.

2 Show that

$$A_1 = \begin{bmatrix} 1 \\ 2 \\ 3 \end{bmatrix} \quad A_2 = \begin{bmatrix} -1 \\ -1 \\ 2 \end{bmatrix} \quad A_3 = \begin{bmatrix} -1 \\ 1 \\ 12 \end{bmatrix}$$

and

$$B_1 = \begin{bmatrix} 0 \\ 1 \\ 5 \end{bmatrix} \quad B_2 = \begin{bmatrix} 1 \\ 3 \\ 8 \end{bmatrix}$$

span the same vector space V. Show that $\{B_1, B_2\}$ is a basis for V. Is $S = \{A_1, A_2, A_3\}$ a basis for V? If not, find a subset of S which is a basis.

3 Let V and W be the vector spaces defined in Examples 1.3.6 and 1.3.23. Find bases for $V + W$ and $V \cap W$.

4 Let the vector space spanned by

$$C_1 = \begin{bmatrix} 1 \\ -1 \\ 1 \end{bmatrix} \quad C_2 = \begin{bmatrix} 6 \\ 6 \\ 1 \end{bmatrix} \quad C_3 = \begin{bmatrix} 8 \\ 4 \\ 3 \end{bmatrix}$$

be denoted by U. Find a basis for U. Find a basis for the intersection of U and the vector space V of Exercise 2. Show that $U + V = \mathbb{F}^3$.

5 Interpret the vectors of Exercises 2 and 4 as points in three-dimensional euclidean space. Sketch U, V, and $U \cap V$. Explain why each of the sets $\{A_1, A_2, A_3\}$ and $\{B_1, B_2\}$ spans V.

6 Find a basis for the vector space $\mathbb{X}$ spanned by the set of matrices

$$\begin{bmatrix} 1 & 0 \\ 1 & 0 \end{bmatrix} \quad \begin{bmatrix} 1 & 1 \\ 0 & 0 \end{bmatrix} \quad \begin{bmatrix} 0 & 1 \\ 0 & 1 \end{bmatrix} \quad \begin{bmatrix} 0 & 0 \\ 1 & 1 \end{bmatrix} \quad \begin{bmatrix} 1 & 0 \\ 0 & 1 \end{bmatrix} \quad \begin{bmatrix} 0 & 1 \\ 1 & 0 \end{bmatrix}$$

Find all subsets of this set of matrices which are bases for $\mathbb{X}$.

7 Show that $\mathbb{F}_n^m$ has a basis consisting of mn matrices each of which has exactly one nonzero entry.

8 Verify that the sets of matrices described in Examples 1.3.4 to 1.3.6 are vector spaces.

9 Verify that S_1 of (1.3.9) and S_2 of (1.3.10) are bases for the vector space V of Example 1.3.6 and that S_3 and S_4 of (1.3.10) are not bases for V.

10 Which of the following sets of matrices form vector spaces?
  (a) For a fixed square matrix A, the set of all matrices which commute with A under multiplication.
  (b) For fixed $m \times n$ and $n \times m$ matrices A and B, the set of all $n \times n$ matrices X such that $AX = XB$.
  (c) For a fixed integer n, the set of all square matrices X of order n such that $X^2 = 0$.
  (d) For fixed matrices A and B, the set of all matrices X such that $AXB = 0$.

(e) The set of all vectors
$$X = [x_1 \ x_2 \ x_3]$$
such that $x_1 + x_2 + x_3 = 1$.
(f) The set of all vectors
$$X = [x_1 \ x_2 \ x_3]$$
such that $x_1 + x_2 + x_3 = 0$.

11 Verify that the sum and intersection of vector spaces as defined in Definition 1.3.12 are vector spaces.

12 Let $\mathbb{U}_1$, $\mathbb{U}_2$, and $\mathbb{U}_3$ be subspaces of the vector space $\mathbb{V}$. Prove that
  (a) $\mathbb{U}_1 + \mathbb{U}_2 = \mathbb{U}_2 + \mathbb{U}_1$.
  (b) $\mathbb{U}_1 + \mathbb{U}_2 = \mathbb{U}_1$ if and only if $\mathbb{U}_2 \subseteq \mathbb{U}_1$.
  (c) $\mathbb{U}_1 + (\mathbb{U}_2 + \mathbb{U}_3) = (\mathbb{U}_1 + \mathbb{U}_2) + \mathbb{U}_3 = \mathbb{U}_1 + \mathbb{U}_2 + \mathbb{U}_3$.
  [Henceforth we write $\mathbb{U}_1 + \mathbb{U}_2 + \mathbb{U}_3$ instead of either $\mathbb{U}_1 + (\mathbb{U}_2 + \mathbb{U}_3)$ or $(\mathbb{U}_1 + \mathbb{U}_2) + \mathbb{U}_3$.]
  (d) $(\mathbb{U}_1 \cap \mathbb{U}_2) + (\mathbb{U}_1 \cap \mathbb{U}_3) \subseteq \mathbb{U}_1 \cap (\mathbb{U}_2 + \mathbb{U}_3)$
  Construct examples to show that both equality and inequality are possible in (d).

13 Let $\mathbb{V}$ be a vector space of $m \times n$ matrices. For fixed matrices B and C of dimensions $p \times m$ and $n \times q$ respectively, we denote the set of all $p \times n$ matrices which can be expressed as BV for some $V \in \mathbb{V}$ by $B\mathbb{V}$ and the set of all $m \times q$ matrices which can be expressed as VC for some $V \in \mathbb{V}$ by $\mathbb{V}C$, that is,
$$B\mathbb{V} = \{BV \mid V \in \mathbb{V}\} \qquad \mathbb{V}C = \{VC \mid V \in \mathbb{V}\}$$
Prove that $B\mathbb{V}$ and $\mathbb{V}C$ are vector spaces.

14 Let
$$A = \begin{bmatrix} 3 & -3 & 1 \\ 3 & 0 & -1 \\ 0 & 3 & -2 \\ 0 & 0 & 0 \end{bmatrix}$$

Find bases for $A\mathbb{F}^3$ and $\mathbb{F}_4 A$. Find bases for the vector spaces $A\mathbb{V}$ and $A\mathbb{U}$, where $\mathbb{V}$ and $\mathbb{U}$ are the vector spaces of Exercises 2 and 4 respectively.

15 Let $\mathbb{S} = \{V_1, V_2, \ldots, V_k\}$ be a basis for a vector space $\mathbb{V}$ of $m \times n$ matrices and let B be a $p \times m$ matrix. Prove that
$$B\mathbb{S} = \{BV_1, BV_2, \ldots, BV_k\}$$
spans $B\mathbb{V}$. Prove that $B\mathbb{S}$ is a basis for $B\mathbb{V}$ if and only if $\mathbb{N}(B) \cap \mathbb{V} = 0$. State and prove the corresponding results for $\mathbb{V}C$.

16 The *row null space*, denoted by $\mathbb{N}_R(A)$, of the $m \times n$ matrix A is defined by

$$\mathbb{N}_R(A) = \{X \mid XA = 0, X \in \mathbb{F}_m\}$$

Prove that $\mathbb{N}_R(A)$ is a vector space. Find a basis for $\mathbb{N}_R(A)$, where A is the matrix of Exercise 14.

17 Let $\mathbb{U}$ and $\mathbb{V}$ be vector spaces and let $\mathbb{S}$ be a spanning set for $\mathbb{V}$. Prove that $\mathbb{V} \subseteq \mathbb{U}$ if and only if $\mathbb{S} \subseteq \mathbb{U}$.

18 Let A and B be $m \times n$ and $p \times n$ matrices respectively. Show that if $\mathbb{N}(A) = \mathbb{N}(B)$, then $\mathbb{N}(AC) = \mathbb{N}(BC)$ for every $n \times q$ matrix C.

19 Let $\mathbb{S}$ be a linearly independent subset of a vector space $\mathbb{V}$. Prove that $\mathbb{S}$ is a basis for $\mathbb{V}$ if and only if every finite subset of $\mathbb{V}$ which contains $\mathbb{S}$ properly is linearly dependent.

20 Let $\mathbb{S}$ be a spanning set for the vector space $\mathbb{V}$. Prove that if there exists an element of $\mathbb{V}$ which can be expressed in exactly one way as a linear combination of the elements of $\mathbb{S}$, then every element of $\mathbb{V}$ can be expressed in exactly one way as a linear combination of the elements of $\mathbb{S}$.

21 Let $\mathbb{S}_1$ and $\mathbb{S}_2$ be bases for the subspaces $\mathbb{U}_1$ and $\mathbb{U}_2$ of the vector space $\mathbb{V}$ respectively. Show that $\mathbb{S}_1 \cup \mathbb{S}_2$ spans $\mathbb{U}_1 + \mathbb{U}_2$. Show that $\mathbb{S}_1 \cup \mathbb{S}_2$ is a basis for $\mathbb{U}_1 + \mathbb{U}_2$ if and only if $\mathbb{U}_1 \cap \mathbb{U}_2 = 0$.[1]

22 Let $\mathbb{U}_1$ and $\mathbb{U}_2$ be subspaces of the vector space $\mathbb{V}$. Prove that $\mathbb{U}_1 \cup \mathbb{U}_2$ is a vector space if and only if either $\mathbb{U}_1 \subseteq \mathbb{U}_2$ or $\mathbb{U}_2 \subseteq \mathbb{U}_1$.

23 Prove that the simultaneous linear homogeneous equations

$$a_{11}x_1 + a_{12}x_2 + \cdots + a_{1n}x_n = 0$$
$$a_{21}x_2 + a_{22}x_2 + \cdots + a_{2n}x_n = 0$$
$$\cdots\cdots\cdots\cdots\cdots\cdots\cdots\cdots\cdots\cdots\cdots$$
$$a_{m1}x_1 + a_{m2}x_2 + \cdots + a_{mn}x_n = 0$$

have a nonzero solution if and only if the n vectors

$$A^{(1)} = \begin{bmatrix} a_{11} \\ a_{21} \\ \vdots \\ a_{m1} \end{bmatrix}, \quad A^{(2)} = \begin{bmatrix} a_{12} \\ a_{22} \\ \vdots \\ a_{m2} \end{bmatrix}, \quad \cdots, \quad A^{(n)} = \begin{bmatrix} a_{1n} \\ a_{2n} \\ \vdots \\ a_{mn} \end{bmatrix}$$

are linearly dependent.

24 Show that the set of all solutions

$$\begin{bmatrix} x_1 \\ x_2 \\ \vdots \\ x_n \end{bmatrix}$$

of the equations of Exercise 23 form a vector space.

---

[1] $\mathbb{S}_1 \cup \mathbb{S}_2$ is the *union* of $\mathbb{S}_1$ and $\mathbb{S}_2$ and is the set which consists of all elements which belong to at least one of the sets $\mathbb{S}_1$ and $\mathbb{S}_2$.

1.4 THE ROW SPACE AND THE COLUMN SPACE OF A MATRIX

We will see that the columns and rows, rather than the individual elements, are the true building blocks of a matrix. It is often essential to think of an $m \times n$ matrix as being composed of an ordered set of m n-dimensional columns or, in the same way, as an ordered set of m n-dimensional rows. For an $m \times n$ matrix A we have denoted the n columns of A by $A^{(1)}, A^{(2)}, \ldots, A^{(n)}$ and the m rows of A by $A_{(1)}, A_{(2)}, \ldots, A_{(m)}$. We then consider A as being partitioned into the n $m \times 1$ submatrices, that is, m-dimensional column vectors, $A^{(1)}, A^{(2)}, \ldots, A^{(n)}$.[1] Consequently we may say that $A^{(1)}, A^{(2)}, \ldots, A^{(n)}$ belong to $\mathbb{F}^m$ and thus the vector space that they span is a subspace of $\mathbb{F}^m$. We call this vector space the *column space* of A and denote it by $\mathbb{C}(A)$. In the same way, we consider A as being partitioned into m $1 \times n$ matrices, that is, n-dimensional row vectors, $A_{(1)}, A_{(2)}, \ldots, A_{(m)}$. These vectors belong to $\mathbb{F}_n$. The vector space spanned by them is a subspace of $\mathbb{F}_n$ called the *row space* of A and denoted by $\mathbb{R}(A)$.

Because most of our discussion of matrices in the remainder of this chapter is in terms of $\mathbb{C}(A)$ and $\mathbb{R}(A)$, row and column vectors (or simply vectors) will be studied separately in some detail. Although our treatment is specifically, at any one moment, in terms either of column or row vectors, it should be remembered that these methods and results are equally applicable to column vectors and row vectors and, in fact, often to vector spaces of matrices of arbitrary but fixed dimensions. In the future, when we have need for some result about column vectors which in an earlier section is stated and proved in terms of row vectors (or vice versa), we will simply refer to the row-vector (or column-vector) result.

EXAMPLE 1.4.1 Find bases for the row and column spaces of the matrix

$$B = \begin{bmatrix} -3 & 0 & 6 & 0 \\ 1 & 0 & -2 & 0 \\ -1 & 0 & 2 & 0 \end{bmatrix}$$

The three row vectors

$$B_{(1)} = [-3 \ 0 \ 6 \ 0] \quad B_{(2)} = [1 \ 0 \ -2 \ 0] \quad B_{(3)} = [-1 \ 0 \ 2 \ 0]$$

span $\mathbb{R}(B)$ but do not form a linearly independent set since

$$B_{(1)} + 2B_{(2)} - B_{(3)} = 0$$

---

[1] We are purposely confusing the jth column of A with the m-dimensional column vector having the same entries. Both are denoted by $A^{(j)}$.

Since $B_{(3)}$ is expressible as a linear combination of $B_{(1)}$ and $B_{(2)}$, it is superfluous and the two vectors $B_{(1)}$ and $B_{(2)}$ also span $\mathbb{R}(B)$. However,

$$B_{(1)} + 3B_{(2)} = 0$$

Thus we can eliminate $B_{(2)}$ and conclude that $\{B_{(1)}\}$ is a basis for $\mathbb{R}(B)$.

In the same way, the four column vectors

$$B^{(1)} = \begin{bmatrix} -3 \\ 1 \\ -1 \end{bmatrix} \quad B^{(2)} = \begin{bmatrix} 0 \\ 0 \\ 0 \end{bmatrix} \quad B^{(3)} = \begin{bmatrix} 6 \\ -2 \\ 2 \end{bmatrix} \quad B^{(4)} = \begin{bmatrix} 0 \\ 0 \\ 0 \end{bmatrix}$$

span $\mathbb{C}(B)$ but do not form a linearly independent set since

$$2B^{(1)} + \alpha B^{(2)} + B^{(3)} + \beta B^{(4)} = 0$$

for every choice of α and β. As before, $B^{(2)}$, $B^{(3)}$, and $B^{(4)}$ can be eliminated, and $\{B^{(1)}\}$ is a basis for $\mathbb{C}(B)$.

Note that $\mathbb{C}(B)$ is the same vector space as $\mathbb{N}(A)$ of Example 1.3.5. ////

Because of the way multiplication of matrices is defined, the row and column spaces of a product are closely related to the row and column spaces of the factors. Specifically:

Lemma 1.4.2 Let A and B be $m \times n$ and $n \times p$ matrices respectively. Then

1. $\mathbb{R}(AB) \subseteq \mathbb{R}(B)$.
2. $\mathbb{C}(AB) \subseteq \mathbb{C}(A)$.

PROOF Set $AB = C$. A careful consideration of Definition 1.2.4 leads to the following relations between the rows of B and C and between the columns of A and C:

$$\begin{aligned} C_{(i)} &= a_{i1} B_{(1)} + a_{i2} B_{(2)} + \cdots + a_{in} B_{(n)} & i &= 1, 2, \ldots, m \\ C^{(j)} &= A^{(1)} b_{1j} + A^{(2)} b_{2j} + \cdots + A^{(n)} b_{nj} & j &= 1, 2, \ldots, p \end{aligned} \quad (1.4.3)$$

(See Exercise 11 of Sec. 1.2.)

Now let $X \in \mathbb{R}(AB) = \mathbb{R}(C)$. Since $\mathbb{R}(C)$ is the vector space spanned by the m rows of C, there exist scalars $\alpha_1, \alpha_2, \ldots, \alpha_m$ such that

$$X = \alpha_1 C_{(1)} + \alpha_2 C_{(2)} + \cdots + \alpha_m C_{(m)}$$

Substituting for each $C_{(i)}$ the expression of equation (1.4.3) and rearranging terms yields

$$\begin{aligned}X = {} & (\alpha_1 a_{11} + \alpha_2 a_{21} + \cdots + \alpha_m a_{m1})B_{(1)} \\ & + (\alpha_1 a_{12} + \alpha_2 a_{22} + \cdots + \alpha_m a_{m2})B_{(2)} \\ & + \cdots + (\alpha_1 a_{1n} + \alpha_2 a_{2n} + \cdots + \alpha_m a_{mn})B_{(n)}\end{aligned}$$

Thus X has been expressed as a linear combination of the rows of B, and it follows that X belongs to the vector space spanned by the rows of B; that is, $X \in \mathbb{R}(B)$. Consequently $\mathbb{R}(AB) = \mathbb{R}(C) \subseteq \mathbb{R}(B)$.

In the same way we can show that $\mathbb{C}(AB) = \mathbb{C}(C) \subseteq \mathbb{C}(A)$. ////

Consideration of the row and column spaces of a matrix permits us to determine when matrix equations of the form $XB = C$ and $BX = C$ have solutions. We have already pointed out that a set of simultaneous linear equations can be written in this form. Unlike the situation which occurs when X, B, and C represent numbers, it is possible that there are no solutions, even when $B \neq 0$. Likewise, it is also possible that there are infinitely many solutions, even when $B \neq 0$. Finally, it will become evident that one of these equations may have a solution while the other does not. (See Exercises 3 and 4 of Sec. 1.2.)

Theorem 1.4.4 (a) Let B and C be $m \times n$ and $p \times n$ matrices respectively. A necessary and sufficient condition that there exist a $p \times m$ matrix A such that $AB = C$ is that $\mathbb{R}(C) \subseteq \mathbb{R}(B)$. Moreover, when such a matrix A does exist, A is unique if and only if the rows of B are linearly independent.

(b) Let B and C be $m \times n$ and $m \times p$ matrices respectively. A necessary and sufficient condition that there exist an $n \times p$ matrix A such that $BA = C$ is that $\mathbb{C}(C) \subseteq \mathbb{C}(B)$. Moreover, when such a matrix A does exist, A is unique if and only if the columns of B are linearly independent.

PROOF It is sufficient to present a proof for (a) since the proof of (b) is essentially identical.

First let $\mathbb{R}(C) \subseteq \mathbb{R}(B)$. We wish to show that the equation $AB = C$ has a solution. Since each of the p rows of C is contained in the row space of B, it can be expressed as a linear combination of the rows of B. Thus, for each i, $i = 1, 2, \ldots, p$, there exist m scalars, a_{ij}, $j = 1, 2, \ldots, m$, such that

$$C_{(i)} = a_{i1} B_{(1)} + a_{i2} B_{(2)} + \cdots + a_{im} B_{(m)} \quad (1.4.5)$$

The p equations of (1.4.5) require pm numbers a_{ij}, $i = 1, 2, \ldots, p$; $j = 1, 2, \ldots, m$. Define the $p \times m$ matrix A by setting $A = [a_{ij}]$. Then it follows from (1.4.5) that $C = AB$.

Conversely, suppose that there is a $p \times m$ matrix $A = [a_{ij}]$ such that $C = AB$. Then the p equations of (1.4.5) are satisfied, and, as in Lemma 1.4.2, we have $\mathbb{R}(AB) = \mathbb{R}(C) \subseteq \mathbb{R}(B)$. We have now shown that $C = AB$ has a solution if and only if $\mathbb{R}(C) \subseteq \mathbb{R}(B)$.

Finally, let us determine when the solution A is unique. Thus we assume that a matrix $A = [a_{ij}]$ exists such that $AB = C$. It follows from Theorem 1.3.25 that for each of the p equations of (1.4.5) the m numbers a_{ij}, $j = 1, 2, \ldots, m$, satisfying the equation are unique if and only if $B_{(1)}, B_{(2)}, \ldots, B_{(m)}$ form a basis for $\mathbb{R}(B)$, that is, if and only if the rows of B are linearly independent. ////

Corollary 1.4.6 Let B and C be $m \times n$ and $p \times n$ matrices respectively. If $\mathbb{R}(C) \subseteq \mathbb{R}(B)$, then $\mathbb{N}(C) \supseteq \mathbb{N}(B)$. Hence, if $\mathbb{R}(C) = \mathbb{R}(B)$, then $\mathbb{N}(B) = \mathbb{N}(C)$.

PROOF Since $\mathbb{R}(C) \subseteq \mathbb{R}(B)$, there is a matrix A such that $AB = C$. Let $X \in \mathbb{N}(B)$. Then

$$CX = ABX = A0 = 0$$

and $X \in \mathbb{N}(C)$. Consequently $\mathbb{N}(C) \supseteq \mathbb{N}(B)$. ////

In Sec. 1.7 we will prove the converse of Corollary 1.4.6; i.e., if $\mathbb{N}(C) \supseteq \mathbb{N}(B)$, then $\mathbb{R}(C) \subseteq \mathbb{R}(B)$. Surprisingly, this is a much deeper result.[1]

An important special case of Theorem 1.4.4 occurs when C is the identity matrix I. The n columns of an identity matrix of order n are the n unit column vectors (1.3.11) and hence form a basis for $\mathbb{F}^n$. Then we have:

Corollary 1.4.7 Let B be an $m \times n$ matrix.

1 A necessary and sufficient condition that there exist a matrix A such that $AB = I$ (where I is necessarily of order n) is that $\mathbb{R}(B) = \mathbb{F}_n$.
2 A necessary and sufficient condition that there exist a matrix A such that $BA = I$ (where I is necessarily of order m) is that $\mathbb{C}(B) = \mathbb{F}^m$.

PROOF Let us apply Theorem 1.4.4 to the equation $AB = I$. A solution exists if and only if $\mathbb{F}_n = \mathbb{R}(I) \subseteq \mathbb{R}(B)$. However, $\mathbb{R}(B) \subseteq \mathbb{F}_n$ for every $m \times n$ matrix B, and hence the condition $\mathbb{F}_n \subseteq \mathbb{R}(B)$ is equivalent to the condition $\mathbb{F}_n = \mathbb{R}(B)$. ////

---

[1] In fact, it is false for arbitrary vector spaces; see, for example, Taylor [9].

EXERCISES

1. Let
$$A = \begin{bmatrix} 1 & 2 & -1 & 1 & 4 \\ 0 & 1 & 2 & -1 & -3 \\ 2 & 7 & 4 & -1 & -1 \end{bmatrix}$$
Find a basis for $\mathbb{R}(A)$ and a basis for $\mathbb{C}(A)$.

2. Let
$$B = \begin{bmatrix} 1 & 2 & -1 \\ 0 & 0 & 1 \\ 1 & 0 & 1 \end{bmatrix} \quad C = \begin{bmatrix} 1 & -1 & 0 \\ 2 & 0 & 1 \end{bmatrix}$$
Find a matrix A such that $AB = C$. Is A unique? Does there exist a matrix A such that $AB = I_3$? Does there exist a matrix A such that $BA = I_3$?

3. Does $[3 \quad -2 \quad -3 \quad 1]$ belong to the row space of
$$\begin{bmatrix} 1 & -1 & -1 & 2 \\ 0 & 1 & 0 & 1 \\ 1 & 0 & -1 & 0 \end{bmatrix}$$

4. Let
$$B = \begin{bmatrix} 1 & 2 & -1 \\ 1 & 1 & 1 \\ 0 & 1 & -2 \end{bmatrix} \quad C = \begin{bmatrix} 1 & 0 & 1 \\ 2 & 1 & 4 \end{bmatrix}$$
Show that there does not exist a matrix A such that $AB = C$ (a) by a direct computation and
(b) by showing that $\mathbb{R}(C) \not\subseteq \mathbb{R}(B)$ and applying Theorem 1.4.4.

5. Find a basis for $\mathbb{C}(A) \cap \mathbb{C}(B)$, where A and B are given in Exercises 1 and 4 respectively.

6. For what values of x is $\mathbb{F}^3$ the column space of the matrix
$$A = \begin{bmatrix} 1 & 0 & 1 \\ -1 & 1 & 2 \\ 0 & 3 & x \end{bmatrix}$$
For what values of x is $\mathbb{F}_3$ the row space of A?

7. Let
$$B = \begin{bmatrix} 1 & 1 & 3 \\ -1 & 2 & 0 \\ 1 & 0 & 2 \end{bmatrix} \quad C = \begin{bmatrix} x & -1 & 1 \\ y & 0 & 1 \end{bmatrix}$$
For what values of x and y do there exist matrices A such that $AB = C$?

8. Let A be an $m \times n$ matrix. Prove that $\mathbb{C}(A) = A\mathbb{F}^n$ and $\mathbb{R}(A) = \mathbb{F}_m A$.

9 Prove Theorem 1.4.4(b).
10 Let A and B be $p \times n$ and $q \times n$ matrices respectively. Show that if $\mathbb{R}(A) = \mathbb{R}(B)$, then $\mathbb{R}(AC) = \mathbb{R}(BC)$ for every $n \times s$ matrix C.
11 Let A and B be $m \times n$ and $p \times n$ matrices respectively and let C be the $(m+p) \times n$ matrix

$$C = \begin{bmatrix} A \\ B \end{bmatrix}$$

Prove:
(a) $\mathbb{R}(A) + \mathbb{R}(B) = \mathbb{R}(C)$.
(b) $\mathbb{R}(A) = \mathbb{R}(C)$ if and only if $\mathbb{R}(B) \subseteq \mathbb{R}(A)$.
(c) $\mathbb{N}(A) \cap \mathbb{N}(B) = \mathbb{N}(C)$.

12 Let A and B be $m \times n$ and $n \times p$ matrices respectively. Prove:
(a) $\mathbb{N}(AB) \supseteq \mathbb{N}(B)$.
(b) $\mathbb{N}(AB) = \mathbb{N}(B)$ if and only if $\mathbb{N}(A) \cap \mathbb{C}(B) = 0$.

13 Let A and B be $m \times n$ and $n \times p$ matrices respectively. Prove that

$$\mathbb{N}(A) \supseteq \mathbb{C}(B) \quad \text{if and only if} \quad \mathbb{R}(A) \subseteq \mathbb{N}_R(B)$$

14 Show that the set of simultaneous linear equations

$$\begin{aligned} a_{11}x_1 + a_{12}x_2 + \cdots + a_{1n}x_n &= d_1 \\ a_{21}x_1 + a_{22}x_2 + \cdots + a_{2n}x_n &= d_2 \\ &\cdots\cdots\cdots\cdots\cdots\cdots\cdots\cdots \\ a_{m1}x_1 + a_{m2}x_2 + \cdots + a_{mn}x_n &= d_m \end{aligned} \quad (1.4.8)$$

has a solution if and only if the vector

$$\begin{bmatrix} d_1 \\ d_2 \\ \vdots \\ d_m \end{bmatrix}$$

is contained in the column space of the $m \times n$ matrix $[a_{ij}]$ formed from the coefficients in the equations.

15 Show that if the rows of the $m \times n$ matrix A are linearly dependent, then $\mathbb{C}(A) \neq \mathbb{F}^m$. [*Hint:* Show that when the rows of A are linearly dependent, the system of equations (1.4.8) does not have a solution for some choice of $d_1, d_2, \ldots, d_m$.]

1.5 THE STEINITZ REPLACEMENT THEOREM

The replacement theorem of Steinitz (Theorem 1.5.1) is of fundamental importance in the study of vector spaces. Although the replacement theorem by itself does not establish the existence of a basis, it will permit us to show (in Theorem

1.5.6) not only that every subspace of $\mathbb{F}_n^m$ has a basis but also that all bases of a given vector space contain the same number of elements. A corollary to the replacement theorem provides a particularly simple means of identifying bases and leads to a method of constructing them. In addition, it will be clear that the matrices which belong to a basis are in no way "special" matrices, as one might think the unit vectors are. In fact, it will follow from the replacement theorem that for any linearly independent set of matrices of a vector space there is some basis which contains it.

Theorem 1.5.1 The Steinitz replacement theorem[1] Let $\mathbb{A} = \{A_1, A_2, \ldots, A_k\}$ be a basis for the vector space $\mathbb{V}$ and let $\mathbb{B} = \{B_1, B_2, \ldots, B_l\}$ be a set of linearly independent matrices belonging to $\mathbb{V}$. Then $l \leq k$. Moreover, there exist $k - l$ of the matrices of $\mathbb{A}$ which, together with the matrices of $\mathbb{B}$, form a basis for $\mathbb{V}$. In particular, if $l = k$, then $\mathbb{B}$ is a basis for $\mathbb{V}$.

PROOF The basic idea underlying the proof is that there is a process for repeatedly replacing one of the elements of $\mathbb{A}$ by one of the elements of $\mathbb{B}$ in such a way that the set which is constructed at each stage spans $\mathbb{V}$. Because we always replace one element of $\mathbb{A}$ with one element of $\mathbb{B}$, the total number of matrices in the set at any time remains unchanged, namely, k.

Since $\mathbb{A}$ spans $\mathbb{V}$, it follows that the $k + 1$ matrices $B_l, A_1, A_2, \ldots, A_k$ also span $\mathbb{V}$. Moreover, this set is linearly dependent since B_l can be expressed as a linear combination of the elements of the spanning set $\mathbb{A}$. Thus there exist numbers $\beta_l, \alpha_1, \alpha_2, \ldots, \alpha_k$, not all of which are zero, such that

$$\beta_l B_l + \alpha_1 A_1 + \alpha_2 A_2 + \cdots + \alpha_k A_k = 0 \qquad (1.5.2)$$

Furthermore, at least one of the numbers $\alpha_1, \alpha_2, \ldots, \alpha_k$ is not zero, for if $\alpha_1 = \alpha_2 = \cdots = \alpha_k = 0$, then $\beta_l \neq 0$ and $\beta_l B_l = 0$, from which it follows that $B_l = 0$. Hence $\{B_l\}$ is a linearly dependent set. However, we know that $\{B_l\}$ is a linearly independent set since it is a subset of the linearly independent set $\mathbb{B}$. To prevent our notation from becoming terribly cumbersome we will assume that the matrices constituting $\mathbb{A}$ are numbered in such a way that we may choose $\alpha_k \neq 0$ in equation (1.5.2) (see the last paragraph of Sec. 1.3). Then A_k can be expressed as a linear combi-

---

[1] Steinitz published this theorem in 1913, undoubtedly not aware that the same result had appeared in a book by Grassmann in 1862.

nation of $B_l, A_1, A_2, \ldots, A_{k-1}$. According to Lemma 1.3.16, A_k can be discarded, and the set

$$\mathbb{A}_1 = \{B_l, A_1, A_2, \ldots, A_{k-1}\}$$

spans $\mathbb{V}$. In other words, A_k has been "replaced" by B_l.

We repeat this process, beginning now with $\mathbb{A}_1$. Since $\mathbb{A}_1$ spans $\mathbb{V}$, it is clear that the $k + 1$ matrices $B_{l-1}, B_l, A_1, A_2, \ldots, A_{k-1}$ also span $\mathbb{V}$. This set is linearly dependent since B_{l-1} can be expressed as a linear combination of the elements of the spanning set $\mathbb{A}_1$. As before, there exist numbers $\delta_{l-1}, \delta_l, \gamma_1, \gamma_2, \ldots, \gamma_{k-1}$, not all zero, such that

$$\delta_{l-1}B_{l-1} + \delta_l B_l + \gamma_1 A_1 + \gamma_2 A_2 + \cdots + \gamma_{k-1}A_{k-1} = 0$$

Moreover, at least one of the numbers $\gamma_1, \gamma_2, \ldots, \gamma_{k-1}$ is not zero, for if $\gamma_1 = \gamma_2 = \cdots = \gamma_{k-1} = 0$, then at least one of δ_{l-1} and δ_l is not zero and we have

$$\delta_{l-1}B_{l-1} + \delta_l B_l = 0$$

which contradicts the linear independence of the set $\{B_{l-1}, B_l\}$. Again, for simplicity of notation, we will assume that the elements of $\mathbb{A}$ are numbered in such a way that we may now choose $\gamma_{k-1} \neq 0$. Then A_{k-1} can be expressed as a linear combination of $B_{l-1}, B_l, A_1, A_2, \ldots, A_{k-2}$. As before, it follows from Lemma 1.3.16 that

$$\mathbb{A}_2 = \{B_{l-1}, B_l, A_1, A_2, \ldots, A_{k-2}\}$$

spans $\mathbb{V}$.

We continue in this way, repeating the process as long as we are able. At each repetition one of the elements of $\mathbb{A}$ is replaced by one of the elements of $\mathbb{B}$. The process can no longer be carried out only when either (1) all the elements of $\mathbb{A}$ have been discarded or (2) all the elements of $\mathbb{B}$ have been brought into the spanning set. The first part of Theorem 1.5.1 merely states that the process can come to an end either by (1) or (2) above happening simultaneously (in which case $l = k$) or by (2) alone occurring (in which case $l < k$). In other words, we wish to show that the third possibility, namely, that (1) occurs before (2) (in which case we would have $l > k$), cannot occur. Let us suppose then that $l > k$ is possible. This means that after k steps of the process we will have shown that

$$\mathbb{A}_k = \{B_{l-k+1}, B_{l-k+2}, \ldots, B_{l-1}, B_l\}$$

spans $\mathbb{V}$. Furthermore, $l > k$ implies that $l - k + 1 \geq 2$, that is, B_1 does not belong to $\mathbb{A}_k$. However, B_1 belongs to $\mathbb{V}$, and thus we must be able to express B_1 as a linear combination of $B_{l-k+1}, B_{l-k+2}, \ldots, B_{l-1}, B_l$. But this is impossible since, by hypothesis, $\mathbb{B}$ is a linearly independent set. This contradiction arose from the supposition that $l > k$, which we now discard. Hence we have shown that $l \leq k$, as required.

The last spanning set we construct using the replacement process described above is $\mathbb{A}_l$. $\mathbb{A}_l$ consists of the l vectors which belonged to $\mathbb{B}$ and $k - l$ of the vectors of $\mathbb{A}$. To prove that the k elements of $\mathbb{A}_l$ form a basis for $\mathbb{V}$ it remains only to show that $\mathbb{A}_l$ is a linearly independent set. By Theorem 1.3.24, some subset of $\mathbb{A}_l$, say $\mathbb{S}$, is a basis for $\mathbb{V}$. Let $\mathbb{S}$ have m elements. Since $\mathbb{S}$ is a subset of $\mathbb{A}_l$, we have $m \leq k$. On the other hand, considering $\mathbb{S}$ as a basis for $\mathbb{V}$ and $\mathbb{A}$ as a linearly independent set of k matrices of $\mathbb{V}$, it follows from the first part of the theorem that $k \leq m$. Combining these results, we have

$$k \leq m \leq k$$

proving that $k = m$ and that $\mathbb{A}_l = \mathbb{S}$ is indeed a basis for $\mathbb{V}$. The last sentence of Theorem 1.5.1 merely points out the special case in which $k - l = 0$. ////

Note that the replacement theorem does not guarantee that a given vector space actually has a spanning set. We will not prove this until Theorem 1.5.6. However, if we assume the existence of bases, the replacement theorem does imply the very important:

Corollary 1.5.3 Any two bases for a vector space contain the same number of elements.

PROOF Let $\mathbb{A} = \{A_1, A_2, \ldots, A_k\}$ and $\mathbb{B} = \{B_1, B_2, \ldots, B_l\}$ be two bases for the same vector space $\mathbb{V}$ and suppose that $l \leq k$. By the replacement theorem there exist $k - l$ matrices of $\mathbb{A}$ which, together with the matrices of $\mathbb{B}$, form a basis for $\mathbb{V}$. However, $\mathbb{B}$ is already a basis for $\mathbb{V}$, and consequently a set consisting of the matrices of $\mathbb{B}$ and any further matrices of $\mathbb{V}$ is linearly dependent. Thus $k - l = 0$. ////

Corollary 1.5.3 gives us a most useful characterization of a basis, namely:

Corollary 1.5.4 Let $\mathbb{V}$ be a vector space which has a basis consisting of k elements. A set of necessary and sufficient conditions that the matrices $B_1, B_2, \ldots, B_l$ of $\mathbb{V}$ form a basis for $\mathbb{V}$ is:

1   $k = l$.
2   $B_1, B_2, \ldots, B_l$ are linearly independent.

In the special case where $\mathbb{V} = \mathbb{F}^m$ we have already found a basis consisting of m elements, namely, the m unit column vectors [see (1.3.11)]. It is the existence of a basis for $\mathbb{F}^m$ which will permit us to prove that every subspace of $\mathbb{F}^m$ has a basis. In order to apply the same methods to $\mathbb{F}_n^m$, we need a basis for $\mathbb{F}_n^m$. Any basis will do. The set of matrices of $\mathbb{F}_n^m$ which can most easily be shown to be a basis is the *mn unit $m \times n$ matrices*

$$\{I(i,j) \mid i = 1, 2, \ldots, m; j = 1, 2, \ldots, n\}$$

where $I(i,j)$ is the $m \times n$ matrix which has a 1 in the (i,j) position and 0s elsewhere. Note that when $n = 1$, these matrices become the m unit column vectors.

According to the replacement theorem, since the mn unit matrices form a basis for $\mathbb{F}_n^m$, every linearly independent subset of $\mathbb{F}_n^m$ contains at most mn elements. In other words:

Lemma 1.5.5   Every set of $mn + 1$ matrices of $\mathbb{F}_n^m$ is linearly dependent.

It is this lemma which provides the means of showing that every subspace of $\mathbb{F}_n^m$ actually possesses a basis.

Theorem 1.5.6   Every subspace of $\mathbb{F}_n^m$ has a basis consisting of not more than mn matrices.

PROOF   For $\mathbb{V} = 0$ we have defined the empty set to be a basis for $\mathbb{V}$. Now assume that $\mathbb{V} \neq 0$. Then $\mathbb{V}$ contains linearly independent sets of matrices. In particular, every set consisting of one nonzero matrix belonging to $\mathbb{V}$ is a linearly independent subset of $\mathbb{V}$. Let $\mathbb{A}$ be a linearly independent set of elements of $\mathbb{V}$ which is not contained in any larger linearly independent set of elements of $\mathbb{V}$. The existence of such a set follows from Lemma 1.5.5, and in fact $\mathbb{A}$ contains at most mn matrices. Thus we may set $\mathbb{A} = \{A_1, A_2, \ldots, A_k\}$. In order to show that $\mathbb{A}$ is a basis for $\mathbb{V}$ it remains only to prove that $A_1, A_2, \ldots, A_k$ span $\mathbb{V}$. Let $X \in \mathbb{V}$. By our method of selection of the set $\mathbb{A}$, the matrices $X, A_1, A_2, \ldots, A_k$ are linearly dependent. Hence there exist scalars, $\beta, \alpha_1, \alpha_2, \ldots, \alpha_k$, not all zero, such that

$$\beta X + \alpha_1 A_1 + \alpha_2 A_2 + \cdots + \alpha_k A_k = 0$$

Moreover, $\beta \neq 0$, for if $\beta = 0$, this equation would state that $A_1, A_2, \ldots,$

A_k are linearly dependent, which would be a contradiction. Hence

$$X = -\frac{\alpha_1}{\beta} A_1 - \frac{\alpha_2}{\beta} A_2 - \cdots - \frac{\alpha_k}{\beta} A_k$$

Thus $A_1, A_2, \ldots, A_k$ span $\mathbb{V}$. ////

Definition 1.5.7 Let $\mathbb{V}$ be a vector space. The number of elements in any basis for $\mathbb{V}$ is called the *dimension*[1] of $\mathbb{V}$ and is written dim $\mathbb{V}$.

Let A be an $m \times n$ matrix. The dimension of the column space of A is called the *column rank* of A and is denoted by $r_C(A)$. Similarly, the dimension of the row space of A is called the *row rank* of A and is denoted by $r_R(A)$. The dimension of the null space of A is the (*column*) *nullity* of A and is denoted by $\nu(A)$.

By Theorem 1.5.6, every subspace $\mathbb{V}$ of $\mathbb{F}_n^m$ has a dimension, and by Corollary 1.5.3, the dimension of $\mathbb{V}$ is well defined; i.e., the dimension of $\mathbb{V}$ is independent of which basis for $\mathbb{V}$ is used.

EXAMPLE 1.5.8 For the 3×4 matrix

$$B = \begin{bmatrix} -3 & 0 & 6 & 0 \\ 1 & 0 & -2 & 0 \\ -1 & 0 & 2 & 0 \end{bmatrix}$$

of Example 1.4.1 we showed that $\{B_{(1)}\}$ and $\{B^{(1)}\}$ are bases for $\mathbb{R}(B)$ and $\mathbb{C}(B)$ respectively. Thus

$$r_R(B) = r_C(B) = 1$$

It was also pointed out that $\mathbb{C}(B) = \mathbb{N}(A)$, where

$$A = \begin{bmatrix} 1 & 2 & -1 \\ 0 & 1 & 1 \end{bmatrix}$$

is the matrix of Example 1.3.4 Thus

$$\nu(A) = 1 \qquad ////$$

---

[1] The use of the word "dimension" in this context may cause some confusion When speaking of a single matrix (and, in particular, of a single vector) *dimension* refers to its physical size, i.e., to the number of rows and columns it has. When speaking about a vector space, dimension refers to the number of elements in a basis. Thus we may have a two-dimensional vector space consisting of three-dimensional column vectors. If we were to emphasize the geometric interpretation of vector spaces, we would call any one-dimensional vector space a *line* and any two-dimensional vector space a *plane*, even if they are subspaces of a three-dimensional vector space.

If $\mathbb{U}$ is a nonzero subspace of the vector space $\mathbb{V}$, then a basis $\mathbb{S}$ for $\mathbb{U}$ is a linearly independent set of matrices which belong to $\mathbb{V}$ and, by Theorem 1.5.1, can be extended to a basis $\mathbb{T}$ for $\mathbb{V}$. Hence $\dim \mathbb{U} \leq \dim \mathbb{V}$. Moreover, according to Corollary 1.5.4, $\mathbb{S}$ is also a basis for $\mathbb{V}$ if and only if the dimension of $\mathbb{V}$ is equal to the dimension of $\mathbb{U}$. Thus, when $\mathbb{U}$ is a subspace of $\mathbb{V}$, we have $\mathbb{U} = \mathbb{V}$ if and only if $\dim \mathbb{U} = \dim \mathbb{V}$.

There is an immediate application of these results to Lemma 1.4.2.

Corollary 1.5.9 Let A and B be $m \times n$ and $n \times p$ matrices respectively. Then:

1 $r_R(AB) \leq r_R(B)$.
2 $r_C(AB) \leq r_C(A)$.

At this point we have no way of comparing the column rank of AB with the column rank of B or the row rank of AB with the row rank of A. These difficulties will be resolved later.

In the remainder of this section we will consider the dimension of a sum of vector spaces. Specifically, let $\mathbb{U}_1, \mathbb{U}_2, \ldots, \mathbb{U}_l$ be vector spaces of $m \times n$ matrices and let

$$\mathbb{A}_1 = \{A_{11}, A_{12}, \ldots, A_{1k_1}\}$$
$$\mathbb{A}_2 = \{A_{21}, A_{22}, \ldots, A_{2k_2}\}$$
$$\cdots\cdots\cdots\cdots\cdots\cdots\cdots$$
$$\mathbb{A}_l = \{A_{l1}, A_{l2}, \ldots, A_{lk_l}\}$$

be bases for $\mathbb{U}_1, \mathbb{U}_2, \ldots, \mathbb{U}_l$ respectively. Clearly the $k_1 + k_2 + \cdots + k_l$ (not necessarily distinct) matrices of the set

$$\mathbb{A} = \{A_{11}, A_{12}, \ldots, A_{1k_1}, A_{21}, \ldots, A_{lk_l}\} \quad (1.5.10)$$

span the vector space $\mathbb{U}_1 + \mathbb{U}_2 + \cdots + \mathbb{U}_l$. By Theorem 1.3.24 some subset of $\mathbb{A}$ is a basis for $\mathbb{U}_1 + \mathbb{U}_2 + \cdots + \mathbb{U}_l$. Thus we have proved

Lemma 1.5.11 Let $\mathbb{U}_1, \mathbb{U}_2, \ldots, \mathbb{U}_l$ be vector spaces of $m \times n$ matrices. Then

$$\dim \mathbb{U}_1 + \mathbb{U}_2 + \cdots + \mathbb{U}_l \leq \dim \mathbb{U}_1 + \dim \mathbb{U}_2 + \cdots + \dim \mathbb{U}_l \quad (1.5.12)$$

Moreover, we have equality in (1.5.12) if and only if $\mathbb{A}$ is a linearly independent set. Let us give this a name.

Definition 1.5.13 Let $\mathbb{U}_1, \mathbb{U}_2, \ldots, \mathbb{U}_l$ be vector spaces of $m \times n$ matrices and set $\mathbb{U} = \mathbb{U}_1 + \mathbb{U}_2 + \cdots + \mathbb{U}_l$. If

$$\dim \mathbb{U} = \dim \mathbb{U}_1 + \dim \mathbb{U}_2 + \cdots + \dim \mathbb{U}_l$$

we say that $\mathbb{U}$ is a *direct sum* of $\mathbb{U}_1, \mathbb{U}_2, \ldots, \mathbb{U}_l$ and write

$$\mathbb{U} = \mathbb{U}_1 \oplus \mathbb{U}_2 \oplus \cdots \oplus \mathbb{U}_l$$

A natural example of a direct sum occurs when a vector space is identified by means of some basis. In that case the vector space is immediately expressable as a direct sum of one-dimensional subspaces. Specifically, let $\mathbb{A} = \{A_1, A_2, \ldots, A_k\}$ be any basis for the k-dimensional vector space $\mathbb{V}$. For each i, $i = 1, 2, \ldots, k$, denote by $\mathbb{V}_i$ the one-dimensional vector space spanned by the single matrix A_i. Then

$$\dim \mathbb{V} = k = \dim \mathbb{V}_1 + \dim \mathbb{V}_2 + \cdots + \dim \mathbb{V}_k$$

and therefore $\mathbb{V} = \mathbb{V}_1 \oplus \mathbb{V}_2 \oplus \cdots \oplus \mathbb{V}_k$. Note that Theorem 1.3.25 states that every element X of $\mathbb{V}$ has a unique expression of the form

$$X = X_1 + X_2 + \cdots + X_k$$

where $X_i \in \mathbb{V}_i$. This is, in fact, an important property of the direct sum.

Theorem 1.5.14 Let $\mathbb{U}_1, \mathbb{U}_2, \ldots, \mathbb{U}_l$ be vector spaces of $m \times n$ matrices and let $\mathbb{U} = \mathbb{U}_1 + \mathbb{U}_2 + \cdots + \mathbb{U}_l$. Then

$$\mathbb{U} = \mathbb{U}_1 \oplus \mathbb{U}_2 \oplus \cdots \oplus \mathbb{U}_l \quad (1.5.15)$$

if and only if every element X belonging to $\mathbb{U}$ has a unique representation as

$$X = X_1 + X_2 + \cdots + X_l \quad (1.5.16)$$

where $X_i \in \mathbb{U}_i$, $i = 1, 2, \ldots, l$.

PROOF For each i, $i = 1, 2, \ldots, l$, select a basis

$$\mathbb{A}_i = \{A_{i1}, A_{i2}, \ldots, A_{ik_i}\}$$

for $\mathbb{U}_i$ and let $\mathbb{A}$ be the set of $k_1 + k_2 + \cdots + k_l$ (not necessarily distinct) matrices

$$\mathbb{A} = \{A_{11}, A_{12}, \ldots, A_{1k_1}, A_{21}, \ldots, A_{lk_l}\}$$

Clearly $\mathbb{A}$ spans $\mathbb{U}$. According to the discussion preceding Definition 1.5.13, equation (1.5.15) holds if and only if the elements of $\mathbb{A}$ are (distinct and) linearly independent and hence form a basis for $\mathbb{U}$. Thus, in order to prove the theorem we will show that $\mathbb{A}$ is a linearly independent set if and only if every matrix X belonging to $\mathbb{U}$ has a unique representation of the form of equation (1.5.16).

First, let us assume that $\mathbb{U} = \mathbb{U}_1 \oplus \mathbb{U}_2 \oplus \cdots \oplus \mathbb{U}_l$. Then the elements of $\mathbb{A}$ are linearly independent. Let $X \in \mathbb{U}$ and suppose that

$$X = X_1 + X_2 + \cdots + X_l = Y_1 + Y_2 + \cdots + Y_l \quad (1.5.17)$$

where $X_i, Y_i \in \mathbb{U}_i$, $i = 1, 2, \ldots, l$. We wish to show that the two representations for X given in equation (1.5.17) are in fact the same representation. It follows from equation (1.5.17) that

$$0 = X - X = (X_1 - Y_1) + (X_2 - Y_2) + \cdots + (X_l - Y_l) \quad (1.5.18)$$

Setting $X_i - Y_i = Z_i$, equation (1.5.18) becomes

$$0 = Z_1 + Z_2 + \cdots + Z_l \quad (1.5.19)$$

and $Z_i \in \mathbb{U}_i$. Since $\mathbb{A}_i$ is a basis for $\mathbb{U}_i$, there exist scalars $\alpha_{i1}, \alpha_{i2}, \ldots, \alpha_{ik_i}$ such that

$$Z_i = \alpha_{i1} A_{i1} + \alpha_{i2} A_{i2} + \cdots + \alpha_{ik_i} A_{ik_i} \quad (1.5.20)$$

Substituting each of the l equations of (1.5.20) into equation (1.5.19), we have

$$0 = \alpha_{11} A_{11} + \alpha_{12} A_{12} + \cdots + \alpha_{1k_1} A_{1k_1} + \alpha_{21} A_{21} + \cdots + \alpha_{lk_l} A_{lk_l}$$

But we previously assumed that $\mathbb{A}$ is a linearly independent set. Therefore

$$\alpha_{11} = \alpha_{12} = \cdots = \alpha_{1k_1} = \alpha_{21} = \cdots = \alpha_{lk_l} = 0$$

from which it immediately follows that

$$Z_1 = Z_2 = \cdots = Z_l = 0$$

Consequently $X_i = Y_i$, $i = 1, 2, \ldots, l$, as required.

Conversely, let us now assume that the representation (1.5.16) is unique for every element X of $\mathbb{U}$. In particular, it is unique for $X = 0$. We wish to show that $\mathbb{A}$ is a linearly independent set. Let

$$0 = \alpha_{11} A_{11} + \alpha_{12} A_{12} + \cdots + \alpha_{1k_1} A_{1k_1} + \alpha_{21} A_{21} + \cdots + \alpha_{lk_l} A_{lk_l}$$

and set

$$Z_i = \alpha_{i1} A_{i1} + \alpha_{i2} A_{i2} + \cdots + \alpha_{ik_i} A_{ik_i} \qquad i = 1, 2, \ldots, l$$

Then we have

$$0 = Z_1 + Z_2 + \cdots + Z_l$$

But certainly

$$0 = 0 + 0 + \cdots + 0$$

is one representation for 0 in the form of (1.5.16). Since we have assumed that the representation of 0 is unique, it follows that

$$Z_1 = Z_2 = \cdots = Z_l = 0$$

However, $\mathbb{A}_i = \{A_{i1}, A_{i2}, \ldots, A_{ik_i}\}$, being a basis for $\mathbb{U}_i$, is a linearly independent set. Consequently $Z_i = 0$ implies that $\alpha_{i1} = \alpha_{i2} = \cdots = \alpha_{ik_i} = 0$. We can now conclude that $\mathbb{A}$ is a linearly independent set. ////

EXERCISES

1 Let

(a) $A = \begin{bmatrix} 1 & 3 & -2 \\ 2 & 6 & -4 \\ 0 & 0 & 0 \end{bmatrix}$ (b) $A = \begin{bmatrix} 1 & 2 & -1 \\ 0 & 1 & 1 \\ 2 & 1 & 0 \end{bmatrix}$

(c) $A = \begin{bmatrix} 1 & 2 & -1 \\ 0 & 1 & 1 \end{bmatrix}$

Find bases for $\mathbb{R}(A)$, $\mathbb{C}(A)$, and $\mathbb{N}(A)$.

2 Find the row rank, the column rank, and the nullity of:

(a) $\begin{bmatrix} 1 & 2 & 0 & -1 & -1 & 1 \\ -1 & 0 & 1 & 1 & 0 & -2 \\ 0 & 2 & 1 & 0 & -1 & -1 \end{bmatrix}$

(b) $\begin{bmatrix} 1 & 2 & 0 & -1 & -1 & 1 \\ -1 & 0 & 1 & 1 & 0 & -2 \end{bmatrix}$

3 Extend the linearly independent set of vectors

$$\begin{bmatrix} 0 \\ 2 \\ -1 \\ 0 \end{bmatrix} \begin{bmatrix} 0 \\ 1 \\ 1 \\ 0 \end{bmatrix}$$

to a basis for $\mathbb{F}^4$ using the method of the replacement theorem and the basis for $\mathbb{F}^4$ consisting of the unit column vectors.

4 Extend the basis for each $\mathbb{N}(A)$ of Exercise 1 to a basis for $\mathbb{F}^3$ using the method of the replacement theorem and the basis for $\mathbb{F}^3$ consisting of the unit column vectors.

5 Let

$$A = \begin{bmatrix} 2 & 1 \\ 4 & 2 \end{bmatrix} \qquad B = \begin{bmatrix} 3 & 0 \\ 0 & 4 \end{bmatrix}$$

Find bases for each of the following vector spaces:
(a) $\{X \mid XA = AX\}$ (b) $\{X \mid XA = BX\}$ (c) $\{X \mid AX = XB\}$

6 Let

$$A = \begin{bmatrix} 1 & 1 & 0 \\ 0 & 0 & 0 \\ 0 & 1 & 0 \end{bmatrix} \qquad B = \begin{bmatrix} 1 & 0 & 1 \\ 0 & 2 & 1 \end{bmatrix}$$

Find bases for:
(a) $\mathbb{R}(A) \cap \mathbb{R}(B)$.
(b) $\mathbb{R}(A) + \mathbb{R}(B)$. Is the sum direct?
(c) $\mathbb{N}(A) \cap \mathbb{N}(B)$.
(d) $\mathbb{N}(A) + \mathbb{N}(B)$. Is the sum direct?
(e) $\mathbb{N}(B) + \mathbb{C}(A)$. Is the sum direct?

7 Let $\mathbb{V}$ and $\mathbb{U}$ be the vector spaces of Exercises 2 and 4 of Sec. 1.3. Find a basis $\mathbb{S}$ for $\mathbb{V} \cap \mathbb{U}$. Extend $\mathbb{S}$ to a basis for $\mathbb{V}$ using the method of the replacement theorem and the basis $\{B_1, B_2\}$ for $\mathbb{V}$. Extend $\mathbb{S}$ to a basis for $\mathbb{U}$ using the basis $\{C_2, C_3\}$ for $\mathbb{U}$.

8 Let $\{X_1, X_2, X_3\}$ be a basis for the vector space $\mathbb{V}$. Show that $\{X_1 + X_2, X_2 + X_3, X_3\}$ is also a basis for $\mathbb{V}$.

9 Show that if $m < n$, then m simultaneous linear homogeneous equations in n unknowns have a nonzero solution. (*Hint:* Prove that the columns of the matrix of coefficients are linearly dependent.)

10 Let $\mathbb{U}$ and $\mathbb{V}$ be vector spaces, $\mathbb{U} \subseteq \mathbb{V}$. Show that there exists a subspace $\mathbb{W}$ of $\mathbb{V}$ such that $\mathbb{V} = \mathbb{U} \oplus \mathbb{W}$.

11 Let $\mathbb{V}_1, \mathbb{V}_2$ be subspaces of $\mathbb{V}$ for which $\mathbb{V} = \mathbb{V}_1 + \mathbb{V}_2$. Prove that $\mathbb{V} = \mathbb{V}_1 \oplus \mathbb{V}_2$ if and only if $\mathbb{V}_1 \cap \mathbb{V}_2 = 0$. (This is a very important characterization of a direct sum.)

12 Let $\mathbb{V}_1, \mathbb{V}_2$ be vector spaces of $m \times n$ matrices. Prove that

$$\dim(\mathbb{V}_1 + \mathbb{V}_2) = \dim \mathbb{V}_1 + \dim \mathbb{V}_2 - \dim(\mathbb{V}_1 \cap \mathbb{V}_2)$$

Show that Exercise 11 is a direct consequence of this formula. (*Hint:* Begin with a basis for $\mathbb{V}_1 \cap \mathbb{V}_2$. Apply Exercise 10 with $\mathbb{U} = \mathbb{V}_1 \cap \mathbb{V}_2$, $\mathbb{V} = \mathbb{V}_1$ and also with $\mathbb{U} = \mathbb{V}_1 \cap \mathbb{V}_2$, $\mathbb{V} = \mathbb{V}_2$.)

13 Let $\mathbb{V}_1, \mathbb{V}_2$ be subspaces of $\mathbb{F}_n^m$. Prove that

$$\dim(\mathbb{V}_1 \cap \mathbb{V}_2) \geq \dim \mathbb{V}_1 + \dim \mathbb{V}_2 - mn$$

14 Let $\mathbb{V}_1, \mathbb{V}_2, \ldots, \mathbb{V}_l$ be subspaces of the vector space $\mathbb{V}$ such that $\mathbb{V} = \mathbb{V}_1 + \mathbb{V}_2 + \cdots + \mathbb{V}_l$. For each j, $j = 1, 2, \ldots, l$, set

$$\mathbb{W}_j = \mathbb{V}_1 + \mathbb{V}_2 + \cdots + \mathbb{V}_{j-1} + \mathbb{V}_{j+1} + \cdots + \mathbb{V}_l$$

that is, $\mathbb{W}_j$ is the sum of all of the $\mathbb{V}_i$'s except $\mathbb{V}_j$. Show that

$\mathbb{V} = \mathbb{V}_1 \oplus \mathbb{V}_2 \oplus \cdots \oplus \mathbb{V}_l$ if and only if
$$\mathbb{W}_j \cap \mathbb{V}_j = 0 \quad j = 1, 2, \ldots, l$$
Show that Exercise 11 is a special case of this result.

15 Let A be a square matrix of order n. Prove that there exists a positive integer p and $p+1$ scalars $\alpha_p, \alpha_{p-1}, \ldots, \alpha_1, \alpha_0$ belonging to $\mathbb{F}$, not all zero, such that
$$\alpha_p A^p + \alpha_{p-1} A^{p-1} + \cdots + \alpha_1 A + \alpha_0 I = 0$$
Prove that p can always be chosen to be less than $n^2 + 1$.

16 Let A and B be $m \times n$ and $p \times q$ matrices, respectively, and denote by C the $(m+p) \times (n+q)$ matrix
$$C = \begin{bmatrix} A & 0 \\ 0 & B \end{bmatrix}$$
Prove that
(a) $\mathbb{R}(C) = \mathbb{R}([A \ 0]) \oplus \mathbb{R}([0 \ B])$.
(b) $\mathbb{N}(C) = \mathbb{N}([A \ 0]) \cap \mathbb{N}([0 \ B])$.
(c) $\mathbb{C}(C) = \mathbb{C}\left(\begin{bmatrix} A \\ 0 \end{bmatrix}\right) \oplus \mathbb{C}\left(\begin{bmatrix} 0 \\ B \end{bmatrix}\right)$.

Use these results to give relations between the row ranks of A, B, C, between the column ranks of A, B, C, and between the nullities of A, B, C.

17 Let $\mathbb{V}_1, \mathbb{V}_2$, and $\mathbb{V}$ be subspaces of $\mathbb{F}^n$ such that $\mathbb{V} = \mathbb{V}_1 \oplus \mathbb{V}_2$ and let A be an $m \times n$ matrix. Show that if $\mathbb{N}(A) \cap \mathbb{V} = 0$, then
$$A\mathbb{V} = A\mathbb{V}_1 \oplus A\mathbb{V}_2$$

18 Let $\mathbb{V}_1$ and $\mathbb{V}_2$ be subspaces of $\mathbb{F}^n$ and let A be an $m \times n$ matrix. Prove:
(a) $A(\mathbb{V}_1 + \mathbb{V}_2) = A\mathbb{V}_1 + A\mathbb{V}_2$ (b) $A(\mathbb{V}_1 \cap \mathbb{V}_2) \subseteq A\mathbb{V}_1 \cap A\mathbb{V}_2$
Give an example to show that we do not always have equality in (b). Further, prove that if $\mathbb{N}(A) \cap (\mathbb{V}_1 + \mathbb{V}_2) = 0$, then
$$A(\mathbb{V}_1 \cap \mathbb{V}_2) = A\mathbb{V}_1 \cap A\mathbb{V}_2$$

19 Let $\mathbb{V}$ be a subspace of $\mathbb{F}^n$ and let A be an $m \times n$ matrix. Prove that
$$\dim A\mathbb{V} = \dim \mathbb{V} - \dim[\mathbb{N}(A) \cap \mathbb{V}]$$
[*Hint:* Extend a basis for $\mathbb{N}(A) \cap \mathbb{V}$ to a basis for $\mathbb{V}$. Multiply the elements of this basis by A.]

20 Let $\mathbb{V}_1, \mathbb{V}_2$, and $\mathbb{V}$ be subspaces of $\mathbb{F}^n$ such that $\mathbb{V} = \mathbb{V}_1 \oplus \mathbb{V}_2$ and let A be an $m \times n$ matrix. Prove that $A\mathbb{V} = A\mathbb{V}_1 \oplus A\mathbb{V}_2$ if and only if
$$\mathbb{N}(A) \cap \mathbb{V} = [\mathbb{N}(A) \cap \mathbb{V}_1] \oplus [\mathbb{N}(A) \cap \mathbb{V}_2]$$

Show that Exercise 17 follows directly from this result. (*Hint:* It is always true that $\mathbb{N}(A) \cap \mathbb{V} \supseteq [\mathbb{N}(A) \cap \mathbb{V}_1] \oplus [\mathbb{N}(A) \cap \mathbb{V}_2]$. Thus we are concerned only with the inclusion in the opposite direction.)

1.6 NONSINGULAR MATRICES

We have associated three vector spaces with a given matrix A, namely, the column space $\mathbb{C}(A)$, the row space $\mathbb{R}(A)$, and the column null space $\mathbb{N}(A)$. [The row null space $\mathbb{N}_R(A)$ has also been introduced, but it does not play a prominent role in our treatment of the subject.] There are two relations between the dimensions of these vector spaces which are basic to all the work that follows. First, we will show, in Theorem 1.6.1, that for any matrix the sum of the column rank and the nullity is equal to the number of columns. The second relation which we will prove between the dimensions of these vector spaces is contained in Theorem 1.7.11 and states that for any matrix the row rank and the column rank are equal. As a result it will then not be necessary to distinguish the row rank from the column rank, and we will speak simply of the rank of a matrix.

For an $m \times n$ matrix the dimensions of the vectors belonging to the column space and the null space are generally different; that is, $\mathbb{C}(A) \subseteq \mathbb{F}^m$, $\mathbb{N}(A) \subseteq \mathbb{F}^n$. Thus, unless $m = n$, it is impossible to speak of the sum of these vector spaces. However, the dimensions of these vector spaces can be added.

Theorem 1.6.1 For any $m \times n$ matrix A

$$r_C(A) + v(A) = n$$

PROOF We begin by choosing any basis $\mathbb{S}_1$ for $\mathbb{N}(A)$. According to the definition of the nullity, $\mathbb{S}_1$ contains $v(A)$ elements, say

$$\mathbb{S}_1 = \{X_1, X_2, \ldots, X_{v(A)}\}$$

The elements of $\mathbb{S}_1$ are n-dimensional column vectors. By Theorem 1.5.1, $\mathbb{S}_1$ can be extended to a basis for $\mathbb{F}^n$; that is, there exist $n - v(A)$ vectors in $\mathbb{F}^n$, say $X_{v(A)+1}, X_{v(A)+2}, \ldots, X_n$ such that

$$\mathbb{S} = \{X_1, X_2, \ldots, X_{v(A)}, X_{v(A)+1}, X_{v(A)+2}, \ldots, X_n\}$$

is a basis for $\mathbb{F}^n$.

In order to prove that $r_C(A) = n - v(A)$, we will show that the $n - v(A)$ m-dimensional column vectors $AX_{v(A)+1}, AX_{v(A)+2}, \ldots, AX_n$ form a basis for $\mathbb{C}(A) = A\mathbb{F}^n \subseteq \mathbb{F}^m$. Set

$$\mathbb{S}_2 = \{X_{v(A)+1}, X_{v(A)+2}, \ldots, X_n\}$$

so that $S = S_1 \cup S_2$. Since S is a basis for $\mathbb{F}^n$, it follows that $A\mathbb{F}^n$ is spanned by

$$AS = \{AX_1, AX_2, \ldots, AX_{v(A)}, AX_{v(A)+1}, AX_{v(A)+2}, \ldots, AX_n\}$$

But

$$AX_1 = AX_2 = \cdots = AX_{v(A)} = 0$$

because $X_1, X_2, \ldots, X_{v(A)}$ belong to $\mathbb{N}(A)$. Thus

$$AS_2 = \{AX_{v(A)+1}, AX_{v(A)+2}, \ldots, AX_n\}$$

spans $A\mathbb{F}^n$ and contains $n - v(A)$ elements. It remains only to show that AS_2 is a linearly independent set.

Denote the vector space which has S_2 as a basis by $\mathbb{V}$. Then it is easily seen that

$$\mathbb{F}^n = \mathbb{N}(A) \oplus \mathbb{V}$$

and consequently (see Exercise 11 of Sec. 1.5)

$$\mathbb{N}(A) \cap \mathbb{V} = 0$$

Now let $\alpha_{v(A)+1}, \alpha_{v(A)+2}, \ldots, \alpha_n$ be scalars for which

$$\alpha_{v(A)+1} A X_{v(A)+1} + \alpha_{v(A)+2} A X_{v(A)+2} + \cdots + \alpha_n A X_n = 0$$

and set $\quad Y = \alpha_{v(A)+1} X_{v(A)+1} + \alpha_{v(A)+2} X_{v(A)+2} + \cdots + \alpha_n X_n$

Then $\quad AY = \alpha_{v(A)+1} A X_{v(A)+1} + \alpha_{v(A)+2} A X_{v(A)+2} + \cdots + \alpha_n A X_n = 0$

and consequently $\quad Y \in \mathbb{N}(A) \cap \mathbb{V} = 0$

But $X_{v(A)+1}, X_{v(A)+2}, \ldots, X_n$ are linearly dependent, and therefore

$$\alpha_{v(A)+1} = \alpha_{v(A)+2} = \cdots = \alpha_n = 0$$

as required. ////

There is a theorem similar to Theorem 1.6.1 connecting the row rank, the row nullity, and the number of rows of a matrix, the statement and proof of which are left as an exercise.

In Sec. 1.7 we will prove that the row rank and the column rank of any matrix are equal. As a preliminary we prove a lemma which is, in fact, a special case of the general theorem. As we will see, the machinery needed to prove the general theorem depends heavily on this lemma.

Lemma 1.6.2 Let A be a square matrix. The rows of A are linearly independent if and only if the columns of A are linearly independent.

PROOF It is sufficient to show that when the rows of A are linearly independent, the columns of A are also linearly independent. Hence, assume that $A_{(1)}, A_{(2)}, \ldots, A_{(n)}$ are linearly independent and let $\alpha_1, \alpha_2, \ldots, \alpha_n$ be numbers for which

$$\alpha_1 A^{(1)} + \alpha_2 A^{(2)} + \cdots + \alpha_n A^{(n)} = 0 \qquad (1.6.3)$$

If we denote the vector

$$\begin{bmatrix} \alpha_1 \\ \alpha_2 \\ \vdots \\ \alpha_n \end{bmatrix}$$

by X, then equation (1.6.3) can be written as

$$AX = 0$$

According to Corollary 1.5.4, $A_{(1)}, A_{(2)}, \ldots, A_{(n)}$ form a basis for $\mathbb{F}_n$, and it follows from part 1 of Corollary 1.4.7 that there exists a square matrix of order n, say C, such that $CA = I$. Then

$$X = I \cdot X = CAX = C \cdot 0 = 0$$

and we have shown that $A^{(1)}, A^{(2)}, \ldots, A^{(n)}$ are linearly independent, as required. ////

Square matrices with the property that their columns (and therefore, their rows) are linearly independent will play a special role in studying general rectangular matrices. Let us gather together some of the characterizations for such matrices which we have established up to this point. In later sections we will find still other characterizations for these matrices and will add them to the list.

Theorem 1.6.4 For a square matrix A of order n the following conditions are equivalent:

1 The columns of A are linearly independent.
2 The rows of A are linearly independent.
3 $\mathbb{C}(A) = \mathbb{F}^n$.
4 $\mathbb{R}(A) = \mathbb{F}_n$.
5 $\mathbb{N}(A) = 0$.
6 There exists a square matrix B of order n such that

$$AB = BA = I \qquad (1.6.5)$$

PROOF

$$\begin{array}{ll} 1 \Leftrightarrow 2 & \text{Lemma 1.6.2} \\ 1 \Leftrightarrow 3 & \text{Corollary 1.5.4} \\ 2 \Leftrightarrow 4 & \text{Corollary 1.5.4} \\ 3 \Leftrightarrow 5 & \text{Theorem 1.6.1} \\ 6 \Rightarrow 3 & \text{Corollary 1.4.7} \end{array}$$

It remains only to prove that $3 \Rightarrow 6$. By the two parts of Corollary 1.4.7, there exist matrices B and C such that $AB = CA = I$. Then

$$C = CI = CAB = IB = B \qquad ////$$

In proving that $3 \Rightarrow 6$ above, we showed that two matrices, B and C, which satisfy the equations

$$AB = I \qquad CA = I$$

must be equal. In particular, this implies that for a given matrix A, there is at most one solution to (1.6.5) [see part (*e*) of Exercises 3 and 4 of Sec. 1.2].

Definition 1.6.6 A square matrix satisfying any one (and hence all) of the six conditions of Theorem 1.6.4 is said to be *nonsingular*. For a nonsingular matrix A, the (necessarily unique) solution B of

$$AB = BA = I$$

is called the *inverse* of A and is denoted by A^{-1}.

Note that Definition 1.6.6 does not deal at all with matrices which are not square. A matrix which is not nonsingular (and this includes all nonsquare matrices) is called *singular*.

Nonsingular matrices are closely analogous to nonzero numbers, just as the zero matrix is closely analogous to the number zero. We shall see that singular matrices occupy a position somewhere between the two and might be thought of as "semizero" matrices. For example, although we have given examples to demonstrate that the cancellation law for multiplication of matrices is not valid in all cases, e.g., Example 1.2.5, it is valid when the matrix to be canceled is nonsingular. For if $AB = AC$ and A is nonsingular, multiplying on the left by A^{-1}, we have

$$B = IB = A^{-1}AB = A^{-1}AC = IC = C$$

As an important special case, note that if A is nonsingular and $AB = 0 \; (= A0)$, then $B = 0$.

But we must be careful. Since multiplication of matrices is generally not commutative, it does not follow from $AB = CA$ that $B = C$, even when A is nonsingular. In this case we shall say that B and C are *similar*. In Sec. 1.10 this situation is studied in great detail.

Nonsingular matrices have another very important property:

Theorem 1.6.7 Let A be an $m \times n$ matrix and let P and Q be nonsingular matrices of orders m and n respectively. Then A and PAQ have the same row rank and the same column rank.

PROOF It will be sufficient for us to prove that multiplying a matrix on the left by a nonsingular matrix does not change either the row rank or the column rank; that is, PA and A have the same row rank and the same column rank. In exactly the same way it can be shown that multiplication on the right by a nonsingular matrix does not affect either the row rank or the column rank. We can then conclude that PAQ and PA (and hence A) have the same row rank and the same column rank.

Applying part 1 of Lemma 1.4.2 twice, we have

$$\mathbb{R}(PA) \subseteq \mathbb{R}(A)$$

and $$\mathbb{R}(A) = \mathbb{R}(IA) = \mathbb{R}(P^{-1}PA) \subseteq \mathbb{R}(PA)$$

Thus $\mathbb{R}(A) = \mathbb{R}(PA)$ and consequently A and PA have the same row rank.

Now let us prove that A and PA have the same column rank. Since P is a square matrix of order m, PA and A have the same dimensions. Thus, making use of Theorem 1.6.1, in order to show that PA and A have the same column rank it is sufficient to show that they have the same nullity. Clearly

$$\mathbb{N}(A) \subseteq \mathbb{N}(PA)$$

since if $AX = 0$, then $PAX = 0$. In the same way,

$$\mathbb{N}(PA) \subseteq \mathbb{N}(P^{-1}PA) = \mathbb{N}(A)$$

Combining these relations, we have

$$\mathbb{N}(A) = \mathbb{N}(PA)$$

and consequently PA and A have the same nullity. ////

The proof of Theorem 1.6.7 provides a typical example of the special properties of nonsingular matrices. The existence of the inverse gives us much more freedom in moving matrices from one side of an equation to the other.

In the next section we will prove that the row rank and the column rank of any matrix are equal. The proof consists of little more than repeated applications of Theorem 1.6.7. Beginning with an arbitrary $m \times n$ matrix, we multiply it on the left and right by carefully chosen nonsingular matrices. This, of course, does not change either the row rank or the column rank. Finally, we arrive at an $m \times n$ matrix which has a particularly simple form and in which the row rank and column rank are equal.

EXERCISES

1 Show that the matrix

$$A = \begin{bmatrix} 1 & 0 & 1 \\ 1 & 1 & 0 \\ 1 & 1 & 1 \end{bmatrix}$$

is nonsingular by showing that:
(a) The rows of A form a basis for $\mathbb{F}_3$,
(b) The columns of A form a basis for $\mathbb{F}^3$,
(c) $\mathbb{N}(A) = 0$.

2 Find the inverse of

$$A = \begin{bmatrix} 1 & 3 \\ -1 & 0 \end{bmatrix}$$

by solving for B in the equation:

(a) $AB = I_2$ \qquad (b) $BA = I_2$

3 Repeat Exercise 2 using the matrix of Exercise 1 and replacing I_2 by I_3.

4 Verify Theorem 1.6.1 for the matrices:

(a) $\begin{bmatrix} 1 & 2 & 1 & 1 & 0 & 1 \\ 2 & 0 & 1 & 1 & 0 & -1 \end{bmatrix}$ \qquad (b) $\begin{bmatrix} 1 & 2 \\ 2 & 0 \\ 1 & 1 \\ 1 & 1 \\ 0 & 0 \\ 1 & -1 \end{bmatrix}$

5 Let

$$A = \begin{bmatrix} 1 & 1 & 2 \\ 0 & -1 & 0 \end{bmatrix}$$

Show that $AB = I_2$ has a solution but that $CA = I_3$ does not. Is B unique?

6 Find the inverse of the matrix of Exercise 6 of Sec. 1.4. For which values of x does the inverse exist?

7 Prove that the $(m + n) \times (m + n)$ matrix
$$\begin{bmatrix} I_n & 0 \\ A & I_m \end{bmatrix}$$
is nonsingular for every $m \times n$ matrix A. Find the inverse.

8 Prove that the $n \times n$ matrix
$$\begin{bmatrix} 1 & 0 & 0 & \cdots & 0 \\ a_{21} & 1 & 0 & \cdots & 0 \\ a_{31} & a_{32} & 1 & \cdots & 0 \\ \cdots\cdots\cdots\cdots\cdots\cdots\cdots \\ a_{n1} & a_{n2} & a_{n3} & \cdots & 1 \end{bmatrix}$$
is nonsingular for every choice of $a_{21}, a_{31}, a_{32}, \ldots, a_{n1}, a_{n2}, \ldots$.

9 State and prove the row analog of Theorem 1.6.1 and apply it to the two matrices of Exercise 4.

10 Show that if A is nonsingular, then A^{-1} is also nonsingular and its inverse is given by
$$(A^{-1})^{-1} = A$$

11 Prove that if A and B are nonsingular matrices of the same order, then AB is nonsingular and its inverse is given by
$$(AB)^{-1} = B^{-1}A^{-1}$$

12 Show that if A and B are square matrices of the same order and either A or B is singular, then AB is singular.

13 Show that if A and B are $m \times n$ and $n \times m$ matrices and if either A or B is singular, then at least one of AB and BA is singular. Find an example to show that one of AB and BA may be nonsingular.

14 Let A be an $m \times n$ matrix ($m < n$) whose rows are linearly independent. Prove that there exists an $(n - m) \times n$ matrix B such that the $n \times n$ matrix
$$\begin{bmatrix} A \\ B \end{bmatrix}$$
is nonsingular.

15 Show that if A and B are square matrices and $AB = I$, then $BA = I$.

16 Let A be an $m \times n$ matrix and let P and Q be nonsingular matrices of order m and n respectively. Prove that
$$\mathsf{N}(PAQ) = Q^{-1}\mathsf{N}(A)$$
Use this relation, Theorem 1.6.1, and Exercise 9 to give a different proof of Theorem 1.6.7.

17 Let A be a square matrix of order n. Prove that
$$\mathbb{F}^n = \mathbb{N}(A) \oplus \mathbb{C}(A)$$
if and only if $\mathbb{N}(A^2) = \mathbb{N}(A)$. (*Hint:* Use Exercise 11 of Sec. 1.5.)

18 Let A be a square matrix such that $A^2 = A$. Prove that $A = I$ or A is singular.

19 Let A be a square matrix of order n. Prove that a necessary and sufficient condition for there to exist numbers $\alpha_p, \alpha_{p-1}, \ldots, \alpha_1, \alpha_0$ belonging to $\mathbb{F}$ for which $\alpha_p A^p + \alpha_{p-1} A^{p-1} + \cdots + \alpha_1 A + \alpha_0 I = 0$ and $\alpha_0 \neq 0$ is that A be nonsingular. (See Exercise 15 of Sec. 1.5.)

1.7 THE ELEMENTARY OPERATIONS

When a matrix is multiplied on either side by a nonsingular matrix, neither its row rank nor its column rank is changed. This section is devoted to showing how a given matrix can be transformed by a sequence of multiplications by nonsingular matrices into a matrix having a particularly simple form (1.7.9). It will be clear that the row rank and the column rank of the resulting matrix, and hence of the original matrix, are equal. This sequence of multiplications can best be described in terms of *operations* on the rows and columns of the matrix. We will also show how each of these operations can be carried out by means of a multiplication by a properly chosen nonsingular matrix.

In this section we define two sets of operations on a matrix. One set consists of three column operations, and the other consists of the corresponding three row operations. These six operations will be called the *elementary operations*. We will also determine the nonsingular matrices which effect these operations. They will be called the *elementary matrices*.

The First Elementary Operation

Let A and B be $m \times n$ matrices and let r and s be distinct integers, $1 \leq r < s \leq m$. Further suppose that the rows of A and B are identical except for their rth and sth rows. Specifically, the rth row of A is identical with the sth row of B, and the sth row of A is identical with the rth row of B; that is,

$$B_{(i)} = A_{(i)} \quad i \neq r, s$$
$$B_{(r)} = A_{(s)}$$
$$B_{(s)} = A_{(r)}$$

We say that B has been obtained from A by the operation of interchanging the

rth and sth rows of A. We denote the operation of interchanging the rth and sth rows of a matrix by $\mathscr{R}_{rs}$ and write

$$\mathscr{R}_{rs} : A \to B$$

$\mathscr{R}_{rs}$ is called the *elementary row operation of the first kind*. Clearly we also have

$$\mathscr{R}_{rs} : B \to A$$

and thus the operation $\mathscr{R}_{rs}$ when performed twice on a matrix leaves it unchanged.

We have said that the elementary operations can be carried out by multiplying the given matrix by properly chosen nonsingular matrices. Let us construct a nonsingular matrix corresponding to the operation $\mathscr{R}_{rs}$. For distinct integers r and s, $1 \leq r < s \leq m$, we define the square matrix of order m, which we call R_{rs}, by

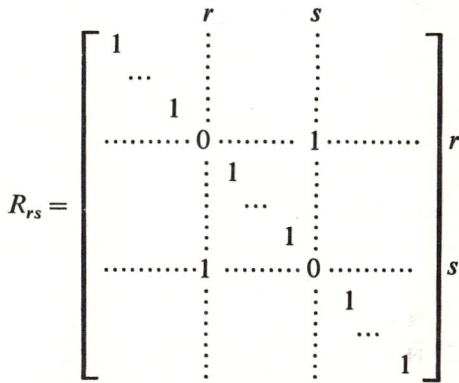

that is, R_{rs} has 1 in all the diagonal positions except the (r, r) and (s, s), positions, 1 in the (r, s) and (s, r) positions, and 0 elsewhere. Note that

$$\mathscr{R}_{rs} : I \to R_{rs}$$

R_{rs} is called the *elementary row matrix of the first kind*, corresponding to the elementary row operation $\mathscr{R}_{rs}$. In order to show that this definition is justified we must prove that the operation $\mathscr{R}_{rs}$ can be carried out by means of a multiplication on the left by R_{rs}, that is:

Lemma 1.7.1 Let $\mathscr{R}_{rs}$ be an elementary row operation of the first kind and let R_{rs} be the corresponding elementary row matrix of the first kind of order m. Then

$$\mathscr{R}_{rs} : A \to R_{rs} A$$

for every $m \times n$ matrix A.

PROOF The proof consists of comparing the entries of the matrix product $R_{rs}A$ with the result obtained by performing the operation $\mathscr{R}_{rs}$ on A. The details are left as an exercise. ////

EXAMPLE 1.7.2 Let

$$A = \begin{bmatrix} 1 & 2 \\ 3 & 4 \\ 5 & 6 \end{bmatrix} \quad \begin{matrix} r = 2 \\ s = 3 \end{matrix}$$

Then

$$R_{23} = \begin{bmatrix} 1 & 0 & 0 \\ 0 & 0 & 1 \\ 0 & 1 & 0 \end{bmatrix}$$

and it can easily be verified that

$$\mathscr{R}_{23}: \begin{bmatrix} 1 & 2 \\ 3 & 4 \\ 5 & 6 \end{bmatrix} \to \begin{bmatrix} 1 & 2 \\ 5 & 6 \\ 2 & 3 \end{bmatrix} = \begin{bmatrix} 1 & 0 & 0 \\ 0 & 0 & 1 \\ 0 & 1 & 0 \end{bmatrix} \begin{bmatrix} 1 & 2 \\ 3 & 4 \\ 5 & 6 \end{bmatrix} \quad ////$$

The Second Elementary Operation

Let A and B be $m \times n$ matrices, r an integer, $1 \leq r \leq m$, and λ a nonzero number. Suppose that the rows of A and B are related as follows:

$$B_{(i)} = A_{(i)} \quad i \neq r$$
$$B_{(r)} = \lambda A_{(r)}$$

B has been obtained from A by the operation of multiplying all the entries of the rth row of A by λ. We denote this operation by $\mathscr{R}_r(\lambda)$ and write

$$\mathscr{R}_r(\lambda): A \to B$$

$\mathscr{R}_r(\lambda)$ is called the *elementary row operation of the second kind*.

Let us construct a nonsingular matrix corresponding to the operation $\mathscr{R}_r(\lambda)$. For an integer r, $1 \leq r \leq m$, and a nonzero number λ, we define the square matrix of order m, which we call $R_r(\lambda)$, by

$$R_r(\lambda) = \begin{bmatrix} 1 & & & \vdots & & & \\ & \ddots & & \vdots & & & \\ & & 1 & \vdots & & & \\ \cdots & \cdots & \cdots & \lambda & \cdots & \cdots & \cdots \\ & & & \vdots & 1 & & \\ & & & \vdots & & \ddots & \\ & & & \vdots & & & 1 \end{bmatrix} r$$

that is, $R_r(\lambda)$ has 1 in all the diagonal positions except the (r, r) position, λ in the (r, r) position, and 0 elsewhere. Note that

$$\mathcal{R}_r(\lambda) : I \to R_r(\lambda)$$

$R_r(\lambda)$ is the *elementary row matrix of the second kind*, corresponding to the elementary row operation $\mathcal{R}_r(\lambda)$. It is important to remember that λ must be a nonzero number.

The row operation $\mathcal{R}_r(\lambda)$ can be carried out by means of a multiplication on the left by $R_r(\lambda)$, that is:

Lemma 1.7.3 Let $\mathcal{R}_r(\lambda)$ be an elementary row operation of the second kind and let $R_r(\lambda)$ be the corresponding elementary row matrix of the second kind of order m. Then

$$\mathcal{R}_r(\lambda) : A \to R_r(\lambda)A$$

for every $m \times n$ matrix A.

PROOF The proof is left as an exercise.   ////

EXAMPLE 1.7.4 Let

$$A = \begin{bmatrix} 1 & 2 \\ 3 & 4 \\ 5 & 6 \end{bmatrix} \quad \begin{array}{l} r = 2 \\ \lambda = -7 \end{array}$$

Then

$$R_2(-7) = \begin{bmatrix} 1 & 0 & 0 \\ 0 & -7 & 0 \\ 0 & 0 & 1 \end{bmatrix}$$

and it can easily be verified that

$$\mathcal{R}_2(-7): \begin{bmatrix} 1 & 2 \\ 3 & 4 \\ 5 & 6 \end{bmatrix} \to \begin{bmatrix} 1 & 2 \\ -21 & -28 \\ 5 & 6 \end{bmatrix} = \begin{bmatrix} 1 & 0 & 0 \\ 0 & -7 & 0 \\ 0 & 0 & 1 \end{bmatrix} \begin{bmatrix} 1 & 2 \\ 3 & 4 \\ 5 & 6 \end{bmatrix} \quad ////$$

The Third Elementary Operation

Let A and B be $m \times n$ matrices, r and s distinct integers, $1 \leq r, s \leq m$, and α any number. Suppose that the rows of A and B are related as follows:

$$B_{(i)} = A_{(i)} \quad i \neq r$$
$$B_{(r)} = A_{(r)} + \alpha A_{(s)}$$

Note that we require r and s to be distinct. B has been obtained from A by the operation of adding the sth row of A, multiplied by α, to the rth row of A. We denote this operation by $\mathscr{R}_{rs}(\alpha)$ and write

$$\mathscr{R}_{rs}(\alpha) : A \to B$$

$\mathscr{R}_{rs}(\alpha)$ is called the *elementary row operation of the third kind*.

For distinct integers r and s, $1 \leq r, s \leq m$, and a number α we define the square matrix of order m, which we call $R_{rs}(\alpha)$, by

$$R_{rs}(\alpha) = \begin{bmatrix} 1 & & & & \vdots & & \\ & \ddots & & & \vdots & & \\ & & 1 & & \vdots & & \\ & & \cdots & 1 & \cdots \alpha & \cdots & \\ & & & & \vdots & & \\ & & & & 1 & & \\ & & & & \vdots & \ddots & \\ & & & & \vdots & & 1 \end{bmatrix} \begin{matrix} \\ \\ \\ r \\ \\ \\ \\ \end{matrix}$$

that is, $R_{rs}(\alpha)$ has 1 in all the diagonal positions, α in the (r, s) position, and 0 elsewhere. Note that

$$\mathscr{R}_{rs}(\alpha) : I \to R_{rs}(\alpha)$$

$R_{rs}(\alpha)$ is the *elementary row matrix of the third kind*, corresponding to the elementary row operation $\mathscr{R}_{rs}(\alpha)$.

The row operation $\mathscr{R}_{rs}(\alpha)$ can be carried out by means of a multiplication on the left by $R_{rs}(\alpha)$, that is:

Lemma 1.7.5 Let $\mathscr{R}_{rs}(\alpha)$ be an elementary row operation of the third kind and let $R_{rs}(\alpha)$ be the corresponding elementary row matrix of the third kind of order m. Then

$$\mathscr{R}_{rs}(\alpha) : A \to R_{rs}(\alpha)A$$

for every $m \times n$ matrix A.

PROOF. The proof is left as an exercise. ////

EXAMPLE 1.7.6 Let

$$A = \begin{bmatrix} 1 & 2 \\ 3 & 4 \\ 5 & 6 \end{bmatrix} \quad \begin{matrix} r = 2 \\ s = 3 \\ \alpha = -2 \end{matrix}$$

Then
$$R_{23}(-2) = \begin{bmatrix} 1 & 0 & 0 \\ 0 & 1 & -2 \\ 0 & 0 & 1 \end{bmatrix}$$

and it can easily be verified that

$$\mathscr{R}_{23}(-2): \begin{bmatrix} 1 & 2 \\ 3 & 4 \\ 5 & 6 \end{bmatrix} \rightarrow \begin{bmatrix} 1 & 2 \\ -7 & -8 \\ 5 & 6 \end{bmatrix} = \begin{bmatrix} 1 & 0 & 0 \\ 0 & 1 & -2 \\ 0 & 0 & 1 \end{bmatrix} \begin{bmatrix} 1 & 2 \\ 3 & 4 \\ 5 & 6 \end{bmatrix} \quad ////$$

We have now defined and illustrated the three elementary row operations $\mathscr{R}_{rs}$, $\mathscr{R}_r(\lambda)$, and $\mathscr{R}_{rs}(\alpha)$. The second set of three operations on a matrix consists of the *elementary column operations*:

1 $\mathscr{C}_{rs}$, interchange the rth and sth columns of the matrix
2 $\mathscr{C}_r(\lambda)$, multiply all the entries of the rth column of the matrix by λ
3 $\mathscr{C}_{rs}(\alpha)$, add the sth column of the matrix, multiplied by α, to the rth column

To these operations there correspond the elementary column matrices, C_{rs}, $C_r(\lambda)$, and $C_{rs}(\alpha)$. It is left to the reader to construct these matrices and show that the elementary column operations can be carried out by multiplying the matrix on the right by the corresponding elementary matrices.

When the elementary operations and matrices were introduced, it was claimed that any matrix could be reduced to a simple form by means of them and without changing either the row rank or the column rank of the matrix. As we will see, this process depends on Theorem 1.6.7 and hence we must first show that:

Lemma 1.7.7 The elementary matrices are nonsingular.

PROOF. It is sufficient to show that each elementary matrix has an inverse.

$$[R_{rs}]^{-1} = R_{rs} \qquad [C_{rs}]^{-1} = C_{rs}$$
$$[R_r(\lambda)]^{-1} = R_r\!\left(\frac{1}{\lambda}\right)^{*} \qquad [C_r(\lambda)]^{-1} = C_r\!\left(\frac{1}{\lambda}\right)$$
$$[R_{rs}(\alpha)]^{-1} = R_{rs}(-\alpha) \qquad [C_{rs}(\alpha)]^{-1} = C_{rs}(-\alpha) \qquad ////$$

Thus an elementary operation performed on a matrix changes neither the row rank nor the column rank. Let us now demonstrate how a given matrix can be reduced to one having a simpler form, namely, (1.7.9), by means of elementary operations.

---

* It is for this reason that we insist that $\lambda \neq 0$.

Theorem 1.7.8 Let A be an $m \times n$ matrix. There is a finite sequence of elementary row and column operations which transforms A into the $m \times n$ matrix.[1]

$$\begin{bmatrix} I & 0 \\ 0 & 0 \end{bmatrix} \quad (1.7.9)$$

PROOF The proof consists of describing the process of reducing A to the desired form. Since we would quickly run out of letters if we changed the notation for the matrix we are operating on after each operation, it will be convenient to refer to it as A throughout the reduction process.

If $A = 0$, then A already has the form (1.7.9) and there is nothing to prove. Henceforth we assume $A \neq 0$. First we wish to obtain a nonzero number in the (1, 1) position of A. If a_{11} is already nonzero, this step is unnecessary. Since $A \neq 0$, the entry in some position of A, say the entry in the (r, s) position, is not zero. This number can be brought into the (1, 1) position by at most two elementary operations of the first kind. If $r \neq 1$, we interchange the first and rth rows of A. If $s \neq 1$, we interchange the first and sth columns of A. In this way we arrive at a matrix (still called A) for which $a_{11} \neq 0$. Now multiply the entries of the first row of A by $1/a_{11}$. This is an elementary row operation of the second kind, and in the resulting matrix A we have $a_{11} = 1$.

By at most $m + n - 2$ elementary operations of the third kind we can obtain zeros in all positions of the first row and first column of A except for the (1, 1) position and thereby reduce A to a matrix of the form

$$A = \begin{bmatrix} 1 & 0 \\ 0 & \tilde{A} \end{bmatrix} \quad (1.7.10)$$

where $\tilde{A}$ is an $(m - 1) \times (n - 1)$ matrix. This is accomplished in the following way:

1 For $i = 2, 3, \ldots, m$, add the first row of A, multiplied by $-a_{i1}$, to the ith row of A,
2 For $j = 2, 3, \ldots, n$, add the first column of A, multiplied by $-a_{1j}$, to the jth column of A.

The resulting matrix has the form (1.7.10).

If $\tilde{A} = 0$, we are finished. If $\tilde{A} \neq 0$, we repeat the entire process.

---

[1] If the row rank of A is equal to the number of rows of A, the row of zero matrices of (1.7.9) does not actually appear. Similarly, if the column rank of A is equal to the number of columns of A, the column of zero matrices of (1.7.9) does not actually appear.

All the elementary operations on the matrix A of (1.7.10) needed to reduce the submatrix $\tilde{A}$ can easily be chosen in such a way that at no time are the first row and first column of A disturbed. In this way we reduce A to the form

$$A = \begin{bmatrix} 1 & 0 & 0 \\ 0 & 1 & 0 \\ 0 & 0 & \tilde{A} \end{bmatrix}$$

where $\tilde{A}$ is now an $(m-2) \times (n-2)$ matrix.

Continuing this process, after a finite number of elementary operations we arrive at the required form

$$\begin{bmatrix} I & 0 \\ 0 & 0 \end{bmatrix} \qquad ////$$

Both the row rank and the column rank of a matrix of the form of (1.7.9) are equal to the order of I. Applying Theorem 1.6.7, we have now proved:

Theorem 1.7.11 *The row rank and the column rank of a matrix are equal.*

It is no longer necessary to distinguish between the row rank and the column rank of a matrix, and henceforth we use the term *rank* to refer to either one. The rank of a matrix A will be denoted by $r(A)$.

When the elementary row and column operations are carried out by means of elementary row and column matrices, Theorem 1.7.8 becomes:

Corollary 1.7.12 *Let A be an $m \times n$ matrix of rank r. There exist elementary matrices $R_1, R_2, \ldots, R_k$ and $C_1, C_2, \ldots, C_l$ which transform A into the form (1.7.9), that is,*

$$R_k \cdots R_2 R_1 A C_1 C_2 \cdots C_l = \begin{bmatrix} I_r & 0 \\ 0 & 0 \end{bmatrix} \qquad (1.7.13)$$

Let us multiply both sides of equation (1.7.13) on the left by $R_1^{-1} R_2^{-1} \cdots R_k^{-1}$ and on the right by $C_l^{-1} \cdots C_2^{-1} C_1^{-1}$. Then

$$A = R_1^{-1} R_2^{-1} \cdots R_k^{-1} \begin{bmatrix} I_r & 0 \\ 0 & 0 \end{bmatrix} C_l^{-1} \cdots C_2^{-1} C_1^{-1}$$

However, we have shown that the inverse of an elementary matrix is again an elementary matrix, and so we have proved:

Corollary 1.7.14 Let A be an $m \times n$ matrix of rank r. There exist elementary matrices $R_1, R_2, \ldots, R_k$ and $C_1, C_2, \ldots, C_l$ such that

$$A = R_1 R_2 \cdots R_k \begin{bmatrix} I_r & 0 \\ 0 & 0 \end{bmatrix} C_l \cdots C_2 C_1 \quad (1.7.15)$$

Note that although we use the same notation in both cases, the elementary matrices of equation (1.7.15) are the inverses of the elementary matrices of equation (1.7.13) and are in the opposite order. It is in the form of equation (1.7.15) that the reduction of a matrix by elementary matrices is most often used.

Theorem 1.7.11 is fundamental to all of matrix theory. The rank of a matrix is one of its most important attributes. In Sec. 1.9 we will see another way in which the rank can be defined. Now, however, let us show how Theorem 1.7.11 leads to a proof of the converse of Corollary 1.4.6. If B and C are $m \times n$ and $p \times n$ matrices respectively, then $\mathbb{R}(B)$ and $\mathbb{R}(C)$ are both subspaces of $\mathbb{F}_n$ while $\mathbb{N}(B)$ and $\mathbb{N}(C)$ are subspaces of $\mathbb{F}^n$. In Corollary 1.4.6 we showed that if $\mathbb{R}(C) \subseteq \mathbb{R}(B)$, then $\mathbb{N}(C) \supseteq \mathbb{N}(B)$. It follows from Theorem 1.6.1 (without Theorem 1.7.11) that if $\mathbb{N}(C) \supseteq \mathbb{N}(B)$, then $r_C(C) \leq r_C(B)$. With Theorem 1.7.11 we have:

Theorem 1.7.16 Let B and C be $m \times n$ and $p \times n$ matrices respectively. If $\mathbb{N}(C) \supseteq \mathbb{N}(B)$, then $\mathbb{R}(C) \subseteq \mathbb{R}(B)$. Hence, if $\mathbb{N}(C) = \mathbb{N}(B)$, then $\mathbb{R}(C) = \mathbb{R}(B)$.

PROOF Let $\mathbb{N}(C) \supseteq \mathbb{N}(B)$ and set

$$D = \begin{bmatrix} B \\ C \end{bmatrix}$$

According to part (c) of Exercise 11 in Sec. 1.4,

$$\mathbb{N}(B) \cap \mathbb{N}(C) = \mathbb{N}(D) \quad (1.7.17)$$

Also, it is clear that

$$\mathbb{R}(B) \subseteq \mathbb{R}(D) \quad \mathbb{R}(C) \subseteq \mathbb{R}(D) \quad (1.7.18)$$

However, since $\mathbb{N}(C) \supseteq \mathbb{N}(B)$, we have

$$\mathbb{N}(B) \cap \mathbb{N}(C) = \mathbb{N}(B)$$

Comparing this equation with equation (1.7.17) yields

$$\mathbb{N}(B) = \mathbb{N}(D)$$

Thus $v(B) = v(D)$, and it follows from Theorems 1.6.1 and 1.7.11 that $r(B) = r(D)$. Combining this with the first part of (1.7.18), we have shown that $\mathbb{R}(B)$ is a subspace of $\mathbb{R}(D)$ whose dimension is equal to the dimension of $\mathbb{R}(D)$. Applying Corollary 1.5.4, we conclude that

$$\mathbb{R}(B) = \mathbb{R}(D)$$

The second part of (1.7.18) now yields

$$\mathbb{R}(C) \subseteq \mathbb{R}(B) \qquad ////$$

Thus the row space and the null space of a matrix determine each other, though neither one determines the matrix itself.

EXERCISES

1. Reduce the matrices

   (a) $A = \begin{bmatrix} 0 & 0 & 1 & 1 \\ 2 & -4 & 2 & 0 \\ -1 & 2 & 0 & 1 \end{bmatrix}$ (b) $B = \begin{bmatrix} 0 & 1 & 1 & 0 \\ 2 & 0 & 2 & -4 \\ -2 & 3 & 1 & 4 \end{bmatrix}$

   to the form (1.7.9) by means of elementary operations.

2. Transform one of the matrices of Exercise 1 into the other by means of elementary operations by first reducing both matrices to form (1.7.9). Show that it is possible to transform one matrix into the other without transforming either into form (1.7.9).

3. Find the rank of the matrix

   $$A = \begin{bmatrix} 1 & 2 & -1 & 0 & 1 \\ 2 & 0 & 0 & 1 & 1 \\ 1 & -2 & 1 & 1 & 0 \end{bmatrix}$$

   (a) By finding a basis for $\mathbb{R}(A)$.
   (b) By finding a basis for $\mathbb{C}(A)$.
   (c) By finding a basis for $\mathbb{N}(A)$ and applying Theorem 1.6.1.
   (d) By transforming A to form (1.7.9).

4. Reduce the matrix A of the examples of this section to the form (1.7.9).

5. Let

   $$A = \begin{bmatrix} 0 & 1 & 2 \\ -1 & 1 & 0 \\ 1 & 0 & 0 \end{bmatrix}$$

Show that A can be reduced to the form (1.7.9) using only elementary row operations. Show also that A can be reduced to the form (1.7.9) using only elementary column operations. Can this be done with the matrices of Exercise 1? Why?

6 Describe each of the elementary column matrices C_{rs}, $C_r(\lambda)$, and $C_{rs}(\alpha)$. Show that every elementary column matrix is equal to an elementary row matrix of the same kind.

7 Let A be a $4 \times n$ matrix. Find a matrix R such that RA can be obtained from A by performing all the following operations:
(a) Interchanging the first and third rows of A.
(b) Adding a fifth row to A which is equal to the fourth row of A.
(c) Multiplying the fourth row of A by 3.
(d) Deleting the second row of A.

8 Prove that a matrix is nonsingular if and only if it can be expressed as a product of elementary matrices.

9 We wish to perform one elementary row operation $\mathscr{R}$ and one elementary column operation $\mathscr{C}$ on a given matrix. Show that the same matrix is obtained irrespective of whether $\mathscr{R}$ or $\mathscr{C}$ is performed first. Show that this result is no longer true if both operations are row operations or both are column operations. (*Hint:* Use the elementary matrices.)

10 Supply proofs for Lemmas 1.7.1, 1.7.3, and 1.7.5 and Corollary 1.7.14.

11 Let A and B be $m \times n$ and $p \times q$ matrices respectively and let C be the $(m + p) \times (n + q)$ matrix

$$C = \begin{bmatrix} A & 0 \\ 0 & B \end{bmatrix}$$

Prove that $r(C) = r(A) + r(B)$ and $v(C) = v(A) + v(B)$.

12 Two $m \times n$ matrices A and B are said to be *equivalent* if there exist nonsingular matrices P and Q such that $B = PAQ$. Prove that equivalence of matrices is an equivalence relation.

13 Prove that two matrices are equivalent if and only if one of them can be transformed into the other by a finite number of elementary operations.

14 Prove that two $m \times n$ matrices are equivalent if and only if they have the same rank.

15 Prove that the first elementary row operation can be expressed in terms of the second and third elementary row operations.

16 Prove that a nonsingular matrix can be reduced to the identity matrix using only elementary row operations or only elementary column operations (see Exercise 5).

17 Let A and B be $m \times n$ matrices for which there exist matrices C and D such that $A = CB$, $B = AD$. Prove that there exist matrices F and G such that $A = BF$, $B = GA$.

18 Let B and C be $m \times n$ and $p \times n$ matrices respectively. Prove that a necessary and sufficient condition that there exist a $p \times m$ matrix A such that $AB = C$ is that $\mathbb{N}(C) \supseteq \mathbb{N}(B)$.

19 Let A be an $m \times n$ matrix. Show that A can be expressed as

$$A = XY$$

where $X \in \mathbb{F}^m$, $Y \in \mathbb{F}_n$, if and only if $r(A) \leq 1$. Use the methods of this exercise to give an independent proof of the fact that if the row rank of a matrix is 1, then the column rank of the matrix is 1.

20 Let $A = [a_{ij}]$ be an $m \times n$ matrix. We define the $n \times m$ matrix $C = [c_{ij}]$ by

$$a_{ij} = c_{ji} \quad \begin{matrix} i = 1, 2, \ldots, m \\ j = 1, 2, \ldots, n \end{matrix}$$

that is, the entry in the (i, j) position of A is equal to the entry in the (j, i) position of C. Then C is called the *transpose* of A and is denoted by A^T. Prove:

(a) $(A + B)^T = A^T + B^T$.
(b) $(\alpha A)^T = \alpha A^T$.
(c) $(AB)^T = B^T A^T$.
(d) $(A^T)^T = A$.
(e) $r(A) = r(A^T)$.
(f) If A is nonsingular, then A^T is nonsingular and $(A^T)^{-1} = (A^{-1})^T$.

21 Prove that if A and C are $m \times n$ matrices such that $\mathbb{R}(A) = \mathbb{R}(C)$, there is a nonsingular matrix B of order m such that $A = BC$. [*Hint:* Show that $\mathbb{R}(PAQ) = \mathbb{R}(PCQ)$ for all nonsingular matrices P and Q of orders m and n respectively. Choose P and Q so that PCQ has the form (1.7.9) and use Exercise 14 of Sec. 1.6.]

22 Let A be a square singular matrix. Prove that there exists a nonzero matrix B such that $AB = BA = 0$. [*Hint:* First prove the result for matrices which have the form (1.7.9).]

23 Let A be a square matrix. Prove that $r(A) = r(A^2)$ if and only if

$$\mathbb{F}^n = \mathbb{C}(A) \oplus \mathbb{N}(A)$$

24 Show that when $\mathbb{F}$ is the real field,

$$\mathbb{F}^n = \mathbb{C}(A^T) \oplus \mathbb{N}(A)$$

for every $m \times n$ matrix A. Show that this equation is no longer valid when $\mathbb{F}$ is the complex field.

25 Let A and B be $m \times n$ and $n \times p$ matrices respectively. Prove:
 (a) $r(AB) \leq \min\{r(A), r(B)\}$.
 (b) $v(AB) \leq v(A) + v(B)$. [*Hint:* Extend a basis for $\mathbb{N}(B)$ to a basis for $\mathbb{N}(AB)$.]
 (c) $r(A) + r(B) - n \leq r(AB)$.

26 Let A and B be $m \times n$ matrices. Prove that

$$r(A + B) \leq r(A) + r(B)$$

27 Let A and B be $m \times n$ and $n \times p$ matrices respectively. Prove that $\mathbb{N}(A) = \mathbb{C}(B)$ if and only if $\mathbb{R}(A) = \mathbb{N}_R(B)$. (*Hint:* See Exercise 13 of Sec. 1.4. Use an argument based on ranks and nullities.)

28 Let A be an $m \times n$ matrix of rank r in which the first r rows and the first r columns are linearly independent sets. Let $\tilde{A}$ be the $r \times r$ submatrix of A formed by deleting all rows and all columns of A except the first r. Prove that $\tilde{A}$ is nonsingular. Construct an example to show that if the rank of A is greater than the number of linearly independent rows and columns used to form $\tilde{A}$, then $\tilde{A}$ need not be nonsingular. (*Hint:* Instead of deleting these rows and columns from A, change them into rows and columns of zeros by multiplying A on the left and on the right by properly chosen nonsingular matrices.)

29 Prove that a set of necessary and sufficient conditions that a matrix have rank r is:
 (a) There is a nonsingular submatrix of order r.
 (b) For each s, $s > r$, every submatrix of order s is singular.

30 Prove that a set of necessary and sufficient conditions that a matrix A have rank r is:
 (a) A has a nonsingular submatrix $\tilde{A}$ of rank r.
 (b) Every square submatrix of A of order $r + 1$ which has $\tilde{A}$ as a submatrix is singular.
 (*Hint:* Show that each row of A which does not meet $\tilde{A}$ can be expressed uniquely as a linear combination of the rows of A which do meet $\tilde{A}$.)

31 Let A, X, Y, and α be $n \times n$, $n \times 1$, $1 \times n$, and 1×1 matrices respectively and let A be singular. Prove that if the $(n + 1) \times (n + 1)$ matrix

$$\begin{bmatrix} A & X \\ Y & \alpha \end{bmatrix}$$

is nonsingular for some value of α, then it is nonsingular for all values of α.

1.8 PERMUTATIONS[1]

In the next section we will introduce the determinant of a square matrix. The definition of the determinant we will use requires a knowledge of some of the basic facts about the set of permutations of the set $\mathbb{N} = \{1, 2, \ldots, n\}$. Let us begin with a definition.

Definition 1.8.1 A *permutation* of a set $\mathbb{N}$ is a one-to-one mapping of $\mathbb{N}$ onto itself. The set of all permutations of $\mathbb{N}$ is denoted by $\mathbb{S}_n$.

We will be concerned only with permutations of a set $\mathbb{N}$ consisting of the first n positive integers. In order to avoid having to make exceptions for the special case when $n = 1$, we assume throughout this section (and only this section) that $n \geq 2$. A permutation σ which maps 1 onto a, 2 onto b, 3 onto c, ..., n onto d is denoted by

$$\sigma = \begin{pmatrix} 1 & 2 & 3 & \cdots & n \\ a & b & c & \cdots & d \end{pmatrix} \quad (1.8.2)$$

The mapping which maps every element of $\mathbb{N}$ onto itself is a permutation and is called the *identity permutation*. It is denoted either by I or by

$$\begin{pmatrix} 1 & 2 & 3 & \cdots & n \\ 1 & 2 & 3 & \cdots & n \end{pmatrix}$$

There are n different possibilities for the entry a of (1.8.2); that is, a can be any one of the numbers $1, 2, 3, \ldots, n$. Once a has been chosen, there are $n - 1$ choices for b. Continuing in this way, we have:

Lemma 1.8.3 $\mathbb{S}_n$ contains $n!$ elements.

Since the domain and range of a permutation are identical, for any two elements of $\mathbb{S}_n$, say σ_1 and σ_2, we may define the product $\sigma_1 \sigma_2$ by

$$(\sigma_1 \sigma_2)(x) = \sigma_1(\sigma_2(x)) \quad x \in \mathbb{N}$$

Then $\sigma_1 \sigma_2$ is again a permutation of $\mathbb{N}$. It is easy to show that with this definition, multiplication of permutations of $\mathbb{N}$ is associative. It will soon become apparent that, in general, multiplication of permutations is not commutative.

---

[1] Most of this section forms one of the two digressions from matrix theory in this chapter. The other one is Sec. 1.11. Each is self-contained and independent of the remainder of the book.

EXAMPLE 1.8.4 Let $\mathbb{N} = \{1, 2, 3, 4, 5\}$ (so that $n = 5$) and set

$$\sigma_1 = \begin{pmatrix} 1 & 2 & 3 & 4 & 5 \\ 2 & 1 & 5 & 3 & 4 \end{pmatrix} \quad \sigma_2 = \begin{pmatrix} 1 & 2 & 3 & 4 & 5 \\ 3 & 5 & 2 & 4 & 1 \end{pmatrix}$$

According to the definition of a product of permutations, we have

$$\sigma_1 \sigma_2: \begin{array}{l} 1 \to (\sigma_1 \sigma_2)(1) = \sigma_1(\sigma_2(1)) = \sigma_1(3) = 5 \\ 2 \to (\sigma_1 \sigma_2)(2) = \sigma_1(\sigma_2(2)) = \sigma_1(5) = 4 \\ 3 \to (\sigma_1 \sigma_2)(3) = \sigma_1(\sigma_2(3)) = \sigma_1(2) = 1 \\ 4 \to (\sigma_1 \sigma_2)(4) = \sigma_1(\sigma_2(4)) = \sigma_1(4) = 3 \\ 5 \to (\sigma_1 \sigma_2)(5) = \sigma_1(\sigma_2(5)) = \sigma_1(1) = 2 \end{array}$$

and hence

$$\sigma_1 \sigma_2 = \begin{pmatrix} 1 & 2 & 3 & 4 & 5 \\ 5 & 4 & 1 & 3 & 2 \end{pmatrix} \quad ////$$

It will be convenient for us to identify the permutation τ of $\mathbb{N}' = \{1, 2, \ldots, n-1\}$ with the permutation σ of $\mathbb{N} = \{1, 2, \ldots, n\}$ when

$$\tau(x) = \sigma(x) \quad x = 1, 2, \ldots, n-1$$

and

$$\sigma(n) = n \quad (1.8.5)$$

Henceforth we will consider τ and σ satisfying (1.8.5) to be the same permutation. In particular, all identity permutations are the same permutation.

Definition 1.8.6 Let σ be a permutation of $\mathbb{N} = \{1, 2, \ldots, n\}$. If there exist $a, b \in \mathbb{N}$, $a \neq b$, such that

$$\begin{array}{l} \sigma(a) = b \\ \sigma(b) = a \\ \sigma(x) = x \quad x \neq a, b \end{array}$$

then σ is a *transposition* and is denoted by $(a \ b)$.

In other words, a transposition is a permutation of $\mathbb{N}$ which interchanges two elements of $\mathbb{N}$ and leaves all other elements of $\mathbb{N}$ fixed.

Theorem 1.8.7 Every permutation can be expressed as a product of transpositions.

PROOF The proof is by an induction on n, beginning with $n = 2$. Clearly $\mathbb{S}_2$ consists of the two elements

$$I = \begin{pmatrix} 1 & 2 \\ 1 & 2 \end{pmatrix} \quad (1 \ 2) = \begin{pmatrix} 1 & 2 \\ 2 & 1 \end{pmatrix}$$

Since (1 2) is already the product of one transposition and

$$I = (1\ 2)(1\ 2)$$

the theorem is true for $n = 2$.

Now suppose that every permutation of $\mathbb{N}' = \{1, 2, \ldots, n-1\}$ is expressible as a product of transposition and let σ be a permutation of $\mathbb{N} = \{1, 2, \ldots, n\}$. If $\sigma(n) = n$, we may consider σ to be a permutation of $\mathbb{N}'$ and hence σ can be expressed as a product of transpositions. On the other hand, suppose

$$\sigma(n) = a \neq n$$

In this case we set

$$\tau = (a\ n)\sigma$$

and then

$$\tau(n) = (a\ n)\sigma(n) = (a\ n)(a) = n$$

that is, τ maps n onto itself. Consequently τ can be considered as a permutation of N' and thus can be expressed as a product of transpositions, say

$$\tau = \rho_1 \rho_2 \cdots \rho_k$$

Clearly

$$\sigma = (a\ n)(a\ n)\sigma = (a\ n)\tau = (a\ n)\rho_1\rho_2 \cdots \rho_k$$

completing the induction. ////

The number of transpositions into which a permutation can be factored is not unique. Indeed, to any factorization we can always add two more (identical) transpositions. We can show, however, that for a given permutation the number of transpositions in its various factorizations is either always odd or always even. The proof is most easily described using the following polynomial in the n variables $x_1, x_2, \ldots, x_n$:

$$f(x_1, x_2, \ldots, x_n) = \prod_{i<j}^{n} (x_i - x_j)$$
$$= (x_1 - x_2)(x_1 - x_3) \cdots (x_1 - x_n)(x_2 - x_3) \cdots (x_{n-1} - x_n)$$

For any permutation σ of $\mathbb{N}$ we define the polynomial σf in the same n variables, $x_1, x_2, \ldots, x_n$, by

$$(\sigma f)(x_1, x_2, \ldots, x_n) = \prod_{i<j}^{n} (x_{\sigma(i)} - x_{\sigma(j)})$$

i.e., the permutation σ acts on the subscripts of the variables. When $\sigma(i) < \sigma(j)$, the term $x_{\sigma(i)} - x_{\sigma(j)}$ is one of the factors of $f(x_1, x_2, \ldots, x_n)$. On the other hand, when $\sigma(i) > \sigma(j)$, then $x_{\sigma(i)} - x_{\sigma(j)} = -(x_{\sigma(j)} - x_{\sigma(i)})$ and $x_{\sigma(j)} - x_{\sigma(i)}$ is one of the factors of $f(x_1, x_2, \ldots, x_n)$. Thus, aside from possible differences of sign, the factors of $f(x_1, x_2, \ldots, x_n)$ and $(\sigma f)(x_1, x_2, \ldots, x_n)$ are identical. The exact relation between f and σf will follow from:

Lemma 1.8.8 Let $\rho = (a\ b)$ be a transposition of $\mathbb{N}$. Then

$$(\rho f)(x_1, x_2, \ldots, x_n) = -f(x_1, x_2, \ldots, x_n)$$

PROOF In order to see more easily the effect of ρ on f we must first express $f(x_1, x_2, \ldots, x_n)$ in the form

$$f(x_1, x_2, \ldots, x_n) = \Theta(x_a - x_b) \prod_{\substack{c \neq a,b \\ d \neq a,b \\ c<d}} (x_c - x_d) \prod_{c \neq a,b} (x_a - x_c) \prod_{c \neq a,b} (x_b - x_c) \quad (1.8.9)$$

where $\Theta = +1$ or $\Theta = -1$. That this can be done follows from the discussion which precedes the lemma. Clearly Θ is uniquely determined by the choice of a and b. For a given pair of distinct integers, a and b, it is possible (but not at all necessary) to determine whether $\Theta = +1$ or $\Theta = -1$.

Now let us investigate the effect of ρ on each of the terms on the right-hand side of (1.8.9). Clearly ρ leaves Θ and $\prod_{\substack{c \neq a,b \\ d \neq a,b \\ c<d}} (x_c - x_d)$ unchanged and has the effect of interchanging the two terms $\prod_{c \neq a,b} (x_a - x_c)$ and $\prod_{c \neq a,b} (x_b - x_c)$. Finally, $\rho(x_a - x_b) = x_b - x_a = -(x_a - x_b)$. Thus the total effect of ρ is to change the sign of $f(x_1, x_2, \ldots, x_n)$, that is,

$$(\rho f)(x_1, x_2, \ldots, x_n) = -f(x_1, x_2, \ldots, x_n) \qquad ////$$

Let σ be a permutation of $\mathbb{N} = \{1, 2, \ldots, n\}$ for which some factorization into transpositions contains an odd number of transpositions. It follows from repeated applications of Lemma 1.8.8 that

$$(\sigma f)(x_1, x_2, \ldots, x_n) = -f(x_1, x_2, \ldots, x_n)$$

On the other hand, if some factorization of σ were to contain an even number of transpositions, then, in the same way, we would have

$$(\sigma f)(x_1, x_2, \ldots, x_n) = f(x_1, x_2, \ldots, x_n)$$

Clearly this would be a contradiction, and consequently we have proved:

Theorem 1.8.10 Every permutation of $\mathbb{N} = \{1, 2, \ldots, n\}$, $n \geq 2$, can be expressed as a product of a finite number of transpositions. If the number of transpositions in one such factorization of the permutation σ is odd (even), then the number of transpositions in every factorization of σ is odd (even). In this case we call σ an *odd (even) permutation*.

Because of Theorem 1.8.10 we are able to make the following definition:

Definition 1.8.11 The *signature* of a permutation σ (written sgn σ) is defined by

$$\operatorname{sgn} \sigma = \begin{cases} +1 & \text{if } \sigma \text{ is an even permutation} \\ -1 & \text{if } \sigma \text{ is an odd permutation} \end{cases}$$

The operation $\mathscr{R}_{rs}$ is a transposition of the rows of a matrix, interchanging the rth and sth rows. Similarly $\mathscr{C}_{rs}$ is a transposition of the columns of a matrix. We can generalize this to any permutation of $\mathbb{N} = \{1, 2, \ldots, n\}$. Let A be a matrix with n rows and let $\mathscr{R}_\sigma$ be the operation of performing the permutation σ on the rows of A. As in Sec. 1.7, the operation $\mathscr{R}_\sigma$ can be carried out by multiplying the given matrix on the left by a properly chosen matrix nonsingular R_σ. As before, R_σ is obtained by performing the operation $\mathscr{R}_\sigma$ on the identity matrix:

$$\mathscr{R}_\sigma : I \to R_\sigma$$

where R_σ has 1 in the $(\sigma(1), 1)$, $(\sigma(2), 2)$, $\ldots$, $(\sigma(n), n)$ positions and 0 elsewhere.

Theorem 1.8.12 Let σ be a permutation of $\mathbb{N} = \{1, 2, \ldots, n\}$. Then

$$\mathscr{R}_\sigma : A \to R_\sigma A$$

for every matrix A with n rows.

PROOF According to Theorem 1.8.7, there exist transpositions $\rho_1 = (r_1 \ s_1)$, $\rho_2 = (r_2 \ s_2)$, $\ldots$, $\rho_k = (r_k \ s_k)$ such that

$$\sigma = \rho_1 \rho_2 \cdots \rho_k$$

Then
$$\mathscr{R}_\sigma = \mathscr{R}_{\rho_1} \mathscr{R}_{\rho_2} \cdots \mathscr{R}_{\rho_k} = \mathscr{R}_{r_1 s_1} \mathscr{R}_{r_2 s_2} \cdots \mathscr{R}_{r_k s_k}$$

Set
$$R_\sigma = R_{r_1 s_1} R_{r_2 s_2} \cdots R_{r_k s_k}$$

It now follows by repeated applications of Lemma 1.7.1 that

$$\mathscr{R}_\sigma(A) = \mathscr{R}_{r_1 s_1} \mathscr{R}_{r_2 s_2} \cdots \mathscr{R}_{r_k s_k}(A)$$
$$= R_{r_1 s_1} R_{r_2 s_2} \cdots R_{r_k s_k} A$$
$$= R_\sigma A$$

for every matrix A with n rows. ////

We call R_σ the *permutation matrix* corresponding to the permutation σ. Thus there is a one-to-one correspondence between permutations of $\mathbb{N} = \{1, 2, \ldots, n\}$ and permutation matrices of order n. Moreover,

$$R_\sigma R_\tau = R_{\sigma\tau}$$

for all $\sigma, \tau \in \mathbb{S}_n$.

Because we have defined multiplication of permutation by

$$(\sigma_1 \sigma_2)(x) = \sigma_1[\sigma_2(x)]$$

rather than by

$$(x)(\sigma_1 \sigma_2) = [(x)\sigma_1]\sigma_2$$

the treatment of permutations of the columns of a matrix is not as straightforward as the discussion above (see Exercises 18 and 19).

EXERCISES

1 For the permutations σ_1 and σ_2 of Example 1.8.4, find:

(a) $\sigma_2 \sigma_1$ (b) $(\sigma_1)^2$ (c) $(\sigma_1)^3$
(d) $(\sigma_1)^4$ (e) $(\sigma_1)^5$ (f) $(\sigma_2)^2$

2 Express

$$\sigma = \begin{pmatrix} 1 & 2 & 3 & 4 & 5 & 6 & 7 & 8 & 9 \\ 7 & 9 & 1 & 6 & 5 & 2 & 3 & 4 & 8 \end{pmatrix}$$

as a product of transpositions. Is σ odd or even?

3 Let σ be a permutation of $\mathbb{N}$. Prove that there is a unique permutation $\tilde{\sigma}$ of $\mathbb{N}$ such that

$$\tilde{\sigma}\sigma = \sigma\tilde{\sigma} = I$$

$\tilde{\sigma}$ is called the *inverse* of σ and is denoted by σ^{-1}.

4 For the permutations σ_1 and σ_2 of Example 1.8.4, find:

(a) σ_1^{-1} (b) σ_2^{-1}
(c) $\sigma_1 \sigma_2 \sigma_1^{-1}$ (d) $\sigma_2 \sigma_1 \sigma_2^{-1}$

5 Find the inverse of the permutation σ of Exercise 2. Express σ^{-1} as a product of transpositions. Is σ^{-1} odd or even?

6 Show that the inverse of the permutation σ of (1.8.2) can be found by interchanging the two rows of σ (and then rearranging the columns so that the first row is again $1, 2, \ldots, n$).

7 Prove that if a permutation is expressed as a product of transpositions, then the inverse of the permutation is the product of the transpositions in the opposite order.
8 Prove that exactly half of the permutations of $\mathbb{N}$ are even.
9 Prove the assertion that the multiplication of permutations is associative.
10 Let σ and τ be permutations of $\mathbb{N}$. Prove that $(\sigma\tau)^{-1} = \tau^{-1}\sigma^{-1}$.
11 Let σ be a permutation of $\mathbb{N}$. Show that

$$\{(i, \sigma(i)) \mid i \in \mathbb{N}\} = \{(\sigma^{-1}(j), j) \mid j \in \mathbb{N}\}$$

12 Show that $\mathbb{S}_n = \{\sigma^{-1} \mid \sigma \in \mathbb{S}_n\}$.
13 Let σ and τ be permutations of $\mathbb{N}$. Prove:
   (a) $\text{sgn}(\sigma\tau) = (\text{sgn } \sigma)(\text{sgn } \tau)$ (b) $\text{sgn } \sigma = \text{sgn } \sigma^{-1}$
14 Prove that

$$\begin{pmatrix} 1 & 2 & \cdots & a-1 & a & a+1 & \cdots & n-1 & n \\ 1 & 2 & \cdots & a-1 & n & a & \cdots & n-2 & n-1 \end{pmatrix}$$
$$= (a \; a+1)(a \; a+2) \cdots (a \; n-1)(a \; n)$$

15 Let $\mathbb{N}' = \{1, 2, \ldots, n-1\}$ and $\mathbb{N} = \{1, 2, \ldots, n\}$. A permutation σ of $\mathbb{N}$ which leaves n fixed is considered to be a permutation of $\mathbb{N}'$. Denote the set of permutations of $\mathbb{N}'$ by $\mathbb{S}_{n-1}$ and let $a \in \mathbb{N}$, $a \neq n$. Prove that

$$\mathbb{S}_{n-1}(a \; n) = \{\sigma(a \; n) \mid \sigma \in \mathbb{S}_{n-1}\}$$

is the set of all permutations of $\mathbb{N}$ which map a onto n and that

$$(a \; n)\mathbb{S}_{n-1} = \{(a \; n)\sigma \mid \sigma \in \mathbb{S}_{n-1}\}$$

is the set of all permutations of $\mathbb{N}$ which map n onto a. How many permutations are there in each set?

16 For the permutations σ_1 and σ_2 of Example 1.8.4, find:

   (a) R_{σ_1} (b) R_{σ_2}
   (c) $R_{\sigma_1 \sigma_2}$ (d) $R_{\sigma_2 \sigma_1}$

17 Let $\mathscr{C}_\sigma$ be the operation of performing the permutation σ on the columns of a matrix and let

$$\mathscr{C}_\sigma : I \to C_\sigma$$

Find:

   (a) C_{σ_1} (b) C_{σ_2}
   (c) $C_{\sigma_1 \sigma_2}$ (d) $C_{\sigma_2 \sigma_1}$

where σ_1 and σ_2 are the permutations of Example 1.8.4.

18 Prove:

(a) $C_{\sigma_1}C_{\sigma_2} = C_{\sigma_2\sigma_1}$
(b) $C_\sigma = R_\sigma^{-1}$
(c) $R_{\sigma^{-1}} = R_\sigma^{-1}$
(d) $C_{\sigma^{-1}} = C_\sigma^{-1}$
(e) $R_\sigma^{-1} = R_\sigma^T$
(f) $C_\sigma^{-1} = C_\sigma^T$

19 Formulate and prove a theorem corresponding to Theorem 1.8.12 for column operations. Verify that the matrices C_{σ_1} and C_{σ_2} of Exercise 17 conform with these results.

20 Let $\mathbb{S}_1$ be a subset of $\mathbb{N} = \{1, 2, \ldots, n\}$ consisting of m integers, $\mathbb{S}_1 = \{s_1, s_2, \ldots, s_m\}$, $s_1 < s_2 < \cdots < s_m$. Let $\mathbb{S}_2$ be the set of $n - m$ elements of $\mathbb{N}$ which do not belong to $\mathbb{S}_1$, $\mathbb{S}_2 = \{s_{m+1}, s_{m+2}, \ldots, s_n\}$, $s_{m+1} < s_{m+2} < \cdots < s_n$. Set

$$\sigma_{\mathbb{S}} = \begin{pmatrix} 1 & 2 & \cdots & m & m+1 & m+2 & \cdots & n \\ s_1 & s_2 & \cdots & s_m & s_{m+1} & s_{m+2} & \cdots & s_n \end{pmatrix}$$

Show that

$$\operatorname{sgn} \sigma_{\mathbb{S}} = (-1)\exp\left(\sum_{j=1}^{m} s_j - j\right) = (-1)\exp\left(\sum_{j=m+1}^{n} s_j - j\right)$$

(*Hint:* Use an induction on m.) Can this result be extended to three subsets $\mathbb{S}_1, \mathbb{S}_2, \mathbb{S}_3$ of $\mathbb{N}$?

1.9 DETERMINANTS[1]

In this section we will define the determinant of a square matrix and investigate many of its properties. Actually, we will present two quite different definitions of the determinant (Definition 1.9.1 and Theorem 1.9.20) and show that they are equivalent. The determinant function can be developed not only for matrices whose entries are from a field but also for matrices with entries from any set in which there are "reasonable" operations of addition, subtraction, and multiplication. This can be done since the operation of division is never mentioned.

---

[1] In his four-volume treatise, "The Theory of Determinants in the Historical Order of Development," (Macmillan, London, 1906 et seq.) Thomas Muir quotes extensively from a letter written by Leibnitz in 1693 to de L'Hospital. In this letter Leibnitz announced a new method for displaying and solving a system of linear equations in several unknowns. Essentially he had formulated a double-subscript notation similar to the one we use today, and he used it to obtain what is now known as Cramer's rule (see Exercise 19 of Sec. 1.9). Cramer discovered the same results independently, and his book, which was published in 1750 (unlike Leibnitz' letters, which were not published until 1850), attracted considerable attention. The concept of a matrix not merely as an array of numbers but as an object which could be added and multiplied is due primarily to Hamilton (1853) and Cayley (1858).

By reasonable operations we will mean that all the properties of a field as given in Definition 0.1.1, except for item 10, are satisfied.[2] In order to keep our notation simple, *in this section* we will denote a set having reasonable operations of addition, subtraction, and multiplication by $\mathbb{F}$. The reader will lose very little (at least until Sec. 1.10) by continuing to think of $\mathbb{F}$ as a field.

The determinant is a function whose domain is the set of all square matrices with entries from $\mathbb{F}$ and whose range is $\mathbb{F}$ itself. The determinant is not defined for matrices which are not square.

Definition 1.9.1 Let $A = [a_{ij}]$ be a square matrix of order n. We define the *determinant* of A to be

$$\sum_{\sigma \in \mathbb{S}_n} \operatorname{sgn} \sigma \, a_{1\sigma(1)} \, a_{2\sigma(2)} \cdots a_{n\sigma(n)} \quad (1.9.2)$$

and denote it by *det A*.

For any positive integer n we have denoted the set of all permutations of $\mathbb{N} = \{1, 2, \ldots, n\}$ by $\mathbb{S}_n$ and showed that $\mathbb{S}_n$ contains $n!$ elements. For $n = 1$, $\mathbb{S}_1$ consists of the identity permutation I alone. For $n \geq 2$, exactly half of the elements of $\mathbb{S}_n$ are even permutations, and exactly half are odd permutations. (See Exercise 8 of Sec. 1.8.)

Thus, for $n \geq 2$, $\operatorname{sgn} \sigma = +1$ for $\frac{1}{2}n!$ permutations σ belonging to $\mathbb{S}_n$ and $\operatorname{sgn} \sigma = -1$ for the remaining $\frac{1}{2}n!$ permutations.

A careful consideration of Definition 1.9.1 will show that the determinant of a matrix of order n is a homogeneous polynomial of degree n in the entries of the matrix, consisting of $n!$ terms, each with a coefficient of either $+1$ or -1. Moreover, each of these terms contains exactly one entry from each row and exactly one entry from each column of the matrix. Conversely, because Definition 1.9.1 calls for us to use all permutations of $\mathbb{N} = \{1, 2, \ldots, n\}$, every product of n entries of the matrix which consists of exactly one entry from each row and exactly one entry from each column is contained in (1.9.2).

EXAMPLE 1.9.3 Set $n = 1$ and $A = [a_{11}]$. Then $\mathbb{S}_1 = \{I\}$, and hence

$$\det A = \operatorname{sgn} I \, a_{1, I(1)} = a_{11} \quad ////$$

---

[2] We have already pointed out that such a system is called a ring. In Sec. 1.10 we will apply the results of this section to matrices whose entries are from the ring of polynomials in a single variable x with coefficients from a field.

EXAMPLE 1.9.4 Set $n = 2$ and $A = [a_{ij}]$, $i, j = 1, 2$. Then $\mathbb{S}_2 = \{I, (1\ 2)\}$, and hence

$$\det A = \operatorname{sgn} I\, a_{1, I(1)} a_{2, I(2)} + \operatorname{sgn}(1\ 2)\, a_{1, (1\ 2)(1)} a_{2, (1\ 2)(2)}$$
$$= a_{11} a_{22} - a_{12} a_{21} \qquad\qquad ////$$

EXAMPLE 1.9.5 For $n = 3$ we have

$$\mathbb{S}_3 = \left\{ I, (1\ 2), (1\ 3), (2\ 3), \begin{pmatrix} 1 & 2 & 3 \\ 2 & 3 & 1 \end{pmatrix}, \begin{pmatrix} 1 & 2 & 3 \\ 3 & 1 & 2 \end{pmatrix} \right\}$$

Since

$$\begin{pmatrix} 1 & 2 & 3 \\ 2 & 3 & 1 \end{pmatrix} = (1\ 2)(2\ 3) \qquad \begin{pmatrix} 1 & 2 & 3 \\ 3 & 1 & 2 \end{pmatrix} = (2\ 3)(1\ 2)$$

it follows that

$$\operatorname{sgn} \begin{pmatrix} 1 & 2 & 3 \\ 2 & 3 & 1 \end{pmatrix} = \operatorname{sgn} \begin{pmatrix} 1 & 2 & 3 \\ 3 & 1 & 2 \end{pmatrix} = +1$$

and therefore

$$\det A = \operatorname{sgn} I\, a_{11} a_{22} a_{22} + \operatorname{sgn}(1\ 2)\, a_{12} a_{21} a_{33} + \operatorname{sgn}(1\ 3)\, a_{13} a_{22} a_{31}$$
$$+ \operatorname{sgn}(2\ 3)\, a_{11} a_{23} a_{32} + \operatorname{sgn}\begin{pmatrix} 1 & 2 & 3 \\ 2 & 3 & 1 \end{pmatrix} a_{12} a_{23} a_{31}$$
$$+ \operatorname{sgn}\begin{pmatrix} 1 & 2 & 3 \\ 3 & 1 & 2 \end{pmatrix} a_{13} a_{21} a_{32}$$
$$= a_{11} a_{22} a_{33} - a_{12} a_{21} a_{33} - a_{13} a_{22} a_{31} - a_{11} a_{23} a_{32}$$
$$+ a_{12} a_{23} a_{31} + a_{13} a_{21} a_{32} \qquad\qquad ////$$

These formulas agree with the definition of the determinant for 1×1, 2×2, and 3×3 matrices that the reader has met before. The determinant of larger matrices can be evaluated in the same way, but the number of terms becomes unpleasantly large very soon. In this section we will find several other ways to evaluate the determinant of a matrix.

In Definition 1.9.1 the row and column subscripts of each entry of A appear to play different roles. Let us demonstrate that in fact this is not the case.

Lemma 1.9.6 Let $A = [a_{ij}]$ be a matrix of order n. Then

$$\det A = \sum_{\tau \in \mathbb{S}_n} \operatorname{sgn} \tau\, a_{\tau(1)1} a_{\tau(2)2} \cdots a_{\tau(n)n} \qquad (1.9.7)$$

PROOF. We wish to show that each of the $n!$ terms of (1.9.2) is identical with one of the $n!$ terms on the right-hand side of equation (1.9.7). More specifically, we wish to show that the term in (1.9.2) corresponding to the permutation σ is equal to the term of (1.9.7) corresponding to the permutation $\tau = \sigma^{-1}$; that is,

$$\operatorname{sgn} \sigma \, a_{1\sigma(1)} a_{2\sigma(2)} \cdots a_{n\sigma(n)} = \operatorname{sgn} \sigma^{-1} \, a_{\sigma^{-1}(1)1} a_{\sigma^{-1}(2)2} \cdots a_{\sigma^{-1}(n)n}$$

for every $\sigma \in \mathbb{S}_n$. Let j be one of the integers of $\mathbb{N} = \{1, 2, \ldots, n\}$. There exists $i \in \mathbb{N}$ such that $\sigma(i) = j$. In fact, $i = \sigma^{-1}(j)$. Then

$$a_{i\sigma(i)} = a_{ij} = a_{\sigma^{-1}(j)j}$$

Since

$$\{(i, \sigma(i)) \mid i \in \mathbb{N}\} = \{(\sigma^{-1}(j), j) \mid j \in \mathbb{N}\}$$

(see Exercise 11 of Sec. 1.8) it follows that

$$\operatorname{sgn} \sigma \, a_{1\sigma(1)} a_{2\sigma(2)} \cdots a_{n\sigma(n)} = \operatorname{sgn} \sigma \, a_{\sigma^{-1}(1)1} a_{\sigma^{-1}(2)2} \cdots a_{\sigma^{-1}(n)n}$$

But $\operatorname{sgn} \sigma = \operatorname{sgn} \sigma^{-1}$, and thus

$$\det A = \sum_{\sigma \in \mathbb{S}_n} \operatorname{sgn} \sigma^{-1} \, a_{\sigma^{-1}(1)1} a_{\sigma^{-1}(2)2} \cdots a_{\sigma^{-1}(n)n} \quad (1.9.8)$$

The set of elements of $\mathbb{S}_n$ is exactly the same as the set of inverses of the elements of $\mathbb{S}_n$ (see Exercise 12 of Sec. 1.8). Consequently we may take the summation in equation (1.9.8) for all σ^{-1} belonging to $\mathbb{S}_n$. Thus equation (1.9.8) becomes

$$\det A = \sum_{\sigma^{-1} \in \mathbb{S}_n} \operatorname{sgn} \sigma^{-1} a_{\sigma^{-1}(1)1} a_{\sigma^{-1}(2)2} \cdots a_{\sigma^{-1}(n)n} \quad (1.9.9)$$

Set $\sigma^{-1} = \tau$. Then equation (1.9.9) becomes equation (1.9.7). ////

A very important property of the determinant function is proved in Theorem 1.9.18, namely, that the determinant of a product of matrices of order n is equal to the product of the determinants of the matrices. The main tool used in the proof of this result is the expression of a matrix as a product of elementary matrices and a matrix of the form (1.7.9), as given in Corollary 1.7.14. Thus we first investigate the effect each of the elementary row and column operations has on the determinant. The three elementary operations are considered in Lemmas 1.9.10, 1.9.14, and 1.9.15. The second part of each of these lemmas shows that Theorem 1.9.18 holds in the special case when one of the matrices of the product is an elementary matrix. Because of Lemma 1.9.6 it will be sufficient to consider only the three row operations.

Lemma 1.9.10 (a) det $R_{rs} = -1$.
(b) det $R_{rs}A = -\det A = (\det R_{rs})(\det A)$
for every square matrix A.

PROOF (a) R_{rs} has exactly n entries which are not zero. These occur in the (r, s) and (s, r) positions and in all the diagonal positions except the (r, r) and (s, s) positions, and they are all equal to 1. Thus, only one term of the expression (1.9.2) for det R_{rs} is not zero, namely, the term containing exactly the n nonzero entries of R_{rs}. This is the term of the expression (1.9.2) corresponding to the transposition $\rho = (r \; s)$ of $\mathbb{S}_n$. Thus

$$\det R_{rs} = \operatorname{sgn} \rho \, 1 \cdot 1 \cdots 1 = -1$$

(b) Let $A = [a_{ij}]$ and set $R_{rs}A = B = [b_{ij}]$. Then A and B are related by

$$b_{ij} = a_{ij} \quad i \neq r, s \quad j = 1, 2, \ldots, n \quad (1.9.11)$$
$$b_{rj} = a_{sj}$$
$$b_{sj} = a_{rj}$$

We can express equations (1.9.11) more simply by using permutations. Let σ be any element of $\mathbb{S}_n$ and set $(r \; s)\sigma = \tau$. Then equations (1.9.11) reduce to the single equation

$$b_{\tau(j)j} = a_{\sigma(j)j} \quad j = 1, 2, \ldots, n$$

For if $\sigma(j) = i \neq r, s$, then

$$\tau(j) = (r \; s)[\sigma(j)] = (r \; s)(i) = i$$

and hence
$$b_{\tau(j)j} = b_{ij} = a_{ij} = a_{\sigma(j)j}$$

while if $\sigma(j) = r$, then

$$\tau(j) = (r \; s)[\sigma(j)] = (r \; s)(r) = s$$

and hence
$$b_{\tau(j)j} = b_{sj} = a_{rj} = a_{\sigma(j)j}$$

and if $\sigma(j) = s$, then

$$\tau(j) = (r \; s)[\sigma(j)] = (r \; s)(s) = r$$

and hence
$$b_{\tau(j)j} = b_{rj} = a_{sj} = a_{\sigma(j)j}$$

Thus, using Lemma 1.9.6, we have

$$\det B = \sum_{\tau \in \mathbb{S}_n} \operatorname{sgn} \tau \, b_{\tau(1)1} b_{\tau(2)2} \cdots b_{\tau(n)n}$$
$$= \sum_{(r \; s)\sigma \in \mathbb{S}_n} \operatorname{sgn}(r \; s)\sigma \, a_{\sigma(1)1} a_{\sigma(2)2} \cdots a_{\sigma(n)n}$$
$$= - \sum_{(r \; s)\sigma \in \mathbb{S}_n} \operatorname{sgn} \sigma \, a_{\sigma(1)1} a_{\sigma(2)2} \cdots a_{\sigma(n)n}$$

since $\text{sgn}(r\ s)\sigma = -\text{sgn}\ \sigma$ for every $\sigma \in \mathbb{S}_n$. It is clear that the set of $n!$ permutation

$$\{(r\ s)\sigma \mid \sigma \in \mathbb{S}_n\}$$

is exactly the set of $n!$ elements of $\mathbb{S}_n$. Thus the last summation above is simply the sum for all permutations contained in $\mathbb{S}_n$. Consequently we have shown that $\det B = -\det A$. ////

Part (a) of Lemma 1.9.10 can be stated as

$$\det R_{rs} = \text{sgn}(r\ s)$$

Since any permutation can be expressed as a product of transpositions, it follows that:

Corollary 1.9.12 For every permutation σ

$$\det R_\sigma = \text{sgn}\ \sigma = \det C_\sigma$$

Lemma 1.9.10 has another very useful corollary.

Corollary 1.9.13 If two rows (or two columns) of a square matrix are identical, then the determinant of the matrix is zero.[1]

PROOF Let the rth and sth rows of A be identical, $r \neq s$. Then

$$A = R_{rs}A$$

and hence, by Lemma 1.9.10,

$$\det A = \det(R_{rs}A) = -\det A$$

Thus $\det A = 0$. ////

Corollary 1.9.13 is a special case of Corollary 1.9.17, which states that the determinant of any square matrix whose rows or columns are linearly dependent is zero.

We return to the investigation of the effect on the determinant of the elementary operations.

Lemma 1.9.14 (a) $\det R_{rs}(\alpha) = 1$.
(b) $\det R_{rs}(\alpha)A = \det A = [\det R_{rs}(\alpha)](\det A)$

for every square matrix A.

---

[1] This proof fails when $\mathbb{F}$ has the property $1 + 1 = 0$ (as, for example, when $\mathbb{F} = \mathbb{Z}_2$). The corollary is still true in this case, but a different proof is necessary. See Exercise 23 and set $m = 2$.

PROOF (a) $R_{rs}(\alpha)$ contains exactly $n+1$ nonzero entries. These occur in the n diagonal positions, where we have 1s, and in the (r, s) position, where we have α. First, using Definition 1.9.1 for the determinant of $R_{rs}(\alpha)$, we wish to show that every term of expression (1.9.2) which contains α as a factor is equal to zero. Clearly α appears in the term of (1.9.2) corresponding to the permutation σ if and only if $\sigma(r) = s$. In fact, such a term is equal to sgn σ multiplied by all the entries in the $(j, \sigma(j))$ positions of $R_{rs}(\alpha)$, $j = 1, 2, \ldots, n$. In particular, let us look at the entry in the $(s, \sigma(s))$ position of $R_{rs}(\alpha)$. Set $\sigma(s) = t$. Since we already have $\sigma(r) = s$ and $r \neq s$, it follows that $\sigma(s) \neq s$, that is, $t \neq s$. But the only nonzero entry in the sth row of $R_{rs}(\alpha)$ is the 1 in the (s, s) position. Thus the entry in the $(s, t) = (s, \sigma(s))$ position of $R_{rs}(\alpha)$ is zero, and consequently the term of (1.9.2) corresponding to σ is also zero.

With the elimination of the possibility that α is a factor of some term of (1.9.2), it becomes clear that the only nonzero term in (1.9.2) is the one which corresponds to the permutation I of $\mathbb{S}_n$. Hence

$$\det R_{rs}(\alpha) = (\text{sgn } I) \, 1 \cdot 1 \cdots 1 = 1$$

(b) Let $A = [a_{ij}]$ and set $R_{rs}(\alpha)A = B = [b_{ij}]$. Then A and B are related by

$$b_{ij} = a_{ij} \quad i \neq r \quad j = 1, 2, \ldots, n$$
$$b_{rj} = a_{rj} + \alpha a_{sj}$$

As before, it will be useful to express this relation in terms of permutations. Let σ be any element of $\mathbb{S}_n$. Then we have

$$b_{i\sigma(i)} = a_{i\sigma(i)} \quad i \neq r$$
$$b_{r\sigma(r)} = a_{r\sigma(r)} + \alpha a_{s\sigma(r)}$$

Thus, by Definition 1.9.1, we have

$$\det B = \sum_{\sigma \in \mathbb{S}_n} \text{sgn } \sigma \, b_{1\sigma(1)} b_{2\sigma(2)} \cdots b_{r\sigma(r)} \cdots b_{n\sigma(n)}$$
$$= \sum_{\sigma \in \mathbb{S}_n} \text{sgn } \sigma \, a_{1\sigma(1)} a_{2\sigma(2)} \cdots (a_{r\sigma(r)} + \alpha a_{s\sigma(r)}) \cdots a_{n\sigma(n)}$$
$$= \sum_{\sigma \in \mathbb{S}_n} \text{sgn } \sigma \, a_{1\sigma(1)} a_{2\sigma(2)} \cdots a_{r\sigma(r)} \cdots a_{n\sigma(n)}$$
$$\quad + \alpha \sum_{\sigma \in \mathbb{S}_n} \text{sgn } \sigma \, a_{1\sigma(1)} a_{2\sigma(2)} \cdots a_{s\sigma(r)} \cdots a_{n\sigma(n)}$$
$$= \det A + \alpha \det C$$

where C is a matrix in which the rth and sth rows are equal. In fact,

$$C_{(i)} = A_{(i)} \quad i \neq r$$
$$C_{(r)} = A_{(s)} = C_{(s)}$$

According to Corollary 1.9.13, det $C = 0$ and thus det $B = $ det A. ////

Lemma 1.9.15 (a) det $R_r(\lambda) = \lambda$.
(b) det $R_r(\lambda)A = \lambda$ det $A = [\det R_r(\lambda)](\det A)$
for every square matrix A.

PROOF The proof is left as an exercise. ////

In defining the third elementary operation we were careful to specify that $\lambda \neq 0$. According to the first parts of Lemmas 1.9.10, 1.9.14, and 1.9.15, the determinant of an elementary matrix is never zero.

Corollary 1.7.14, together with Lemmas 1.9.10, 1.9.14, and 1.9.15, leads to an expression, (1.9.16), for the determinant of any square matrix in terms of determinants of elementary matrices and a matrix of the form (1.7.9). Since the determinant of each of these matrices is readily obtainable, equation (1.9.16) describes a method for evaluating the determinant which is often more practical than using Definition 1.9.1 directly. For a given matrix A, Corollary 1.7.14 states that there exist elementary matrices $R_1, R_2, \ldots, R_k$ and $C_1, C_2, \ldots, C_l$ such that

$$A = R_1 R_2 \cdots R_k \begin{bmatrix} I_r & 0 \\ 0 & 0 \end{bmatrix} C_l C_{l-1} \cdots C_1$$

where I_r is an identity matrix whose order is equal to the rank of A. By repeated use of Lemmas 1.9.10, 1.9.14, and 1.9.15 and their column-matrix counterparts we have

$$\det A = \det\left(R_1 R_2 \cdots R_k \begin{bmatrix} I_r & 0 \\ 0 & 0 \end{bmatrix} C_l C_{l-1} \cdots C_1 \right)$$
$$= (\det R_1)\left[\det\left(R_2 \cdots R_k \begin{bmatrix} I_r & 0 \\ 0 & 0 \end{bmatrix} C_l C_{l-1} \cdots C_1 \right)\right]$$
$$= (\det R_1)(\det R_2)\left[\det\left(R_3 \cdots R_k \begin{bmatrix} I_r & 0 \\ 0 & 0 \end{bmatrix} C_l C_{l-1} \cdots C_1 \right)\right]$$
$$\cdots\cdots\cdots\cdots\cdots\cdots\cdots\cdots\cdots\cdots\cdots\cdots\cdots\cdots\cdots\cdots$$
$$= (\det R_1)(\det R_2) \cdots (\det R_k)\left[\det\left(\begin{bmatrix} I_r & 0 \\ 0 & 0 \end{bmatrix} C_l C_{l-1} \cdots C_1 \right)\right]$$
$$= (\det R_1)(\det R_2) \cdots (\det R_k)\left[\det\left(\begin{bmatrix} I_r & 0 \\ 0 & 0 \end{bmatrix} C_l C_{l-1} \cdots C_2 \right)\right](\det C_1)$$
$$\cdots\cdots\cdots\cdots\cdots\cdots\cdots\cdots\cdots\cdots\cdots\cdots\cdots\cdots\cdots\cdots$$
$$= (\det R_1)(\det R_2) \cdots (\det R_k)\left(\det\begin{bmatrix} I_r & 0 \\ 0 & 0 \end{bmatrix}\right)(\det C_l)(\det C_{l-1}) \cdots (\det C_1)$$

(1.9.16)

Since the determinant of each elementary matrix is nonzero, it follows that $\det A = 0$ if and only if $\det \begin{bmatrix} I_r & 0 \\ 0 & 0 \end{bmatrix} = 0$. But it is clear from Definition 1.9.1 that

$$\det \begin{bmatrix} I_r & 0 \\ 0 & 0 \end{bmatrix} = \begin{cases} 1 & \text{if the rank of } A \text{ is } n \text{ (in which case the zero matrices do not actually appear)} \\ 0 & \text{if the rank of } A \text{ is less than } n \text{ (in which case the zero matrices do actually appear)} \end{cases}$$

Thus we have another characterization of nonsingular matrices to add to those of Theorem 1.6.4.

Corollary 1.9.17 Let A be a square matrix. Then A is nonsingular if and only if $\det A \neq 0$.

Corollary 1.7.14 shows that a nonsingular matrix A can be expressed as a product of elementary matrices; for when A is nonsingular, the matrix of (1.7.9) becomes an identity matrix. Since each elementary row matrix is also an elementary column matrix, and conversely, we can consider A to be expressed either as a product of elementary row matrices or as a product of elementary column matrices. By applying the same method that was used to obtain equation (1.9.16) to a product of two matrices of the same order we can prove the main theorem of this section, namely:

Theorem 1.9.18 For any two matrices, A and B, of the same order,

$$\det AB = (\det A)(\det B)$$

PROOF If either A or B is singular, then AB is also singular (see Exercise 12 of Sec. 1.6), and, by Corollary 1.9.17, we have

$$\det AB = 0 = (\det A)(\det B)$$

Now suppose that A and B are both nonsingular. Then the theorem follows from the second parts of Lemmas 1.9.10, 1.9.14, and 1.9.15. Since A is nonsingular, A can be expressed as a product of elementary row matrices, say

$$A = R_1 R_2 \cdots R_k$$

As before,

$$\begin{aligned} \det A &= \det(R_1 R_2 \cdots R_k B) \\ &= (\det R_1)(\det R_2) \cdots (\det R_k)(\det B) \\ &= (\det R_1 R_2 \cdots R_k)(\det B) \\ &= (\det A)(\det B) \end{aligned}$$
////

There is an alternative definition for the determinant of a matrix which emphasizes the functional rather than the computational aspects of determinants. This definition is in terms of three elementary properties of the determinant function. One drawback of this type of definition is that it is often necessary to prove that such a function actually exists. By presenting Definition 1.9.1 first we avoid this step. However, in Theorem 1.9.20 we will prove that these properties define a unique function whose domain is the set of square matrices with entries from $\mathbb{F}$ and whose range is $\mathbb{F}$. Clearly this function must be the determinant.

Two of the three properties of the determinant function that we will use have already been explicitly stated. We will also need:

Lemma 1.9.19 The determinant is an *n-linear* function of its rows, that is, if A, B and C are $n \times n$ matrices which are identical except for the rth row, and if

$$C_{(r)} = \alpha A_{(r)} + \beta B_{(r)}$$

for some α and β in $\mathbb{F}$ then

$$\det C = \alpha \det A + \beta \det B$$

PROOF   The proof is left as an exercise.   ////

Theorem 1.9.20 There is exactly one function D whose domain is the set of square matrices with entries from $\mathbb{F}$ and whose range is $\mathbb{F}$ which satisfies the three conditions:

1   If two rows of A are equal, then $D(A) = 0$.
2   D is an n-linear function of the rows of A.
3   $D(I) = 1$.

Consequently, $D(A) = \det A$.

PROOF   We have already shown that the determinant, as defined in Definition 1.9.1, has properties 1, 2, and 3.

Conversely, let D be a mapping which satisfies conditions 1, 2, and 3. First, we wish to show that if B is a matrix obtained from A by interchanging the rth and sth rows of A, then

$$D(B) = -D(A) \qquad (1.9.21)$$

Let C be defined by

$$C_{(r)} = C_{(s)} = A_{(r)} + A_{(s)} = B_{(r)} + B_{(s)}$$
$$C_{(i)} = A_{(i)} = B_{(i)} \qquad i \neq r, s$$

By condition 1,
$$D(C) = 0$$

However, applying condition 2 to both the rth and sth rows of C, we have
$$D(C) = D(A) + D(B) + D(X) + D(Y)$$

where
$$X_{(r)} = X_{(s)} = A_{(r)}$$
$$Y_{(r)} = Y_{(s)} = A_{(s)}$$
$$X_{(i)} = Y_{(i)} = A_{(i)} \qquad i \neq r, s$$

Thus $D(X) = D(Y) = 0$, and we have established (1.9.21).

The analog of Corollary 1.9.12 follows directly from (1.9.21), that is,
$$D(R_\sigma) = \text{sgn } \sigma \qquad (1.9.22)$$

for every permutation σ.

Now let $A = [a_{ij}]$ be an $n \times n$ matrix. By repeated use of condition 2 we have

$$D(A) = \sum_\alpha a_{1\alpha(1)} a_{2\alpha(2)} \cdots a_{n\alpha(n)} D(I_{\alpha(1), \alpha(2), \cdots, \alpha(n)}) \qquad (1.9.23)$$

where α is any map of $\mathbb{N} = \{1, 2, \ldots, n\}$ into itself and $I_{\alpha(1), \alpha(2), \cdots, \alpha(n)}$ represents the $n \times n$ matrix whose rows are $I_{\alpha(1)}, I_{\alpha(2)}, \ldots, I_{\alpha(n)}$. (Note that there are n^n such maps, of which only $n!$ are permutations.) Making use of condition *1*, we see that

$$D(I_{\alpha(1), \alpha(2), \cdots, \alpha(n)}) = 0$$

unless α is a permutation of $\mathbb{N}$. When α is a permutation, $I_{\alpha(1), \alpha(2), \cdots, \alpha(n)}$ has 1 in the $(1, \alpha(1)), (2, \alpha(2)), \ldots, (n, \alpha(n))$ positions and 0 elsewhere, that is,

$$I_{\alpha(1), \alpha(2), \cdots, \alpha(n)} = C_\alpha = R_{\alpha^{-1}}$$

[see part (*b*) of Exercise 18 in Sec. 1.8]. Since sgn α = sgn α^{-1} for every permutation α, the theorem follows by substituting (1.9.22) into (1.9.23).

////

As we have pointed out, the direct use of Definition 1.9.1 is not a convenient way of computing the determinant of a given matrix except when its order is very small. The determinant can also be obtained by applying Theorem 1.9.18 to (1.7.15). This method is not much better. However, several practical schemes have been devised for computing the determinant. We will present one which, in addition, leads to a very useful expression for the inverse of a matrix.

Let $A = [a_{ij}]$ be a square matrix of order n. According to Definition 1.9.1, the determinant of A is a linear homogeneous polynomial in the entries of any one row of A each of whose coefficients is a homogeneous polynomial of degree $n - 1$ in the entries of the remaining $n - 1$ rows. More specifically, if we select one row of A, say the pth row, and group together those terms of the determinant which contain a_{p1}, those which contain a_{p2}, etc., we have

$$\det A = a_{p1} A_{p1} + a_{p2} A_{p2} + \cdots + a_{pn} A_{pn} \qquad (1.9.24)$$

where
$$A_{pq} = \sum_{\substack{\sigma \in \mathbb{S}_n \\ \sigma(p)=q}} \operatorname{sgn} \sigma \, a_{1\sigma(1)} a_{2\sigma(2)} \cdots a_{p-1, \sigma(p-1)} a_{p+1, \sigma(p+1)} \cdots a_{n\sigma(n)}$$

$$= \sum_{\substack{\sigma \in \mathbb{S}_n \\ \sigma(p)=q}} \operatorname{sgn} \sigma \prod_{\substack{r \in \mathbb{N} \\ r \neq p}} a_{r(r)} \qquad (1.9.25)$$

Equation (1.9.24) is called the pth *row expansion* of det A, and A_{pq} is the *cofactor* of a_{pq} in A. The form of A_{pq} in equation (1.9.25) is strongly suggestive of Definition 1.9.1 for the determinant. Theorem 1.9.26 shows that A_{pq} is, up to a possible difference in sign, the determinant of a submatrix of A of order $n - 1$.

Theorem 1.9.26 Let $A = [a_{ij}]$ be a matrix of order n and let B_{pq} be the submatrix of A formed by deleting the pth row and the qth column of A. Then

$$A_{pq} = (-1)^{p+q} \det B_{pq}$$

PROOF Let
$$B_{pq} = [b_{rs}] \qquad r, s = 1, 2, \ldots, n-1$$

Then

$$B_{pq} = \begin{bmatrix} a_{11} & a_{12} & \cdots & a_{1,q-1} & a_{1,q+1} & \cdots & a_{1n} \\ a_{21} & a_{22} & \cdots & a_{2,q-1} & a_{2,q+1} & \cdots & a_{2n} \\ \cdot & \cdot & & \cdot & \cdot & & \cdot \\ a_{p-1,1} & a_{p-1,2} & \cdots & a_{p-1,q-1} & a_{p-1,q+1} & \cdots & a_{p-1,n} \\ a_{p+1,1} & a_{p+1,2} & \cdots & a_{p+1,q-1} & a_{p+1,q+1} & \cdots & a_{p+1,n} \\ \cdot & \cdot & & \cdot & \cdot & & \cdot \\ a_{n1} & a_{n2} & \cdots & a_{n,q-1} & a_{n,q+1} & \cdots & a_{nn} \end{bmatrix}$$

Hence, for $1 \leq r, s \leq n$, $r \neq p$, $s \neq q$,

$$a_{rs} = \begin{cases} b_{rs} & r \leq p-1, s \leq q-1 \\ b_{r-1,s} & r \geq p+1, s \leq q-1 \\ b_{r,s-1} & r \leq p-1, s \geq q+1 \\ b_{r-1,s-1} & r \geq p+1, s \geq q+1 \end{cases} \qquad (1.9.27)$$

The four parts of equation (1.9.27) can be expressed in a single equation using permutations. Define the permutations ρ and τ by

$$\rho = \begin{pmatrix} 1 & 2 & \cdots & p-1 & p & p+1 & \cdots & n-1 & n \\ 1 & 2 & \cdots & p-1 & n & p & \cdots & n-2 & n-1 \end{pmatrix}$$
$$= (p \;\; p+1)(p \;\; p+2) \cdots (p \;\; n-1)(p \;\; n)$$
$$\tau = \begin{pmatrix} 1 & 2 & \cdots & q-1 & q & q+1 & \cdots & n-1 & n \\ 1 & 2 & \cdots & q-1 & n & q & \cdots & n-2 & n-1 \end{pmatrix}$$
$$= (q \;\; q+1)(q \;\; q+2) \cdots (q \;\; n-1)(q \;\; n)$$

Then equation (1.9.27) becomes

$$a_{rs} = b_{\rho(r)\tau(s)} \qquad \begin{matrix} r \neq p \\ s \neq q \end{matrix} \qquad (1.9.28)$$

For if $r \leq p-1$, then $\rho(r) = r$, while if $r \geq p+1$, then $\rho(r) = r-1$. Note that we have already excluded the possibility that $r = p$. It then follows that $\rho(r)$ is never equal to n. A similar argument shows that $\tau(s)$ is the correct second subscript for b in equation (1.9.28).

Let σ be any permutation of $\mathbb{N} = \{1, 2, \ldots, n\}$ such that $\sigma(p) = q$. By equation (1.9.28), we have

$$a_{r\sigma(r)} = b_{\rho(r)\tau\sigma(r)} \qquad r \neq p$$

Taking the product of these terms for all $r \in \mathbb{N}$, $r \neq p$, yields

$$\prod_{\substack{r \in \mathbb{N} \\ r \neq p}} a_{r\sigma(r)} = \prod_{\rho(r) \in \mathbb{N}'} b_{\rho(r)\tau\sigma(r)} \qquad (1.9.29)$$

where $\mathbb{N}' = \{1, 2, \ldots, n-1\}$. We now set $\rho(r) = t$. Then $r = \rho^{-1}(t)$ and

$$\tau\sigma(r) = \tau\sigma\rho^{-1}(t)$$

If we set $\tau\sigma\rho^{-1} = \omega$, then equation (1.9.29) becomes

$$\prod_{\substack{r \in \mathbb{N} \\ r \neq p}} a_{r\sigma(r)} = \prod_{t \in \mathbb{N}'} b_{t\omega(t)} = b_{1\omega(1)} b_{2\omega(2)} \cdots b_{n-1, \omega(n-1)}$$

Moreover, since ρ and τ are fixed permutations of $\mathbb{N}$, it follows that as σ runs through the $(n-1)!$ permutations of $\mathbb{N}$ for which $\sigma(p) = q$, then $\omega = \tau\sigma\rho^{-1}$ runs through the $(n-1)!$ distinct permutations of $\mathbb{N}'$. Hence

$$\det B_{pq} = \sum_{\omega \in S_{n-1}} \operatorname{sgn} \omega \; b_{1\omega(1)} b_{2\omega(2)} \cdots b_{n-1, \omega(n-1)}$$
$$= \sum_{\substack{\sigma \in S_n \\ \sigma(p)=q}} \operatorname{sgn} \tau\sigma\rho^{-1} \prod_{\substack{r \in \mathbb{N} \\ r \neq p}} a_{r\sigma(r)}$$
$$= \operatorname{sgn} \tau \; \operatorname{sgn} \rho^{-1} \sum_{\substack{\sigma \in S_n \\ \sigma(p)=q}} \operatorname{sgn} \sigma \prod_{\substack{r \in \mathbb{N} \\ r \neq p}} a_{r\sigma(r)}$$
$$= \operatorname{sgn} \tau \; \operatorname{sgn} \rho^{-1} A_{pq} = (-1)^{2n-p-q} A_{pq} = (-1)^{p+q} A_{pq}$$

since $\qquad \operatorname{sgn} \tau = (-1)^{n-q} \qquad \operatorname{sgn} \rho^{-1} = \operatorname{sgn} \rho = (-1)^{n-q} \qquad \text{////}$

Thus we have shown that in the expression for the determinant of a matrix the coefficient of each entry is, up to sign, the determinant of some $(n-1) \times (n-1)$ submatrix of the matrix. Let us give these submatrices and their determinants names.

Definition 1.9.30 Let $A = [a_{ij}]$ be an $n \times n$ matrix. The submatrix B_{pq} of order $n-1$ formed from A by deleting the pth row and the qth column of A is called the *minor* of a_{pq} in A. The number

$$A_{pq} = (-1)^{p+q} \det B_{pq}$$

is called the *cofactor* of a_{pq}. Finally, the matrix of order n which has the cofactor A_{pq} of a_{pq} in the (q, p) position, $p, q, = 1, 2, \ldots, n$, is called the *adjoint*[1] of A and is denoted[2] by *adj A*.

According to equation (1.9.24), the cofactor of a_{pq} is the coefficient of a_{pq} in the pth row expansion for the determinant of A. This entire discussion can be repeated expressing the determinant of A as a linear homogeneous polynomial in the entries of the qth column of A. The coefficient of a_{pq} is the same in both expressions for det A, namely, the cofactor A_{pq} of a_{pq} in A. Thus, in addition to equation (1.9.24) we have

$$\det A = a_{1q} A_{1q} + a_{2q} A_{2q} + \cdots + a_{nq} A_{nq} \quad (1.9.31)$$

which we call the qth *column expansion* of det A.

The adjoint of a matrix leads to an interesting and often useful construction of the inverse of a nonsingular matrix.

Theorem 1.9.32 For any square matrix A,

$$A \cdot \operatorname{adj} A = \operatorname{adj} A \cdot A = (\det A) I$$

In particular, when A is nonsingular,

$$A^{-1} = \frac{1}{\det A} \operatorname{adj} A$$

PROOF We will first show that

$$A \cdot \operatorname{adj} A = (\det A) I \quad (1.9.33)$$

is a direct consequence of equation (1.9.24) and Corollary 1.9.13. Set

$$A \cdot \operatorname{adj} A = C = [c_{ij}]$$

---

[1] The adjoint of a matrix as defined here should not be confused with the adjoint of a linear transformation of an abstract vector space (see, for example, Halmos [6] or Taylor [9]).
[2] It will be convenient to define the adjoint of any 1×1 matrix to be the 1×1 matrix [1].

According to Definition 1.9.30 and the definition of multiplication of matrices [allowing for the fact that A_{pq} is in the (q, p) position of adj A], we have

$$c_{ij} = a_{i1} A_{j1} + a_{i2} A_{j2} + \cdots + a_{in} A_{jn} \quad i, j = 1, 2, \ldots, n \quad (1.9.34)$$

But, by equation (1.9.24), the right-hand side of equation (1.9.34) is equal to the determinant of the matrix formed from A by replacing $A_{(j)}$ by $A_{(i)}$. When $i = j$, this matrix is simply A and hence

$$c_{ii} = \det A \quad i = 1, 2, \ldots, n$$

This is equation (1.9.24) again. On the other hand, when $i \neq j$, the right-hand side of equation (1.9.34) is equal to the determinant of a matrix whose ith and jth rows are identical, i.e., they are both equal to $A_{(i)}$. By Corollary 1.9.13,

$$c_{ij} = 0 \quad i \neq j$$

completing the proof of equation (1.9.33).

A proof that adj $A \cdot A = (\det A) I$ would be a repetition of the above argument beginning with equation (1.9.31) instead of equation (1.9.24).

When A is nonsingular, it follows from Corollary 1.9.17 that $\det A \neq 0$ and consequently

$$A \cdot \frac{1}{\det A} \operatorname{adj} A = \frac{1}{\det A} \operatorname{adj} A \cdot A = I$$

Thus
$$A^{-1} = \frac{1}{\det A} \operatorname{adj} A \quad ////$$

Note that when A is singular, then $\det A = 0$ and hence

$$A \cdot \operatorname{adj} A = \operatorname{adj} A \cdot A = 0$$

EXAMPLE 1.9.35 Let

$$A = \begin{bmatrix} 1 & -2 & 0 \\ 2 & -1 & 5 \\ 1 & 0 & 4 \end{bmatrix}$$

Find (a) the minor of $-2 \ (= a_{12})$, (b) the cofactor of -2, (c) adj A, and (d) A^{-1}.

(a) We will use the notation of Definition 1.9.30. The minor of $a_{12} = -2$ is

$$B_{12} = \begin{bmatrix} 2 & 5 \\ 1 & 4 \end{bmatrix}$$

and is obtained by deleting the first row and the second column of A.

(b) the cofactor of -2 is

$$A_{12} = (-1)^{1+2} \det \begin{bmatrix} 2 & 5 \\ 1 & 4 \end{bmatrix} = -3$$

(c) Continuing in this way, we find that

$$\operatorname{adj} A = \begin{bmatrix} -4 & 8 & -10 \\ -3 & 4 & -5 \\ 1 & -2 & 3 \end{bmatrix}$$

(d) Since $\det A = 2$, we have

$$A^{-1} = \frac{1}{2} \begin{bmatrix} -4 & 8 & -10 \\ -3 & 4 & -5 \\ 1 & -2 & 3 \end{bmatrix} = \begin{bmatrix} -2 & 4 & -5 \\ -\frac{3}{2} & 2 & -\frac{5}{2} \\ \frac{1}{2} & -1 & \frac{3}{2} \end{bmatrix} \quad ////$$

We conclude this section with a brief introduction to a special type of matrix which will be used extensively in later sections of this chapter.

Definition 1.9.36 For $i = 1, 2, \ldots, k$, let A_i be a square matrix of order n_i. The square matrix

$$A = \begin{bmatrix} A_1 & 0 & 0 & \cdots & 0 \\ 0 & A_2 & 0 & \cdots & 0 \\ 0 & 0 & A_3 & \cdots & 0 \\ \multicolumn{5}{c}{\dotfill} \\ 0 & 0 & 0 & \cdots & A_k \end{bmatrix} \quad (1.9.37)$$

of order $n_1 + n_2 + \cdots + n_k$ is called the *direct sum* of the matrices $A_1, A_2, \ldots, A_k$. A more convenient notation for (1.9.37) is

$$A = A_1 \oplus A_2 \oplus \cdots \oplus A_k$$

We have already encountered the direct sum. In terms of Definition 1.9.36, Theorem 1.7.8 states that a square matrix of order n and rank r can be transformed by elementary operations into the direct sum of an identity matrix of order r and a zero matrix of order $n - r$. (Of course, Theorem 1.7.8 applies to nonsquare matrices as well.) The direct sum of matrices will play a prominent role in Secs. 1.10, 1.12, and 1.13.

For any positive integer n and any square matrix B it can easily be shown that
$$\det(I_n \oplus B) = \det B = \det(B \oplus I_n) \quad (1.9.38)$$
The first-row expansion of $\det(I_n \oplus B)$ yields
$$\det(I_n \oplus B) = 1 \cdot \det(I_{n-1} \oplus B) = \det(I_{n-1} \oplus B)$$
After repeating this process n times, we have the first equation of (1.9.38). The proof of the second equation of (1.9.38) is similar. This is a simple special case of:

Theorem 1.9.39 Let
$$A = A_1 \oplus A_2 \oplus \cdots \oplus A_k$$
Then
$$\det A = (\det A_1)(\det A_2) \cdots (\det A_k)$$

PROOF The proof is by an induction on k. When $k = 1$, we have $A = A_1$ and the theorem is immediate. Now assume the theorem is true for direct sums of $k-1$ matrices and write

$$A = \begin{bmatrix} A_1 & 0 & \cdots & 0 & 0 \\ 0 & A_2 & \cdots & 0 & 0 \\ \cdots & \cdots & \cdots & \cdots & \cdots \\ 0 & 0 & \cdots & A_{k-1} & 0 \\ 0 & 0 & \cdots & 0 & A_k \end{bmatrix}$$

$$= \begin{bmatrix} A_1 & 0 & \cdots & 0 & 0 \\ 0 & A_2 & \cdots & 0 & 0 \\ \cdots & \cdots & \cdots & \cdots & \cdots \\ 0 & 0 & \cdots & A_{k-1} & 0 \\ 0 & 0 & \cdots & 0 & I \end{bmatrix} \begin{bmatrix} I & 0 & \cdots & 0 & 0 \\ 0 & I & \cdots & 0 & 0 \\ \cdots & \cdots & \cdots & \cdots & \cdots \\ 0 & 0 & \cdots & I & 0 \\ 0 & 0 & \cdots & 0 & A_k \end{bmatrix}$$

Using (1.9.38) twice, we see that
$$\det A = [\det(A_1 + A_2 + \cdots + A_{k-1})](\det A_k)$$
and, applying the induction hypothesis,
$$\det A = [(\det A_1)(\det A_2) \cdots (\det A_{k-1})](\det A_k) \quad ////$$

EXERCISES

1 Find the determinant of each of the following matrices:

(a) $[-3]$ (b) $\begin{bmatrix} 1 & 2 \\ 2 & -5 \end{bmatrix}$ (c) $\begin{bmatrix} 1 & 2 & -1 \\ -7 & 0 & 5 \\ -4 & 1 & 3 \end{bmatrix}$

(d) $\begin{bmatrix} 1 & 2 & -1 \\ 2 & 1 & 3 \\ 3 & 3 & 2 \end{bmatrix}$ (e) $\begin{bmatrix} 1 & 2 & -2 & 0 \\ 3 & -3 & 5 & -1 \\ 1 & 3 & 5 & 12 \\ -3 & 0 & 1 & 4 \end{bmatrix}$

(1) Directly from Definition 1.9.1.
(2) By transforming the matrix to the form (1.7.9) using elementary operations.
(3) By using the first-row expansion.
(4) By using the last-column expansion.

2 Find the determinant of each of the following matrices:

(a) $\begin{bmatrix} 1 & 2 & 0 & 0 \\ -1 & 2 & 0 & 0 \\ 0 & 0 & 1 & 2 \\ 0 & 0 & 2 & -5 \end{bmatrix}$ (b) $\begin{bmatrix} 2 & 0 & 0 & 0 \\ 0 & 1 & 2 & -1 \\ 0 & 2 & 0 & 5 \\ 0 & -4 & 1 & 3 \end{bmatrix}$

(1) Directly from Definition 1.9.1 and (2) By using Theorem 1.9.39.

3 For what values of x is

$$\begin{bmatrix} 1 & 2 & -1 \\ x & 1 & 3 \\ 3 & 3 & x \end{bmatrix}$$

nonsingular?

4 Find the inverse of the matrix in part (c) of Exercise 1:
   (a) By using Theorem 1.9.32.
   (b) By expressing the matrix as a product of elementary matrices and using Exercise 11 of Sec. 1.6.
   (c) By treating the columns of the matrix as a basis for $\mathbb{F}^3$.

5 Find the adjoint of the matrix in part (d) of Exercise 1. Verify Theorem 1.9.32.

6 Let $A = [a_{ij}]$ be a square matrix of order n of the form

$$A = \begin{bmatrix} 0 & 0 & 0 & \cdots & 0 & a_{1n} \\ a_{21} & 0 & 0 & \cdots & 0 & 0 \\ 0 & a_{32} & 0 & \cdots & 0 & 0 \\ \multicolumn{6}{c}{\dotfill} \\ 0 & 0 & 0 & \cdots & a_{n,n-1} & 0 \end{bmatrix}.$$

Find det A (a) by an induction on n and (b) by showing that A is related to a permutation matrix.

7 Prove Lemma 1.9.15.

8 Let A be a square matrix of order n and let $\alpha \in \mathbb{F}$. Find (a) det αA and (b) adj αA.

9 Let A be a square matrix of order n. Show that the rank of adj A is:
   (a) n when A has rank n.
   (b) 1 when A has rank $n-1$.
   (c) 0 when A has rank less than $n-1$.

10 Show that
$$\text{adj } A^T = (\text{adj } A)^T$$
for every square matrix A.

11 Show that
$$\text{adj } A^{-1} = (\text{adj } A)^{-1}$$
for every nonsingular matrix A.

12 Show that
$$\det A = \det A^T$$
for every square matrix A by:
   (a) Using Lemma 1.9.6.
   (b) Using a first-row expansion of A, a first-column expansion of A^T, and an induction on the order of A.
   (c) Reducing A to the form (1.7.9) using elementary operations.

13 A matrix is *symmetric* if $A^T = A$ and *skew-symmetric* if $A^T = -A$. Prove that if $1 + 1 \neq 0$, then every skew-symmetric matrix with entries from $\mathbb{F}$ which is of odd order is singular. (See Exercise 20 of Sec. 1.7.)

14 Show that the adjoint of a symmetric matrix is symmetric. Show that the adjoint of a skew-symmetric matrix is skew-symmetric if its order is even and symmetric if its order is odd.

15 Show that for every square matrix A, the matrices $A + A^T$, AA^T, and A^TA are symmetric and $A - A^T$ is skew-symmetric.

16 A square matrix $A = [a_{ij}]$ is *lower triangular* if $a_{ij} = 0$ whenever $i < j$. Prove that the determinant of a lower-triangular matrix is equal to the product of its diagonal entries.

17 Let A, B, C and D be $m \times m$, $m \times n$, $n \times m$, and $n \times n$ matrices respectively. Show that
$$\det \begin{bmatrix} A & 0 \\ C & D \end{bmatrix} = \det \begin{bmatrix} A & B \\ 0 & D \end{bmatrix} = (\det A)(\det D)$$

Hint: Show that
$$\begin{bmatrix} A & 0 \\ C & D \end{bmatrix} = \begin{bmatrix} A & 0 \\ 0 & I \end{bmatrix} \begin{bmatrix} I & 0 \\ C & I \end{bmatrix} \begin{bmatrix} I & 0 \\ 0 & D \end{bmatrix}$$

18 Let A be a square matrix of order n, $n \geq 2$. Show that

$$\text{adj}(\text{adj } A) = (\det A)^{n-2} A$$

[*Hint:* Consider three cases: (*a*) A is nonsingular (use Theorem 1.9.32 and Exercises 8 and 11); (*b*) A is singular, $n \geq 3$ (use Exercise 9); (*c*) $n = 2$.]

19 *Cramer's rule* (1750). Let

$$a_{11}x_1 + a_{12}x_2 + \cdots + a_{1n}x_n = b_1$$
$$a_{21}x_1 + a_{22}x_2 + \cdots + a_{2n}x_n = b_2$$
$$\cdots\cdots\cdots\cdots\cdots\cdots\cdots\cdots\cdots$$
$$a_{n1}x_1 + a_{n2}x_2 + \cdots + a_{nn}x_n = b_n$$

be a system of n simultaneous linear equations in n unknowns for which the matrix of coefficients

$$A = [a_{ij}] \qquad i, j = 1, 2, \ldots, n$$

is nonsingular. Show that the unique solution of this system of equations is given by

$$x_k = \frac{\det B_k}{\det A}$$

where B_k is the $n \times n$ matrix formed from A by replacing the kth column by the column vector

$$\begin{bmatrix} b_1 \\ b_2 \\ \vdots \\ b_n \end{bmatrix}$$

(*Hint:* Express the system of equations in the matrix form $AX = B$ and multiply both sides of the equation by adj A.)

20 Let A be an $m \times n$ matrix of rank r. Show that for each value of k, $k = 1, 2, \ldots, r$, A has a nonsingular submatrix of order k. (*Hint:* Show that if every $k \times k$ submatrix of A is singular, then, for all $l > k$, every $l \times l$ submatrix of A is singular. Also, see Exercises 28 to 30 of Sec. 1.7.)

21 Let $A = [a_{ij}]$ be a square matrix of order n. Treat det A as a quadratic polynomial in the entries of the first two rows of A. What is the coefficient of $a_{11}a_{22}$? What is the coefficient of $a_{12}a_{21}$? In this way, find row and column expansions of det A, expressing det A as a sum of $\frac{1}{2}n(n-1)$ terms, where each term consists of a product of a determinant of a 2×2 matrix and a determinant of an $(n-2) \times (n-2)$ matrix. (This is a special case of Exercise 23 below.)

22 Show that for any $a_1, a_2, b_1, b_2, c_1, c_2, d_1, d_2 \in \mathbb{F}$

$$\det\begin{bmatrix} a_1 & c_1 \\ a_2 & c_2 \end{bmatrix}\det\begin{bmatrix} b_1 & d_1 \\ b_2 & d_2 \end{bmatrix} - \det\begin{bmatrix} b_1 & c_1 \\ b_2 & c_2 \end{bmatrix}\det\begin{bmatrix} a_1 & d_1 \\ a_2 & d_2 \end{bmatrix}$$
$$= \det\begin{bmatrix} a_1 & b_1 \\ a_2 & b_2 \end{bmatrix}\det\begin{bmatrix} c_1 & d_1 \\ c_2 & d_2 \end{bmatrix}$$

[*Hint:* Use Exercises 17 and 21 and the matrix

$$\begin{bmatrix} a_1 & 0 & c_1 & d_1 \\ a_2 & 0 & c_2 & d_2 \\ 0 & b_1 & c_1 & d_1 \\ 0 & b_2 & c_2 & d_2 \end{bmatrix}$$

23 *The Laplace expansion of a determinant* (1772). Let A be an $n \times n$ matrix and let m be a fixed integer, $1 \leq m \leq n - 1$. Partition A as

$$A = \begin{bmatrix} B \\ C \end{bmatrix}$$

where B has m rows and C has $n - m$ rows. Let Θ be a subset of $\mathbb{N} = \{1, 2, \ldots, n\}$ consisting of m integers, $\Theta = \{\theta_1, \theta_2, \ldots, \theta_m\}$, $\theta_1 < \theta_2 < \cdots < \theta_m$. Further let $\overline{\Theta}$ be the set of $n - m$ integers of $\mathbb{N}$ which do not belong to Θ, $\overline{\Theta} = \{\bar{\theta}_1, \bar{\theta}_2, \ldots, \bar{\theta}_{n-m}\}$, $\bar{\theta}_1 < \bar{\theta}_2 < \cdots < \bar{\theta}_{n-m}$. Define B_Θ to be the square matrix of order m formed from B by deleting all columns $B^{(j)}$ for $j \notin \Theta$. Similarly, define $C_{\overline{\Theta}}$ to be the square matrix of order $n - m$ formed from C by deleting all columns $C^{(j)}$ for $j \notin \overline{\Theta}$. Finally, let σ_Θ be the permutation

$$\sigma_\Theta = \begin{pmatrix} 1 & 2 & \cdots & m & m+1 & \cdots & n \\ \theta_1 & \theta_2 & \cdots & \theta_m & \bar{\theta}_1 & \cdots & \bar{\theta}_{n-m} \end{pmatrix}$$

Prove that

$$\det A = \sum_\Theta \operatorname{sgn} \sigma_\Theta (\det B_\Theta)(\det C_{\overline{\Theta}})$$

where the sum is taken over the $n!/[m!(n-m)!]$ subsets Θ of $\mathbb{N}$. What does this theorem become when $m = 1$?

24 *The Cauchy-Binet theorem*[1] (1812) Let A and B be $m \times n$ and $n \times m$ matrices respectively, $m \leq n$, and let $C = AB$. For each subset Θ of $\mathbb{N} = \{1, 2, \ldots, n\}$ consisting of m elements of $\mathbb{N}$, define A_Θ to be the square matrix of order m formed from A by deleting all columns $A^{(j)}$ for $j \notin \Theta$.

---

[1] This theorem was discovered and published independently by both Cauchy and Binet in 1812.

Similarly, define B_Θ to be the square matrix of order m formed from B by deleting all rows $B_{(j)}$ for $j \notin \Theta$. Prove that

$$\det C = \sum_\Theta (\det A_\Theta)(\det B_\Theta)$$

where the sum is taken over the $n!/[m!(n-m)!]$ subsets Θ of $\mathbb{N}$. (*Hint:* Set

$$D = \begin{bmatrix} 0 & -A \\ B & I \end{bmatrix} = \begin{bmatrix} AB & -A \\ 0 & I \end{bmatrix} \begin{bmatrix} I & 0 \\ B & I \end{bmatrix}$$

Apply Exercise 20 of Sec. 1.8 and Exercise 23 above.)

1.10 LINEAR TRANSFORMATIONS AND INVARIANT SPACES

In Sec. 1.1 we defined a matrix to be a rectangular array of numbers. There is, however, another approach to the concept of a matrix which is of prime importance throughout mathematics. A matrix is thought of not so much in terms of what it is but rather in terms of what it does.

For a (fixed) $m \times n$ matrix A and a (variable) vector $X \in \mathbb{F}^n$, a vector

$$Y = AX \quad (1.10.1)$$

belonging to $\mathbb{F}^m$ is determined. We can consider equation (1.10.1) as a mapping

$$\mathscr{A} : X \to Y \quad (= AX) \quad (1.10.2)$$

of the vector X (belonging to $\mathbb{F}^n$) to the vector Y (belonging to $\mathbb{F}^m$). The effect on X of the mapping $\mathscr{A}$ is obtained by multiplying X on the left by the matrix A.

Similarly, if $\mathbb{V}$ is a subspace of $\mathbb{F}^n$, then

$$\mathbb{W} = A\mathbb{V} = \{AX \mid X \in \mathbb{V}\}$$

is a subspace of $\mathbb{F}^m$ (see Exercise 13 of Sec. 1.3). The mapping $\mathscr{A}$ of (1.10.2) induces a mapping

$$\mathscr{A} : \mathbb{V} \to \mathbb{W} \quad (= A\mathbb{V})$$

of the subspace $\mathbb{V}$ of $\mathbb{F}^n$ to the subspace $\mathbb{W}$ of $\mathbb{F}^m$. As before, $\mathscr{A}$ is carried out by multiplying the elements of $\mathbb{V}$ on the left by the matrix A.

Not all mappings of $\mathbb{F}^n$ into $\mathbb{F}^m$ can be carried out by means of a matrix multiplication. For example, it is certainly necessary (though not sufficient) that $\mathscr{A}$ map the zero vector of $\mathbb{F}^n$ into the zero vector of $\mathbb{F}^m$. We will see that the mappings of one vector space into another which can be carried out by a matrix multiplication are characterized in Definition 1.10.3.

Definition 1.10.3[1] A mapping $\mathscr{A}$ of a vector space $\mathbb{V}$ into a vector space $\mathbb{W}$ which satisfies the two conditions

1. $\mathscr{A}(X + Y) = \mathscr{A}(X) + \mathscr{A}(Y) \qquad X, Y \in \mathbb{V}$
2. $\mathscr{A}(\alpha X) = \mathscr{A}(X\alpha) = \alpha \mathscr{A}(X) \qquad \alpha \in \mathbb{F}, X \in \mathbb{V}$

(1.10.4)

is called a *linear transformation* of $\mathbb{V}$ into $\mathbb{W}$.

EXAMPLE 1.10.5 It is not difficult to verify that the mapping

$$\mathscr{A}: \begin{bmatrix} x_1 \\ x_2 \\ \vdots \\ x_n \end{bmatrix} \to \begin{bmatrix} x_n \\ x_{n-1} \\ \vdots \\ x_1 \end{bmatrix}$$

is a linear transformation of $\mathbb{F}^n$ into $\mathbb{F}^n$. ////

EXAMPLE 1.10.6 The mapping

$$\mathscr{D}: \begin{bmatrix} x_1 \\ x_2 \\ x_3 \\ \vdots \\ x_{n-1} \\ x_n \end{bmatrix} \to \begin{bmatrix} x_2 \\ 2x_3 \\ 3x_4 \\ \vdots \\ (n-1)x_n \end{bmatrix}$$

is a linear transformation of $\mathbb{F}^n$ into $\mathbb{F}^{n-1}$. $\mathscr{D}$ is called the *derivative*. Why? ////

For every $m \times n$ matrix A, the mapping $\mathscr{A}$ defined by (1.10.2) provides an example of a linear transformation of $\mathbb{F}^n$ into $\mathbb{F}^m$. Moreover, two $m \times n$ matrices yield the same linear transformation of $\mathbb{F}^n$ into $\mathbb{F}^m$ if and only if they are equal. Conversely, we will prove (in Theorem 1.10.9) that every linear transformation of $\mathbb{F}^n$ into $\mathbb{F}^m$ is of this type; i.e., for each linear transformation

$$\mathscr{A}: X \to Y$$

of $\mathbb{F}^n$ into $\mathbb{F}^m$ there exists exactly one $m \times n$ matrix A for which

$$Y = AX$$

---

[1] All the material of this chapter can be formulated in terms of linear transformations rather than in terms of matrices. The theorems concerning matrices then follow for the corresponding theorems about linear transformations. Indeed, this is very often done; see, for example, Halmos [6].

for all $X \in \mathbb{F}^n$. This will establish a one-to-one correspondence between the set of all linear transformations of $\mathbb{F}^n$ into $\mathbb{F}^m$ and the set $\mathbb{F}_n^m$ of all $m \times n$ matrices with entries in $\mathbb{F}$. Because of this correspondence it is seldom necessary to study matrices and linear transformations separately. Information about one can usually be translated into information about the other. We have chosen matrices rather than linear transformations as a framework for this chapter primarily because matrices are more suitable for the subjects to be discussed in succeeding chapters.

Because of Equations (1.10.4), a linear transformation of one vector space into another is completely determined by its effect on a basis. More specifically:

Lemma 1.10.7 Let $\mathbb{V}$ and $\mathbb{W}$ be vector spaces, let $\mathbb{X} = \{X_1, X_2, \ldots, X_k\}$ be a basis for $\mathbb{V}$ and let $Y_1, Y_2, \ldots, Y_k$ be any k (distinct or not) elements of $\mathbb{W}$. Then there is exactly one linear transformation $\mathscr{A}$ of $\mathbb{V}$ into $\mathbb{W}$ for which

$$\mathscr{A} : X_i \to Y_i \qquad i = 1, 2, \ldots, k \qquad (1.10.8)$$

PROOF Let us first show that there is at least one linear transformation $\mathscr{A}$ of $\mathbb{V}$ into $\mathbb{W}$ for which (1.10.8) holds. Since $\mathbb{X}$ is a basis for $\mathbb{V}$, for each $X \in \mathbb{V}$ there exist unique numbers $\alpha_1, \alpha_2, \ldots, \alpha_k$ belonging to $\mathbb{F}$ such that

$$X = \alpha_1 X_1 + \alpha_2 X_2 + \cdots + \alpha_k X_k$$

We now define the mapping $\mathscr{A}$ of $\mathbb{V}$ into $\mathbb{W}$ by

$$\mathscr{A} : X \to \alpha_1 Y_1 + \alpha_2 Y_2 + \cdots + \alpha_k Y_k$$

Clearly this implies

$$\mathscr{A} : X_i \to Y_i \qquad i = 1, 2, \ldots, k$$

It only remains to verify that the mapping $\mathscr{A}$ defined in this way is a linear transformation, i.e., that $\mathscr{A}$ satisfies equations (1.10.4). This is left as an exercise. (Note that if $\alpha_1, \alpha_2, \ldots, \alpha_k$ were not uniquely defined by X, then this definition of $\mathscr{A}$ might be ambiguous.)

Conversely, in order to show that $\mathscr{A}$ is unique, let us suppose that $\mathscr{A}_1$ and $\mathscr{A}_2$ are linear transformations of $\mathbb{V}$ into $\mathbb{W}$ satisfying (1.10.8). Let $X \in \mathbb{V}$,

$$X = \alpha_1 X_1 + \alpha_2 X_2 + \cdots + \alpha_k X_k$$

Then, by Definition 1.10.3,

$$\begin{aligned} \mathscr{A}_1(X) &= \mathscr{A}_1(\alpha_1 X_1 + \alpha_2 X_2 + \cdots + \alpha_k X_k) \\ &= \alpha_1 \mathscr{A}_1(X_1) + \alpha_2 \mathscr{A}_1(X_2) + \cdots + \alpha_k \mathscr{A}_1(X_k) \\ &= \alpha_1 Y_1 + \alpha_2 Y_2 + \cdots + \alpha_k Y_k \end{aligned}$$

In the same way,

$$\begin{aligned}\mathscr{A}_2(X) &= \mathscr{A}_2(\alpha_1 X_1 + \alpha_2 X_2 + \cdots + \alpha_k X_k)\\ &= \alpha_1 \mathscr{A}_2(X_1) + \alpha_2 \mathscr{A}_2(X_2) + \cdots + \alpha_k \mathscr{A}_2(X_k)\\ &= \alpha_1 Y_1 + \alpha_2 Y_2 + \cdots + \alpha_k Y_k\end{aligned}$$

Consequently $\mathscr{A}_1(X) = \mathscr{A}_2(X)$ for every $X \in \mathbb{V}$. Thus $\mathscr{A}_1$ and $\mathscr{A}_2$ are the same mapping. ////

Theorem 1.10.9 Let $\mathscr{A}$ be a linear transformation of $\mathbb{F}^n$ into $\mathbb{F}^m$. Then there exists exactly one $m \times n$ matrix A such that

$$\mathscr{A}: X \to AX \quad (1.10.10)$$

for all $X \in \mathbb{F}^n$.

PROOF We can prove that there exists at least one $m \times n$ matrix A satisfying (1.10.10) simply by constructing it. According to Lemma 1.10.7, it is sufficient to show that (1.10.10) holds for the elements of some basis for $\mathbb{F}^n$. We will select the basis consisting of the n unit column vectors, $I^{(1)}, I^{(2)}, \ldots, I^{(n)}$. Denote the image of $I^{(i)}$ under the mapping $\mathscr{A}$ by Y_i, that is,

$$\mathscr{A}: I^{(i)} \to Y_i \quad i = 1, 2, \ldots, n$$

Clearly $Y_i \in \mathbb{F}^m$. Define A to be the $m \times n$ matrix whose ith column is Y_i. Then

$$AI^{(i)} = A^{(i)} = Y_i \quad i = 1, 2, \ldots, n$$

and hence $\quad \mathscr{A}: I^{(i)} \to Y_i \ (= AI^{(i)}) \quad i = 1, 2, \ldots, n$

Thus A satisfies (1.10.10).

In order to show that A is unique, let us suppose that A_1 and A_2 are $m \times n$ matrices which satisfy (1.10.10). Then

$$A_1 X = A_2 X$$

for all $X \in \mathbb{F}^n$, and hence $\mathbb{N}(A_1 - A_2) = \mathbb{F}^n$. It follows from Theorem 1.6.1 that the rank of $A_1 - A_2$ is zero and therefore

$$A_1 - A_2 = 0 \qquad ////$$

Because of Theorem 1.10.9 we can say that A is the *matrix of the linear transformation* $\mathscr{A}$ and that $\mathscr{A}$ is the *linear transformation of the matrix A*.

There is a completely analogous development of linear transformations of $\mathbb{F}_n$ into $\mathbb{F}_m$. In this case there is a one-to-one correspondence between linear transformations $\mathscr{A}$ and $n \times m$ matrices A such that

$$\mathscr{A}: X \to Y \quad (= XA)$$

for all $X \in \mathbb{F}_n$. The details are left for the exercises.

The most important linear transformations are those which map a vector space into itself, and we will study these in more detail. When a linear transformation maps $\mathbb{F}^n$ into itself, its matrix is square and has order n. We will be able to obtain a great deal of information about a square matrix by studying the subspaces which are mapped into themselves by the linear transformation corresponding to the matrix.

Definition 1.10.11 Let A be a square matrix of order n. A subspace $\mathbb{V}$ of $\mathbb{F}^n$ is *A-invariant* (or when there is no doubt about the choice of A, simply *invariant*) under the linear transformation

$$\mathscr{A}: X \to AX$$

of $\mathbb{F}^n$ into itself if $AX \in \mathbb{V}$ for all $X \in \mathbb{V}$.

For every matrix A of order n, both $\mathbb{F}^n$ and 0 are A-invariant subspaces of $\mathbb{F}^n$. Another example of an A-invariant subspace of $\mathbb{F}^n$ is $\mathbb{N}(A)$. It is a simple matter to show that the sum and the intersection of any number of A-invariant subspaces of $\mathbb{F}^n$ are again A-invariant.

The most important invariant subspaces of $\mathbb{F}^n$ are those of dimension 1, and we will consider these in this section before turning our attention in Sec. 1.12 to the more complicated theory of invariant subspaces in general. If X is a nonzero vector of $\mathbb{F}^n$ and the (one-dimensional) vector space which has X as a basis is denoted by $\mathbb{V}$, then $\mathbb{V}$ is A-invariant if and only if there exists a scalar λ in $\mathbb{F}$ such that

$$AX = \lambda X \quad (1.10.12)$$

Note that although X is not permitted to be the zero vector, there is no restriction against λ being zero. In particular, then, every nonzero vector contained in $\mathbb{N}(A)$ is a basis for a one-dimensional A-invariant subspace of $\mathbb{F}^n$ [with $\lambda = 0$ in equation (1.10.12)].

Definition 1.10.13 Let A be a square matrix of order n. If X is a nonzero vector of $\mathbb{F}^n$ and λ is a number such that equation (1.10.12) is satisfied, then λ is called a *characteristic root* of A and X is called a *characteristic vector* of A corresponding to the characteristic root λ.

Clearly equation (1.10.12) is equivalent to

$$(A - \lambda I)X = 0$$

Thus X is a characteristic vector of A corresponding to the characteristic root λ if and only if X is a nonzero element of $\mathbb{N}(A - \lambda I)$. However, applying Theorem 1.6.1 and Corollary 1.9.17, we see that $\mathbb{N}(A - \lambda I) \neq 0$ if and only if $A - \lambda I$ is singular or, equivalently, if and only if

$$\det(A - \lambda I) = 0$$

Thus the characteristic roots of A are precisely the solutions of the equation

$$\det(A - xI) = \det \begin{bmatrix} a_{11} - x & a_{12} & \cdots & a_{1n} \\ a_{21} & a_{22} - x & \cdots & a_{2n} \\ \vdots & & & \vdots \\ a_{n1} & a_{n2} & \cdots & a_{nn} - x \end{bmatrix} = 0 \quad (1.10.14)$$

Evaluating this determinant, we find that $\det(A - xI)$ is a polynomial of degree n (where n is the order of A). The coefficient of x^n is $(-1)^n$, and the constant term of $\det(A - xI)$, obtained by setting $x = 0$ in equation (1.10.14), is $\det A$.

Definition 1.10.15 Let A be a square matrix. Then the polynomial $\det(A - xI)$ is called the *characteristic polynomial* of A. The equation

$$\det(A - xI) = 0$$

is called the *characteristic equation* of A.

EXAMPLE 1.10.16 Let

$$A = \begin{bmatrix} 2 & 1 \\ 3 & 4 \end{bmatrix}$$

The characteristic polynomial of A is

$$\det \begin{bmatrix} 2 - x & 1 \\ 3 & 4 - x \end{bmatrix} = x^2 - 6x + 5 = (x - 1)(x - 5)$$

the characteristic equation of A is

$$x^2 - 6x + 5 = 0$$

and the characteristic roots of A are

$$\lambda_1 = 1 \qquad \lambda_2 = 5$$

The characteristic vectors of A can be found by examining $A - I$ and $A - 5I$,

$$A - I = \begin{bmatrix} 1 & 1 \\ 3 & 3 \end{bmatrix} \qquad A - 5I = \begin{bmatrix} -3 & 1 \\ 3 & -1 \end{bmatrix}$$

Clearly

$$X_1 = \begin{bmatrix} 1 \\ -1 \end{bmatrix} \in \mathbb{N}(A - I) \qquad X_2 = \begin{bmatrix} 1 \\ 3 \end{bmatrix} \in \mathbb{N}(A - 5I) \qquad (1.10.17)$$

Moreover X_1 is a basis for $\mathbb{N}(A - I)$, and X_2 is a basis for $\mathbb{N}(A - 5I)$. Thus X_1 is the unique (up to multiplication by a nonzero scalar) characteristic vector of A corresponding to the characteristic root $\lambda_1 = 1$ and X_2 is the unique (up to multiplication by a nonzero scalar) characteristic vector of A corresponding to the characteristic root $\lambda_2 = 5$. The two equations of (1.10.17) can also be written in the form of equation (1.10.12)

$$A \begin{bmatrix} 1 \\ -1 \end{bmatrix} = 1 \begin{bmatrix} 1 \\ -1 \end{bmatrix} \qquad A \begin{bmatrix} 1 \\ 3 \end{bmatrix} = 5 \begin{bmatrix} 1 \\ 3 \end{bmatrix}$$

The only A-invariant subspaces of $\mathbb{F}^2$ are 0, $\mathbb{F}^2$, $\mathbb{V}_1$, and $\mathbb{V}_2$, where $\mathbb{V}_1$ and $\mathbb{V}_2$ are the one-dimensional vector spaces spanned by X_1 and X_2, respectively.

////

Let us pause here to point out that we can define A-invariant subspaces of $\mathbb{F}_n$ in an analogous way and that one-dimensional A-invariant subspaces of $\mathbb{F}_n$ correspond to nonzero vectors X such that $XA = \lambda X$. It is evident that we are then led to define *row characteristic roots*, *row characteristic vectors*, the *row characteristic polynomial*, and the *row characteristic equation*. However, it follows from equation (1.10.14) and from Definition 1.10.15 that the (column) characteristic roots, polynomial, and equation are equal to their row counterparts. *Warning:* The corresponding characteristic vectors are quite different.

The roots of the characteristic equation of A are the characteristic roots of A. Clearly 0 is a characteristic root of A if and only if the constant term of its characteristic polynomial (which, as we pointed out above, is equal to det A) is 0. We can now add another useful necessary and sufficient condition that a matrix be nonsingular to the seven we already have (see Theorem 1.6.4 and Corollary 1.9.17).

Corollary 1.10.18 Let A be a square matrix. Then A is nonsingular if and only if 0 is not a characteristic root of A.

We have shown that to each characteristic root of A there corresponds at least one characteristic vector. It is clear from equation (1.10.12) that characteristic vectors corresponding to distinct characteristic roots are distinct. We can, in fact, prove that they are linearly independent.

Lemma 1.10.19 If $X_1, X_2, \ldots, X_k$ are characteristic vectors of A corresponding to the distinct characteristic roots $\lambda_1, \lambda_2, \ldots, \lambda_k$, then $X_1, X_2, \ldots, X_k$ are linearly independent.

PROOF Let V be the vector space spanned by $X_1, X_2, \ldots, X_k$. We may assume that $\lambda_1, \lambda_2, \ldots, \lambda_k$ are ordered in such a way that $X_1, X_2, \ldots, X_l$ form a basis for V. In order to show that $X_1, X_2, \ldots, X_k$ are linearly independent we will prove that $l = k$. Hence let us suppose that $l < k$. Then there exist numbers $\beta_1, \beta_2, \ldots, \beta_l$ such that

$$X_{l+1} = \beta_1 X_1 + \beta_2 X_2 + \cdots + \beta_l X_l \quad (1.10.20)$$

Multiply both sides of equation (1.10.20) on the left by A. Then

$$AX_{l+1} = \beta_1 AX_1 + \beta_2 AX_2 + \cdots + \beta_l AX_l$$

from which it follows that

$$\lambda_{l+1} X_{l+1} = \beta_1 \lambda_1 X_1 + \beta_2 \lambda_2 X_2 + \cdots + \beta_l \lambda_l X_l \quad (1.10.21)$$

On the other hand, when we multiply both sides of equation (1.10.20) by λ_{l+1}, we have

$$\lambda_{l+1} X_{l+1} = \beta_1 \lambda_{l+1} X_1 + \beta_2 \lambda_{l+1} X_2 + \cdots + \beta_l \lambda_{l+1} X_l \quad (1.10.22)$$

Subtracting equation (1.10.22) from equation (1.10.21) yields

$$0 = \beta_1(\lambda_1 - \lambda_{l+1})X_1 + \beta_2(\lambda_2 - \lambda_{l+1})X_2 + \cdots + \beta_l(\lambda_l - \lambda_{l+1})X_l$$

Since $X_1, X_2, \ldots, X_l$ are linearly independent, it follows that

$$\beta_1(\lambda_1 - \lambda_{l+1}) = \beta_2(\lambda_2 - \lambda_{l+1}) = \cdots = \beta_l(\lambda_l - \lambda_{l+1}) = 0$$

But $\lambda_1, \lambda_2, \ldots, \lambda_l, \lambda_{l+1}$ are distinct, and consequently

$$\beta_1 = \beta_2 = \cdots = \beta_l = 0$$

We have now shown that $X_{l+1} = 0$. This is a contradiction since X_{l+1} is a characteristic vector. Since the assumption that $l < k$ leads to a contradiction, we conclude that $l = k$; that is, $X_1, X_2, \ldots, X_k$ are linearly independent. ////

The simplest situation that can arise is for the characteristic polynomial of A to be expressible as a product of n distinct linear factors. Then A has n distinct characteristic roots, and we can prove:

Theorem 1.10.23 If A is a square matrix of order n having n distinct characteristic roots, $\lambda_1, \lambda_2, \ldots, \lambda_n$, then there is a nonsingular matrix P such that

$$P^{-1}AP = \begin{bmatrix} \lambda_1 & 0 & \cdots & 0 \\ 0 & \lambda_2 & \cdots & 0 \\ \multicolumn{4}{c}{\dotfill} \\ 0 & 0 & \cdots & \lambda_n \end{bmatrix} \quad (1.10.24)$$

PROOF Let $X_1, X_2, \ldots, X_k$ be characteristic vectors of A corresponding to the characteristic roots $\lambda_1, \lambda_2, \ldots, \lambda_k$ respectively and let P be the $n \times n$ matrix whose columns are $X_1, X_2, \ldots, X_k$, that is

$$P^{(i)} = X_i \quad i = 1, 2, \ldots, n$$

Then $\quad AP^{(i)} = AX_i = \lambda_i X_i = \lambda_i P^{(i)} \quad i = 1, 2, \ldots, n \quad (1.10.25)$

We can combine the n equations of (1.10.25) into the single matrix equation

$$AP = P \begin{bmatrix} \lambda_1 & 0 & \cdots & 0 \\ 0 & \lambda_2 & \cdots & 0 \\ \multicolumn{4}{c}{\dotfill} \\ 0 & 0 & \cdots & \lambda_n \end{bmatrix} \quad (1.10.26)$$

Lemma 1.10.19 implies that P is nonsingular, and the theorem follows by multiplying equation (1.10.26) on the left by P^{-1}. ////

The hypothesis of Theorem 1.10.23 may be relaxed somewhat. If A has n linearly independent characteristic vectors, corresponding to distinct characteristic roots or not, then the same construction yields a nonsingular matrix P for which $P^{-1}AP$ is a diagonal matrix whose distinct diagonal entries[1] are the characteristic roots of A. Conversely, if $P^{-1}AP$ is a diagonal matrix for some nonsingular matrix P, then the columns of P are a set of linearly independent characteristic vectors of A and the distinct diagonal entries of $P^{-1}AP$ are characteristic roots of A.

The corresponding theorem for row characteristic vectors states that a necessary and sufficient condition that there exist a nonsingular matrix Q such

---

[1] At this point of our discussion it is not at all clear what happens either when the characteristic polynomial of A has multiple roots or when the diagonal entries of $P^{-1}AP$ are not distinct. We will consider this problem shortly.

that QAQ^{-1} is a diagonal matrix is that A have n linearly independent row characteristic vectors. In this case, the rows of Q are the row characteristic vectors of A, and the distinct diagonal entries of QAQ^{-1} are characteristic roots of A. Setting $Q = P^{-1}$, we see that A has n linearly independent row characteristic vectors if and only if it has n linearly independent (column) characteristic vectors.

There is another way of looking at the row characteristic vectors of a matrix. We can identify them with the (column) characteristic vectors of A^T; that is, since

$$XA = \lambda X \quad \text{implies} \quad A^T X^T = \lambda X^T$$

(see Exercise 20 of Sec. 1.7) we see that X is a row characteristic vector of A if and only if X^T is a (column) characteristic vector of A^T. The characteristic polynomial, equation, and roots of A^T are identical to the characteristic polynomial, equation, and roots of A. However, the characteristic vectors may be quite different.

EXAMPLE 1.10.27 In Example 1.10.16 we showed that the matrix

$$A = \begin{bmatrix} 2 & 1 \\ 3 & 4 \end{bmatrix}$$

has characteristic vectors

$$\begin{bmatrix} 1 \\ -1 \end{bmatrix} \quad \text{and} \quad \begin{bmatrix} 1 \\ 3 \end{bmatrix}$$

corresponding to the characteristic roots

$$\lambda = 1 \quad \text{and} \quad \lambda = 5$$

respectively. As in Theorem 1.10.23, we form

$$P = \begin{bmatrix} 1 & 1 \\ -1 & 3 \end{bmatrix}$$

Then

$$AP = \begin{bmatrix} 1 & 5 \\ -1 & 15 \end{bmatrix} = \begin{bmatrix} 1 & 1 \\ -1 & 3 \end{bmatrix} \begin{bmatrix} 1 & 0 \\ 0 & 5 \end{bmatrix} = P \begin{bmatrix} 1 & 0 \\ 0 & 5 \end{bmatrix}$$

Moreover P is nonsingular, and

$$P^{-1} = \frac{1}{4} \begin{bmatrix} 3 & -1 \\ 1 & 1 \end{bmatrix}$$

Thus

$$P^{-1}AP = \frac{1}{4} \begin{bmatrix} 3 & -1 \\ 1 & 1 \end{bmatrix} \begin{bmatrix} 2 & 1 \\ 3 & 4 \end{bmatrix} \begin{bmatrix} 1 & 1 \\ -1 & 3 \end{bmatrix} = \begin{bmatrix} 1 & 0 \\ 0 & 5 \end{bmatrix}$$

as expected.

In the same way we see that

$$[3 \quad -1] \quad \text{and} \quad [1 \quad 1]$$

are the row characteristic vectors of A corresponding to the characteristic roots

$$\lambda_1 = 1 \quad \text{and} \quad \lambda_2 = 5$$

respectively. Set

$$Q = \begin{bmatrix} 3 & -1 \\ 1 & 1 \end{bmatrix}$$

Then Q is nonsingular, and

$$Q^{-1} = \frac{1}{4}\begin{bmatrix} 1 & 1 \\ -1 & 3 \end{bmatrix}$$

Thus we again have

$$QAQ^{-1} = \frac{1}{4}\begin{bmatrix} 3 & -1 \\ 1 & 1 \end{bmatrix}\begin{bmatrix} 2 & 1 \\ 3 & 4 \end{bmatrix}\begin{bmatrix} 1 & 1 \\ -1 & 3 \end{bmatrix} = \begin{bmatrix} 1 & 0 \\ 0 & 5 \end{bmatrix} \qquad ////$$

It may be difficult at this time to see the significance of equation (1.10.24). Equation (1.10.24) is reminiscent of equation (1.7.13), reducing a given matrix to a simple form by means of left and right multiplications. However, in Theorem 1.10.23 we consider only square matrices, and the type of reduction that is permitted is severely restricted.

Definition 1.10.28 Let A and B be square matrices of the same order. If there exists a nonsingular matrix P such that

$$B = P^{-1}AP$$

we say that B is *similar* to A.[1]

---

[1] The similarity of matrices has many applications. For example, it is useful in solving a system of linear homogeneous first-order differential equations

$$X' = AX \qquad \text{(I)}$$

where A is a square matrix of real numbers, X is a column vector of unknown functions in the variable t, and X' is the vector of the first derivatives of these functions with respect to t. If we introduce a column vector Y of new unknowns related to the original unknowns by

$$Y = PX$$

where P is a nonsingular matrix of real numbers, then the system of differential equations (I) becomes

$$Y' = PAP^{-1}Y$$

For a more geometric interpretation of similarity of matrices, see Halmos [6, Secs. 46 and 47].

It follows from Exercises 10 and 11 of Sec. 1.6 that similarity of matrices is an equivalence relation. Clearly similar matrices are also equivalent (see Exercise 12 of Sec. 1.7), and consequently they have the same rank. In Corollary 1.10.29 we present a number of other properties shared by similar matrices.

Corollary 1.10.29 Similar matrices have the same characteristic polynomial and consequently the same characteristic equation, the same characteristic roots, and the same determinant.

PROOF Let $B = P^{-1}AP$. Then
$$\det(B - xI) = \det[P^{-1}AP - P^{-1}(xI)P]$$
$$= \det[P^{-1}(A - xI)P]$$
$$= (\det P^{-1})[\det(A - xI)](\det P)$$
$$= \det(A - xI) \qquad \qquad ////$$

It is not true, however, that similar matrices have the same characteristic vectors (see Exercise 20). Also the converse of Corollary 1.10.29 (namely, that matrices having the same characteristic polynomial are similar) is false. Essentially, Secs. 1.12 and 1.13 are devoted to determining necessary and sufficient conditions for two matrices to be similar and to selecting, from all the matrices similar to a given matrix, the one which has the "simplest" form. In the case where A is of order n and has n linearly independent characteristic vectors we have already solved the problem, for then A is similar to a diagonal matrix.

When a matrix does not have n linearly independent characteristic vectors (either because the characteristic polynomial has multiple roots or because it is not expressible as a product of linear factors), the simple form of equation (1.10.24) is no longer obtainable. In order to prepare for a discussion of this more complicated situation, let us interpret Lemma 1.10.19 and Theorem 1.10.23 in the language of invariant subspaces. In Lemma 1.10.19, let us denote by $\mathbb{V}_i$ the one-dimensional subspace of $\mathbb{F}^n$ having the characteristic vector X_i of A as a basis. When $k = n$, Lemma 1.10.19 states that

$$\mathbb{F}^n = V_1 \oplus V_2 \oplus \cdots \oplus V_n \qquad (1.10.30)$$

and it follows in Theorem 1.10.23 that A is similar to the direct sum of n 1×1 matrices,

$$P^{-1}AP = [\lambda_1] \oplus [\lambda_2] \oplus \cdots \oplus [\lambda_n]$$

In general, whenever $\mathbb{F}^n$ can be expressed as a direct sum of A-invariant subspaces [as in (1.10.30), though the subspaces need not be one-dimensional], the same basic argument as that used in Theorem 1.10.23 yields the same basic result, namely, that A is similar to the direct sum of square matrices, one for each of the A-invariant subspaces $\mathbb{V}_i$, and that the square matrix corresponding to the subspace $\mathbb{V}_i$ has order equal to the dimension of $\mathbb{V}_i$. These results are contained in:

Theorem 1.10.31 The fundamental theorem of direct sums $\mathbb{F}^n$ can be expressed as

$$\mathbb{F}^n = \mathbb{V}_1 \oplus \mathbb{V}_2 \oplus \cdots \oplus \mathbb{V}_k$$

where $\mathbb{V}_i$ is an A-invariant subspace of $\mathbb{F}^n$ of dimension n_i if and only if there exists a nonsingular matrix P such that

$$P^{-1}AP = B_1 \oplus B_2 \oplus \cdots \oplus B_k \quad (1.10.32)$$

where B_i is a square matrix of order n_i.

PROOF First, let us suppose that

$$\mathbb{F}^n = \mathbb{V}_1 \oplus \mathbb{V}_2 \oplus \cdots \oplus \mathbb{V}_k \quad (1.10.33)$$

We wish to construct a nonsingular matrix P such that $P^{-1}AP$ is given by (1.10.32). For each i, $i = 1, 2, \ldots, k$, let

$$\mathbb{X}_i = \{X_{i1}, X_{i2}, \ldots, X_{in_i}\}$$

be a basis for $\mathbb{V}_i$. Since $\mathbb{F}^n$ is a direct sum of $\mathbb{V}_1, \mathbb{V}_2, \ldots, \mathbb{V}_k$, it follows from Definition 1.5.13 that

$$\mathbb{X} = \mathbb{X}_1 \cup \mathbb{X}_2 \cup \cdots \cup \mathbb{X}_k = \{X_{11}, X_{12}, \ldots, X_{1n_1}, X_{21}, \ldots, X_{kn_k}\}$$

is a basis for $\mathbb{F}^n$. Let P be the matrix whose columns are the n vectors of $\mathbb{X}$, in the order $X_{11}, X_{12}, \ldots, X_{1n_1}, X_{21}, \ldots, X_{kn_k}$. (Note that this imposes an order on the bases $\mathbb{X}_1, \mathbb{X}_2, \ldots, \mathbb{X}_k$ but not on the vectors within each basis.) Then P is a square matrix of order n and, in fact, is nonsingular. The columns of AP are $AX_{11}, AX_{12}, \ldots, AX_{1n_1}, AX_{21}, \ldots, AX_{kn_k}$. Since each of the vector spaces $\mathbb{V}_1, \mathbb{V}_2, \ldots, \mathbb{V}_k$ is A-invarient, the first n_1 columns of AP are elements of $\mathbb{V}_1$, the next n_2 columns of AP are elements of $\mathbb{V}_2$, etc. Hence each of the first n_1 columns of AP can be expressed as a linear combination of the elements of $\mathbb{X}_1$, each of the next n_2 columns of AP can be expressed as a linear combination of the elements of $\mathbb{X}_2$, and so on. Therefore, for each j, $j = 1, 2, \ldots, n_1$, there exist numbers

$$\alpha_{1j}, \alpha_{2j}, \ldots, \alpha_{n_1 j}$$

such that

$$AX_{1j} = \alpha_{1j}X_{11} + \alpha_{2j}X_{12} + \cdots + \alpha_{n_1 j}X_{1n_1}$$
$$= \alpha_{1j}X_{11} + \alpha_{2j}X_{12} + \cdots + \alpha_{n_1 j}X_{1n_1} + 0X_{12}$$
$$+ \cdots + 0X_{kn_k} \quad (1.10.34)$$

Similarly, for each j, $j = 1, 2, \ldots, n_2$, there exist numbers

$$\beta_{1j}, \beta_{2j}, \ldots, \beta_{n_2 j}$$

such that

$$AX_{2j} = \beta_{1j}X_{21} + \beta_{2j}X_{22} + \cdots + \beta_{n_2 j}X_{2n_2}$$
$$= 0X_{11} + 0X_{12} + \cdots + 0X_{1n_1} + \beta_{1j}X_{21} + \cdots$$
$$+ \beta_{n_2 j}X_{2n_2} + \cdots + 0X_{kn_k} \quad (1.10.35)$$

Continuing in this way for each of the sets $\mathbb{X}_3, \mathbb{X}_4, \ldots, \mathbb{X}_k$ and combining the resulting n equations into a single matrix equation, we have

$$AP = P \begin{bmatrix} B_1 & 0 & \cdots & 0 \\ 0 & B_2 & \cdots & 0 \\ \multicolumn{4}{c}{\dotfill} \\ 0 & 0 & \cdots & B_k \end{bmatrix} \quad (1.10.36)$$

where each B_i is a square matrix of order n_i. In particular, the n_1 equations of (1.10.34) and the n_2 equations of (1.10.35) yield the $n_1 \times n_1$ and $n_2 \times n_2$ matrices

$$B_1 = \begin{bmatrix} \alpha_{11} & \alpha_{12} & \cdots & \alpha_{1n_1} \\ \alpha_{21} & \alpha_{22} & \cdots & \alpha_{2n_1} \\ \multicolumn{4}{c}{\dotfill} \\ \alpha_{n_1 1} & \alpha_{n_1 2} & \cdots & \alpha_{n_1 n_1} \end{bmatrix} \quad B_2 = \begin{bmatrix} \beta_{11} & \beta_{12} & \cdots & \beta_{1n_2} \\ \beta_{21} & \beta_{22} & \cdots & \beta_{2n_2} \\ \multicolumn{4}{c}{\dotfill} \\ \beta_{n_2 1} & \beta_{n_2 2} & \cdots & \beta_{n_2 n_2} \end{bmatrix}$$

respectively. Since P is nonsingular, we may multiply equation (1.10.36) on the left by P^{-1}, completing the proof.

Conversely, suppose P is a nonsingular matrix for which

$$P^{-1}AP = B_1 \oplus B_2 \oplus \cdots \oplus B_k$$

where B_i is a square matrix of order n_i. By expressing this equation in the form (1.10.36) we see that the first n_1 columns of P form a basis for an A-invariant subspace $\mathbb{V}_1$ of $\mathbb{F}^n$, the next n_2 columns of P form a basis for an A-invariant subspace $\mathbb{V}_2$ of $\mathbb{F}^n$, etc., and finally,

$$\mathbb{F}^n = \mathbb{V}_1 \oplus \mathbb{V}_2 \oplus \cdots \oplus \mathbb{V}_k \qquad ////$$

EXAMPLE 1.10.37 Let

$$A = \begin{bmatrix} 3 & 1 & -2 \\ -1 & -1 & 4 \\ 0 & -1 & 3 \end{bmatrix}$$

The characteristic polynomial of A is

$$\det \begin{bmatrix} 3-x & 1 & -2 \\ -1 & -1-x & 4 \\ 0 & -1 & 3-x \end{bmatrix} = -x^3 + 5x^2 - 8x + 4 = -(x-2)^2(x-1)$$

and hence the characteristic roots of A are

$$\lambda_1 = \lambda_2 = 2 \qquad \lambda_3 = 1$$

The characteristic vectors of A can be found by examining the matrices

$$A - 2I = \begin{bmatrix} 1 & 1 & -2 \\ -1 & -3 & 4 \\ 0 & -1 & 1 \end{bmatrix} \qquad A - I = \begin{bmatrix} 2 & 1 & -2 \\ -1 & -2 & 4 \\ 0 & -1 & 2 \end{bmatrix}$$

Clearly $\left\{ \begin{bmatrix} 1 \\ 1 \\ 1 \end{bmatrix} \right\}$ is a basis for $\mathbb{N}(A - 2I)$, and hence $\begin{bmatrix} 1 \\ 1 \\ 1 \end{bmatrix}$ is the only characteristic vector of A (up to multiplication by a scalar) corresponding to the characteristic root 2. In the same way we see that $\left\{ \begin{bmatrix} 0 \\ 2 \\ 1 \end{bmatrix} \right\}$ is a basis for $\mathbb{N}(A - I)$ and $\begin{bmatrix} 0 \\ 2 \\ 1 \end{bmatrix}$ is the only characteristic vector of A (up to multiplication by a scalar) corresponding to the characteristic root 1. Moreover, A has no other characteristic roots (or characteristic vectors). Thus $\mathbb{F}^3$ cannot be expressed as a direct sum of three one-dimensional A-invariant subspaces, and A is not similar to a diagonal matrix. However, if we let $\mathbb{V}_1$ and $\mathbb{V}_2$ be the vector spaces which have the sets

$$\mathbb{X}_1 = \left\{ X_{11} = \begin{bmatrix} 1 \\ 1 \\ 1 \end{bmatrix}, X_{12} = \begin{bmatrix} 2 \\ -1 \\ 0 \end{bmatrix} \right\} \qquad \mathbb{X}_2 = \left\{ \begin{bmatrix} 0 \\ 2 \\ 1 \end{bmatrix} \right\}$$

as bases, then $\mathbb{V}_1$ and $\mathbb{V}_2$ are A-invariant (since

$$AX_{12} = \begin{bmatrix} 5 \\ -1 \\ 1 \end{bmatrix} = 2X_{12} + X_{11} \in \mathbb{V}_1) \qquad \text{and} \qquad \mathbb{F}^3 = \mathbb{V}_1 \oplus \mathbb{V}_2$$

We now set
$$P = \begin{bmatrix} 1 & 2 & 0 \\ 1 & -1 & 2 \\ 1 & 0 & 1 \end{bmatrix}$$

Then
$$AP = \begin{bmatrix} 2 & 5 & 0 \\ 2 & -1 & 2 \\ 2 & 1 & 1 \end{bmatrix} = \begin{bmatrix} 1 & 2 & 0 \\ 1 & -1 & 2 \\ 1 & 0 & 1 \end{bmatrix} \begin{bmatrix} 2 & 1 & 0 \\ 0 & 2 & 0 \\ 0 & 0 & 1 \end{bmatrix} = P \begin{bmatrix} 2 & 1 & 0 \\ 0 & 2 & 0 \\ 0 & 0 & 1 \end{bmatrix}$$

and hence
$$P^{-1}AP = \begin{bmatrix} 2 & 1 & 0 \\ 0 & 2 & 0 \\ 0 & 0 & 1 \end{bmatrix} = B_1 \oplus B_2$$

where
$$B_1 = \begin{bmatrix} 2 & 1 \\ 0 & 2 \end{bmatrix} \qquad B_2 = [1] \qquad ////$$

Definition 1.10.38 Let λ be a characteristic root of A. The multiplicity of λ as a root of the characteristic polynomial of A is called the *algebraic multiplicity* of λ, and $v(A - \lambda I)$ is called the *geometric multiplicity* of λ.

In Example 1.10.37, 2 is a characteristic root of algebraic multiplicity 2, since $\lambda_1 = \lambda_2 = 2$, and geometric multiplicity 1, since $\begin{bmatrix} 1 \\ 1 \\ 1 \end{bmatrix}$ is the only characteristic vector of A corresponding to the characteristic root 2.

Let A and B be similar matrices, $B = P^{-1}AP$. Then
$$B - \lambda I = P^{-1}(A - \lambda I)P$$
for every $\lambda \in \mathbb{F}$, and it follows from Theorem 1.6.7 that the geometric multiplicities of the characteristic roots of A and B are equal. Moreover, since similar matrices have the same characteristic polynomial, the algebraic multiplicities of their characteristic roots are equal. In Exercise 24 we outline a proof that the algebraic multiplicity of a characteristic root of a matrix is never less than its geometric multiplicity. In Secs. 1.12 and 1.13 we will investigate further the properties of the algebraic and geometric multiplicities of the characteristic roots of a matrix.

As we have seen, similar matrices have many properties in common, and much information about a matrix A can be obtained by finding a matrix B, similar to A, which is the direct sum of smaller matrices $B_1, B_2, \ldots, B_k$. According to Theorem 1.10.31, this will be accomplished when we have expressed $\mathbb{F}^n$ as a direct sum of A-invariant subspaces $\mathbb{V}_1, \mathbb{V}_2, \ldots, \mathbb{V}_k$. Theorem 1.10.31 raises two problems:

1 How to express $\mathbb{F}^n$ as a direct sum of A-invariant subspaces $\mathbb{V}_1, \mathbb{V}_2, \ldots, \mathbb{V}_k$.
2 How to make the form of the matrices $B_1, B_2, \ldots, B_k$ as simple as possible.

The first question is answered in Sec. 1.12. The second question requires us to decide which of the possible forms for $B_1, B_2, \ldots, B_k$ we want. Two somewhat different "simple" forms are given in Sec. 1.13.

EXERCISES

1 Find the matrix of the linear transformation of Example 1.10.5.
2 Find the matrix of the linear transformation of Example 1.10.6.
3 Let $\mathscr{B}$ be the linear transformation of $\mathbb{F}^3$ into itself for which

$$\mathscr{B}: \begin{array}{c} \begin{bmatrix} 1 \\ 2 \\ 0 \end{bmatrix} \to \begin{bmatrix} -1 \\ 3 \\ 2 \end{bmatrix} \\ \begin{bmatrix} 0 \\ 2 \\ 1 \end{bmatrix} \to \begin{bmatrix} 3 \\ 4 \\ -2 \end{bmatrix} \\ \begin{bmatrix} -1 \\ 1 \\ 1 \end{bmatrix} \to \begin{bmatrix} 3 \\ 1 \\ -3 \end{bmatrix} \end{array}$$

What vector is $\begin{bmatrix} 1 \\ 1 \\ 1 \end{bmatrix}$ mapped into by $\mathscr{B}$? What vector is mapped into $\begin{bmatrix} 1 \\ 1 \\ 1 \end{bmatrix}$ by $\mathscr{B}$? Find the matrix B of $\mathscr{B}$. Show that the subspace of $\mathbb{F}^3$ which has

$$\left\{ \begin{bmatrix} -1 \\ 2 \\ 1 \end{bmatrix}, \begin{bmatrix} 0 \\ 1 \\ 0 \end{bmatrix} \right\}$$

as a basis is B-invariant.

4 Show that there is a nonzero vector X such that

$$\mathscr{A}(X) = \mathscr{B}(X)$$

where $\mathscr{A}$ and $\mathscr{B}$ are the linear transformations of Example 1.10.5 (with $n=3$) and Exercise 3 respectively.

5 Find all nonsingular matrices P such that

$$P^{-1} \begin{bmatrix} 1 & 1 \\ 0 & 2 \end{bmatrix} P = \begin{bmatrix} 1 & 0 \\ 0 & 2 \end{bmatrix}$$

Find all nonsingular matrices Q such that
$$Q^{-1}\begin{bmatrix} 1 & 1 \\ 0 & 2 \end{bmatrix} Q = \begin{bmatrix} 1 & 0 \\ 1 & 2 \end{bmatrix}$$

6 Show that
$$\begin{bmatrix} 1 & y \\ 0 & 1 \end{bmatrix} \text{ and } \begin{bmatrix} 1 & z \\ 0 & 1 \end{bmatrix}$$
are similar to each other for all nonzero numbers y and z. What happens when y or z is zero?

7 Show that
$$\begin{bmatrix} 1 & 0 \\ w & 1 \end{bmatrix}$$
is similar to the matrices of Exercise 6 for all nonzero numbers w.

8 Find the characteristic roots and all the characteristic vectors of
$$A = \begin{bmatrix} 3 & 4 & 5 \\ 0 & 1 & -1 \\ -1 & -2 & -1 \end{bmatrix}$$

Find a matrix P such that $P^{-1}AP$ is a diagonal matrix. Find the row characteristic vectors of A and use them to construct a nonsingular matrix Q such that QAQ^{-1} is a diagonal matrix.

9 Repeat Exercise 8 using the matrix
$$A = \begin{bmatrix} 2 & 3 & 0 \\ -4 & -5 & 0 \\ 3 & 3 & -1 \end{bmatrix}$$

10 Find the characteristic roots and all the corresponding row and column characteristic vectors of
$$A = \begin{bmatrix} 3 & 2 & -1 \\ 0 & -1 & 0 \\ 1 & -4 & 1 \end{bmatrix}$$

Is A similar to a diagonal matrix?

11 Let $\mathcal{A}$ be a linear transformation of the vector space $\mathbb{V}$ into the vector space $\mathbb{W}$. We define the *image* of $\mathcal{A}$, denoted by im $\mathcal{A}$, as follows:

$Y \in$ im $\mathcal{A}$ if there exists $X \in \mathbb{V}$ such that $\mathcal{A}: X \to Y$

Prove that im $\mathcal{A}$ is a subspace of $\mathbb{W}$.

12 Let $\mathscr{A}$ be a linear transformation of the vector space $\mathbb{V}$ into the vector space $\mathbb{W}$. We define the *kernel* of $\mathscr{A}$, denoted by ker $\mathscr{A}$, as follows:

$$\ker \mathscr{A} = \{X \mid X \in \mathbb{V} \text{ and } \mathscr{A} : X \to 0 \in \mathbb{W}\}$$

Prove that ker $\mathscr{A}$ is a subspace of $\mathbb{V}$.

13 Let $\mathscr{A}$ be a linear transformation of $\mathbb{V}$ into $\mathbb{W}$. Prove that

$$\dim \mathbb{V} = \dim(\ker \mathscr{A}) + \dim(\operatorname{im} \mathscr{A})$$

Interpret this result when $\mathbb{V} = \mathbb{F}^n$, $\mathbb{W} = \mathbb{F}^m$, and A is the matrix of $\mathscr{A}$.

14 Show that if B is a matrix which commutes with A, then $\mathbb{N}(B)$ is A-invariant.

15 Show that $\mathbb{C}(A)$ is A-invariant for every square matrix A.

16 Prove that the sum and intersection of A-invariant vector spaces are A-invariant.

17 Prove that the characteristic polynomial of a direct sum of matrices is equal to the product of the characteristic polynomials of the individual matrices.

18 Prove that if X is a characteristic vector of A corresponding to the characteristic root λ, then, for every positive integer k, X is a characteristic vector of A^k corresponding to the characteristic root λ^k.

19 Show that if X is a characteristic vector of A corresponding to the characteristic root λ, then, for all $\alpha, \beta \in \mathbb{F}$, X is a characteristic vector of $\alpha A + \beta I$ corresponding to the characteristic root $\alpha\lambda + \beta$.

20 Let $B = P^{-1}AP$. Show that X is a characteristic vector of A corresponding to the characteristic root λ if and only if $P^{-1}X$ is a characteristic vector of B corresponding to the characteristic root λ.

21 Let A and B be square matrices of the same order and let A be nonsingular. Prove that AB is similar to BA. Give an example to show that if A and B are singular, then AB need not be similar to BA.

22 Let A and B be $m \times n$ and $n \times m$ matrices respectively. Prove that if X is a characteristic vector of AB corresponding to the characteristic root $\lambda \neq 0$, then BX is a characteristic vector of BA corresponding to the characteristic root λ. Why is it necessary to specify that $\lambda \neq 0$?

23 Let A be a square matrix of order n and let V be an A-invariant subspace of $\mathbb{F}^n$ of dimension m. Show that A is similar to a matrix of the form

$$\begin{bmatrix} B_{11} & B_{12} \\ 0 & B_{22} \end{bmatrix}$$

where B_{11}, B_{12}, and B_{22} are $m \times m$, $m \times (n-m)$, and $(n-m) \times (n-m)$ matrices respectively.

24 Prove that the algebraic multiplicity of a characteristic root of a matrix is greater than or equal to its geometric multiplicity. (*Hint:* Let $X_1, X_2, \ldots, X_l$ be linearly independent characteristic vectors of the matrix A corresponding to the characteristic root λ. Construct a nonsingular matrix P whose first l columns are $X_1, X_2, \ldots, X_l$ and consider the characteristic polynomial of $P^{-1}AP$. Apply Exercise 17 of Sec. 1.9.)

1.11 POLYNOMIALS

Let
$$f(x) = \alpha_m x^m + \alpha_{m-1} x^{m-1} + \cdots + \alpha_1 x + \alpha_0$$
be a polynomial with coefficients from $\mathbb{F}$. For a square matrix A, we define the *matrix polynomial* $f(A)$ by
$$f(A) = \alpha_m A^m + \alpha_{m-1} A^{m-1} + \cdots + \alpha_1 A + \alpha_0 I$$
Then $f(A)$ is again a square matrix whose order is the same as the order of A. In the next section we will show that for a given square matrix A, we can find the A-invariant subspaces of $\mathbb{F}^n$ by considering polynomials in A with coefficients from $\mathbb{F}$. For this reason we first need to develop a number of theorems dealing with polynomials in one variable.

Theorem 1.11.1 The division algorithm Let $f(x)$ and $g(x)$ be polynomials, $g(x) \neq 0$. Then there exist polynomials $q(x)$ and $r(x)$ such that
$$f(x) = q(x)g(x) + r(x)$$
and the degree of $r(x)$ is less than the degree of $g(x)$,[1] that is
$$\deg r(x) < \deg g(x)$$

PROOF When $f(x) = 0$, then $q(x) = r(x) = 0$ satisfies the theorem. Thus we may henceforth assume that $f(x) \neq 0$. Let

$$f(x) = \alpha_m x^m + \alpha_{m-1} x^{m-1} + \cdots + \alpha_0 \qquad m \geqq 0 \qquad \alpha_m \neq 0$$
$$g(x) = \beta_l x^l + \beta_{l-1} x^{l-1} + \cdots + \beta_0 \qquad l \geqq 0 \qquad \beta_l \neq 0$$

The proof is by an induction on m, the degree of $f(x)$. It will be convenient for us to think of $g(x)$ as being fixed throughout the proof. First, suppose $m = 0$. Then $f(x) = \alpha_0$, and two distinct cases arise. If

---
[1] We define the degree of the zero polynomial to be $-\infty$.

$l = 0$, so that $g(x) = \beta_0 \neq 0$, then we set $q(x) = \alpha_0/\beta_0$, $r(x) = 0$. If $l \geq 1$, then $q(x) = 0$, $r(x) = f(x)$ satisfies the theorem. Thus we have established the theorem for $m = 0$.

Now suppose that for a fixed nonzero polynomial $g(x)$ the theorem has been proved for all polynomials $f(x)$ of degree less than k. Let $f(x)$ be a polynomial of degree k. If $k < l$, we again set $q(x) = 0$, $r(x) = f(x)$. On the other hand, if $k \geq l$, then we define the polynomial $\tilde{f}(x)$ by

$$\tilde{f}(x) = f(x) - \frac{\alpha_k}{\beta_l} x^{k-l} g(x) \qquad (1.11.2)$$

The degree of $\tilde{f}(x)$ is definitely less than k and hence, by the induction hypothesis, there exist polynomials $\tilde{q}(x)$ and $\tilde{r}(x)$ such that

$$\tilde{f}(x) = \tilde{q}(x) g(x) + \tilde{r}(x) \qquad (1.11.3)$$

and
$$\deg \tilde{r}(x) < \deg g(x)$$

Substituting equation (1.11.3) into equation (1.11.2), we have

$$f(x) = \tilde{f}(x) + \frac{\alpha_k}{\beta_l} x^{k-l} g(x)$$

$$= \left[\tilde{q}(x) + \frac{\alpha_k}{\beta_l} x^{k-l}\right] g(x) + \tilde{r}(x)$$

The required polynomials, $q(x)$ and $r(x)$, are obtained by setting

$$q(x) = \tilde{q}(x) + \frac{\alpha_k}{\beta_l} x^{k-l}, \qquad r(x) = \tilde{r}(x) \qquad ////$$

EXAMPLE 1.11.4   Let

$$f(x) = x^5 - 4x^4 + 2x^3 + 5x^2 - 4x - 2 \qquad g(x) = -x^4 + 3x^3 - 4x$$

If we set

$$q(x) = -x + 1 \qquad r(x) = -x^3 + x^2 - 2$$

then
$$f(x) = q(x) g(x) + r(x)$$

and
$$\deg r(x) = 3 < 4 = \deg g(x) \qquad ////$$

The division algorithm is usually thought of as a division process, $q(x)$ being the quotient and $r(x)$ being the remainder after dividing $f(x)$ by $g(x)$.

When $r(x) = 0$, we have

$$f(x) = q(x)g(x)$$

and we say that $g(x)$ is a *divisor* of $f(x)$ and that $f(x)$ is a *multiple* of $g(x)$.

Definition 1.11.5 The polynomial $d(x)$ is a *greatest common divisor* of the polynomials $f_1(x), f_2(x), \ldots, f_k(x)$ provided:
1 $d(x)$ is a divisor of each of the polynomials $f_1(x), f_2(x), \ldots, f_k(x)$.
2 If $h(x)$ is a divisor of each of the polynomials $f_1(x), f_2(x), \ldots, f_k(x)$, then $h(x)$ is a divisor of $d(x)$.

Definition 1.11.5 does not tell us how many, if any, greatest common divisors a set of polynomials has. In fact, if $d_1(x)$ is a greatest common divisor of a set of polynomials and α is a nonzero element of $\mathbb{F}$, then

$$d_2(x) = \alpha d_1(x)$$

is also a greatest common divisor of these polynomials. Conversely, suppose that $d_1(x)$ and $d_2(x)$ are both greatest common divisors of the same set of polynomials. Since $d_2(x)$ is a divisor of each of the polynomials and $d_1(x)$ is a greatest common divisor, it follows that $d_1(x)$ divides $d_2(x)$. Interchanging the roles of $d_1(x)$ and $d_2(x)$ shows that $d_2(x)$ divides $d_1(x)$. Thus $d_1(x)$ and $d_2(x)$ divide each other, and consequently they differ by at most a multiplication by a constant; i.e., there exists $\alpha \in \mathbb{F}$ such that

$$d_2(x) = \alpha d_1(x) \qquad \alpha \neq 0 \qquad (1.11.6)$$

Any two polynomials $d_1(x)$ and $d_2(x)$ which satisfy equation (1.11.6) are called *associates*. Associateness of polynomials is an equivalence relation, and every nonzero polynomial has exactly one associate whose leading coefficient is 1. A polynomial whose leading coefficient is 1 is said to be *monic*. If we also define the zero polynomial to be monic, then every polynomial has a unique monic associate and we have proved:

Corollary 1.11.7 If a set of polynomials has any greatest common divisors, then they have a unique monic greatest common divisor.

The monic greatest common divisor of a set of polynomials is referred to as *the greatest common divisor*. We express the statement that $d(x)$ is the greatest common divisor of $f_1(x), f_2(x), \ldots, f_k(x)$ by writing

$$d(x) = (f_1(x), f_2(x), \ldots, f_k(x))$$

Note that we have not yet established the existence of any greatest common divisors. This will be done in Theorem 1.11.8, but first we will consider the case where some of the polynomials are the zero polynomial. Since

$$h(x) \cdot 0 = 0$$

for every polynomial $h(x)$, it follows that every polynomial $h(x)$, including $h(x) = 0$, is a divisor of 0. Moreover, 0 is the only polynomial which has every polynomial as a divisor. Thus, if

$$f_1(x) = f_2(x) = \cdots = f_k(x) = 0$$

then 0 is their only greatest common divisor. Conversely, the only polynomial which 0 divides is the zero polynomial. Thus if 0 is a greatest common divisor of $f_1(x), f_2(x), \ldots, f_k(x)$, then

$$f_1(x) = f_2(x) = \cdots = f_k(x) = 0$$

Finally, if $f_1(x), f_2(x), \ldots, f_l(x)$ are nonzero polynomials and $f_{l+1}(x) = f_{l+2}(x) = \cdots = f_k(x) = 0$, then the set of greatest common divisors of $f_1(x), f_2(x), \ldots, f_l(x)$ is exactly the same as the set of greatest common divisors of $f_1(x), f_2(x), \ldots, f_k(x)$.

Theorem 1.11.8 The euclidean algorithm[1] Every pair of polynomials has a greatest common divisor.

PROOF When $f(x) = g(x) = 0$, then 0 is their greatest common divisor. When $f(x) = 0$ and $g(x) \neq 0$, then the set of greatest common divisors of $f(x)$ and $g(x)$ is identical to the set of greatest common divisors of $g(x)$ alone, which is the set of associates of $g(x)$.

Finally, let $f(x)$ and $g(x)$ be nonzero polynomials. By Theorem 1.11.1 there exist polynomials $q(x)$ and $r(x)$ such that

$$f(x) = q(x)g(x) + r(x)$$

and
$$\deg r(x) < \deg g(x)$$

Moreover, the set of greatest common divisors, if any, of $f(x)$ and $g(x)$ is exactly the same as the set of greatest common divisors of $g(x)$ and $r(x)$. If $r(x) = 0$, then $g(x)$ is a greatest common divisor of $g(x)$ and $r(x)$, and hence of $f(x)$ and $g(x)$. If $r(x) \neq 0$, we apply Theorem 1.11.1 again. There exist polynomials $q_1(x)$ and $r_1(x)$ such that

$$g(x) = q_1(x)r(x) + r_1(x)$$

and
$$\deg r_1(x) < \deg r(x) < \deg g(x)$$

---

[1] This algorithm was discovered by Euclid, who used it to obtain the greatest common divisor of two integers. The theory of the greatest common divisor and the least common multiple are virtually identical for polynomials and for integers.

As before, if $r_1(x) = 0$, then $r(x)$ is a greatest common divisor of $r(x)$ and $r_1(x)$ and hence of $g(x)$ and $r(x)$ and, in the same way, of $f(x)$ and $g(x)$. If $r_1(x) \neq 0$, then we continue this process

$$r(x) = q_2(x)r_1(x) + r_2(x)$$
$$\cdots\cdots\cdots\cdots\cdots\cdots\cdots$$
$$r_{l-3}(x) = q_{l-1}(x)r_{l-2}(x) + r_{l-1}(x) \quad (1.11.9)$$
$$r_{l-2}(x) = q_l(x)r_{l-1}(x) + r_l(x)$$

and $\quad \deg r_l(x) < \deg r_{l-1}(x) < \cdots < \deg r(x) < \deg g(x)$

Since the degree of the remainder is decreasing at each step, it follows that for some value of l we have $r_l(x) = 0$. Then $r_{l-1}(x)$ is a greatest common divisor of $f(x)$ and $g(x)$. ////

Corollary 1.11.10 For each pair of polynomials the greatest common divisor exists and is unique.

The proof of Theorem 1.11.8 is constructive; i.e., it provides us with a method for actually constructing a greatest common divisor of two polynomials. Later we will see that it is not the only method.

EXAMPLE 1.11.11 Let us use the euclidean algorithm to find the greatest common divisor of the pair of polynomials

$$f(x) = x^5 - 4x^4 + 2x^3 + 5x^2 - 4x - 2 \quad g(x) = -x^4 + 3x^3 - 4x$$

of Example 1.11.4. According to Example 1.11.4,

$$x^5 - 4x^4 + 2x^3 + 5x^2 - 4x - 2 = (-x + 1)(-x^4 + 3x^3 - 4x) + (-x^3 + x^2 - 2)$$

and, continuing this process, we have

$$-x^4 + 3x^3 - 4x = (x - 2)(-x^3 + x^2 - 2) + (2x^2 - 2x - 4)$$
$$-x^3 + x^2 - 2 = (-\tfrac{1}{2}x)(2x^2 - 2x - 4) + (-2x - 2)$$
$$2x^2 - 2x - 4 = (-x + 2)(-2x - 2)$$

Thus $-2x - 2$ is a greatest common divisor of $x^5 - 4x^4 + 2x^3 + 5x^2 - 4x - 2$ and $-x^4 + 3x^3 - 4x$. The monic associate of $-2x - 2$ is $x + 1$, and hence

$$x + 1 = (x^5 - 4x^4 + 2x^3 + 5x^2 - 4x - 2, -x^4 + 3x^3 - 4x) \quad ////$$

There is an important consequence of this construction. Working backward through equations (1.11.9), we can express the greatest common divisor of any two polynomials as a linear combination of them with polynomial coefficients; that is:

Theorem 1.11.12 If
$$d(x) = (f(x), g(x))$$
then there exist polynomials $s(x)$ and $t(x)$ such that
$$d(x) = s(x)f(x) + t(x)g(x) \quad (1.11.13)$$

PROOF In equation (1.11.9) [with $r_l(x) = 0$], $d(x)$ is the monic associate of $r_{l-1}(x)$. Therefore it is sufficient to prove that there exist polynomials $\tilde{s}(x)$ and $\tilde{t}(x)$ such that
$$r_{l-1}(x) = \tilde{s}(x)f(x) + \tilde{t}(x)g(x)$$
The proof consists of successively eliminating $r_{l-2}(x), r_{l-3}(x), \ldots, r_1(x)$ and $r(x)$ from equations (1.11.9). Initially we have
$$r_{l-1}(x) = r_{l-3}(x) - q_{l-1}(x)r_{l-2}(x)$$
and
$$r_{l-2}(x) = r_{l-4}(x) - q_{l-2}(x)r_{l-3}(x)$$
Substituting for $r_{l-2}(x)$ yields
$$\begin{aligned} r_{l-1}(x) &= r_{l-3}(x) - q_{l-1}(x)[r_{l-4}(x) - q_{l-2}(x)r_{l-3}(x)] \\ &= -q_{l-1}(x)r_{l-4}(x) + [q_{l-1}(x)q_{l-2}(x) + 1]r_{l-3}(x) \end{aligned}$$
and thus $r_{l-2}(x)$ has been eliminated. At the next stage we eliminate $r_{l-3}(x)$, and so on, until finally $r(x)$ is eliminated, leaving $r_{l-1}(x)$ expressed in terms of $f(x)$ and $g(x)$. ////

EXAMPLE 1.11.14 Let us apply Theorem 1.11.12 to the polynomials of Examples 1.11.4 and 1.11.11. We have already shown that the greatest common divisor is $x + 1$.

$$\begin{aligned} -2x - 2 &= (-x^3 + x^2 - 2) - (-\tfrac{1}{2}x)(2x^2 - 2x - 4) \\ &= (-x^3 + x^2 - 2) \\ &\quad - (-\tfrac{1}{2}x)[(-x^4 + 3x^3 - 4x) - (x - 2)(-x^3 + x^2 - 2)] \\ &= -\tfrac{1}{2}(x^2 - 2x - 2)(-x^3 + x^2 - 2) + (\tfrac{1}{2}x)(-x^4 + 3x^3 - 4x) \\ &= -\tfrac{1}{2}(x^2 - 2x - 2)[(x^5 - 4x^4 + 2x^3 + 5x^2 - 4x - 2) \\ &\quad - (-x + 1)(-x^4 + 3x^3 - 4x)] + (\tfrac{1}{2}x)(-x^4 + 3x^3 - 4x) \\ &= -\tfrac{1}{2}(x^2 - 2x - 2)(x^5 - 4x^4 + 2x^3 + 5x^2 - 4x - 2) \\ &\quad + \tfrac{1}{2}(-x^3 + 3x^2 + x - 2)(-x^4 + 3x^3 - 4x) \end{aligned}$$

Thus we have equation (1.11.13), where
$$s(x) = \tfrac{1}{4}(x^2 - 2x - 2) \qquad t(x) = \tfrac{1}{4}(x^3 - 3x^2 - x + 2) \qquad ////$$

Thus far we have largely restricted our attention to a pair of polynomials. Now, for any finite set of polynomials we can prove that there exists a greatest common divisor and that it has an expression similar to the form of equation (1.11.13). To do this, we show that in the construction of a greatest common divisor of more than two polynomials we may successively replace two polynomials by their greatest common divisor. It will be helpful to prove the result first for a set of three polynomials.

Lemma 1.11.15 Let $f(x)$, $g(x)$, and $j(x)$ be polynomials and set

$$\tilde{d}(x) = (f(x), g(x)) \qquad d(x) = (\tilde{d}(x), j(x))$$

Then
$$d(x) = (f(x), g(x), j(x))$$

PROOF By Corollary 1.11.10, $d(x)$ and $\tilde{d}(x)$ exist and are unique. Since $d(x)$ is a divisor of $\tilde{d}(x)$ and $\tilde{d}(x)$ is a divisor of $f(x)$ and $g(x)$, it follows that $d(x)$ is a divisor of $f(x)$ and $g(x)$. Since $d(x)$ also divides $j(x)$, $d(x)$ satisfies the first part of Definition 1.11.5.

To establish the second part of the definition we must show that any polynomial $h(x)$ which is a divisor of $f(x)$, $g(x)$, and $j(x)$ is also a divisor of $d(x)$. Since $h(x)$ divides $f(x)$ and $g(x)$, it follows that $h(x)$ divides $\tilde{d}(x)$. Also, since $h(x)$ divides $\tilde{d}(x)$ and $j(x)$, it divides $d(x)$. Thus $d(x)$ is a monic greatest common divisor of $f(x)$, $g(x)$, and $j(x)$, and therefore

$$d(x) = (f(x), g(x), j(x)) \qquad ////$$

By repeated applications of Lemma 1.11.15 we have:

Theorem 1.11.16 Let $f_1(x), f_2(x), \ldots, f_k(x)$ be polynomials and set

$$d_1(x) = (f_1(x), f_2(x))$$
$$d_2(x) = (d_1(x), f_3(x))$$
$$\cdots\cdots\cdots\cdots\cdots\cdots\cdots\cdots\cdots$$
$$d_{k-1}(x) = (d_{k-2}(x), f_k(x))$$

Then
$$d_{k-1} = (f_1(x), f_2(x), \ldots, f_k(x))$$

We can obtain an expression, similar to equation (1.11.13), for the greatest common divisor of a set of polynomials by means of Theorem 1.11.16 and repeated use of Theorem 1.11.12.

Theorem 1.11.17 If

$$d(x) = (f_1(x), f_2(x), \ldots, f_k(x))$$

then there exist polynomials $s_1(x), s_2(x), \ldots, s_k(x)$ such that

$$d(x) = s_1(x)f_1(x) + s_2(x)f_2(x) + \cdots + s_k(x)f_k(x) \quad (1.11.18)$$

PROOF Define $d_1(x), d_2(x), \ldots, d_{k-1}(x)$ as in Theorem 1.11.16. Then $d(x) = d_{k-1}(x)$. By Theorem 1.11.12, there exist polynomials $s(x)$ and $t(x)$ such that

$$d_1(x) = s(x)f_1(x) + t(x)f_2(x)$$

Similarly there exist polynomials $u(x)$ and $v(x)$ such that

$$d_2(x) = u(x)d_1(x) + v(x)f_3(x)$$

Substituting the first of these equations into the second yields

$$d_2(x) = u(x)s(x)f_1(x) + u(x)t(x)f_2(x) + v(x)f_3(x) \quad (1.11.19)$$

As before, there exist polynomials $w(x)$ and $y(x)$ such that

$$d_3(x) = w(x)d_2(x) + y(x)f_4(x) \quad (1.11.20)$$

Substituting (1.11.19) into (1.11.20) yields an expression for $d_3(x)$ in terms of $f_1(x), f_2(x), f_3(x)$, and $f_4(x)$. Continuing this process, we eventually arrive at an expression for $d(x)$ in the form of equation (1.11.18). ////

EXAMPLE 1.11.21 Find the greatest common divisor of

$$f(x) = x^5 - 4x^4 + 2x^3 + 5x^2 - 4x - 2$$
$$g(x) = -x^4 + 3x^3 - 4x \qquad h(x) = 2x^2 + 6x + 5$$

and express it in the form of equation (1.11.18).

Set

$$d_1(x) = (f(x), g(x))$$

In Example 1.11.11 we showed that

$$d_1(x) = x + 1$$

By Theorem 1.11.17,

$$d(x) = (x + 1, 2x^2 + 6x + 5)$$

is the greatest common divisor of $f(x), g(x)$, and $h(x)$. The euclidean algorithm can be used to show that

$$d(x) = 1$$

and, applying the method of Theorem 1.11.12, we have

$$1 = -\tfrac{1}{2}(x^3 - 6x - 4)(x^5 - 4x^4 + 2x^3 + 5x^2 - 4x - 2)$$
$$- \tfrac{1}{2}(x^4 - x^3 - 7x^2 + 4)(-x^4 + 3x^3 - 4x) + 1(2x^2 + 6x + 5) \qquad ////$$

Although the euclidean algorithm provides a constructive method for obtaining greatest common divisors, this is not the method commonly used when greatest common divisors are to be computed. Instead, each polynomial is expressed as a product of its divisors, and the greatest common divisor is immediately apparent.

Definition 1.11.22 The polynomials $f_1(x), f_2(x), \ldots, f_k(x)$ are *relatively prime* if

$$(f_1(x), f_2(x), \ldots, f_k(x)) = 1$$

and are *relatively prime in pairs* if

$$(f_i(x), f_j(x)) = 1 \qquad i, j = 1, 2, \ldots, k \qquad i \neq j$$

Lemma 1.11.23 If $f(x)$ is relatively prime to each of the polynomials $f_1(x), f_2(x), \ldots, f_k(x)$, then $f(x)$ is relatively prime to the product

$$f_1(x)f_2(x) \cdots f_k(x)$$

PROOF According to Theorem 1.11.12, for each $i, i = 1, 2, \ldots, k$, there exist polynomials $s_i(x)$ and $t_i(x)$ such that

$$t_i(x)f_i(x) = 1 - s_i(x)f(x)$$

Multiplying these k equations together yields

$$t_1(x)t_2(x) \cdots t_k(x)f_1(x)f_2(x) \cdots f_k(x) = \prod_{i=1}^{k} (1 - s_i(x)f(x))$$
$$= 1 - u(x)f(x)$$

where $u(x)$ is a function of $f(x), s_1(x), s_2(x), \ldots, s_k(x)$. We can write this equation as

$$1 = u(x)f(x) + t_1(x)t_2(x) \cdots t_k(x)f_1(x)f_2(x) \cdots f_k(x)$$

Now suppose that $h(x)$ divides both $f(x)$ and the product $f_1(x)f_2(x) \cdots f_k(x)$. Then $h(x)$ also divides 1 and hence is a nonzero number. Thus

$$(f(x), f_1(x)f_2(x) \cdots f_k(x)) = 1 \qquad ////$$

Definition 1.11.24 A polynomial is *irreducible* if it is not a constant and if its only divisors are its associates and the nonzero elements of $\mathbb{F}$.

The irreducibility of a polynomial may depend on the choice of $\mathbb{F}$. Although all polynomials of degree 1 are irreducible no matter what field $\mathbb{F}$ is, $x^2 + 1$ is irreducible when $\mathbb{F}$ is the real field but not when $\mathbb{F}$ is the complex field.

If a polynomial is irreducible, then all its associates are irreducible. Moreover, if $f(x)$ is a monic irreducible polynomial, then

$$(f(x), g(x)) = \begin{cases} f(x) & \text{when } f(x) \text{ divides } g(x) \\ 1 & \text{when } f(x) \text{ does not divide } g(x) \end{cases}$$

Lemma 1.11.25 If $p(x)$ is an irreducible polynomial which divides the product $f_1(x)f_2(x) \cdots f_k(x)$, then $p(x)$ divides at least one of the polynomials $f_1(x), f_2(x), \ldots, f_k(x)$.

PROOF By Lemma 1.11.23, $p(x)$ is not relatively prime to at least one of the polynomials $f_i(x)$, that is,

$$d(x) = (p(x), f_i(x)) \neq 1$$

for some i. Since $p(x)$ is irreducible, $d(x)$ is the monic associate of $p(x)$ and $p(x)$ divides $f_i(x)$. ////

Theorem 1.11.26 The unique factorization theorem Every polynomial of positive degree can be expressed in exactly one way (up to associates) as a product of irreducible polynomials.

PROOF The proof that every polynomial of positive degree has at least one expression as a product of irreducible polynomials is by an induction on the degree. Let $f(x)$ be a polynomial of degree n, $n \geq 1$. If $n = 1$, then $f(x)$ is itself irreducible. Now suppose that every polynomial of degree less than n can be expressed as a product of irreducible polynomials and consider $f(x)$. If $f(x)$ is irreducible, then the required expression is simply $f(x)$ itself and we are finished. On the other hand, if $f(x)$ is not irreducible, then, by Definition 1.11.24, $f(x)$ has a divisor, say $g(x)$, of degree k, where $1 \leq k \leq n - 1$. Hence there exists a polynomial $h(x)$ such that

$$f(x) = g(x)h(x) \qquad 1 \leq \deg g(x), \deg h(x) \leq n - 1$$

By the induction hypothesis, both $g(x)$ and $h(x)$ can be expressed as a product of irreducible polynomials. Consequently $f(x)$ is also expressible as a product of irreducible polynomials. It remains only to show that such an expression is unique up to associates. Suppose that $f(x)$ can be expressed in two ways as a product of irreducible polynomials, say

$$p_1(x)p_2(x) \cdots p_r(x) = f(x) = q_1(x)q_2(x) \cdots q_s(x) \quad (1.11.27)$$

By Lemma 1.11.25, $p_1(x)$ is a divisor of $q_i(x)$ for some i, $1 \leq i \leq s$. We may assume that the terms of (1.11.27) are arranged so that $i = 1$; that is, $p_1(x)$ is a divisor of $q_1(x)$. But $q_1(x)$ is irreducible, and hence its only divisors are the nonzero numbers [and $p_1(x)$ is not one of these] and its associates. Thus there is a nonzero number α such that

$$p_1(x) = \alpha q_1(x)$$

It then follows from equation (1.11.27) that

$$\alpha p_2(x) \cdots p_r(x) = q_2(x) \cdots q_s(x)$$

Continuing this process, we see that $r = s$ and the polynomials $p_1(x)$, $p_2(x), \ldots, p_r(x)$ and $q_1(x), q_2(x), \ldots, q_s(x)$ are equal in some order, up to associates. ////

By requiring that each of the irreducible polynomials be monic we eliminate the ambiguity caused by associates. Thus:

Theorem 1.11.28 Every monic polynomial of positive degree can be expressed in exactly one way as a product of irreducible monic polynomials of positive degree.

The unique factorization theorem provides an alternate method for finding the greatest common divisor of a set of polynomials. First we express each nonzero polynomial of the set as a product of monic irreducible polynomials and a nonzero number. We use these expressions to find the highest power of each irreducible polynomial which divides all the polynomials of the set. The product of these is the greatest common divisor.

EXAMPLE 1.11.29   Let

$$f(x) = x^5 - 4x^3 - 2x^2 + 3x + 2 \qquad g(x) = -x^4 + 3x^3 - 4x$$
$$h(x) = 2x^2 - 8x + 6$$

Find (a) $(f(x), g(x))$ and (b) $(f(x), g(x), h(x))$ using the method described in the last paragraph.

Since

$$f(x) = 1 \cdot (x+1)^3(x-1)(x-2)$$
$$g(x) = -1 \cdot (x+1) \qquad (x-2)^2 x$$
$$h(x) = 2 \cdot \qquad (x-1) \qquad (x-3)$$

we immediately have

$$(f(x), g(x)) = (x+1)(x-2) = x^2 - x - 2$$
$$(f(x), g(x), h(x)) = 1 \qquad ////$$

One drawback of this method is that it does not provide a means for obtaining the expression (1.11.18) for the greatest common divisor. Another drawback is that not all polynomials factor as nicely as those in Example 1.11.29. It would not be an easy task to find the greatest common divisor of the polynomials in Example 1.11.4 in this manner.

There is a theory for the multiples of a set of polynomials which in many ways is analogous to the one we have just developed for divisors. Let us begin with a definition.

Definition 1.11.30 The polynomial $n(x)$ is a *least common multiple* of the polynomials $f_1(x), f_2(x), \ldots, f_k(x)$ provided:

1    $n(x)$ is a multiple of each of the polynomials $f_1(x), f_2(x), \ldots, f_k(x)$.
2   If the polynomial $h(x)$ is a multiple of $f_1(x), f_2(x), \ldots, f_k(x)$, then $h(x)$ is a multiple of $n(x)$.

If $n_1(x)$ and $n_2(x)$ are least common multiples of the same set of polynomials, then they are multiples of each other and hence are associates. Thus:

Corollary 1.11.31 If a set of polynomials has any least common multiples, then it has a unique monic least common multiple.

The monic least common multiple of a set of polynomials is called *the least common multiple*. We express the fact that $n(x)$ is the least common multiple of $f_1(x), f_2(x), \ldots, f_k(x)$ by writing

$$n(x) = [f_1(x), f_2(x), \ldots, f_k(x)]$$

Theorem 1.11.32 For every finite set of polynomials the least common multiple exists and is unique.

PROOF   Since the only multiple of 0 is 0, it follows that 0 is the unique least common multiple of any set of polynomials which contains 0.

Now, assume that $f_1(x), f_2(x), \ldots, f_k(x)$ are nonzero polynomials. By the unique factorization theorem there exist irreducible monic polynomials $p_1(x), p_2(x), \ldots, p_l(x)$ and nonzero numbers $\alpha_1, \alpha_2, \ldots, \alpha_l$ such that

$$f_1(x) = \alpha_1 p_1(x)^{e_{11}} p_2(x)^{e_{12}} \cdots p_l(x)^{e_{1l}}$$
$$f_2(x) = \alpha_2 p_1(x)^{e_{21}} p_2(x)^{e_{22}} \cdots p_l(x)^{e_{2l}}$$
$$\cdots \cdots \cdots \cdots \cdots \cdots \cdots \cdots \cdots \cdots \cdots$$
$$f_k(x) = \alpha_k p_1(x)^{e_{k1}} p_2(x)^{e_{k2}} \cdots p_l(x)^{e_{kl}}$$

where $e_{11}, e_{12}, \ldots, e_{kl}$ are nonnegative (possibly zero) integers. For each $j, j = 1, 2, \ldots, l$, we set

$$r_j = \max_{i = 1, 2, \cdots, k} e_{ij}$$

Then

$$n(x) = p_1(x)^{r_1} p_2(x)^{r_2} \cdots p_l(x)^{r_l}$$

is the least common multiple of $f_1(x), f_2(x), \ldots, f_k(x)$. ////

EXAMPLE 1.11.33 Find (a) $[f(x), g(x)]$ and (b) $[f(x), g(x), h(x)]$ where $f(x), g(x)$, and $h(x)$ are the polynomials of Example 1.11.29.

In Example 1.11.29 we expressed $f(x), g(x)$, and $h(x)$ as products of a nonzero number and irreducible monic polynomials. Applying the method of construction of the least common multiple which we used to prove Theorem 1.11.32, we have

$$[f(x), g(x)] = (x + 1)^3 (x - 1)(x - 2)^2 x$$
$$[f(x), g(x), h(x)] = (x + 1)^3 (x - 1)(x - 2)^2 x(x - 3) \quad ////$$

In the next two sections we will apply the results of this section to matrix polynomials. For a square matrix A of order n, the matrices $I, A, A^2, \ldots, A^{n^2-1}, A^{n^2}$ are $n^2 + 1$ elements of the n^2-dimensional vector space $\mathbb{F}_n^n$ and hence are linearly dependent. Thus, there exist scalars $\alpha_{n^2}, \alpha_{n^2-1}, \ldots, \alpha_2, \alpha_1, \alpha_0$, not all zero, such that

$$\alpha_{n^2} A^{n^2} + \alpha_{n^2-1} A^{n^2-1} + \cdots + \alpha_2 A^2 + \alpha_1 A + \alpha_0 I = 0$$

In other words,

$$f(A) = 0$$

where $f(x)$ is the polynomial

$$f(x) = \alpha_{n^2} x^{n^2} + \alpha_{n^2-1} x^{n^2-1} + \cdots + \alpha_2 x^2 + \alpha_1 x + \alpha_0$$

In such a case we say that *A satisfies* the polynomial $f(x)$. Thus we have proved that a matrix of order n satisfies at least one nonzero polynomial of degree no greater than n^2. (See Exercise 15 of Sec. 1.5.) In the next section we will improve on this considerably.

If A and B are matrices which commute, then any power of A commutes with any power of B. More generally, when A and B commute, we have

$$f(A)g(B) = g(B)f(A)$$

for any polynomials $f(x)$ and $g(x)$. Since A always commutes with itself, it follows that

$$f(A)g(A) = g(A)f(A)$$

Likewise, if

$$f(x) = g(x)h(x)$$

then

$$f(A) = g(A)h(A)$$

EXERCISES

1 Find

    (a) $(x^5 - x, 2x^5 - x^4 + 2x^3 - x^2)$
    (b) $(x^5 - x, x^5 - 2x^4 + x^3 - 2x^2)$
    (c) $(x^5 - x, x^4 - 4x^2 + 4)$

(1) by means of the euclidean algorithm and (2) by expressing each polynomial as a product or irreducible polynomials.

2 Express each of the greatest common divisors in Exercise 1 in the form of equation (1.11.13).

3 Find:

    (a) $(x^5 - x, 2x^5 - x^4 + 2x^3 - x^2, x^5 - 2x^4 + x^3 - 2x^2)$
    (b) $(x^3 + x, x^3 - x, 4x^2 - 1)$

(1) by means of the euclidean algorithm and Lemma 1.11.15 and (2) by expressing each polynomial as a product of irreducible polynomials. Express each greatest common divisor in the form of equation (1.11.18).

4 Let $f(x)$, $g(x)$, and $h(x)$ be polynomials. Prove that:

    (a) $(f(x) + g(x), g(x)) = (f(x), g(x))$.
    (b) $(h(x)f(x), h(x)g(x)) = h(x)(f(x), g(x))$.

(c) $(f(x) + g(x), f(x) - g(x)) = (f(x), g(x))$ provided $1 + 1 \neq 0$.

(d) $[f(x), g(x), h(x)] = [[f(x), g(x)], h(x)]$.

(e) $[f(x), g(x)] = \dfrac{f(x)g(x)}{(f(x), g(x))}$ provided $(f(x), g(x)) \neq 0$.

(f) $\dfrac{[f(x), g(x), h(x)]}{(f(x), g(x), h(x))} = \dfrac{[f(x), g(x)][f(x), h(x)][g(x), h(x)]}{f(x)g(x)h(x)}$

provided $f(x)g(x)h(x) \neq 0$.

5 Find the least common multiple of each of the sets of polynomials in Exercises 1 and 3
(1) by expressing each polynomial as a product of irreducible polynomials and (2) by using parts (e) and (f) of Exercise 4.

6 Prove that the polynomials $q(x)$ and $r(x)$ of Theorem 1.11.1 are unique.

7 Let $f(x)$ and $g(x)$ be polynomials where the degree of $g(x)$ is at least 1. Show that there exist polynomials $r_0(x), r_1(x), r_2(x), \ldots, r_k(x)$, each of degree less than the degree of $g(x)$, such that

$$f(x) = r_0(x) + r_1(x)g(x) + r_2(x)[g(x)]^2 + \cdots + r_k(x)[g(x)]^k \quad (1.11.34)$$

[*Hint:* Apply the euclidean algorithm to $f(x)$ and $g(x)$, obtaining $q(x)$ and $r(x)$. Next, apply the euclidean algorithm to $q(x)$ and $g(x)$.]

8 Express $f(x)$ in the form of equation (1.11.34) when

(a) $f(x) = x^7 + 2x^3 - x^2 + x + 3 \quad g(x) = x^2 - 1$.

(b) $f(x) = 2x^4 - 3x^3 - 2x^2 + 2x - 7 \quad g(x) = x - 2$.

9 Prove that the polynomials $r_0(x), r_1(x), r_2(x), \ldots, r_k(x)$ of Exercise 7 are unique. [What corresponds to equation (1.11.34) when the euclidean algorithm is applied to the rational integers rather than to polynomials; in particular, when $g(x)$ is replaced by 10?]

10 Find a polynomial of degree 2 satisfied by the matrix:

(a) $\begin{bmatrix} 2 & 1 \\ 3 & 4 \end{bmatrix}$

(b) $\begin{bmatrix} 2 & 1 \\ 0 & 2 \end{bmatrix}$

(c) $\begin{bmatrix} 3 & -3 \\ 3 & -3 \end{bmatrix}$

(d) $\begin{bmatrix} 2 & -2 & 1 \\ 3 & -3 & 1 \\ 0 & 0 & -1 \end{bmatrix}$

11 Find a polynomial of degree 3 satisfied by the matrix:

(a) $\begin{bmatrix} 2 & -4 & 4 \\ 1 & -2 & 3 \\ 0 & 0 & 1 \end{bmatrix}$

(b) $\begin{bmatrix} 3 & 1 & -2 \\ -1 & -1 & 4 \\ 0 & -1 & 3 \end{bmatrix}$

and show that the matrix does not satisfy any polynomial of lesser degree.

12 Let $f_1(x), f_2(x), \ldots, f_k(x)$ be polynomials satisfied by the matrix A. Prove that A also satisfies $d(x) = (f_1(x), f_2(x), \ldots, f_k(x))$.

13 Let A be a square matrix of order n, let $Z \in \mathbb{F}^n$, and let $\mathbb{U}$ be an A-invariant subspace of $\mathbb{F}^n$. Show that if $f(x)$ and $g(x)$ are polynomials for which

$$f(A)Z \in \mathbb{U} \qquad g(A)Z \in \mathbb{U}$$

and

$$d(x) = (f(x), g(x))$$

then

$$d(A)Z \in \mathbb{U}$$

14 Let A be a square matrix of order n and let $Z \in \mathbb{F}^n, Z \neq 0$. Let $p(x)$ and $f(x)$ be polynomials for which

$$p(A)Z = f(A)Z = 0$$

Show that if $p(x)$ is irreducible, then $p(x)$ divides $f(x)$.

15 Let A be a square matrix of order n and let $Z \in \mathbb{F}^n$. Let $p(x)$ and $f(x)$ be polynomials and let $p(x)$ be irreducible. Show that if e is a positive integer for which

$$p(A)^e Z = f(A)Z = 0 \qquad p(A)^{e-1}Z \neq 0$$

then $p(x)^e$ divides $f(x)$.

16 Let A be a square matrix of order n and let $Z \in \mathbb{F}^n$. Let $\mathbb{U}$ be an A-invariant subspace of $\mathbb{F}^n$ and let $p(x)$ and $f(x)$ be polynomials, $p(x)$ irreducible. Show that if e is a positive integer for which

$$p(A)^e Z \in \mathbb{U} \qquad f(A)Z \in \mathbb{U} \qquad p(A)^{e-1}Z \notin \mathbb{U}$$

then $p(x)^e$ divides $f(x)$.

17 Let $f(x)$ and $g(x)$ be relatively prime polynomials and let A be a square matrix of order n. Prove that

$$r(f(A)) + r(g(A)) \geq n$$

(*Hint:* Use Theorem 1.11.12).

18 Prove that for a given set of polynomials $f_1(x), f_2(x), \ldots, f_k(x)$ the polynomial $h(x)$ can be expressed in the form

$$h(x) = s_1(x)f_1(x) + s_2(x)f_2(x) + \cdots + s_k(x)f_k(x)$$

if and only if $d(x) = (f_1(x), f_2(x), \ldots, f_k(x))$ divides $h(x)$.

1.12 THE DECOMPOSITION THEOREMS

In the fundamental theorem of direct sums (Theorem 1.10.31) we proved that a decomposition of $\mathbb{F}^n$ into a direct sum of k A-invariant subspaces leads to a matrix B, similar to A, which is the direct sum of k submatrices. Moreover, we indicated how the matrices of the direct sum correspond to these invariant subspaces and showed that the order of a submatrix is equal to the dimension of the invariant subspace corresponding to it. However, the problem of actually decomposing $\mathbb{F}^n$ into a direct sum of invariant subspaces was not attacked. The main purpose of this section is to show how such a decomposition can be obtained. Our main tool will be the information about polynomials developed in the last section.

It was pointed out in Sec. 1.10 that when A is a square matrix, $\mathbb{N}(A)$ is an A-invariant vector space. Moreover, if B is any matrix which commutes with A, then $\mathbb{N}(B)$ is also A-invariant. (See Exercise 14 of Sec. 1.10.) In particular, when B can be expressed as a polynomial in A, say $B = f(A)$, then B commutes with A and $\mathbb{N}(B) = \mathbb{N}[f(A)]$ is A-invariant. In Theorem 1.12.6 we will show that $\mathbb{F}^n$ can be expressed as a direct sum of just such invariant subspaces.

Let A be a fixed $n \times n$ matrix. At the end of the last section we proved that A satisfies some nonzero polynomial of degree no greater than n^2. We also indicated that this result could be improved. Let h be the smallest integer for which there exists a nonzero polynomial of degree h satisfied by A. The last theorem of this section states that $h \leq n$. First, we need to establish that this polynomial is unique (up to associates). Hence let us suppose that

$$m(x) = a_h x^h + a_{h-1} x^{h-1} + \cdots + a_0 \qquad a_h \neq 0$$
$$\tilde{m}(x) = \tilde{a}_h x^h + \tilde{a}_{h-1} x^{h-1} + \cdots + \tilde{a}_0 \qquad \tilde{a}_h \neq 0$$

are both polynomials of degree h satisfied by A. Set

$$q(x) = \tilde{a}_h m(x) - a_h \tilde{m}(x)$$

Clearly

$$q(A) = 0$$

and the degree of $q(x)$ is less than h. By the definition of h, we see that $q(x) = 0$. Thus

$$\tilde{m}(x) = \frac{\tilde{a}_h}{a_h} m(x)$$

that is, $m(x)$ and $\tilde{m}(x)$ are associates. It follows that there is exactly one monic polynomial of degree h satisfied by A. We call this polynomial the *minimal polynomial* of A and denote it by $m(x)$.

Although we have shown that the minimal polynomial of a square matrix exists, we have not given any hints of how to find it (other than the obvious method, which involves determining which subsets of the collection of matrices $I, A, A^2, \ldots, A^{n^2}$ are linearly dependent). In fact, this is a rather nasty problem. However, at the end of this section we will obtain another characterization of the minimal polynomial, and in Sec. 1.13 we will derive a number of important and useful relations between the minimal polynomial and the characteristic polynomial of a matrix. In particular, we will show that the minimal polynomial of a matrix divides the characteristic polynomial. This fact will cut down considerably the work required to find the minimal polynomial of a given matrix.

Lemma 1.12.1 Let A be a square matrix whose minimal polynomial is $m(x)$ and let $f(x)$ be any polynomial. Then $f(A) = 0$ if and only if $m(x)$ divides $f(x)$.

PROOF First, suppose that $m(x)$ divides $f(x)$; that is, there is a polynomial $q(x)$ such that

$$f(x) = m(x)q(x)$$

Substituting A for x, we have

$$f(A) = m(A)q(A) = 0$$

Conversely, suppose that $f(x)$ is a polynomial which is satisfied by A. By the division algorithm, there exist polynomials $q(x)$ and $r(x)$ such that

$$f(x) = q(x)m(x) + r(x) \qquad \deg r(x) < \deg m(x)$$

Then $\qquad 0 = f(A) = q(A)m(A) + r(A) = r(A)$

Since the degree of $r(x)$ is less than the degree of $m(x)$ and $m(x)$ is the minimal polynomial of A, it follows that $r(x) = 0$. Thus $f(x) = q(x)m(x)$, and $m(x)$ is a divisor of $f(x)$. ////

We will use the minimal polynomial of A to obtain a decomposition of $\mathbb{F}^n$ into the direct sum of A-invariant subspaces. First, let us factor the minimal polynomial into its irreducible factors. According to Theorem 1.11.28, $m(x)$ can be expressed as

$$m(x) = p_1(x)^{e_1}, p_2(x)^{e_2} \cdots p_k(x)^{e_k} \qquad (1.12.2)$$

where $p_1(x), p_2(x), \ldots, p_k(x)$ are distinct irreducible monic polynomials and $e_1, e_2, \ldots, e_k$ are positive integers. Clearly $p_1(x)^{e_1}$, $p_2(x)^{e_2}, \ldots, p_k(x)^{e_k}$ are relatively prime in pairs, i.e.,

$$(p_i(x)^{e_i}, p_j(x)^{e_j}) = 1 \qquad i \neq j$$

Moreover, if we define the polynomials $q_i(x)$ by[1]

$$q_i(x) = \frac{m(x)}{p_i(x)^{e_i}} = \prod_{j \neq i} p_j(x)^{e_j} \qquad i = 1, 2, \ldots, k \qquad (1.12.3)$$

then

$$(p_i(x), q_i(x)) = 1 \qquad i = 1, 2, \ldots, k$$

and

$$(q_1(x), q_2(x), \ldots, q_k(x)) = 1$$

By Theorem 1.11.17, there exist polynomials $s_1(x), s_2(x), \ldots, s_k(x)$ such that

$$1 = s_1(x)q_1(x) + s_2(x)q_2(x) + \cdots + s_k(x)q_k(x) \qquad (1.12.4)$$

Observe also that for each i there exist polynomials $u_i(x)$ and $v_i(x)$ such that

$$1 = u_i(x)p_i(x)^{e_i} + v_i(x)q_i(x) \qquad i = 1, 2, \ldots, k \qquad (1.12.5)$$

Now that all the polynomials we will need have been defined, we can begin to apply some of the machinery built up here and in the last section.

Theorem 1.12.6 The primary decomposition theorem Let A be a square matrix of order n whose minimal polynomial is given by equation (1.12.2). Then

$$\mathbb{F}^n = \mathbb{N}[p_1(A)^{e_1}] \oplus \mathbb{N}[p_2(A)^{e_2}] \oplus \cdots \oplus \mathbb{N}[p_k(A)^{e_k}] \qquad (1.12.7)$$

PROOF Let $q_i(x)$ and $s_i(x)$, $i = 1, 2, \ldots, k$, be defined by equations (1.12.3) and (1.12.4) respectively. Substituting A for x in equation (1.12.4) yields

$$I = s_1(A)q_1(A) + s_2(A)q_2(A) + \cdots + s_k(A)q_k(A) \qquad (1.12.8)$$

Let X be an arbitrary element of $\mathbb{F}^n$. Multiplying equation (1.12.8) on the right by X, we have

$$X = s_1(A)q_1(A)X + s_2(A)q_2(A)X + \cdots + s_k(A)q_k(A)X \qquad (1.12.9)$$

If we set

$$Y_i = s_i(A)q_i(A)X \qquad i = 1, 2, \ldots, k$$

---

[1] When $k = 1$, we set $q_1(x) = 1$ and the remainder of the discussion proceeds without difficulty. However, in this case, the primary decomposition theorem (Theorem 1.12.6) does not yield a meaningful decomposition of $\mathbb{F}^n$.

then equation (1.12.9) becomes
$$X = Y_1 + Y_2 + \cdots + Y_k \quad (1.12.10)$$
Moreover, $Y_i \in \mathbb{N}[p_i(A)^{e_i}]$
since $\quad p_i(A)^{e_i} Y_i = p_i(A)^{e_i} s_i(A) q_i(A) X = m(A) s_i(A) X = 0$

Because X is an arbitrary element of $\mathbb{F}^n$, it follows from equation (1.12.10) that
$$\mathbb{F}^n = \mathbb{N}[p_1(A)^{e_1}] + \mathbb{N}[p_2(A)^{e_2}] + \cdots + \mathbb{N}[p_k(A)^{e_k}] \quad (1.12.11)$$

According to Theorem 1.5.14, in order to show that the sum on the right hand side of equation (1.12.11) is actually a direct sum we must prove that the expression for X of equation (1.12.10), namely,
$$X = Y_1 + Y_2 + \cdots + Y_k \quad Y_i \in \mathbb{N}[p_i(A)^{e_i}]$$
is unique. Suppose we also have
$$X = Z_1 + Z_2 + \cdots + Z_k \quad Z_i \in \mathbb{N}[p_i(A)^{e_i}]$$
Then
$$0 = (Y_1 - Z_1) + (Y_2 - Z_2) + \cdots + (Y_k - Z_k) \quad (1.12.12)$$

Now, let i be fixed, $1 \le i \le k$. Because of the way in which $q_i(x)$ was defined, $p_j(x)^{e_j}$ divides $q_i(x)$ for all $j \ne i$. Therefore
$$q_i(A)(Y_j - Z_j) = 0 \quad j \ne i$$
and consequently if we multiply both sides of equation (1.12.12) on the left by $q_i(A)$, we have
$$0 = q_i(A)(Y_i - Z_i)$$

Moreover, equation (1.12.5) states that there exist polynomials $u_i(x)$ and $v_i(x)$ such that
$$I = u_i(A) p_i(A)^{e_i} + v_i(A) q_i(A) \quad (1.12.13)$$
Multiplying equation (1.12.13) on the right by $Y_i - Z_i$, we see that
$$Y_i - Z_i = u_i(A) p_i(A)^{e_i} (Y_i - Z_i) + v_i(A) q_i(A)(Y_i - Z_i) = 0$$
Thus $\quad Y_i = Z_i$

We have now proved that for each $X \in \mathbb{F}^n$ the expression for X which is given by equations (1.12.9) and (1.12.10) is unique, and consequently
$$\mathbb{F}^n = \mathbb{N}[p_1(A)^{e_1}] \oplus \mathbb{N}[p_2(A)^{e_2}] \oplus \cdots \oplus \mathbb{N}[p_k(A)^{e_k}] \quad ////$$

EXAMPLE 1.12.14 Let

$$A = \begin{bmatrix} 2 & 1 \\ 3 & 4 \end{bmatrix}$$

In Example 1.10.16 we showed that the characteristic polynomial of A is

$$x^2 - 6x + 5 = (x-1)(x-5)$$

It is easily verified that the minimal polynomial of A is also $x^2 - 6x + 5$.[1] [See part (a) of Exercise 10 in Sec. 1.11.] Thus

$$p_1(x)^{e_1} = x - 1 \qquad p_2(x)^{e_2} = x - 5$$

We have already shown that

$\left\{ \begin{bmatrix} 1 \\ -1 \end{bmatrix} \right\}$ is a basis for $\mathbb{N}(A - I)$ $\qquad \left\{ \begin{bmatrix} 1 \\ 3 \end{bmatrix} \right\}$ is a basis for $\mathbb{N}(A - 5I)$

and that

$$\mathbb{F}^2 = \mathbb{N}(A - I) \oplus \mathbb{N}(A - 5I) \qquad\qquad ////$$

EXAMPLE 1.12.15 Let

$$A = \begin{bmatrix} 3 & 1 & -2 \\ -1 & -1 & 4 \\ 0 & -1 & 3 \end{bmatrix}$$

In Example 1.10.37 we showed that the characteristic polynomial of A is

$$-(x^3 - 5x^2 + 8x - 4) = -(x-2)^2(x-1)$$

It is easily verified that the minimal polynomial of A is $x^3 - 5x^2 + 8x - 4$.[1] [See part (b) Exercise 11 in Sec. 1.11.] Thus

$$p_1(x)^{e_1} = (x-2)^2 \qquad p_2(x)^{e_2} = x - 1$$

and we have already shown that

$\left\{ \begin{bmatrix} 0 \\ 2 \\ 1 \end{bmatrix} \right\}$ is a basis for $\mathbb{N}(A - I)$

It remains to find a basis for $\mathbb{N}[(A - 2I)^2]$ and to verify the primary decomposition theorem. First we calculate

$$(A - 2I)^2 = \begin{bmatrix} 0 & 0 & 0 \\ 2 & 4 & -6 \\ 1 & 2 & -3 \end{bmatrix}$$

---

[1] It is not entirely a coincidence that the minimal polynomial and the characteristic polynomial of A are equal. The underlying reasons for this will be made clear in Sec. 1.13.

It is not difficult to show that

$$\left\{ \begin{bmatrix} 1 \\ 1 \\ 1 \end{bmatrix}, \begin{bmatrix} 2 \\ -1 \\ 0 \end{bmatrix} \right\} \text{ is a basis for } \mathbb{N}[(A-2I)^2]$$

Note that $\mathbb{N}[(A-2I)^2]$ is the vector space which we denoted by $\mathbb{V}_1$ in Example 1.10.37. Thus

$$\mathbb{F}^3 = \mathbb{N}[(A-2I)^2] \oplus \mathbb{N}(A-I) \qquad ////$$

EXAMPLE 1.12.16 Let

$$A = \begin{bmatrix} 2 & -2 & 1 \\ 3 & -3 & 1 \\ 0 & 0 & -1 \end{bmatrix}$$

The minimal polynomial of A is

$$m(x) = x^2 + x = x(x+1)$$

[See part (d) of Exercise 10 in Sec. 1.11.] Thus

$$p_1(x)^{e_1} = x \qquad p_2(x)^{e_2} = x+1$$

Clearly

$$\left\{ \begin{bmatrix} 1 \\ 1 \\ 0 \end{bmatrix} \right\} \text{ is a basis for } \mathbb{N}(A)$$

Since

$$A + I = \begin{bmatrix} 3 & -2 & 1 \\ 3 & -2 & 1 \\ 0 & 0 & 0 \end{bmatrix}$$

it follows that

$$\left\{ \begin{bmatrix} 1 \\ 2 \\ 1 \end{bmatrix}, \begin{bmatrix} 0 \\ 1 \\ 2 \end{bmatrix} \right\} \text{ is a basis for } \mathbb{N}(A+I)$$

Again,

$$\mathbb{F}^3 = \mathbb{N}(A) \oplus \mathbb{N}(A+I) \qquad ////$$

EXAMPLE 1.12.17 Let

$$A = \begin{bmatrix} 3 & 4 & 5 \\ 0 & 1 & -1 \\ -1 & -2 & -1 \end{bmatrix}$$

The minimal polynomial of A is

$$m(x) = x^3 - 3x^2 + 2x = x(x-1)(x-2)$$

and therefore $\quad p_1(x)^{e_1} = x \quad p_2(x)^{e_2} = x - 1 \quad p_3(x)^{e_3} = x - 2$

Since

$$A - I = \begin{bmatrix} 2 & 4 & 5 \\ 0 & 0 & -1 \\ -1 & -2 & -2 \end{bmatrix} \quad A - 2I = \begin{bmatrix} 1 & 4 & 5 \\ 0 & -1 & -1 \\ -1 & -2 & -3 \end{bmatrix}$$

we can verify that

$$\left\{ \begin{bmatrix} 3 \\ -1 \\ -1 \end{bmatrix} \right\} \quad \left\{ \begin{bmatrix} 2 \\ -1 \\ 0 \end{bmatrix} \right\} \quad \left\{ \begin{bmatrix} 1 \\ 1 \\ -1 \end{bmatrix} \right\}$$

are bases for $\mathbb{N}(A)$, $\mathbb{N}(A - I)$, and $\mathbb{N}(A - 2I)$ respectively. As before,

$$\mathbb{F}^3 = \mathbb{N}(A) \oplus \mathbb{N}(A - I) \oplus \mathbb{N}(A - 2I) \qquad ////$$

It can be proved that none of the vector spaces $\mathbb{N}[p_i(A)^{e_i}]$ on the right-hand side of equation (1.12.7) is a zero space. (See Exercise 8.) Thus, unless $k = 1$, equation (1.12.7) splits $\mathbb{F}^n$ nontrivially into a direct sum of invariant subspaces; that is, $\mathbb{F}^n$ is now expressed as a direct sum of invariant subspaces none of which is either 0 or $\mathbb{F}^n$ itself. On the other hand, when $k = 1$, we have

$$m(x) = p_1(x)^{e_1}$$

and the primary decomposition theorem states that

$$\mathbb{F}^n = \mathbb{N}[p_1(A)^{e_1}] = \mathbb{N}(0) = \mathbb{F}^n$$

which, while certainly true, adds no new information, In the secondary decomposition theorem we will determine when an invariant subspace of $\mathbb{F}^n$ of the form $\mathbb{N}[p(A)^e]$ can itself be expressed as the direct sum of invariant subspaces. Then the decomposition of $\mathbb{F}^n$ will be complete.

Recall that our original reason for splitting $\mathbb{F}^n$ into a direct sum of invariant subspaces was that such a decomposition of $\mathbb{F}^n$ enables us to obtain a matrix B, similar to A, having the form of a direct sum. Thus combining Theorems 1.10.31 and 1.12.6 yields:

Corollary 1.12.18 Let A be a square matrix whose minimal polynomial is given by (1.12.2). For each i, $i = 1, 2, \ldots, k$, let

$$\mathbb{X}_i = \{X_{i1}, X_{i2}, \ldots, X_{in_i}\}$$

be a basis for $\mathbb{N}[p_i(A)^{e_i}]$ and let P be the $n \times n$ matrix whose columns are $X_{11}, X_{12}, \ldots, X_{1n_1}, X_{21}, \ldots, X_{kn_k}$ in that order. Then

$$P^{-1}AP = B = \begin{bmatrix} B_1 & 0 & \cdots & 0 \\ 0 & B_2 & \cdots & 0 \\ \multicolumn{4}{c}{\dotfill} \\ 0 & 0 & \cdots & B_k \end{bmatrix}$$

where B_i is a square matrix of order n_i.

Examples 1.10.16 and 1.10.37 provide illustrations of Corollary 1.12.18. In Example 1.10.37 there is no indication of how either the two-dimensional A-invariant subspace $\mathbb{V}_1$ or its basis $\mathbb{X}_1$ is obtained. It is easily verified that $\mathbb{V}_1 = \mathbb{N}[(A - 2I)^2]$. The form of $P^{-1}AP$ then follows from the primary decomposition theorem and Corollary 1.12.18. In Example 1.12.33 we will see that when some basis other than $\mathbb{X}_1$ is chosen for $\mathbb{N}[(A - 2I)^2]$, the form of the resulting matrix B_1 is changed.

In Corollary 1.12.18 no attempt was made to obtain any particular matrices for $B_1, B_2, \ldots, B_k$. Nor, in fact, was an attempt made to determine whether any of the matrices $B_1, B_2, \ldots, B_k$ could itself be expressed as a direct sum. If it is possible to split one of the subspaces, say $\mathbb{N}[p_i(A)^{e_i}]$, appearing on the right-hand side of equation (1.12.7) into a direct sum of A-invariant subspaces, then the corresponding matrix B_i can be made to have the form of a direct sum. In Theorem 1.12.27 we will determine when this can be done and, at the same time, exhibit a special basis for a vector space which will lead (in Sec. 1.13) to particularly simple matrices $B_1, B_2, \ldots, B_k$. We begin with:

Definition 1.12.19 Let $\mathbb{U}$ be a nonzero, A-invariant, q-dimensional subspace of $\mathbb{F}^n$. Then $\mathbb{U}$ is a *cyclic space* (with respect to A) if there exists $Y \in \mathbb{U}$ such that the q vectors

$$Y, AY, A^2Y, \ldots, A^{q-1}Y \quad (1.12.20)$$

form a basis for $\mathbb{U}$. The vector Y is called a *generator* of $\mathbb{U}$, and the basis (1.12.20) is called a *cyclic basis* for $\mathbb{U}$.

The generator of a cyclic space is not unique (see Exercise 16). Every nonzero vector of $\mathbb{F}^n$ is a generator of some cyclic space. For, let Y be any nonzero vector, let q be the largest integer for which the vectors $Y, AY, A^2Y,$

..., $A^{q-1}Y$ are linearly independent, and let U be the vector space which has $Y, AY, A^2Y, \ldots, A^{q-1}Y$ as a basis. In order to prove that U is a cyclic space with generator Y, it remains only to show that U is A-invariant. It is sufficient to show that

$$A(A^jY) = A^{j+1}Y \in \mathsf{U} \qquad j = 0, 1, 2, \ldots, q-1$$

For $j = 0, 1, 2, \ldots, q-2$ there is no difficulty since $A^{j+1}Y$ is one of the elements of the basis that we used to define U. Finally, we need to prove that

$$A(A^{q-1}Y) = A^qY \in \mathsf{U}$$

Since $Y, AY, A^2Y, \ldots, A^{q-1}Y, A^qY$ are linearly dependent, there exist scalars, $\beta_0, \beta_1, \beta_2, \ldots, \beta_{q-1}, \beta_q$, not all zero, such that

$$\beta_0 Y + \beta_1 AY + \beta_2 A^2Y + \cdots + \beta_{q-1} A^{q-1}Y + \beta_q A^qY = 0 \quad (1.12.21)$$

However, $Y, AY, A^2Y, \ldots, A^{q-1}Y$ are linearly independent, and it follows that in equation (1.12.21) we must have $\beta_q \neq 0$. Thus

$$A^qY = -\frac{\beta_0}{\beta_q}Y - \frac{\beta_1}{\beta_q}AY - \frac{\beta_2}{\beta_q}A^2Y - \cdots - \frac{\beta_{q-1}}{\beta_q}A^{q-1}Y \in \mathsf{U} \quad (1.12.22)$$

proving that U is A-invariant.

According to equation (1.12.22), if we set

$$f(x) = x^q + \frac{\beta_{q-1}}{\beta_q}x^{q-1} + \cdots + \frac{\beta_2}{\beta_q}x^2 + \frac{\beta_1}{\beta_q}x + \frac{\beta_0}{\beta_q}$$

then

$$f(A)Y = 0 \quad (1.12.23)$$

The (unique) monic nonzero polynomial of least degree, $n(x)$, for which

$$n(A)Y = 0$$

is called the *minimal polynomial* of Y (with respect to A). We have just proved that there is a monic polynomial of degree q satisfying equation (1.12.23). Moreover, since $Y, AY, A^2Y, \ldots, A^{q-1}Y$ are linearly dependent, no nonzero polynomial of degree less than q satisfies equation (1.12.23). Thus we have proved:

Lemma 1.12.24 The degree of the minimal polynomial of a vector is equal to the dimension of the cyclic space it generates.

The minimal polynomial of a vector behaves very much like the minimal polynomial of a matrix.

Lemma 1.12.25 If $n(x)$ is the minimal polynomial of Y (with respect to A) and $f(x)$ is any polynomial for which

$$f(A)Y = 0$$

then $n(x)$ divides $f(x)$. In particular, $n(x)$ divides $m(x)$, the minimal polynomial of A.

PROOF The proof is left as an exercise. ////

In the next theorem we will apply Lemma 1.12.25 to the case where $p(x)$ is an irreducible polynomial of degree l and $Y \in \mathbb{N}[p(A)^e]$, $Y \neq 0$. Then the minimal polynomial of Y has the form

$$n(x) = p(x)^d$$

for some integer d, $1 \leq d \leq l$, and the dimension of the cyclic space generated by Y is dl.

Theorem 1.12.26 Let $p(x)$ be an irreducible monic polynomial of degree l and let $p(x)^e$, $e \geq 1$, be the largest power of $p(x)$ which divides the minimal polynomial, $m(x)$, of A.[1] Then there exists an el-dimensional cyclic subspace $\mathbb{U}$ of $\mathbb{N}[p(A)^e]$.

PROOF Define the polynomials $q(x)$ and $\tilde{q}(x)$ by

$$q(x) = \frac{m(x)}{p(x)^e} \qquad \tilde{q}(x) = \frac{m(x)}{p(x)} = p(x)^{e-1}q(x)$$

Then $q(x)$ and $\tilde{q}(x)$ are polynomials such that

$$q(A) \neq 0 \qquad \tilde{q}(A) \neq 0$$

and

$$p(A)^e q(A) = p(A)\tilde{q}(A) = 0$$

Since $\tilde{q}(A) \neq 0$, there exists a vector Z such that $\tilde{q}(A)Z \neq 0$. Set $Y = q(A)Z$. We wish to show that the cyclic space $\mathbb{U}$ generated by Y is a subspace of $\mathbb{N}[p(A)^e]$ and has dimension el. Because

$$p(A)^e Y = p(A)^e q(A) Z = m(A) Z = 0$$

we have $Y \in \mathbb{N}[p(A)^e]$. Moreover, since

$$p(A)^{e-1} Y = p(A)^{e-1} q(A) Z = \tilde{q}(A) Z \neq 0$$

---

[1] In other words, $p(x)$ is one of the polynomials $p_1(x), p_2(x), \ldots, p_k(x)$ of equation (1.12.2), and e is the corresponding exponent $e_1, e_2, \ldots, e_k$.

the minimal polynomial of Y is $p(x)^e$. By Lemma 1.12.24, the cyclic space U generated by Y has dimension el. ////

According to Theorem 1.12.26

$$\dim \mathsf{U} = el \leq \nu[p(A)^e] = \dim \mathsf{N}[p(A)^e]$$

If $el = \nu[p(A)^e]$, then $\mathsf{N}[p(A)^e]$ is an el-dimensional cyclic space. It is not difficult to prove that a cyclic space is *irreducible*, i.e., it cannot be split into a direct sum of invariant subspaces (see Exercise 17). Hence, if $el = \nu[p(A)^e]$, then the decomposition of $\mathsf{N}[p(A)^e]$ has been carried as far as possible. On the other hand, we will see that if $el < \nu[p(A)^e]$, then $\mathsf{N}[p(A)^e]$ is expressible as a direct sum of (A-invariant) cyclic subspaces. This result is contained in:

Theorem 1.12.27 The secondary decomposition theorem Let $p(x)$ be an irreducible monic polynomial of degree l and let $p(x)^e$, $e \geq 1$, be the largest power of $p(x)$ which divides $m(x)$. Then there exist positive integers $r_1 = e \geq r_2 \geq \cdots \geq r_h \geq 1$ and cyclic subspaces $\mathsf{U}_1 = \mathsf{U}, \mathsf{U}_2, \ldots,$ U_h of $\mathsf{N}[p(A)^e]$ of dimensions $r_1 l = el, r_2 l, \ldots, r_h l$ respectively such that

$$\mathsf{N}[p(A)^e] = \mathsf{U}_1 \oplus \mathsf{U}_2 \oplus \cdots \oplus \mathsf{U}_h$$

PROOF Theorem 1.12.26 states that there exists a vector $Y_1 = Y$ belonging to $\mathsf{N}[p(A)^e]$ which generates a cyclic subspace $\mathsf{U}_1 = \mathsf{U}$ of $\mathsf{N}[p(A)^e]$ of dimension $r_1 l = el$. If $\mathsf{N}[p(A)^e] = \mathsf{U}_1$, then we are finished.

Now suppose that $\mathsf{U}_1 \subsetneq \mathsf{N}[p(A)^e]$. We will construct a cyclic space U_2 with generator Y_2 such that $\mathsf{U}_1 \oplus \mathsf{U}_2 \subseteq \mathsf{N}[p(A)^e]$. Let r_2 be the largest integer for which there exists a vector Z belonging to $\mathsf{N}[p(A)^e]$ such that

$$p(A)^{r_2} Z \in \mathsf{U}_1 \qquad p(A)^{r_2-1} Z \notin \mathsf{U}_1$$

Clearly $r_1 = e \geq r_2 \geq 1$. If $p(A)^{r_2} Z = 0$, we set $Y_2 = Z$. If $p(A)^{r_2} Z \neq 0$, we need to show that there exists a vector Y_2 such that

$$p(A)^{r_2} Y_2 = 0 \in \mathsf{U}_1 \qquad p(A)^{r_2-1} Y_2 \notin \mathsf{U}_1 \qquad (1.12.28)$$

Since $p(A)^{r_2} Z \in \mathsf{U}_1$ and $Y_1, AY_1, A^2 Y_1, \ldots, A^{r_1 l - 1} Y_1$ form a basis for U_1, there exist scalars $\alpha_{r_1 l - 1}, \ldots, \alpha_2, \alpha_1, \alpha_0$ for which

$$p(A)^{r_2} Z = \alpha_{r_1 l - 1} A^{r_1 l - 1} Y_1 + \cdots + \alpha_2 A^2 Y_1 + \alpha_1 A Y_1 + \alpha_0 Y_1 \qquad (1.12.29)$$

Define the polynomial $f(x)$ by

$$f(x) = \alpha_{r_1 l - 1} x^{r_1 l - 1} + \cdots + \alpha_2 x^2 + \alpha_1 x + \alpha_0$$

Equation (1.12.29) then becomes

$$p(A)^{r_2}Z = f(A)Y_1 \quad (1.12.30)$$

Multiplying both sides of this equation by $p(A)^{r_1-r_2}$, we have

$$p(A)^{r_1-r_2}f(A)Y_1 = p(A)^{r_1-r_2}p(A)^{r_2}Z = p(A)^{r_1}Z = p(A)^e Z = 0$$

Since the minimal polynomial of Y_1 is $p(x)^{r_1}$, it follows from Lemma 1.12.25 that $p(x)^{r_1}$ divides $p(x)^{r_1-r_2}f(x)$ and consequently $f(x)$ is divisible by $p(x)^{r_2}$, say

$$f(x) = p(x)^{r_2}g(x)$$

Substituting this expression for $f(x)$ into equation (1.12.30), we have

$$p(A)^{r_2}Z = f(A)Y_1 = p(A)^{r_2}g(A)Y_1$$

If we now set

$$Y_2 = Z - g(A)Y_1$$

then Y_2 satisfies both parts of (1.12.28):

$$\begin{aligned} p(A)^{r_2}Y_2 &= p(A)^{r_2}Z - p(A)^{r_2}g(A)Y_1 = 0 \in \mathbb{U}_1 \\ p(A)^{r_2-1}Y_2 &= p(A)^{r_2-1}Z - p(A)^{r_2-1}g(A)Y_1 \notin \mathbb{U}_1 \end{aligned} \quad (1.12.31)$$

since $p(A)^{r_2-1}g(A)Y_1 \in \mathbb{U}_1$ and $p(A)^{r_2-1}Z \notin \mathbb{U}_1$.

Denote the cyclic space generated by Y_2 by $\mathbb{U}_2$. We wish to prove that $\mathbb{U}_2$ is a subspace of $\mathbb{N}[p(A)^e]$ of dimension $r_2 l$ and that $\mathbb{U}_1 + \mathbb{U}_2$ is a direct sum. Combining the two parts of (1.12.31), we see that $p(x)^{r_2}$ is the minimal polynomial of Y_2, and applying Lemma 1.12.24 shows that the dimension of $\mathbb{U}_2$ is $r_2 l$. Moreover, it follows from the first equation of (1.12.31) that

$$Y_2 \in \mathbb{N}[p(A)^{r_2}] \subseteq \mathbb{N}[p(A)^e]$$

and hence $\mathbb{U}_2 \subseteq \mathbb{N}[p(A)^e]$. Consequently

$$\mathbb{U}_1 + \mathbb{U}_2 \subseteq \mathbb{N}[p(A)^e]$$

and it remains only to prove that this sum is actually a direct sum. Suppose there exists a nonzero vector W contained in $\mathbb{U}_1 \cap \mathbb{U}_2$. Because $W \in \mathbb{U}_2$ and

$$\{Y_2, AY_2, A^2Y_2, \ldots, A^{r_2 l - 1}Y_2\}$$

is a basis for $\mathbb{U}_2$, there exists a nonzero polynomial, say $g(x)$, of degree less that $r_2 l$ such that

$$W = g(A)Y_2$$

If we set
$$d(x) = (g(x), p(x)^{r_2})$$
then
$$d(x) = p(x)^m$$
for some integer m, $0 \leq m \leq r_2$. Moreover, the degree of $g(x)$ is less than the degree of $p(x)^{r_2}$, and we conclude that
$$0 \leq m < r_2$$
By Theorem 1.11.12, there exist polynomials $s(x)$ and $t(x)$ such that
$$d(x) = p(x)^m = s(x)g(x) + t(x)p(x)^{r_2} \quad (1.12.32)$$
Substituting A for x in equation (1.12.32) and multiplying on the right by Y_2 yields
$$p(A)^m Y_2 = s(A)g(A)Y_2 + t(A)p(A)^{r_2} Y_2 = s(A)W \in \mathbb{U}_1$$
since $W \in \mathbb{U}_1$. Then
$$p(A)^{r_2-1} Y_2 = p(A)^{r_2-m-1} p(A)^m Y_2 = p(A)^{r_2-m-1} s(A)W \in \mathbb{U}_1$$
which contradicts the second part of (1.12.31). Thus we have proved that $\mathbb{U}_1 \cap \mathbb{U}_2 = 0$ and consequently that $\mathbb{U}_1 + \mathbb{U}_2$ is a direct sum (see Exercise 11 of Sec. 1.5).

If $\mathbb{U}_1 \oplus \mathbb{U}_2 = \mathbb{N}[p(A)^e]$, we are finished. Otherwise we repeat the entire process. Let r_3 be the largest integer for which there exists a vector Z belonging to $\mathbb{N}[p(A)^e]$ such that
$$p(A)^{r_3} Z \in \mathbb{U}_1 \oplus \mathbb{U}_2 \quad p(A)^{r_3-1} Z \notin \mathbb{U}_1 \oplus \mathbb{U}_2$$
As before, we can demonstrate that Z may be replaced by a vector Y_3 such that
$$p(A)^{r_3} Y_3 = 0 \quad p(A)^{r_3-1} Y_3 \notin \mathbb{U}_1 \oplus \mathbb{U}_2$$
We then define $\mathbb{U}_3$ to be the cyclic space generated by Y_3 and show that $\mathbb{U}_3$ is a subspace of $\mathbb{N}[p(A)^e]$ of dimension $r_3 l$ and that $(\mathbb{U}_1 \oplus \mathbb{U}_2) + \mathbb{U}_3$ is a direct sum. Proceeding in this way, we eventually express $\mathbb{N}[p(A)^e]$ as $\mathbb{U}_1 \oplus \mathbb{U}_2 \oplus \cdots \oplus \mathbb{U}_h$, where each $\mathbb{U}_j$ is a cyclic subspace of $\mathbb{N}[p(A)^e]$ of dimension $r_j l$ with generator Y_j. ////

It certainly cannot be inferred from what we have done that for a given matrix A the cyclic subspaces $\mathbb{U}_1, \mathbb{U}_2, \ldots, \mathbb{U}_h$ of the secondary decomposition theorem are unique. In fact, they are not. Furthermore, it does not even follow that the numbers h, $r_2, r_3, \ldots, r_h$ are uniquely determined. On the other hand, $r_1 = e$ is prescribed by the fact that $p(x)^e$ is the highest power of $p(x)$ which divides $m(x)$. We will consider this problem again in Theorem 1.12.42.

EXAMPLE 1.12.33 Let

$$A = \begin{bmatrix} 3 & 1 & -2 \\ -1 & -1 & 4 \\ 0 & -1 & 3 \end{bmatrix}$$

(See Examples 1.10.37 and 1.12.15.) We have previously shown that

$$m(x) = (x-2)^2(x-1)$$

and that

$$\left\{ \begin{bmatrix} 1 \\ 1 \\ 1 \end{bmatrix}, \begin{bmatrix} 2 \\ -1 \\ 0 \end{bmatrix} \right\}$$

is a basis for $\mathbb{N}[(A-2I)^2]$. However it is not a cyclic basis, though $\mathbb{N}[(A-2I)^2]$ is a cyclic space. If we set

$$Y_1 = \begin{bmatrix} 2 \\ -1 \\ 0 \end{bmatrix}$$

then

$$AY_1 = \begin{bmatrix} 5 \\ -1 \\ 1 \end{bmatrix} \quad A^2 Y_1 = \begin{bmatrix} 12 \\ 0 \\ 4 \end{bmatrix} = 4AY_1 - 4Y_1$$

and

$$\left\{ Y_1 = \begin{bmatrix} 2 \\ -1 \\ 0 \end{bmatrix}, AY_1 = \begin{bmatrix} 5 \\ -1 \\ 1 \end{bmatrix} \right\}$$

is a cyclic basis for $\mathbb{N}[(A-2I)^2]$.
Clearly

$$\left\{ \begin{bmatrix} 0 \\ 2 \\ 1 \end{bmatrix} \right\}$$

is a cyclic basis for $\mathbb{N}(A-I)$. ////

EXAMPLE 1.12.34 Let

$$A = \begin{bmatrix} 2 & -2 & 1 \\ 3 & -3 & 1 \\ 0 & 0 & -1 \end{bmatrix}$$

(See Example 1.12.16.) We have previously shown that

$$m(x) = x(x+1)$$

and that $\quad\quad \dim \mathbb{N}(A) = 1 \quad \dim \mathbb{N}(A+I) = 2$

Then
$$\left\{\begin{bmatrix}1\\1\\0\end{bmatrix}\right\}$$

is a cyclic basis for $\mathbb{N}(A)$. Because the highest power of $x + 1$ which divides $m(x)$ is $x + 1$, we expect the cyclic subspaces of $\mathbb{N}(A + I)$ to be one-dimensional. Indeed, $\mathbb{N}(A + I)$ is not a cyclic space but can be expressed as the direct sum of two one-dimensional cyclic subspaces. If Y_1 and Y_2 are any two linearly independent vectors of $\mathbb{N}(A + I)$, and if we denote the cyclic spaces generated by Y_1 and Y_2 by $\mathbb{U}_1$ and $\mathbb{U}_2$ respectively, then $\mathbb{U}_1$ and $\mathbb{U}_2$ each has dimension 1 and

$$\mathbb{N}(A + I) = \mathbb{U}_1 \oplus \mathbb{U}_2 \qquad ////$$

EXAMPLE 1.12.35 Let

$$A = \begin{bmatrix} -1 & 2 & 2 & 8 \\ 1 & 1 & 2 & 0 \\ 1 & -1 & 0 & -4 \\ -1 & 1 & 1 & 5 \end{bmatrix}$$

It can be verified that the minimal polynomial of A is

$$m(x) = x^3 - 4x^2 + 5x - 2 = (x - 1)^2(x - 2)$$

We compute

$$(A - I)^2 = \begin{bmatrix} 0 & 2 & 6 & 8 \\ 0 & 0 & 0 & 0 \\ 0 & -1 & -3 & -4 \\ 0 & 1 & 3 & 4 \end{bmatrix} \qquad A - 2I = \begin{bmatrix} -3 & 2 & 2 & 8 \\ 1 & -1 & 2 & 0 \\ 1 & -1 & -2 & -4 \\ -1 & 1 & 1 & 3 \end{bmatrix}$$

and we see that

$$\left\{\begin{bmatrix}1\\0\\0\\0\end{bmatrix}, \begin{bmatrix}0\\3\\-1\\0\end{bmatrix}, \begin{bmatrix}0\\4\\0\\-1\end{bmatrix}\right\} \quad \text{and} \quad \left\{\begin{bmatrix}-2\\0\\1\\-1\end{bmatrix}\right\}$$

are bases for $\mathbb{N}[(A - I)^2]$ and $\mathbb{N}(A - 2I)$ respectively. The primary decomposition theorem yields

$$\mathbb{F}^4 = \mathbb{N}[(A - I)^2] \oplus \mathbb{N}(A - 2I)$$

Let us apply the secondary decomposition theorem to $\mathbb{N}[(A - I)^2]$ with $p(x) = x - 1$, $l = 1$, and $r_1 = e = 2$. The largest cyclic subspace of $\mathbb{N}[(A - I)^2]$ has dimension 2. Since the dimension of $\mathbb{N}[(A - I)^2]$ is 3, we expect to be able to express it as

$$\mathbb{N}[(A - I)^2] = \mathbb{U}_1 \oplus \mathbb{U}_2 \quad (1.12.36)$$

where $\mathbb{U}_1$ and $\mathbb{U}_2$ are cyclic spaces of dimension 2 and 1, respectively.

Any vector Y_1 belonging to $\mathbb{N}[(A - I)^2]$ for which Y_1 and AY_1 are linearly independent may be used as a generator for $\mathbb{U}_1$. Specifically, if we set

$$Y_1 = \begin{bmatrix} 1 \\ 0 \\ 0 \\ 0 \end{bmatrix}$$

then

$$AY_1 = \begin{bmatrix} -1 \\ 1 \\ 1 \\ -1 \end{bmatrix} \quad A^2 Y_1 = \begin{bmatrix} -3 \\ 2 \\ 2 \\ -2 \end{bmatrix} = 2AY_1 - Y_1$$

and the vector space which has

$$\left\{ \begin{bmatrix} 1 \\ 0 \\ 0 \\ 0 \end{bmatrix}, \begin{bmatrix} -1 \\ 1 \\ 1 \\ -1 \end{bmatrix} \right\}$$

as a basis can be taken for $\mathbb{U}_1$. In order to find a basis for a cyclic space $\mathbb{U}_2$ which satisfies equation (1.12.36), it suffices to find a vector Y_2 such that

$$Y_2 \notin \mathbb{U}_1 \quad (A - I)Y_2 = 0 \in \mathbb{U}_1$$

Let us try

$$Z = \begin{bmatrix} 0 \\ 3 \\ -1 \\ 0 \end{bmatrix}$$

It is not difficult to show that $Z \notin \mathbb{U}_1$. However,

$$(A - I)Z = \begin{bmatrix} -2 & 2 & 2 & 8 \\ 1 & 0 & 2 & 0 \\ 1 & -1 & -1 & -4 \\ -1 & 1 & 1 & 4 \end{bmatrix} \begin{bmatrix} 0 \\ -3 \\ 1 \\ 0 \end{bmatrix} = \begin{bmatrix} 4 \\ -2 \\ -2 \\ 2 \end{bmatrix} \neq 0$$

At this point we can construct Y_2 as we did in the proof of Theorem 1.12.27.

A simple computation shows that

$$(A - I)Z = \begin{bmatrix} 4 \\ -2 \\ -2 \\ 2 \end{bmatrix} = -2AY_1 + 2Y_1 = -2(A - I)Y_1$$

If we set

$$Y_2 = Z + 2Y_1 = \begin{bmatrix} 2 \\ 3 \\ -1 \\ 0 \end{bmatrix}$$

then we have

$$Y_2 \notin \mathsf{U}_1 \qquad (A - I)Y_2 = 0$$

and the cyclic space generated by Y_2, which we call U_2, satisfies equation (1.12.36).

The dimension of $\mathsf{N}(A - 2I)$ is 1, and hence it is already a cyclic space. Any nonzero element of $\mathsf{N}(A - 2I)$ is a generator. ////

When we combine the primary and secondary decomposition theorems, each of the subspaces $\mathsf{N}[p_i(A)^{e_i}]$ of equation (1.12.7) is expressed as a direct sum of cyclic spaces and thus $\mathbb{F}^n$ itself is expressed as a direct sum of cyclic spaces. Since a cyclic space is irreducible (see Exercise 17), this decomposition can go no further. If a basis is selected for each of the cyclic spaces and a matrix P is formed from them, as in the fundamental theorem of direct sums, then $P^{-1}AP$ is a direct sum of matrices, one corresponding to each of the cyclic spaces. In the next section we will discuss in greater detail the exact form of $P^{-1}AP$ when the columns of P are the vectors in the cyclic bases of the cyclic subspaces.

It was pointed out earlier in discussing the secondary decomposition theorem that there is no reason to suppose that the decomposition of $\mathsf{N}[p(A)^e]$ into the direct sum of cyclic subspaces $\mathsf{U}_1, \mathsf{U}_2, \ldots, \mathsf{U}_h$ is unique. Example 1.12.35 provides a simple illustration of this phenomenon. We chose, rather arbitrarily, the vector

$$Y_1 = \begin{bmatrix} 1 \\ 0 \\ 0 \\ 0 \end{bmatrix}$$

as the generator of the two-dimensional cyclic space U_1. The only requirements placed on Y_1 were that Y_1 belong to $\mathbb{N}[(A-I)^2]$ and that Y_1 and AY_1 be linearly independent. We leave it as an exercise to verify that

$$\tilde{Y}_1 = Y_1 + AY_1 = \begin{bmatrix} 1 \\ 0 \\ 0 \\ 0 \end{bmatrix} + \begin{bmatrix} -1 \\ 1 \\ 1 \\ -1 \end{bmatrix} = \begin{bmatrix} 0 \\ 1 \\ 1 \\ -1 \end{bmatrix}$$

can be used to generate the same cyclic space U_1 and that

$$\hat{Y}_1 = Y_1 + Y_2 = \begin{bmatrix} 1 \\ 0 \\ 0 \\ 0 \end{bmatrix} + \begin{bmatrix} 2 \\ 3 \\ -1 \\ 0 \end{bmatrix} = \begin{bmatrix} 3 \\ 3 \\ -1 \\ 0 \end{bmatrix}$$

can be used to generate a different two-dimensional cyclic subspace of $\mathbb{N}[(A-I)^2]$. It is interesting that we can still choose Y_2 as the generator of U_2.

Since neither the vectors $Y_1, Y_2, \ldots, Y_h$ nor the cyclic spaces $U_1, U_2, \ldots, U_h$ which they generate are unique in the secondary decomposition theorem, it is not unreasonable to expect the values of $h, r_2, \ldots, r_h$ to depend on the choice of $U_1, U_2, \ldots, U_h$. In order to show that this is in fact not the case we introduce a second type of basis for a cyclic space.

Definition 1.12.37 Let $p(x)$ be an irreducible monic polynomial of degree l and let U be a dl-dimensional cyclic subspace of $\mathbb{N}[p(A)^e]$ with generator Y. Then the dl vectors

$$\begin{array}{ccccc}
Y, & AY, & A^2Y, & \ldots, & A^{l-1}Y \\
p(A)Y, & p(A)AY, & p(A)A^2Y, & \ldots, & p(A)A^{l-1}Y \\
\multicolumn{5}{c}{\dotfill} \\
p(A)^{d-1}Y, & p(A)^{d-1}AY, & p(A)^{d-1}A^2Y, & \ldots, & p(A)^{d-1}A^{l-1}Y
\end{array}$$

(1.12.38)

form a basis for U called the *canonical basis*.

When $d=1$, the cyclic basis and the canonical basis generated by Y are identical. For $d>1$ it is necessary to verify that the dl vectors of (1.12.38) do actually form a basis for U. Clearly these vectors belong to U, and since the dimension of U is dl, it is sufficient to show that they span U. For any $X \in U$ there exists a polynomial of degree less than dl, say $f(x)$, such that

$$X = f(A)Y$$

By repeated use of the division algorithm we can show that there exist polynomials $r_0(x), r_1(x), r_2(x), \ldots, r_{d-1}(x)$, each of degree less than l, such that

$$f(x) = r_0(x) + p(x)r_1(x) + p(x)^2 r_2(x) + \cdots + p(x)^{d-1} r_{d-1}(x)$$

(See Exercise 7 of Sec. 1.11.) Then

$$X = f(A)Y = r_0(A)Y + p(A)r_1(A)Y + p(A)^2 r_2(A)Y + \cdots + p(A)^{d-1} r_{d-1}(A)Y$$

and hence X can be expressed as a linear combination of the vectors of (1.12.38).

One advantage the canonical basis has over the cyclic basis for a dl-dimensional space U is that for each value of k, $k = 0, 1, 2, \ldots, d$, we can exhibit a basis for $p(A)^k \mathsf{U}$. The $(d-k)l$ vectors

$$
\begin{array}{lllll}
p(A)^k Y, & p(A)^k A Y, & p(A)^k A^2 Y, & \ldots, & p(A)^k A^{l-1} Y \\
p(A)^{k+1} Y, & p(A)^{k+1} A Y, & p(A)^{k+1} A^2 Y, & \ldots, & p(A)^{k+1} A^{l-1} Y \\
\cdots & \cdots & \cdots & & \cdots \\
p(A)^{d-1} Y, & p(A)^{d-1} A Y, & p(A)^{d-1} A^2 Y, & \ldots, & p(A)^{d-1} A^{l-1} Y
\end{array}
$$

(1.12.39)

are linearly independent elements of $p(A)^k \mathsf{U}$ since they constitute a subset of the canonical basis for U. In order to show that they form a basis for $p(A)^k \mathsf{U}$ it remains only to show that they span $p(A)^k \mathsf{U}$. If $Z \in p(A)^k \mathsf{U}$, then Z can be expressed as

$$Z = p(A)^k X$$

for some $X \in \mathsf{U}$. As before, we can express X as

$$X = f(A)Y = r_0(A)Y + p(A)r_1(A)Y + p(A)^2 r_2(A)Y + \cdots + p(A)^{d-1} r_{d-1}(A)Y$$

and hence

$$Z = p(A)^k X = p(A)^k r_0(A)Y + p(A)^{k+1} r_1(A)Y + \\ p(A)^{k+2} r_2(A)Y + \cdots + p(A)^{k+d-1} r_{d-1}(A)Y \quad (1.12.40)$$

By Lemma 1.12.24, the minimal polynomial of Y is $p(x)^d$, and consequently equation (1.12.40) becomes

$$Z = p(A)^k r_0(A)Y + p(A)^{k+1} r_1(A)Y + p(A)^{k+2} r_2(A)Y + \cdots + p(A)^{d-1} r_{d-k-1}(A)Y$$

(1.12.41)

Because each of the polynomials $r_0(x), r_1(x), \ldots, r_{d-1}(x)$ is of degree less than l, equation (1.12.41) expresses Z as a linear combination of the vectors of (1.12.39). Thus the vectors of (1.12.39) form a basis for $p(A)^k \mathsf{U}$, and we have $p(A)^k \mathsf{U}$ as a $(d-k)$ l-dimensional cyclic space with generator $p(A)^k Y$.

Making use of the canonical basis for a cyclic space, we can now prove:

Theorem 1.12.42 The numbers $h, r_2, r_3, \ldots, r_h$ of the secondary decomposition theorem are uniquely determined by A.

PROOF The method just used to determine the dimension of each of the vector spaces $p(A)^k \mathsf{U}$, where U is a cyclic subspace of $\mathsf{N}[p(A)^e]$, can also be used to find the dimension of the vector spaces $p(A)^k\{\mathsf{N}[p(A)^e]\}$, $k = 0, 1, 2, \ldots, e$. Since

$$\mathsf{N}[p(A)^e] = \mathsf{U}_1 \oplus \mathsf{U}_2 \oplus \cdots \oplus \mathsf{U}_h$$

and $\mathsf{U}_1, \mathsf{U}_2, \ldots, \mathsf{U}_h$ are A-invariant, it follows that for each k we have

$$p(A)^k\{\mathsf{N}[p(A)^e]\} = p(A)^k \mathsf{U}_1 \oplus p(A)^k \mathsf{U}_2 \oplus \cdots \oplus p(A)^k \mathsf{U}_h$$

and hence

$$\dim(p(A)^k\{\mathsf{N}[p(A)^e]\}) = \dim[p(A)^k \mathsf{U}_1] + \dim[p(A)^k \mathsf{U}_2] + \cdots$$
$$+ \dim[p(A)^k \mathsf{U}_h] \quad (1.12.43)$$

In the discussion above we showed that for a cyclic space U of dimension dl,

$$\dim[p(A)^k \mathsf{U}] = \begin{cases} (d-k)l & k = 0, 1, 2, \ldots, d \\ 0 & k \geq d \end{cases} \quad (1.12.44)$$

Let us define $t_e, t_{e-1}, \ldots, t_1$ by

$$e = r_1 \geq r_2 \geq \cdots \geq r_{t_k} \geq k > r_{t_k+1} \geq \cdots \geq r_h \geq 1 \quad (1.12.45)$$

and set $t_{e+1} = 0$; i.e., t_k is the largest integer for which $r_{t_k} \geq k$. Note that $t_1 = h$. If we set

$$s_k = t_k - t_{k+1}$$

then among the cyclic spaces $\mathsf{U}_1, \mathsf{U}_2, \ldots, \mathsf{U}_h$ there are exactly s_k whose dimension is kl. Substituting equation (1.12.44) into equation (1.12.43) yields

$$\dim(p(A)^k\{\mathsf{N}[p(A)^e]\}) = [(r_1 - k) + (r_2 - k) + \cdots + (r_{t_k} - k)]l \quad (1.12.46)$$

Clearly A uniquely determines the dimension of the $e + 1$ vector spaces

$$p(A)^k\{\mathsf{N}[p(A)^e]\} \quad k = 0, 1, 2, \ldots, e$$

and it remains to prove that $h, r_2, r_3, \ldots, r_h$ are uniquely determined by the numbers on the right-hand side of the various equations of (1.12.46). Setting $k = e$ in (1.12.46) produces no information. However, setting $k = e - 1$, we see that

$$\dim(p(A)^{e-1}\{\mathsf{N}[p(A)^e]\}) = t_e l = s_e l$$

which proves that s_e is completely determined by A. Similarly, setting $k = e - 2$ in (1.12.46), we have

$$\dim(p(A)^{e-2}\{\mathbb{N}[p(A)^e]\}) = (2s_e + s_{e-1})l$$

from which we can calculate s_{e-1}. Continuing in this way, we see that the dimensions of the cyclic spaces $\mathbb{U}_1, \mathbb{U}_2, \ldots, \mathbb{U}_h$ can be obtained from the equations of (1.12.46). It follows immediately that h is also uniquely determined by A. ////

EXAMPLE 1.12.47 Let

$$A = \begin{bmatrix} -1 & 2 & 2 & 8 \\ 1 & 1 & 2 & 0 \\ 1 & -1 & 0 & -4 \\ -1 & 1 & 1 & 5 \end{bmatrix}$$

We have seen that the minimal polynomial of A is

$$m(x) = (x - 1)^2(x - 2)$$

(see Example 1.12.35) and that

$$\left\{ \begin{bmatrix} 1 \\ 0 \\ 0 \\ 0 \end{bmatrix}, \begin{bmatrix} 0 \\ 3 \\ -1 \\ 0 \end{bmatrix}, \begin{bmatrix} 0 \\ 4 \\ 0 \\ -1 \end{bmatrix} \right\}$$

is a basis for $\mathbb{N}[(A - I)^2]$. Multiplying each of these vectors by $A - I$, we see that

$$\left\{ \begin{bmatrix} -2 \\ 1 \\ 1 \\ -1 \end{bmatrix} \right\}$$

is a basis for $(A - I)\{\mathbb{N}[(A - I)^2]\}$. Thus, with $p(x) = x - 1$, $e = 2$, and $l = 1$ and setting $k = 1, 0$ we have

$$k = 1: \quad \dim[(A - I)\{\mathbb{N}[(A - I)^2]\}] = 1 = s_2$$
$$k = 0: \quad \dim\{\mathbb{N}[(A - I)^2]\} = 3 = 2s_2 + s_1$$

verifying that $\mathbb{N}[(A - I)^2]$ can be expressed as the direct sum of a two-dimensional cyclic space and a one-dimensional cyclic space. ////

Since the numbers $h, r_1 = e, r_2, \ldots, r_h$ are uniquely determined by A, we can make the following definition.

Definition 1.12.48 Let $p(x)$ be an irreducible monic polynomial which divides the minimal polynomial of A. The polynomials $p(x)^{r_1}, p(x)^{r_2}, \ldots, p(x)^{r_h}$ of the secondary decomposition theorem are called the $p(x)$-*elementary divisors of* A. The totality of $p(x)$-elementary divisors of A for all irreducible monic polynomials $p(x)$ which divide $m(x)$, that is, the polynomials $p_1(x), p_2(x), \ldots, p_k(x)$ of equation (1.12.2), is called the *set of elementary divisors of* A.

The $p(x)$-elementary divisor $p(x)^{r_j}$ arises from the $r_j l$-dimensional cyclic subspace U_j of $\mathsf{N}[p(A)^e]$ of the secondary decomposition theorem.[1] The elementary divisors of the matrix of Example 1.12.47 are $(x-1)^2$, $x-1$, and $x-2$.

Corollary 1.12.49 Similar matrices have the same elementary divisors.

PROOF Let $B = P^{-1}AP$. Then A and B have the same minimal polynomial. Moreover, if

$$\mathsf{N}[p(A)^e] = \mathsf{U}_1 \oplus \mathsf{U}_2 \oplus \cdots \oplus \mathsf{U}_h$$

where U_i is a cyclic space with generator Y_i, then

$$\mathsf{N}[p(B)^e] = P^{-1}\{\mathsf{N}[p(A)^e]\} = P^{-1}\mathsf{U}_1 \oplus P^{-1}\mathsf{U}_2 \oplus \cdots \oplus P^{-1}\mathsf{U}_h$$

and $P^{-1}\mathsf{U}_i$ is a cyclic space with generator $P^{-1}Y_i$. (See Exercise 16 of Sec. 1.6.) The cyclic spaces U_i and $P^{-1}\mathsf{U}_i$ have the same dimension and therefore give rise to the same elementary divisor. ////

Corollary 1.12.50 Let A be a square matrix. Then A and A^T have the same elementary divisors.

PROOF A and A^T have the same minimal polynomial. It follows from the proof of Theorem 1.12.42 that in order to prove that A and A^T have the same elementary divisors it is sufficient to establish

$$\dim(p(A)^k\{\mathsf{N}[p(A)^e]\}) = \dim(p(A^T)^k\{\mathsf{N}[p(A^T)^e]\}) \qquad k = 1, 2, \ldots, e$$

Clearly

$$\mathsf{N}[p(A)^k] \cap \mathsf{N}[p(A)^e] = \mathsf{N}[p(A)^k]$$

so that $\qquad \dim\{\mathsf{N}[p(A)^k] \cap \mathsf{N}[p(A)^e]\} = \nu[p(A)^k] \qquad (1.12.51)$

---

[1] For an entirely different approach to the study of elementary divisors see Albert [1].

Making use of Exercise 19 of Sec. 1.5 (with $\mathbb{V} = \mathbb{N}[p(A)^e]$) and equation (1.12.51), we have

$$\dim(p(A)^k\{\mathbb{N}[p(A)^e]\}) = v[p(A)^e] - v[p(A)^k]$$
$$= v[p(A^T)^e] - v[p(A^T)^k] = \dim(p(A^T)^k\{\mathbb{N}[p(A^T)^e]\})$$

as required. ////

The minimal polynomial of A is equal to the least common multiple of the elementary divisors of A. It follows from the secondary decomposition theorem that

$$v[p(A)^e] = \dim(\mathbb{N}[p(A)^e]) = \dim \mathbb{U}_1 + \dim \mathbb{U}_2 + \cdots + \dim \mathbb{U}_h$$
$$= \deg p(x)^{r_1} + \deg p(x)^{r_2} + \cdots + \deg p(x)^{r_h}$$
$$= \deg[p(x)^{r_1} p(x)^{r_2} \cdots p(x)^{r_h}] = (r_1 + r_2 + \cdots + r_h)l$$

The degree of the product of the $p(x)$-elementary divisors of A is equal to $v[p(A)^e]$, and, combining this result with equation (1.12.7), we see that the degree of the product of all the elementary divisors of A is equal to the order of A. Since the degree of the least common multiple of a set of polynomials is at most the degree of the product of these polynomials, we can now conclude:

Theorem 1.12.52 The degree of the minimal polynomial of a matrix of order n is at most n.

The product of all the elementary divisors of A is a monic polynomial whose degree is n. In the next section we will show that, up to a possible change in sign, the product of all the elementary divisors of A is equal to the characteristic polynomial of A.

EXERCISES

1 Find the minimal polynomial and the elementary divisors of each of the following matrices:

(a) $\begin{bmatrix} 1 & 0 \\ 0 & 1 \end{bmatrix}$ (b) $\begin{bmatrix} 1 & 0 \\ 0 & 2 \end{bmatrix}$ (c) $\begin{bmatrix} 1 & 1 \\ 0 & 1 \end{bmatrix}$ (d) $\begin{bmatrix} 1 & 0 \\ 1 & 1 \end{bmatrix}$

(e) $\begin{bmatrix} 1 & 1 \\ 0 & 2 \end{bmatrix}$ (f) $\begin{bmatrix} 1 & 1 \\ 1 & 1 \end{bmatrix}$ (g) $\begin{bmatrix} 1 & -1 \\ -1 & 1 \end{bmatrix}$ (h) $\begin{bmatrix} 1 & 2 \\ 4 & 3 \end{bmatrix}$

In each case express $\mathbb{F}^2$ as a direct sum of irreducible cyclic spaces.

2   For each of the following matrices

(a) $\begin{bmatrix} -1 & 1 \\ -2 & 1 \end{bmatrix}$ (b) $\begin{bmatrix} 1 & -2 \\ 2 & 1 \end{bmatrix}$

show that $\mathbb{F}^2$ can be expressed as a direct sum of two one-dimensional cyclic spaces when $\mathbb{F}$ is the complex field but cannot be so expressed when $\mathbb{F}$ is the real field.   Show that $\mathbb{F}^2$ is a cyclic space with generator

$$Y = \begin{bmatrix} 1 \\ 0 \end{bmatrix}$$

when $\mathbb{F}$ is either the real or complex field. (Note that when $\mathbb{F}$ is the complex field, $\mathbb{F}^2$ is a two-dimensional cyclic space even though it can be expressed as the direct sum of two one-dimensional cyclic spaces; see Exercises 15 and 17.)

3   Find the minimal polynomial and the elementary divisors of the matrix

$$A = \begin{bmatrix} -3 & 1 & 3 \\ 3 & -1 & -3 \\ -5 & 1 & 5 \end{bmatrix}$$

Show that A is similar to a diagonal matrix and use Corollary 1.12.49 to find the minimal polynomial of A. Express $\mathbb{F}^3$ as a direct sum of three one-dimensional cyclic spaces.

4   Find the minimal polynomial and the elementary divisors of the matrix

$$A = \begin{bmatrix} 0 & 0 & 1 \\ 1 & -1 & 1 \\ -1 & 0 & -2 \end{bmatrix}$$

Show that A is similar to the direct sum of a 1×1 matrix and a 2×2 matrix but is not similar to a diagonal matrix. Express $\mathbb{F}^3$ as a direct sum of cyclic spaces. (*Hint:* Use Theorem 1.13.16 to find the minimal polynomial of A.)

5   Let

$$A = \begin{bmatrix} 2 & -2 & 1 \\ 1 & 5 & -1 \\ 2 & 3 & 2 \end{bmatrix}$$

Show that the minimal polynomial of A is $(x - 3)^3$. Show that $\mathbb{F}^3$ is a cyclic space with respect to A and that

$$Y = \begin{bmatrix} 1 \\ 0 \\ 0 \end{bmatrix}$$

is a generator. Show also that

$$\tilde{Y} = \begin{bmatrix} 0 \\ 0 \\ 1 \end{bmatrix}$$

does not generate $\mathbb{F}^3$. Find the canonical and cyclic bases for $\mathbb{F}^3$ using Y and obtain $P^{-1}AP$ where:

(a) P is the 3×3 matrix whose columns are the vectors in the canonical basis.

(b) P is the 3×3 matrix whose columns are the vectors in the cyclic basis.

6 Let

$$A = \begin{bmatrix} 1 & 1 & 0 & 0 \\ -2 & 0 & 1 & 0 \\ 2 & 0 & 0 & 1 \\ -2 & -1 & -1 & -1 \end{bmatrix}$$

Show that the minimal polynomial of A is $(x^2 + 1)^2$. Show that if $\mathbb{F}$ is the real field then $\mathbb{F}^4$ is a cyclic space. Find a generator for $\mathbb{F}^4$. Find a nonzero vector of $\mathbb{F}^4$ which does *not* generate $\mathbb{F}^4$. Using the generator previously found, obtain the cyclic and canonical bases for $\mathbb{F}^4$. Apply the method of (1.12.39) to construct a canonical basis for $(A^2 + I)\mathbb{F}^4$. Find $P^{-1}AP$ when:

(a) P is the 4×4 matrix whose columns are the vectors of the canonical basis of $\mathbb{F}^4$.

(b) P is the 4×4 matrix whose columns are the vectors of the cyclic basis of $\mathbb{F}^4$.

Show that if $\mathbb{F}$ is the complex field then $\mathbb{F}^4$ can be expressed as the direct sum of two two-dimensional cyclic subspaces.

7 Let A be a diagonal matrix and let the distinct diagonal entries of A be $\lambda_1, \lambda_2, \ldots, \lambda_k$. Prove that

$$m(x) = (x - \lambda_1)(x - \lambda_2) \cdots (x - \lambda_k)$$

8 Let $f(x)$ be a factor of the minimal polynomial of A of degree at least 1. Show that $f(A)$ is singular.

9 Let $p(x)$ be an irreducible polynomial such that $p(A)$ is singular. Show that $p(x)$ is a factor of the minimal polynomial of A.

10 Prove that a matrix and its transpose have the same minimal polynomial.

11 Prove that similar matrices have the same minimal polynomial.

12 Prove that the minimal polynomial of a direct sum of matrices is equal to the least common multiple of the minimal polynomials of the individual matrices. Prove that the elementary divisors of a direct sum of matrices is equal to the union of the elementary divisors of the individual matrices.

13 Let
$$d(x) = (f_1(x), f_2(x), \ldots, f_k(x))$$

Prove that
$$\mathbb{N}[d(A)] = \mathbb{N}[f_1(A)] \cap \mathbb{N}[f_2(A)] \cap \cdots \cap \mathbb{N}[f_k(A)]$$

14 Let $p(x)$ be an irreducible polynomial and let $p(x)^e$, $e \geq 1$, be the highest power of $p(x)$ which divides the minimal polynomial of A. Prove that
$$\mathbb{N}[p(A)^f] = \mathbb{N}[p(A)^{f+1}]$$
if and only if $f \geq e$.

15 Let $p_1(x)$ and $p_2(x)$ be distinct irreducible monic polynomials of degree l_1 and l_2 respectively. Let $\mathbb{U}_1$ and $\mathbb{U}_2$ be cyclic subspaces of $\mathbb{N}[p_1(A)^{e_1}]$ and $\mathbb{N}[p_2(A)^{e_2}]$ of dimension $d_1 l_1$ and $d_2 l_2$ with generators Y_1 and Y_2 respectively. Prove that $Y_1 + Y_2$ generates a cyclic subspace of $\mathbb{F}^n$ of dimension $d_1 l_1 + d_2 l_2$.

16 Let $\mathbb{U}$ be a cyclic subspace of $\mathbb{N}[p(A)^e]$ with generator Y and let $q(x)$ be a polynomial which is relatively prime to $p(x)$. Show that $Z = q(A)Y$ is also a generator of $\mathbb{U}$.

17 Let $p(x)$ be an irreducible polynomial, let $\mathbb{U}$ be a cyclic subspace of $\mathbb{N}[p(A)^e]$ with generator Y, and let $\mathbb{V}$ be a nonzero A-invariant subspace of $\mathbb{U}$. Prove that $\mathbb{V}$ has the form
$$\mathbb{V} = p(A)^k \mathbb{U}$$
for some k, $0 \leq k < e$, and that $\mathbb{V}$ is a cyclic space and is generated by $W = p(A)^k Y$. Show how it now follows that $\mathbb{U}$ is irreducible. [*Hint:* Let k be defined to be the largest integer for which there exists a vector $Z \in \mathbb{V}$ such that $p(A)^k Z = 0$ and $p(A)^{k-1} Z \neq 0$.] Is this result still true when $p(x)$ is not assumed to be irreducible? (See Exercise 15.)

18 Let A be a square matrix of order n. Prove that $\mathbb{F}^n$ is a cyclic space with respect to A if and only if the minimal polynomial of A has degree n.

19 Let A be a square matrix of order n. Prove that there exists a vector in $\mathbb{F}^n$ whose minimal polynomial is equal to the minimal polynomial of A.

1.13 THE CANONICAL FORMS

At the end of Sec. 1.10 we raised two problems. An answer to the first, namely, how to decompose $\mathbb{F}^n$ into a direct sum of A-invariant subspaces, was provided by the two decomposition theorems of Sec. 1.12. We have now expressed $\mathbb{F}^n$ as a direct sum of cyclic spaces corresponding to the elementary divisors of A. Since the elementary divisors are powers of irreducible polynomials, it follows from Exercise 17 of Sec. 1.12 that the cyclic spaces which correspond to them are irreducible. Also in Sec. 1.12 we introduced two special types of bases for cyclic subspaces of $\mathbb{N}[p(A)^e]$. In this section we will see that each of these bases provides an answer to the second problem raised at the end of Sec. 1.10, namely, how to make the forms of the matrices $B_1, B_2, \ldots, B_k$ of equation (1.10.32) as simple as possible.

Let us assume that for a given matrix A, each of the subspaces $\mathbb{N}[p_i(A)^{e_i}]$ of the primary decomposition theorem has been expressed as in the secondary decomposition theorem

$$\mathbb{N}[p_i(A)^{e_i}] = \mathbb{U}_{i1} \oplus \mathbb{U}_{i2} \oplus \cdots \oplus \mathbb{U}_{ih_i}$$

If for each of the subspaces $\mathbb{U}_{ij}$, $j = 1, 2, \ldots, h_i$, of $\mathbb{N}[p_i(A)^{e_i}]$ we select some basis, say $\mathbb{Y}_{ij}$, then the union of these bases

$$\mathbb{X}_i = \mathbb{Y}_{i1} \cup \mathbb{Y}_{i2} \cup \cdots \cup \mathbb{Y}_{ih_i} \qquad (1.13.1)$$

is a basis for $\mathbb{N}[p_i(A)^{e_i}]$. Similarly, if for each $\mathbb{N}[p_i(A)^{e_i}]$, $i = 1, 2, \ldots, k$, we select the basis $\mathbb{X}_i$ of (1.13.1), then their union

$$\mathbb{X} = \mathbb{X}_1 \cup \mathbb{X}_2 \cup \cdots \cup \mathbb{X}_k$$

is a basis for $\mathbb{F}^n$. Finally, if we form the matrix P from the vectors of $\mathbb{X}$, then P is nonsingular and $P^{-1}AP$ is a direct sum of k matrices $B_1, B_2, \ldots, B_k$, one for each of the subspaces $\mathbb{N}[p_i(A)^{e_i}]$. Moreover, each of the B_i's is itself the direct sum of h_i matrices, say

$$B_i = C_{i1} \oplus C_{i2} \oplus \cdots \oplus C_{ih_i}$$

one for each of the cyclic subspaces of $\mathbb{N}[p_i(A)^{e_i}]$. Thus $P^{-1}AP$ has the form

$$\begin{aligned} P^{-1}AP &= B_1 \oplus B_2 \oplus \cdots \oplus B_k \\ &= \{C_{11} \oplus C_{12} \oplus \cdots \oplus C_{1h_1}\} \oplus \{C_{21} \oplus C_{22} \oplus \cdots \oplus C_{2h_2}\} \\ &\quad \oplus \cdots \oplus \{C_{k1} \oplus C_{k2} \oplus \cdots \oplus C_{kh_k}\} \end{aligned} \qquad (1.13.2)$$

where the matrices $C_{i1}, C_{i2}, \ldots, C_{ih_i}$ correspond to the $p_i(x)$-elementary divisors of A. We will investigate the form of each matrix C_{ij} of equation (1.13.2) when $\mathbb{Y}_{ij}$ is the cyclic basis for $\mathbb{U}_{ij}$ and when $\mathbb{Y}_{ij}$ is the canonical basis for $\mathbb{U}_{ij}$.

We begin with a dl-dimensional cyclic subspace $\mathbb{U}$ of $\mathbb{N}[p(A)^e]$ with generator Y. The matrix C which corresponds to $\mathbb{U}$ [and to the elementary divisor $p(x)^d$] will have the simplest form when we use the cyclic basis

$$\mathbb{Y} = \{Y, AY, A^2Y, \ldots, A^{dl-1}Y\} \quad (1.13.3)$$

for $\mathbb{U}$. Let us set

$$p(x)^d = x^{dl} + \alpha_{dl-1}x^{dl-1} + \alpha_{dl-2}x^{dl-2} + \cdots + \alpha_1 x + \alpha_0 \quad (1.13.4)$$

Multiplying each vector belonging to the basis $\mathbb{Y}$ of (1.13.3) on the left by A and expressing the resulting vector (which, of course, is contained in $\mathbb{U}$) as a linear combination of the vectors of $\mathbb{Y}$, we have

$$\begin{aligned} A \cdot Y &= 0Y + 1AY + 0A^2Y + \cdots + 0A^{dl-1}Y \\ A \cdot AY &= 0Y + 0AY + 1A^2Y + \cdots + 0A^{dl-1}Y \\ &\cdots \\ A \cdot A^{dl-2}Y &= 0Y + 0AY + 0A^2Y + \cdots + 1A^{dl-1}Y \\ A \cdot A^{dl-1}Y &= -\alpha_0 Y - \alpha_1 AY - \alpha_2 A^2 Y - \cdots - \alpha_{dl-1}A^{dl-1}Y \end{aligned} \quad (1.13.5)$$

The entries of the matrix C which corresponds to $\mathbb{U}$ are the coefficients in the equations of (1.13.5). Thus

$$C = \begin{bmatrix} 0 & 0 & 0 & \cdots & 0 & -\alpha_0 \\ 1 & 0 & 0 & \cdots & 0 & -\alpha_1 \\ 0 & 1 & 0 & \cdots & 0 & -\alpha_2 \\ \cdots & & & & & \\ 0 & 0 & 0 & \cdots & 1 & -\alpha_{dl-1} \end{bmatrix} \quad (1.13.6)$$

Each of the matrices C_{ij} of equation (1.13.2) has the form (1.13.6).

Definition 1.13.7 The matrix C of (1.13.6) is called the *companion matrix* of the monic polynomial

$$x^{dl} + \alpha_{dl-1}x^{dl-1} + \alpha_{dl-2}x^{dl-2} + \cdots + \alpha_1 x + \alpha_0$$

A matrix consisting of the direct sum of the companion matrices of the elementary divisors of A, in any order, is called the *rational canonical form* of A.

The order of the companion matrix of a polynomial is equal to the degree of the polynomial. Equation (1.13.6) and Definition 1.13.7 can be used to construct the companion matrix of any monic polynomial and hence the rational canonical form of any matrix. According to the discussion leading up to Definition 1.13.7, we have proved:

Theorem 1.13.8 A matrix is similar to its rational canonical form.

EXAMPLE 1.13.9 Let

$$A = \begin{bmatrix} 2 & 1 \\ 3 & 4 \end{bmatrix}$$

We have already shown (in Examples 1.10.16, 1.10.27, and 1.12.14) that the elementary divisors of A are $x - 1$ and $x - 5$, corresponding to the cyclic subspaces

$$\mathbb{U}_{11} = \mathbb{N}(A - I) \qquad \mathbb{U}_{21} = \mathbb{N}(A - 5I)$$

with generators

$$Y_{11} = \begin{bmatrix} 1 \\ -1 \end{bmatrix} \qquad Y_{21} = \begin{bmatrix} 1 \\ 3 \end{bmatrix}$$

respectively. The companion matrices of $x - 1$ and $x - 5$ are the 1×1 matrices [1] and [5] respectively. We then set

$$P = \begin{bmatrix} 1 & 1 \\ -1 & 3 \end{bmatrix}$$

and obtain

$$P^{-1}AP = \begin{bmatrix} 1 & 0 \\ 0 & 5 \end{bmatrix} \qquad ////$$

EXAMPLE 1.13.10 Let

$$A = \begin{bmatrix} 3 & 1 & -2 \\ -1 & -1 & 4 \\ 0 & -1 & 3 \end{bmatrix}$$

In Examples 1.10.37, 1.12.15, and 1.12.33 we showed that the elementary divisors of A are

$$(x - 2)^2 = x^2 - 4x + 4 \qquad x - 1$$

corresponding to the cyclic subspaces

$$\mathbb{U}_{11} = \mathbb{N}[(A - 2I)^2] \qquad \mathbb{U}_{21} = \mathbb{N}(A - I)$$

with generators

$$Y_{11} = \begin{bmatrix} 2 \\ -1 \\ 0 \end{bmatrix} \qquad Y_{21} = \begin{bmatrix} 0 \\ 2 \\ 1 \end{bmatrix}$$

respectively. The cyclic basis generated by Y_{11} is

$$\mathbb{Y}_1 = \left\{ \begin{bmatrix} 2 \\ -1 \\ 0 \end{bmatrix}, \begin{bmatrix} 5 \\ -1 \\ 1 \end{bmatrix} \right\}$$

The companion matrices of $(x-2)^2$ and $x-1$ are

$$\begin{bmatrix} 0 & -4 \\ 1 & 4 \end{bmatrix} \qquad [1]$$

and it is not difficult to verify that if we set

$$P = \begin{bmatrix} 2 & 5 & 0 \\ -1 & -1 & 2 \\ 0 & 1 & 1 \end{bmatrix}$$

then
$$P^{-1}AP = \begin{bmatrix} 0 & -4 & 0 \\ 1 & 4 & 0 \\ 0 & 0 & 1 \end{bmatrix} = \begin{bmatrix} 0 & -4 \\ 1 & 4 \end{bmatrix} \oplus [1] \qquad ////$$

In Corollary 1.12.49 we proved that similar matrices have the same elementary divisors. Conversely, according to Theorem 1.13.8, two matrices which have the same elementary divisors are each similar to the same rational canonical form and hence are similar to each other. Combining these results, we have a characterization of similar matrices; that is:

Theorem 1.13.11 Two matrices are similar if and only if they have the same elementary divisors.

Corollary 1.13.12 now follows from Corollary 1.12.50:

Corollary 1.13.12 Let A be a square matrix. Then A is similar to A^T.

Surprisingly, the computational problem of transforming A into A^T by means of a similarity is generally no easier than transforming A into any other matrix similar to it.

We can use Theorem 1.13.8 to obtain a great deal of information about any square matrix. Since the rational canonical form is a direct sum of matrices, we can investigate each of these individually. In particular, because of the simple form of the companion matrix we can easily compute its minimal polynomial and its characteristic polynomial.

Theorem 1.13.13 If C is the companion matrix of the qth degree monic polynomial $g(x)$, then the minimal polynomial of C is $g(x)$ and the characteristic polynomial of C is $(-1)^q g(x)$.

PROOF If we set

$$g(x) = x^q + \alpha_{q-1}x^{q-1} + \alpha_{q-2}x^{q-2} + \cdots + \alpha_1 x + \alpha_0$$

then
$$C = \begin{bmatrix} 0 & 0 & 0 & \cdots & -\alpha_0 \\ 1 & 0 & 0 & \cdots & -\alpha_1 \\ 0 & 1 & 0 & \cdots & -\alpha_2 \\ \cdots & \cdots & \cdots & \cdots & \cdots \\ 0 & 0 & 0 & \cdots & -\alpha_{q-1} \end{bmatrix}$$

$\mathbb{F}^q$ is a cyclic space with respect to C, and the q unit column vectors form a cyclic basis with generator $I^{(1)}$ since

$$CI^{(1)} = I^{(2)}, \quad C^2 I^{(1)} = CI^{(2)} = I^{(3)}, \quad \ldots, \quad C^{q-1}I^{(1)} = I^{(q)}$$

$$C^q I^{(1)} = CI^{(q)} = \begin{bmatrix} -\alpha_0 \\ -\alpha_1 \\ -\alpha_2 \\ \vdots \\ -\alpha_{q-1} \end{bmatrix} = (-\alpha_0 I - \alpha_1 C - \alpha_2 C^2 - \cdots - \alpha_{q-1}C^{q-1})I^{(1)}$$

Consequently the minimal polynomial of $I^{(1)}$ with respect to C is $g(x)$. By Lemma 1.12.25, $g(x)$ divides $m(x)$, the minimal polynomial of C. On the other hand, it follows from Theorem 1.12.52 that the degree of $m(x)$ is at most q. Thus we have shown that $m(x) = g(x)$.

We use an induction on q to prove that the characteristic polynomial of C is $(-1)^q g(x)$. First, let us set $q = 1$, say

$$g(x) = x + \alpha_0$$

Then $\qquad C = [-\alpha_0] \qquad C - xI = [-\alpha_0 - x]$

and the result is immediate. Now assume that the result has been established for all polynomials of degree $q - 1$ and that

$$g(x) = x^q + \alpha_{q-1}x^{q-1} + \cdots + \alpha_2 x^2 + \alpha_1 x + \alpha_0$$

When C is the companion matrix of $g(x)$, we have

$$C - xI = \begin{bmatrix} -x & 0 & 0 & \cdots & -\alpha_0 \\ 1 & -x & 0 & \cdots & -\alpha_1 \\ 0 & 1 & -x & \cdots & -\alpha_2 \\ \cdots & \cdots & \cdots & \cdots & \cdots \\ 0 & 0 & 0 & \cdots & -x-\alpha_{q-1} \end{bmatrix}$$

and, expanding the determinant of $C - xI$ along the first row, we see that

$$\det(C - xI) = (-1)^2(-x) \det \begin{bmatrix} -x & 0 & \cdots & -\alpha_1 \\ 1 & -x & \cdots & -\alpha_2 \\ \cdots\cdots\cdots\cdots\cdots\cdots\cdots\cdots \\ 0 & 0 & \cdots & -x-\alpha_{q-1} \end{bmatrix}$$

$$+ 0 + \cdots + 0 + (-1)^{q+1}(-\alpha_0) \det \begin{bmatrix} 1 & -x & 0 & \cdots & 0 \\ 0 & 1 & -x & \cdots & 0 \\ \cdots\cdots\cdots\cdots\cdots\cdots\cdots\cdots \\ 0 & 0 & 0 & \cdots & 1 \end{bmatrix} \quad (1.13.14)$$

The first determinant on the right-hand side of equation (1.13.14) is the determinant of a matrix of the form $\tilde{C} - xI$, where $\tilde{C}$ is the companion matrix of

$$\tilde{g}(x) = x^{q-1} + \alpha_{q-1} x^{q-2} + \cdots + \alpha_2 x + \alpha_1$$

By the induction hypothesis,

$$\det(C - xI) = -x[(-1)^{q-1}\tilde{g}(x)] + (-1)^q \alpha_0 = (-1)^q g(x) \qquad ////$$

Corollary 1.13.15 If $p(x)$ is an irreducible monic polynomial, then the companion matrix of $p(x)^e$ has the single elementary divisor $p(x)^e$.

PROOF We have just shown that $I^{(1)}$ generates a cyclic space (with respect to C) whose dimension is equal to the degree of $p(x)^e$. ////

In Sec. 1.12 we proved that the minimal polynomial of a matrix is the least common multiple of its elementary divisors. Theorem 1.13.13 permits us to express the characteristic polynomial of a matrix in terms of its elementary divisors.

Theorem 1.13.16 The characteristic polynomial of a matrix of order n is equal to the product of its elementary divisors multiplied by $(-1)^n$.

PROOF Since similar matrices have the same characteristic polynomial and the same elementary divisors, it is sufficient to find the characteristic polynomial of a matrix A which is in the rational canonical form. According to Exercise 17 of Sec. 1.10, the characteristic polynomial of A is equal to the product of the characteristic polynomials of the companion matrices of the elementary divisors of A. However, by Theorem 1.13.13 and Corollary 1.13.15, the characteristic polynomial of the companion matrix of an elementary divisor of A is equal to the elementary divisor multiplied by $(-1)^q$, where q is the degree of the elementary divisor. ////

Since the minimal polynomial of a matrix is equal to the least common multiple of the elementary divisors, it divides the characteristic polynomial and we have proved the following celebrated theorem.

Theorem 1.13.17   The Hamilton-Cayley theorem[1]   A matrix satisfies its characteristic polynomial.

In this section we also wish to determine the form of the matrix C of (1.13.6) when the canonical basis for a cyclic space is used instead of the cyclic basis. Although the computations involved are somewhat more complicated in this case, the resulting matrix is considerably simpler and consequently more useful. Let

$$p(x) = x^l + \gamma_{l-1} x^{l-1} + \gamma_{l-2} x^{l-2} + \cdots + \gamma_1 x + \gamma_0$$

be an irreducible polynomial and let

$$\mathbb{Y} = \{Y, AY, A^2 Y, \ldots, A^{l-1} Y, p(A)Y, p(A)AY, \ldots, p(A)^{d-1} A^{l-1} Y\} \quad (1.13.18)$$

be the canonical basis for the dl-dimensional cyclic subspace $\mathbb{U}$ of $\mathbb{N}[p(A)^e]$ generated by $\mathbb{Y}$. [Note that $p(x)^d$ is the polynomial of (1.13.4).] Multiplying each vector belonging to the basis $\mathbb{Y}$ of (1.13.18) by A and expressing the resulting vector (which, of course, belongs to $\mathbb{U}$) as a linear combination of the vectors of $\mathbb{Y}$, we have

$$
\begin{aligned}
A \cdot Y = \quad & 0Y + 1AY + 0A^2 Y + \cdots + 0A^{l-1} Y + 0p(A)Y \\
& + 0p(A)AY + \cdots + 0p(A)^{d-1} A^{l-1} Y \\
A \cdot AY = \quad & 0Y + 0AY + 1A^2 Y + \cdots + 0A^{l-1} Y + 0p(A)Y \\
& + 0p(A)AY + \cdots + 0p(A)^{d-1} A^{l-1} Y \\
\cdots \cdots & \cdots \cdots \cdots \cdots \cdots \cdots \cdots \cdots \cdots \cdots \cdots \cdots \cdots \cdots \\
A \cdot A^{l-1} Y = & -\gamma_0 Y - \gamma_1 AY - \gamma_2 A^2 Y - \cdots - \gamma_{l-1} A^{l-1} Y + 1p(A)Y \\
& + 0p(A)AY + \cdots + 0p(A)^{d-1} A^{l-1} Y \\
A \cdot p(A)Y = \quad & 0Y + 0AY + 0A^2 Y + \cdots + 0A^{l-1} Y + 0p(A)Y \\
& + 1p(A)AY + \cdots + 0p(A)^{d-1} A^{l-1} Y \\
\end{aligned}
$$
$$(1.13.19)$$
$$
\begin{aligned}
\cdots \cdots & \cdots \cdots \cdots \cdots \cdots \cdots \cdots \cdots \cdots \cdots \cdots \cdots \cdots \cdots \\
A \cdot p(A)^{d-1} A^{l-2} Y = \quad & 0Y + 0AY + 0A^2 Y + \cdots + 0A^{l-1} Y + 0p(A)Y \\
& + 0p(A)AY + \cdots + 1p(A)^{d-1} A^{l-1} Y \\
A \cdot p(A)^{d-1} A^{l-1} Y = \quad & 0Y + 0AY + 0A^2 Y + \cdots + 0A^{l-1} Y + 0p(A)Y \\
& + 0p(A)AY + \cdots \\
& - \gamma_0 p(A)^{d-1} Y - \gamma_1 p(A)^{d-1} AY - \cdots - \gamma_{l-1} p(A)^{d-1} A^{l-1} Y
\end{aligned}
$$

---

[1] This theorem was first proved for quaternions by Hamilton in 1853. In 1858 Cayley announced the theorem for all square matrices though he presented proofs only for 2×2 and 3×3 matrices. Since then many different proofs have been given.

From the coefficients in equations (1.13.19) we obtain

$$C = \begin{bmatrix} 0 & 0 & 0 & \cdots & 0 & -\gamma_0 & & & & & & & & & & & \\ 1 & 0 & 0 & \cdots & 0 & -\gamma_1 & & & & & & & & & & & \\ 0 & 1 & 0 & \cdots & 0 & -\gamma_2 & & & & & & & & & & & \\ \cdots & \cdots & \cdots & \cdots & \cdots & \cdots & & & & & & & & & & & \\ 0 & 0 & 0 & \cdots & 1 & -\gamma_{l-1} & & & & & & & & & & & \\ & & & & & 1 & 0 & 0 & 0 & \cdots & 0 & -\gamma_0 & & & & & \\ & & & & & & 1 & 0 & 0 & \cdots & 0 & -\gamma_2 & & & & & \\ & & & & & & 0 & 1 & 0 & \cdots & 0 & -\gamma_2 & & & & & \\ & & & & & & \cdots & \cdots & \cdots & \cdots & \cdots & \cdots & & & & & \\ & & & & & & 0 & 0 & 0 & \cdots & 1 & -\gamma_{l-1} & & & & & \\ & & & & & & & & & & & 1 & & & & & \\ & & & & & & & & & & & \cdots & & & & & \\ & & & & & & & & & & & 1 & 0 & 0 & 0 & \cdots & 0 & -\gamma_0 \\ & & & & & & & & & & & & 1 & 0 & 0 & \cdots & 0 & -\gamma_1 \\ & & & & & & & & & & & & 0 & 1 & 0 & \cdots & 0 & -\gamma_2 \\ & & & & & & & & & & & & \cdots & \cdots & \cdots & \cdots & \cdots & \cdots \\ & & & & & & & & & & & & 0 & 0 & 0 & \cdots & 1 & -\gamma_{l-1} \end{bmatrix} \quad (1.13.20)$$

that is, C is a $dl \times dl$ matrix and consists of d copies of the $l \times l$ companion matrix of $p(x)$ in the diagonal block positions, 1s in each of the positions immediately below the last column of the companion matrix (except, of course, for the last of the companion matrices) and 0s elsewhere. Thus C contains 1s in all the $dl - 1$ positions just below the diagonal positions. The matrix C of (1.13.20) is called the *hypercompanion matrix* of the polynomial $p(x)^d$. A matrix consising of the direct sum of the hypercompanion matrices of the elementary divisors of A (in any order) is called the *classical canonical form of A*. Thus we also have proved.

Theorem 1.13.21 Every matrix is similar to its classical canonical form.

EXAMPLE 1.13.22 Let

$$A = \begin{bmatrix} 3 & 1 & -2 \\ -1 & -1 & 4 \\ 0 & -1 & 3 \end{bmatrix}$$

(See Examples 1.10.37, 1.12.15, 1.12.33, and 1.13.10). We have previously shown that

$$m(x) = (x - 2)^2(x - 1)$$

and that

$$\left\{ Y_1 = \begin{bmatrix} 2 \\ -1 \\ 0 \end{bmatrix}, AY_1 = \begin{bmatrix} 5 \\ -1 \\ 1 \end{bmatrix} \right\} \quad \left\{ \begin{bmatrix} 0 \\ 2 \\ 1 \end{bmatrix} \right\}$$

are cyclic bases for $\mathbb{N}[(A - 2I)^2]$ and $\mathbb{N}(A - I)$ respectively. The canonical basis for $\mathbb{N}[(A - 2I)^2]$ generated by Y_1 is given by

$$\left\{ Y_1 = \begin{bmatrix} 2 \\ -1 \\ 0 \end{bmatrix}, (A - 2I)Y_1 = \begin{bmatrix} 1 \\ 1 \\ 1 \end{bmatrix} \right\}$$

To obtain the classical canonical form of A we set

$$P = \begin{bmatrix} 2 & 1 & 0 \\ -1 & 1 & 2 \\ 0 & 1 & 1 \end{bmatrix}$$

and then

$$P^{-1}AP = \begin{bmatrix} 2 & 0 & 0 \\ 1 & 2 & 0 \\ 0 & 0 & 1 \end{bmatrix} = \begin{bmatrix} 2 & 0 \\ 1 & 2 \end{bmatrix} \oplus [1] \qquad ////$$

One special advantage of the classical canonical form occurs when the degree of $p(x)$ is 1. The companion matrix of $x - \lambda$ is the 1×1 matrix $[\lambda]$, and therefore the hypercompanion matrix of $(x - \lambda)^q$ is the $q \times q$ matrix

$$\begin{bmatrix} \lambda & 0 & 0 & \cdots & 0 & 0 \\ 1 & \lambda & 0 & \cdots & 0 & 0 \\ 0 & 1 & \lambda & \cdots & 0 & 0 \\ \multicolumn{6}{c}{\dotfill} \\ 0 & 0 & 0 & \cdots & 1 & \lambda \end{bmatrix} \qquad (1.13.23)$$

When each of the elementary divisors of A is of the form $(x - \lambda)^q$, the classical canonical form of A is the direct sum of matrices of the form (1.13.23) and is called the *Jordan canonical form of A*.

EXERCISES

1 Find the rational canonical form and the classical canonical form of each of the matrices of Exercise 1 of Sec. 1.12. In each case find a matrix P such that $P^{-1}AP$ is the rational canonical form of A. Find a matrix P such that $P^{-1}AP$ is the classical canonical form of A.

2. Repeat Exercise 1 using the matrices of Exercises 3 to 6 of Sec. 1.12.
3. The elementary divisors of A are x, x, x^2, $x^2 - 7x - 2$, and $x - 1$. What is the minimal polynomial of A? What is the characteristic polynomial of A? What is the rational canonical form of A? What is the classical canonical form of A?
4. Repeat Exercise 3 when the elementary divisors of A are $(x - 2)^2$, $(x - 2)^2$, and $(x - 2)^3$.
5. Prove that a matrix is similar to a diagonal matrix if and only if each of its elementary divisors is of the first degree.
6. Prove that the minimal polynomial of a matrix is equal to the characteristic polynomial (up to a possible change in sign) if and only if the elementary divisors are relatively prime in pairs.
7. Let $\mathbb{Y}$ of (1.13.3) be a cyclic basis for a cyclic space $\mathbb{U}$. Find the matrix C of (1.13.6) when the columns of P are the vectors of $\mathbb{Y}$, but in the reversed order. Find C when the columns of P are the vectors of a canonical basis, but in reversed order.
8. Let C be the companion matrix of a monic polynomial of degree n. Show that $\mathbb{F}^n$ is a cyclic space with respect to C with generator $I^{(1)}$ and is a cyclic space with respect to C^T with generator $I_{(n)}$.
9. Let $X \in \mathbb{F}_n$ and $Y \in \mathbb{F}^n$ such that $XY \neq 0$. Set $B = YX$. Prove that the elementary divisors of B are $x - XY$, x, x, $\ldots$, x. Also prove that the characteristic vectors of B are (a) the nonzero elements of $\mathbb{N}(X)$, corresponding to the characteristic root 0 and (b) Y, corresponding to the characteristic root XY.
10. Let
$$A = \begin{bmatrix} 0 & 1 \\ 2 & 1 \end{bmatrix}$$
Show that
$$A^n = \frac{2^n - (-1)^n}{3} A + \frac{2^n + 2(-1)^n}{3} I \qquad n = 0, 1, 2, \ldots \qquad (1.13.24)$$

[*Hint:* Show that for each n, $n \geq 2$, there exist scalars α_n and β_n and a polynomial $q_n(x)$ such that
$$x^n = q_n(x)(x^2 - x - 2) + \alpha_n x + \beta_n$$
Where did $x^2 - x - 2$ come from? It then follows that
$$A^n = \alpha_n A + \beta_n I \qquad n = 0, 1, 2, \ldots$$
Show that α_n and β_n can be evaluated by comparing the characteristic roots of A^n and those of $\alpha_n A + \beta_n I$. See Exercises 18 and 19 of Sec. 1.10].

Note that equation (1.13.24) can easily be verified by means of an induction on n. However, an induction can be used only when the coefficients are known.

11 Let $a_1 = 4$, $a_2 = 2$ and let $a_3, a_4, a_5, \ldots$ be defined recursively by

$$a_{n+2} = a_{n+1} + 2a_n \qquad n = 1, 2, \ldots$$

Prove that

$$a_n = \frac{2^{n-1} - (-1)^{n-1}}{3} 2 + \frac{2^{n-1} + 2(-1)^{n-1}}{3} 4 \qquad n = 1, 2, \ldots$$

[Hint: Set

$$X_n = \begin{bmatrix} a_n \\ a_{n+1} \end{bmatrix} \qquad n = 1, 2, \ldots$$

Show that

$$X_n = A^{n-1} X_1$$

where A is the matrix of Exercise 10.]

12 The Fibonacci numbers

$$1, 1, 2, 3, 5, 8, 13, 21, 34, \ldots$$

are defined recursively by

$$a_1 = a_2 = 1 \qquad a_{n+2} = a_{n+1} + a_n \qquad n = 1, 2, \ldots$$

Show that

$$a_n = \frac{[(1 + \sqrt{5})/2]^n - [(1 - \sqrt{5})/2]^n}{\sqrt{5}} \qquad n = 1, 2, \ldots$$

(a) by the method of Exercise 11 and (b) by an induction on n.

REFERENCES

The standard one-semester course in linear algebra is designed to contain, in addition to all the material of this chapter, a discussion of inner-product spaces; see Halmos [6, chap. 3] or Perlis [8, chap. 9], among many others. This basic course prepares the reader to explore several quite different advanced topics in linear algebra. We will name just a few. Greub [5] presents a detailed and highly abstract account of multilinear algebra. Gantmacher's treatise [2] considers a large number of classical topics, among which are the Perron-Frobenius theorem and the application of matrices to systems of linear differential equations. Graybill [3] presents a careful discussion of the generalized inverse of a matrix and summarizes in textbook form many important research papers of recent years. Taylor [9] develops the theory of vector spaces of arbitrary dimension, pointing out both the similarities and the differences with the finite-dimensional case studied here.

1 Albert, A. A.: "Introduction to Algebraic Theories," University of Chicago Press, Chicago, 1941.
2 Gantmacher, F. R.: "The Theory of Matrices," vols. I and II, trans. from the Russian by K. A. Hirsch, Chelsea Publishing Company, New York, 1959.
3 Graybill, F. A.: "Introduction to Matrices with Applications in Statistics, "Wadsworth Publishing Company, Belmont, Calif. 1969.
4 Greub, W. H.: "Linear Algebra," 3d ed., Springer-Verlag New York Inc., New York, 1967.
5 Greub, W. H.: "Multilinear Algebra," Springer-Verlag New York Inc., New York, 1967.
6 Halmos, P. R.: "Finite-dimensional vector spaces," 2d ed., D. Van Nostrand Company, Inc., Princeton, N.J., 1958.
7 Hoffman, K., and R. Kunze: "Linear Algebra," 2d ed., Prentice-Hall, Inc., Englewood Cliffs, N.J., 1971.
8 Perlis, S.: "Theory of Matrices," Addison-Wesley Press, Inc., Cambridge, Mass. 1952.
9 Taylor, A. E.: "Introduction to Functional Analysis," John Wiley & Sons, Inc., New York, 1958.

2
THE THEORY OF GAMES

2.1 EXAMPLES OF GAMES

This chapter is devoted to an introduction to the theory of games; in particular, to the theory of two-person games, in which there are only a finite number of possible ways in which the game can be played and in which the amount won by one player is equal to the amount lost by the other player. We refer to such games as *two-person finite zero-sum games*.[1] In this section we will present an intuitive discussion of some simple two-person finite zero-sum games, including some very specific and not uncommon children's games. Then we will discuss some other competitive situations which can be treated within the same general format as games. We will speak about the gambling philosophy of our players, their decision-making processes, and, in the case of competitions which are not games and which do not have payoffs in terms of money, their assignment of values.

---

[1] Although earlier attempts had been made to analyze decision-making and competitive situations mathematically, it is fair to say that the theory of games dates from 1928, when von Neumann [6] gave the first complete proof of the fundamental theorem (our Theorem 2.5.6). The subject took a giant leap forward in 1944 with the publication of the book "Theory of Games and Economic Behavior" by von Neumann and Morgenstern and has continued to grow rapidly since then.

[2.1.1]   THE THEORY OF GAMES   169

After this rather informal discussion we will begin the subject afresh, this time giving a careful mathematical treatment. We will then prove rigorously some of the conclusions arrived at earlier in the less formal treatment of this section.

Our attention throughout will be restricted to a game G (or G' or $\tilde{G}$) between two players, known simply as player I and player II (or as player I' and player II'). A game will consist of a succession of "moves" made according to rules set down before the game begins. Every move belongs to one or the other of two distinct types:

1. A choice among a finite number of alternatives made by one or the other of the players, with or without information of the previous moves that have been made but with full knowledge of the rules of the game, e.g., the moves in chess, deciding how many cards to exchange in poker, deciding whether or not to bet in poker, or deciding what to bid or play in games such as bridge.
2. A chance occurrence which neither player can control, e.g., the numbers that appear as a result of throwing dice, the cards received in an exchange in poker, the cards dealt in any card game.

A game consisting only of chance moves is quite uninteresting, at least from the mathematical point of view, since neither player is ever called upon to make a decision, and the theory of games is essentially a device for making decisions. All the games we consider will contain decision-type moves, i.e., type *1* above. After all the moves have been made, the game is over and a payment is made from one player to the other. In accordance with some prearranged set of rules, the amount of the payoff and the direction in which the payoff is to go are determined by the set of moves that has been made.

Our players will maintain an exceptionally conservative philosophy, taking absolutely no unnecessary risks even when there is the possibility of large winnings. They seek not the largest possible windfall but the largest possible *guarantee*.

Let us begin by describing the rules for a few very simple games and the means of *solving* them. A solution of a game will consist of a *best*, or *optimal*, way for each player to play and of the expectation each player will have, using his *optimal strategy*.

EXAMPLE 2.1.1 **Matching pennies (simple form)** Each player is to place one penny on a table. He has the option of putting his penny down with the head side showing or with the tail side showing. Neither player is to know what the other player has done until after his own penny is down. If both pennies show heads, or if both pennies show tails, player I takes both pennies from the table, i.e., player I wins one penny. On the other hand, if one penny shows heads and the other shows tails, player II wins one penny.

In this game, each player makes one move, and no chance moves occur. Each player must make a decision between two alternatives, namely, (1) play heads, H, and (2) play tails, T. We can describe this game in terms of player I's prospects by means of the array

I \ II	H	T
H	+1	−1
T	−1	+1

There are two ways of arriving at the outlook toward the game that we assume each player adopts. We might call them the "susceptibility to espionage" method and the "need for a guarantee" method. It is clear that should either player be able, in some way, to determine the decision his opponent has made before he himself has played, then he would have no difficulty in winning. Thus if a player is fearful that his espionage system is not as good as his opponent's (or that his opponent is a mind reader), he must look for a way of playing which will minimize the effect of his opponent's advantage. In such a case (or for any other reason) each player is anxious to determine how to play in such a way that his expectation is maximized no matter what his opponent does. Clearly a player with no espionage system can do nothing to guarantee that he will fare any better than losing every time the game is played, i.e., every time the pennies are matched. However, we will show that there is a means whereby he can guarantee that his *expectation* is to win half of the games played, no matter how much better his opponent's espionage system is. The method is very simple. He simply throws the coin into the air and permits it to come to rest with either the head or the tail showing, purely by chance (of course, taking care that his opponent does not see which it is). Even with an opponent who is a mind reader, he can expect to win half of the games.[1] Thus, after a series of games, he expects neither to win nor to lose. We say that his expectation is 0. Another way of expressing this is to say that, from player I's point of view, the value of the game is 0. Because of the symmetry of this game, all the above remarks apply equally well to player II. ////

---

[1] Because we have not properly developed the basic notions of probability theory, we must here tacitly assume that the two statements "the probability that player I will win any one game is $\frac{1}{2}$" and "player I expects to win half of the games played" mean the same thing or at least imply each other.

EXAMPLE 2.1.2 **Matching pennies (complex form)** The rules of play for the complex form of matching pennies are the same as the rules for the simple form. The difference lies in the payoff arrangements. Let us suppose that the game is described (again in terms of player I's prospects) by means of the array

I \ II	H	T
H	+1	−2
T	−3	+6

Most of the comments made concerning the simple form of matching pennies apply equally well to the complex form. However, the solution is not the same. Suppose player I adopts the following procedure to decide which alternative (H or T) to choose. He puts three white balls and one black ball in a bag. He then pulls out a ball at random. If it is white, he plays heads. If it is black, he plays tails.

We say that player I has combined the *pure strategies*, H and T, to form the (*mixed*) *strategy*

$$\text{Play H } \tfrac{3}{4} \text{ of the time} \qquad \text{Play T } \tfrac{1}{4} \text{ of the time}$$

Suppose that player II learns of this. How can he use this information? If he constantly plays heads, he can expect to win one game out of every four. However, when he wins, he wins 3, and when he loses, he loses 1. Thus, constantly playing heads, he expects to neither win nor lose but to come out even. On the other hand, if he constantly plays tails, he can expect to win three games out of every four. However, when he wins, he wins only 2, and when he loses, he loses 6. As before, by constantly playing tails, his expectation is 0. Thus, as long as player I continues this procedure for making his move, as far as player II is concerned, it is immaterial whether he plays heads or tails—or, for that matter, any arrangement of heads or tails. Player II's espionage gives him no advantage whatsoever.

We have found a method whereby player I can guarantee that his expectation is 0, irrespective of player II's machinations. What is the best that player II can guarantee for himself, irrespective of player I? Certainly he cannot guarantee that he will expect to win. But does he have a strategy whereby he can guarantee that he will not lose? That is, does there exist a strategy for player II such that his expectation is 0 no matter what player I does? The answer is yes. If he plays heads two-thirds of the time and tails one-third of the time (but in a random fashion, using colored balls), his expectation is 0.

There is still more to be said about this game. Suppose player I plays heads with probability p (for some p, $0 \leq p \leq 1$) and tails with probability

$1 - p$ and player II plays heads with probability q (for some q, $0 \leq q \leq 1$) and tails with probability $1 - q$. Then the probability that both players will show heads is pq, and when this happens, player I wins 1. The probability that player I will play heads and that player II will play tails is $p(1 - q)$, and when this happens, player I loses 2. Similarly, the probability that player I will play tails and player II will play heads is $(1 - p)q$, and then player I loses 3. Finally, the probability that both players will play tails is $(1 - p)(1 - q)$, and then player I wins 6. Thus player I can expect to win

$$(1)pq + (-2)p(1 - q) + (-3)(1 - p)q + (6)(1 - p)(1 - q) =$$
$$12pq - 8p - 9q + 6 = 12(p - \tfrac{3}{4})(q - \tfrac{2}{3}) \qquad (2.1.3)$$

We express this by saying that player I's expectation when he uses the strategy

$$P = [p \quad 1 - p]$$

and player II uses the strategy

$$Q = \begin{bmatrix} q \\ 1 - q \end{bmatrix}$$

is given by (2.1.3). It is clear that (2.1.3) can be expressed as the matrix product

$$P \begin{bmatrix} 1 & -2 \\ -3 & 6 \end{bmatrix} Q$$

More generally, if G is a game with matrix A, and players I and II use strategies P and Q, respectively, then *player I's expectation with respect to P and Q* is denoted by $E(P, Q)$ and is given by

$$E(P, Q) = PAQ$$

In the game we have been discussing, suppose player II decides to use the strategy

$$Q = \begin{bmatrix} q \\ 1 - q \end{bmatrix}$$

for some q, $0 \leq q \leq 1$. It is not difficult to see [from the last expression of (2.1.3)] that unless $q = \tfrac{2}{3}$, player I can improve his expectations. In fact, if $q > \tfrac{2}{3}$, player I would do well to always play heads, i.e., set $p = 1$; for then he would expect to win

$$q(+1) + (1 - q)(-2) = -2 + 3q > 0$$

each game. If $q < \tfrac{2}{3}$, player I would play tails each game, i.e., set $p = 0$; for then his expectation is

$$q(-3) + (1 - q)(6) = 6 - 9q > 0$$

On the other hand, if player II sets $q = \frac{2}{3}$, then player I's expectation is always 0 (no matter what strategy he chooses). Thus the mixed strategy for player II,

$$Q = \begin{bmatrix} \frac{2}{3} \\ \frac{1}{3} \end{bmatrix}$$

(obtained by setting $q = \frac{2}{3}$) is his unique optimal strategy. Similarly we can show that

$$P = \begin{bmatrix} \frac{3}{4} & \frac{1}{4} \end{bmatrix}$$

is the unique optimal strategy for player I. ////

Let us discuss another, still more complex variant of the game of matching pennies, in which player II has three possible pure strategies. In the discussion of this game, we will present another very simple method (the *graphical method*) for solving the games thus far introduced. Because for each player we must be able to draw several accurate graphs in a space of as many dimensions as he has pure strategies, this method has extreme limitations. Later we will show that it is often possible to eliminate some of the obviously unsound strategies each player might (but will not) make and thereby arrive at a "new" game which can be solved by graphical means.

EXAMPLE 2.1.4 **Matching pennies (more complex form)** In this game player I will have the same two alternatives to decide between as in the first two games, namely, to play heads or to play tails. However, player II will have three alternatives: to play heads, to play tails, or, to show no coin at all (which we denote by N). As before, let us suppose that the game is described (in terms of the outcome for player I) using the array

I \ II	H	T	N
H	1	2	4
T	3	1	0

Player I must now decide how often he should play heads and how often he should play tails. From our discussion of the previous games it is clear that once each player has decided how often each of the alternatives open to him should be played, it is always desirable for him to make his moves in a random way (as, for example, by using a bag containing various colored balls). Let us suppose that player I plays heads with probability p ($0 \leq p \leq 1$). Then he must play tails with probability $1 - p$. If player II plays heads, then player I's expectation for each game is $(p)(1) + (1 - p)(3) = 3 - 2p$. If player II plays tails, then player I's expectation for each game is $(p)(2) + (1 - p)(1) = 1 + p$. Finally, if player II shows no coin, then player I's expectation is $(p)(4) + (1 - p)(0) = 4p$. For each of these three situations, let us plot the expectation of player I, using the strategy $P = [p \quad 1 - p]$.

Player I is now able to determine for each value of p the best expectations that he is able to guarantee. This is found by taking, for each value of p, the minimum of the three numbers, $3 - 2p$, $1 + p$, and $4p$. The function

$$m(p) = \min\{3 - 2p, 1 + p, 4p\}$$

is defined for each value of p, $0 \leq p \leq 1$, and is described by the line segments OB, BC, CD in Figs. 2.1.5 and 2.1.6. The best guaranteed expectation for player I is given by

$$\max_{0 \leq p \leq 1} m(p)$$

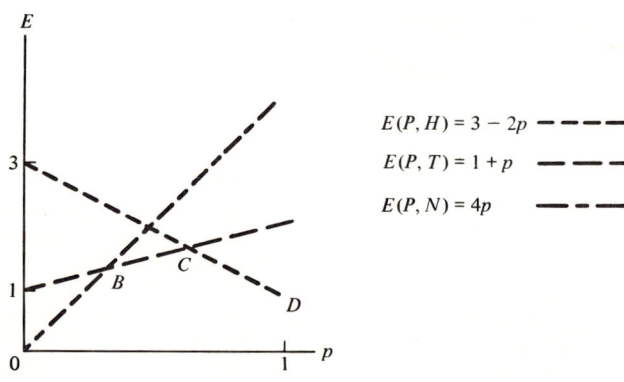

FIGURE 2.1.5

From Figs. 2.1.5 and 2.1.6 it is easily seen that this best guaranteed expectation occurs at the point C, which is the intersection of the lines

$$E = 3 - 2p, \quad E = 1 + p$$

Solving these two equations simultaneously yields $p = \frac{2}{3}$, $E = \frac{5}{3}$. If player I uses the mixed strategy $P = [\frac{2}{3} \quad \frac{1}{3}]$, his expectation is at least $\frac{5}{3}$, irrespective of players II's strategy. Unlike the previous examples, however, it is clear from

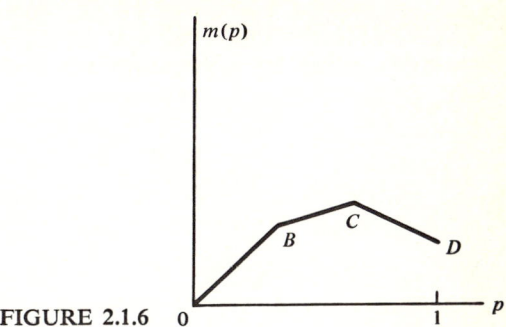

FIGURE 2.1.6

Fig. 2.1.5 that if player II uses the alternative N at all, then player I's expectation becomes greater than $\frac{5}{3}$.

Player II may go through the same process. He must determine with what probability he should play heads, tails, or no coin. Let us denote these probabilities by q_1, q_2, q_3, respectively. Then

$$0 \leq q_i \leq 1 \quad i = 1, 2, 3 \quad q_1 + q_2 + q_3 = 1 \quad (2.1.7)$$

If player I plays heads, then his expectation is

$$q_1 + 2q_2 + 4q_3 = 4 - 3q_1 - 2q_2$$

while if player I plays tails, then his expectation is

$$3q_1 + q_2$$

Let us plot player I's expectation, in each case, assuming that player II uses the strategy

$$Q = \begin{bmatrix} q_1 \\ q_2 \\ q_3 \end{bmatrix}$$

For any choice of q_1, q_2, the least expectation for player I that player II is able to guarantee is

$$n(q_1, q_2) = \max\{4 - 3q_1 - 2q_2, 3q_1 + q_2\}$$

Clearly $n(q_1, q_2)$ is defined for all q_1, q_2 satisfying equations (2.1.7) and is described by the planar segments $RSUT$ and TUV of Figs. 2.1.8 and 2.1.9. Player II wishes to keep player I's expectation as small as possible, and the smallest he can make player I's expectation, solely by means of his own strategy, is

$$\min_{\substack{0 \leq q_1, q_2 \leq 1 \\ q_1 + q_2 \leq 1}} n(q_1, q_2)$$

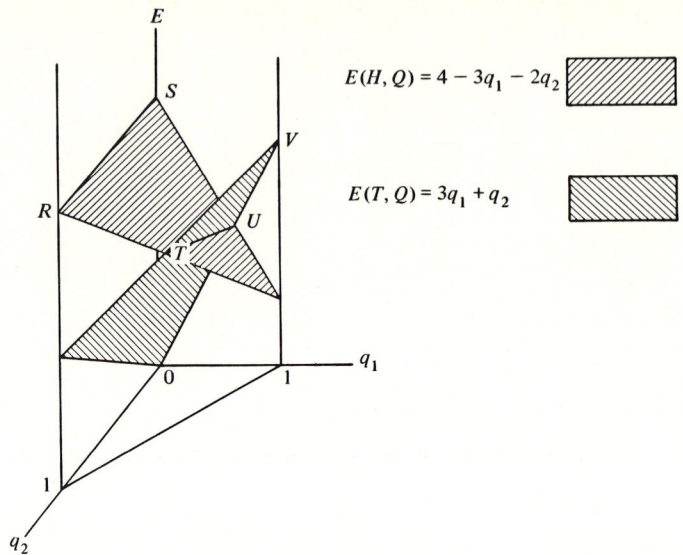

FIGURE 2.1.8

According to Figs. 2.1.8 and 2.1.9, the least guaranteed expectation for player I occurs at T, which is the intersection of the three planes

$$E = 4 - 3q_1 - 2q_2 \qquad E = 3q_1 + q_2 \qquad q_1 + q_2 = 1$$

Solving these equations simultaneously yields

$$q_1 = \tfrac{1}{3} \qquad q_2 = \tfrac{2}{3} \qquad q_3 = 0 \qquad E = \tfrac{5}{3}$$

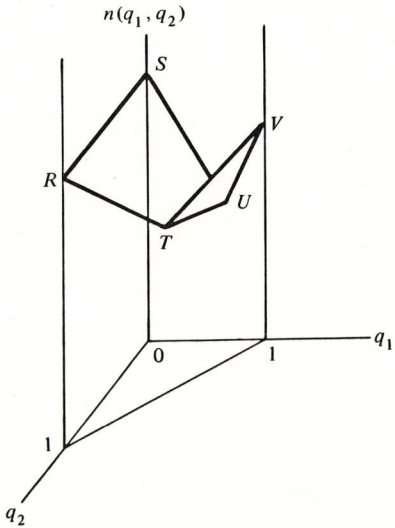

FIGURE 2.1.9

If player II uses the strategy

$$Q = \begin{bmatrix} \frac{1}{3} \\ \frac{2}{3} \\ 0 \end{bmatrix}$$

then player II's expectation is $\frac{5}{3}$, no matter how player I plays. ////

Again we have a situation where this is a number v (in this case, $\frac{5}{3}$) such that:

1. Player I has an optimal strategy (in this case $P = [\frac{2}{3} \quad \frac{1}{3}]$) whereby he is able to guarantee an expectation of winning at least v, irrespective of the play of player II.

2. Player II has an optimal strategy $\left(\text{in this case } Q = \begin{bmatrix} \frac{1}{3} \\ \frac{2}{3} \\ 0 \end{bmatrix}\right)$ whereby he is able to guarantee for himself an expectation of losing no more than v, irrespective of the play of player I.

In such a case, we call v the *value* of the game. The main theorem of this chapter (the von Neumann minimax theorem of Sec. 2.5) states that every game has a value and that both players have optimal strategies.

Note that in Example 2.1.4 player II's (unique) optimal strategy calls upon him never to play N, even though his best possible expectation (namely, 0) occurs when he plays N and player I plays T. Since player II is never going to make use of the pure strategy N, the game would not be changed if this strategy were removed. Thus the game we have just finished discussing is essentially the same game as the one with the array

I \ II	H	T
H	1	2
T	3	1

Unfortunately, it does not become clear that player II never plays N until after the game is solved. In the next game, there will be some alternatives which are so obviously bad that they can be eliminated before we attempt to solve the game.

EXAMPLE 2.1.10 Two-fingered morra There are two players. At some signal, each player puts out either one finger or two fingers and at the same time calls out one of the numbers 2, 3, 4. If the number which one player calls out is equal to the total number of fingers put out and the other player calls out some other number, the first player wins. If both call out incorrect numbers, or if both players call out the correct number, the game is a tie.

Each player has a choice of six alternatives (or pure strategies):

Strategy	Fingers put out	Numbers called
S_1	1	2
S_2	1	3
S_3	1	4
S_4	2	2
S_5	2	3
S_6	2	4

Two of these alternatives have absolutely no chance of winning, namely, S_3 (for player II cannot put out three fingers) and S_4 (since player II must put out some fingers). As far as we are concerned, the game would be unchanged if these strategies were removed. In the same way, if each player were permitted to call any number he wishes (still being restricted to showing one or two fingers), it is clear that no (sensible) player would use any combination of fingers and numbers other than S_1, S_2, S_5, and S_6.

We can represent this game in the matrix form (from player I's point of view) by

I \ II	S_1	S_2	S_5	S_6
S_1	0	1	−1	0
S_2	−1	0	0	1
S_5	1	0	0	−1
S_6	0	−1	1	0

Because of the size of the matrix involved, it would be, at best, extremely cumbersome to try to find a solution to this game by the methods we have previously introduced. However, solutions will be discussed in the exercises.   ////

Each of the games we have examined thus far has consisted solely of one move made by each player. At the time of each move, the players had no information available to them (other than, of course, the rules of the game, including the payoff arrangements). Let us now examine a simple game in which (1) the first move is made by chance, (2) the players move in sequence (rather than simultaneously), and (3) one of the players has available some usable information at the time his move is to be made.

EXAMPLE 2.1.11 **Bluffing** Bluffing is a two-person card game, modeled vaguely after poker, played with a deck of two cards, called high (H) and low (L). Both cards have identical backs and are placed face down on a table. As the first move in the game, player I takes one of the cards. He inspects it (to determine which card he has drawn), making certain that player II does not have an opportunity to see which card has been chosen.

At this point, player I has two alternatives: (1) he may *fold*, in which case he pays one penny to player II and the game is over, or (2) he may *bet* two pennies. Now player II has two alternatives: (1) he may fold, in which case he pays one penny to player I and the game is over, or (2) he may *call*. In the event that player II calls, player I exposes the card he has chosen. If the card is H, player II pays player I two pennies. If the card is L, player I pays player II two pennies. In either case the game is over.

The first move in bluffing is made by chance. Player I receives either H or L. We must assume that these two possibilities are equally likely. We can change this assumption by altering the deck of cards. If the deck consists of two cards marked H and one card marked L, we would assume that the probability that player I's card would be H would be $\frac{2}{3}$ and the probability that it would be L would be $\frac{1}{3}$.

The second move of the game is made by player I. If player I folds, player II has no moves to make. On the other hand, if player I bets, player II has a move to make, either to fold or to bet. In addition, at the time that he is called upon to make his move, player I has some information at his disposal. He knows which card he has drawn.

Suppose player I bets every time he has H and folds every time he has L. He can expect to bet half of the time and fold half of the time. Should player II learn of this, he will never call, since he can never win by calling. In this event, in half of the games, player I will win one penny, and, in the other half of the games, he will lose one penny. Thus his expectation is zero. On the other hand, suppose player I bets no matter what he has drawn. If player II learns of this, he will always call, for then he will lose two pennies half of the time (when player I has H) and he will win two pennies half of the time (when player I has L). Again his expectation (and also player I's expectation) is zero.

Once we have shown that bluffing is a matrix game, it will be immediately clear that there are mixed strategies player I can use which will make his expectation positive, no matter what player II does. We have said that player I has two sensible pure strategies. These are

$$S_1 = \begin{cases} \text{if the card drawn is H, then bet} \\ \text{if the card drawn is L, then bet} \end{cases}$$

$$S_2 = \begin{cases} \text{if the card drawn is H, then bet} \\ \text{if the card drawn is L, then fold} \end{cases}$$

Player I has two other pure strategies. Neither of them is sensible since each would require him to fold when he has drawn H, and so we will ignore them. If we did include them in the array we are going to construct, it would be easy to show that player I would never use them.

We can describe player II's pure strategies more simply. They are

$$T_1 = \text{call (when the option arises)}$$
$$T_2 = \text{fold (when the option arises)}$$

Summarizing the discussion above, player I's expectation is zero either (1) when player I plays S_1 and player II plays T_1 or (2) when player I plays S_2 and player II plays T_2. There are two other possibilities, namely, (3) player I plays S_1, and player II plays T_2, in which case it is clear that player I wins 1, and (4) player I plays S_2, and player II plays T_1. In this last situation, player I will win 2 half of the time (when he draws H), and he will lose 1 half of the time (when he draws L). Thus his expectation is $\frac{1}{2}(2) + \frac{1}{2}(-1) = \frac{1}{2}$. With all of this information, we see that player I's expectation is given by the array

I \ II	T_1	T_2
S_1	0	1
S_2	$\frac{1}{2}$	0

Once we have put this game in matrix form, it is clear that the solution is easily obtained by using the graphical methods introduced in Example 2.1.4.   ////

A strategy for a player is a set of rules which tells him which of several alternatives to choose whenever he has such a decision to make. In the games we have studied thus far each player is called upon to make a decision, i.e., a move, exactly once. In the game of Example 2.1.12, player I may have to make two separate decisions, and his strategies must tell him how to make each one.

EXAMPLE 2.1.12   Three-card Poker   Each player is dealt one card from a deck consisting of the three cards, high (H), medium (M), and low (L). The cards are *ordered* from *strongest* to *weakest* by the ordering $H > M > L$. Player I has two choices:

1 He may *fold*, in which case he pays one penny to player II and the game is over.
2 He may *bet*.

If player I bets, player II has three choices:

1 He may fold, in which case player II pays one penny to player I and the game is over.
2 He may *call*.
3 He may *raise*.

If player II calls, both players expose the card they have been dealt and the player with the stronger card wins two pennies from the other player. The game is then over. On the other hand, if player II raises, player I has two choices:

1 He may fold, in which case he pays player II two pennies and the game is over.
2 He may call.

Should player I now call, the cards are exposed and the player with the stronger card wins three pennies from the other player.
  If player I folds initially, player II needs no strategy since he has no decisions to make. On the other hand, if player I bets, then player II must decide between the three alternatives: (1) fold (F), (2) call (C), and (3) raise (R). The information that player II has available to help him make this decision is the identity of the card he has been dealt. A pure strategy for player II consists

of a plan for all the eventualities he might be called upon to face. In this case there are three such eventualities:

1 Player I has bet, and he (player II) holds H
2 Player I has bet, and he holds M
3 Player I has bet, and he holds L

and for each of these eventualities, player II must assign either F, C, or R. Thus player II has a total of 27 pure strategies. However, some of these strategies are not sensible, and so we will ignore them. If player II has been dealt H, he can gain nothing by folding or by calling. In the same way, if he has been dealt L, he can gain nothing by calling. By folding, he keeps his losings down and by raising, i.e., bluffing, he retains the possibility of winning. Of the 27 pure strategies available to player II, only 6 are sensible. These are tabulated.

PLAYER II

Strategy	Card drawn	Move (if needed)
T_1	H	R
	M	R
	L	R
T_2	H	R
	M	R
	L	F
T_3	H	R
	M	C
	L	R
T_4	H	R
	M	C
	L	F
T_5	H	R
	M	F
	L	R
T_6	H	R
	M	F
	L	F

Player I may have two decisions to make. At the time of these decisions he is aware of which card he has been dealt, and he can make use of this information. When player I bets and player II raises, player I is called upon to make a second decision, namely, whether to call or to fold. Thus player I also has 27 pure strategies available. Clearly, if he has been dealt H, he should never fold (either on the first or on the second move). Also, if he has been dealt L, he should never call. Thus, of the 27 pure strategies available, only 6 are sensible. These are tabulated.

Strategy	Card drawn	First move	Second move (if needed)
S_1	H	B	C
	M	B	C
	L	B	F
S_2	H	B	C
	M	B	F
	L	B	F
S_3	H	B	C
	M	B	C
	L	F	
S_4	H	B	C
	M	F	
	L	F	
S_5	H	B	C
	M	B	F
	L	F	
S_6	H	B	C
	M	F	
	L	B	F

PLAYER I

We are now faced with the prospect of filling in the entries of the 6×6 matrix $A = [a_{ij}]$, where a_{ij} is player I's expectation when he plays S_i and player II plays T_j. There are six possible ways in which the cards can be dealt, namely

	Player I	Player II
1	H	L
2	L	H
3	H	M
4	M	H
5	M	L
6	L	M

We assume that all these possibilities are equally likely. Then

$$a_{11} = \tfrac{1}{6}(3 - 2 + 3 - 3 + 3 - 2) = \tfrac{1}{3},$$
$$a_{21} = \tfrac{1}{6}(1 - 2 + 3 - 3 + 1 - 2) = -\tfrac{1}{3},$$
$$a_{31} = \tfrac{1}{6}(3 - 2 + 2 - 3 + 3 - 2) = \tfrac{1}{6},$$

....

$$a_{12} = \tfrac{1}{6}(3 - 2 + 3 - 2 - 2 - 2) = -\tfrac{1}{3},$$
$$a_{22} = \tfrac{1}{6}(1 - 2 + 3 - 2 + 1 - 2) = -\tfrac{1}{6},$$
$$a_{32} = \tfrac{1}{6}(3 - 2 + 2 - 2 - 2 - 2) = -\tfrac{1}{2},$$

....

////

There are several examples of competitive situations which are not games but which, with the assignment of reasonable numerical values to the possible outcomes, can be analyzed by the methods of this chapter. A simple example of such a nongame competition is:

EXAMPLE 2.1.13 **Duelling** Two players stand facing each other, $2n$ paces apart, where n is a positive integer. Each has a gun containing a single bullet. At a signal from the referee each player may fire at the other if he so wishes. If either player is hit or if both players fire and miss, the "game" is over. Otherwise both players advance one pace, and the procedure is repeated. After taking n paces forward, the players have reached each other and the game is over. It will be convenient to say that a player has taken zero paces forward when he is at his initial position. Also, we will assume that the probability that either player will be able to hit the other increases as the two players approach each other.

At this point we must decide how much information should be made available to each player. A duel in which each player knows when his opponent has shot at him and missed is called a *noisy duel*. A duel in which neither player knows he has been shot at unless he has been hit (as if both guns were equipped with silencers) is called a *silent duel*.

In a noisy duel each player has $n + 1$ sensible pure strategies. They are

$$S_i = \begin{cases} \text{shoot after the } i\text{th pace forward if the other player} \\ \text{has not previously shot and missed} \\ \text{shoot after the } n\text{th pace forward if the other player} \\ \text{has previously shot and missed} \end{cases} \quad i = 0, 1, 2, \ldots, n$$

If the other player has shot and not missed, then no decision need be made.

In a silent duel each player has $n + 1$ pure strategies, all of which appear to be sensible. They are

$S_i =$ shoot after the ith pace forward $\quad i = 0, 1, 2, \ldots, n$

In order to apply the methods of game theory to duelling it is only necessary to assign values to the possible outcomes. It is reasonable to assign the value of $+1$ when player I survives and player II does not and the value -1 when player II survives and player I does not. It is also reasonable to assign the value 0 when both players survive, especially if the duel must then be repeated. Finally, if neither player survives, we must assign a value of 0. If each player feels that his death should be valued at -1 (in terms of his expectation) irrespective of the survival of his opponent, we no longer have a zero-sum game, for if both players are killed, both lose and no one wins. ////

A game in which the rules of play and the payoff arrangements are the same for both players, e.g., two-fingered morra, is called a *symmetric game*.[1] In a symmetric game the set of pure strategies and the set of all (mixed) strategies are the same for both players. An optimal strategy for one player is an optimal strategy for the other player. Moreover, if a symmetric game has a value, it must be zero since it is clear that neither player can find a strategy which will give him a positive expectation against every one of his opponent's strategies.

If, in a symmetric game, S_1 and S_2 are strategies for player I (and hence also for player II), then $E(S_1, S_2) = -E(S_2, S_1)$; for, clearly, the first player's expectation when he plays S_1 and player II plays S_2 is the same as player II's expectation when he plays S_1 and player I plays S_2. But, in any instance, player I's expectation is the negative of player II's expectation. Thus, the matrix of a symmetric game is skew-symmetric, that is, $A^T = -A$.

Any game can be changed into a symmetric game in a number of different ways (see Exercise 16). Let us examine here a particular method of *symmetrization* which requires the addition of a single chance move. In general, symmetrization is itself not a game but a method for transforming any game into a symmetric game.

The symmetrization G' of the game G is a game between the two players, player I' and player II', and is played as follows. A coin is tossed. If it shows heads, then player I' in G' becomes player I in G and player II' becomes player II in G; if the coin shows tails, then player I' becomes player II in G and player II' becomes player I in G. If the pure strategies in G are $S_1, S_2, \ldots, S_m$ for player I and $T_1, T_2, \ldots, T_n$ for player II, then the pure strategies for both of the players in G' are

$$S'_{ij} = \begin{cases} \text{when playing as player I, play } S_i & i = 1, 2, \ldots, m \\ \text{when playing as player II, play } T_j & j = 1, 2, \ldots, n. \end{cases}$$

Thus, in G', each player has mn pure strategies, and the matrix A' of G' is a skew-symmetric $mn \times mn$ matrix.

EXAMPLE 2.1.14 **The symmetrization of bluffing** In the symmetrization of bluffing, players I' and II' each have four pure strategies:

$$S'_{11} = \begin{cases} \text{when playing as player I, play } S_1 \ (=\text{always bet}) \\ \text{when playing as player II, play } T_1 \ (=\text{call}) \end{cases}$$

$$S'_{12} = \begin{cases} \text{when playing as player I, play } S_1 \ (=\text{always bet}) \\ \text{when playing as player II, play } T_2 \ (=\text{fold}) \end{cases}$$

---

[1] Later we shall define a symmetric game in a somewhat different manner.

$$S'_{21} = \begin{cases} \text{when playing as player I, play } S_2 \, (=\text{bet on H only}) \\ \text{when playing as player II, play } T_1 \, (=\text{call}) \end{cases}$$

$$S'_{22} = \begin{cases} \text{when playing as player I, play } S_2 \, (=\text{bet on H only}) \\ \text{when playing as player II, play } T_2 \, (=\text{fold}) \end{cases}$$

Let us fill in some of the entries of the matrix A'. We can find the entry in the (1, 4) position of A' by supposing that player I$'$ selects the strategy S'_{11} and player II$'$ selects the strategy S'_{22}. There are now two possibilities, and the probability of each is $\frac{1}{2}$. When the tossed coin shows heads, player I$'$ (who has become player I and has played S_1, that is, always bet) can expect to win 1 from player II$'$ (who has become player II and has played T_2, that is, fold). When the coin that is tossed shows tails, player I$'$ (who has become player II and has played T_1) will lose $\frac{1}{2}$ to player II$'$ (who has become player I and has played S_2). Thus

$$a'_{14} = E(S'_{11}, S'_{22}) = \tfrac{1}{2}(1 - \tfrac{1}{2}) = \tfrac{1}{4}$$

We can show very quickly that the entry in the (4, 1) position of A' is $-\frac{1}{4}$. For if player I$'$ selects the strategy S'_{22} and player II$'$ selects the strategy S'_{11}, the expectations are the same as in the discussion above except that the roles of the players have been reversed. Thus

$$a'_{41} = E(S'_{22}, S'_{11}) = \tfrac{1}{2}(\tfrac{1}{2} - 1) = -\tfrac{1}{4} = -E(S'_{11}, S'_{22}) = -a'_{14}$$

The entries along the diagonal of A' are all zero; for if players I$'$ and II$'$ both decide to use the strategy $S'_{ij}(1 \leq i, j \leq 2)$, then half of the time (when the coin shows heads) player I$'$ can expect to win a_{ij} and half the time (when it shows tails) player II$'$ can expect to win a_{ij}. Thus

$$E(S'_{ij}, S'_{ij}) = \tfrac{1}{2}(a_{ij} - a_{ij}) = 0$$

We leave the task of finding the remaining entries of A' as an exercise.   ////

In Sec. 2.8 we shall discuss linear-programming problems and show that they can be transformed into games.

EXERCISES

1 Solve the following games:

(a) $A = \begin{bmatrix} 1 & -1 \\ -1 & 3 \end{bmatrix}$ (b) $A = \begin{bmatrix} 1 & 1 \\ 3 & -2 \end{bmatrix}$

(c) $A = \begin{bmatrix} 0 & 3 \\ -2 & 2 \end{bmatrix}$ (d) $A = \begin{bmatrix} 2 & -2 \\ -1 & 3 \end{bmatrix}$

(1) by the method of Example 2.1.2 and (2) by the graphical method.

2 Solve the following games by the methods of this section

(a)

I \ II	T_1	T_2	T_3
S_1	1	−3	4
S_2	4	10	−6

(b)

I \ II	T_1	T_2	T_3
S_1	5	−1	−3
S_2	−3	−1	5

(c)

I \ II	T_1	T_2	T_3
S_1	6	0	−1
S_2	−4	2	4
S_3	−9	−5	−6

3 Show that in Example 2.1.10 playing S_1, S_2, S_5, S_6 with equal probability is an optimal strategy for each player. Are there any other optimal strategies?

4 Change the rules in Example 2.1.10 so that the winner wins an amount equal to the total number of fingers put out. Find the matrix for this game. Show that the solution given in Exercise 3 is no longer a solution. Show that $\bar{P} = [0 \ \ 3/5 \ \ 2/5 \ \ 0]$ is now an optimal strategy for both players. Are there any other optimal strategies?

5 Solve the game of Example 2.1.11.

6 Write out the matrix in Example 2.1.11 showing all of player I's pure strategies (the nonsensible ones as well as the sensible ones). Show that the solutions are the same, with and without taking the nonsensible pure strategies into account.

7 Set up and solve the game of bluffing when the deck used consists of two H's and three L's.

8 Show that, in Example 2.1.12, S_3 is at least as good a strategy as S_4 for player I under all circumstances. (It follows then that player I need never use S_4.)

9 Let G be a noisy duel with the players beginning $2n$ paces apart in which the probability that a player will hit his opponent when firing after his ith pace forward is i/n, $i = 0, 1, 2, \ldots, n$. Show that:
 (a) For n odd, $n = 2k - 1$, S_k is an optimal strategy for each player.
 (b) For n even, $n = 2k$, S_k and S_{k+1} are both optimal strategies for each player.

10 Let G be a silent duel with the players beginning $2n$ paces apart in which the probability that a player will hit his opponent when firing after taking his ith pace forward is i/n, $i = 0, 1, 2, \ldots, n$. Show that
 (a) For $n = 4$, S_2 is an optimal strategy.

(b) For $n = 5$, no pure strategy is optimal, but $\bar{P} = [0 \quad 0 \quad \frac{5}{11} \quad \frac{5}{11} \quad 0 \quad \frac{1}{11}]$ is an optimal strategy for both players.

11 Solve the silent duel in which the players begin six paces apart and the probability of each player's hitting the other is given in the table.

Pace	Player I	Player II
0	1/4	1/5
1	2/4	2/5
2	3/4	4/5
3	4/4	5/5

12 Find the pure strategies and the matrix for the symmetrization of matching pennies (Example 2.1.1). Show that playing each pure strategy with equal probability is an optimal strategy for both players.

13 Find the pure strategies and the matrix for the symmetrization of bluffing. Ignore the nonsensible strategies.

14 Show what happens to the matrix of a game when player I becomes player II and player II becomes player I.

15 *Scissors, paper, and stone* This is an ancient Italian game. At a signal each player puts out either (1) two fingers (scissors), (2) five fingers (paper), or (3) fist (stone). If both players put out the same thing, the game is a draw. Otherwise:

(a) Two fingers beats five fingers (scissors cut paper).

(b) Five fingers beats fist (paper covers stone).

(c) Fist beats two fingers (stone breaks scissors).

If the winner always wins $+1$, set up the matrix of the game and show that playing each strategy with equal probability is an optimal strategy for both players.

16 Let G be a game and form the game G' between the players I' and II' as follows. Players I' and II' can:

(a) Choose a strategy of player I in G.

(b) Choose a strategy of player II in G.

(c) Choose neither.

If players I' and II' both choose a player I strategy or both choose a player II strategy or both choose neither, the game is a tie. If one player chooses a player I strategy and the other chooses a player II strategy, they play that way and the payoff is determined by the payoff in G. If either player selects a player I strategy and the other player chooses no strategy, the player who chose the player I strategy loses 1. If either player selects a player II strategy and the other player chooses no strategy, the player who chose the player II strategy wins 1. Set up the matrix for this game.

2.2 MAXMIN AND MINMAX

It was pointed out at the beginning of Sec. 2.1 that the discussion in that section is very informal and that no attempt would be made there to justify any of the methods introduced. Since we will begin a more formal development of game theory from the very beginning in the next section, where many of the same problems will arise, we devote this section to justifying some of the practices of Sec. 2.1. We will consider here some nonalgebraic fundamentals of game theory. These come primarily from topology and analysis. A thorough discussion of this material, including definitions and proofs for all the theorems stated, would lead us too far astray. In addition, for many readers such a discussion would present little that is new. For these reasons, only those facts which are needed later are presented here, and some of these are given without proof.

Definition 2.2.1 Let $X, X_1, X_2, \ldots, X_k$ be elements of the vector space $\mathbb{V}$.[1] Then X is a *convex combination* of $X_1, X_2, \ldots, X_k$ if there exist real numbers $\alpha_1, \alpha_2, \ldots, \alpha_k$ such that

$$\alpha_1 + \alpha_2 + \cdots + \alpha_k = 1 \qquad \begin{matrix} 0 \leq \alpha_j \leq 1 \\ j = 1, 2, \ldots, k \end{matrix} \qquad (2.2.2)$$

for which

$$X = \alpha_1 X_1 + \alpha_2 X_2 + \cdots + \alpha_k X_k$$

A subset $\mathbb{S}$ of the vector space $\mathbb{V}$ is *convex* if for any k vectors $X_1, X_2, \ldots, X_k$ belonging to $\mathbb{S}$ and any k scalars $\alpha_1, \alpha_2, \ldots, \alpha_k$ satisfying (2.2.2) the vector

$$\alpha_1 X_1 + \alpha_2 X_2 + \cdots + \alpha_k X_k$$

also belongs to $\mathbb{S}$; that is, $\mathbb{S}$ is convex if it contains all the convex combinations of its elements.

The set of all vectors P,

$$P = [p_1, p_2, \ldots, p_m]$$

of $\mathbb{R}_m$ for which

$$p_1 + p_2 + \cdots + p_m = 1 \qquad \begin{matrix} 0 \leq p_i \leq 1 \\ i = 1, 2, \ldots, m \end{matrix} \qquad (2.2.3)$$

---

[1] Throughout this chapter we will consider only vector spaces and matrices whose entries are from the field of real numbers $\mathbb{R}$.

is called the *strategy space for player* I (in the context of this chapter[1]) and is denoted by $\mathbb{P}_m$. Clearly $\mathbb{P}_m$ consists of all convex combinations of the n unit row vectors and hence is convex. In the same way the set of all vectors Q

$$Q = \begin{bmatrix} q_1 \\ q_2 \\ \vdots \\ q_n \end{bmatrix}$$

of $\mathbb{R}^n$ for which

$$q_1 + q_2 + \cdots + q_n = 1 \qquad \begin{array}{c} 0 \leq q_j \leq 1 \\ j = 1, 2, \ldots, n \end{array}$$

is a convex set called the *strategy space for player* II and is denoted by $\mathbb{Q}^n$.

Note that neither $\mathbb{P}_m$ nor $\mathbb{Q}^n$ is a vector space. However, since all the unit vectors of $\mathbb{R}_m$ belong to $\mathbb{P}_m$ and all of the unit vectors of $\mathbb{R}^n$ belong to $\mathbb{Q}^n$, the only vector space which contains $\mathbb{P}_m$ is $\mathbb{R}_m$ and the only vector space which contains $\mathbb{Q}^n$ is $\mathbb{R}^n$. Equations (2.2.2) describe the entries of an element of $\mathbb{P}_m$. Thus, for $A \in \mathbb{R}_n^m$, $Z \in \mathbb{R}_n$, Z is a convex combination of the rows of A if and only if there exists a vector P belonging to $\mathbb{P}_m$ such that $Z = PA$. (Compare with Theorem 1.4.4.)

The vector spaces $\mathbb{R}_n$ and $\mathbb{R}^n$ have an important geometric interpretation, as n-dimensional euclidean space.[2] The vector $X = [x_1, x_2, \ldots, x_n]$ is interpreted as the point whose coordinates are $x_1, x_2, \ldots, x_n$. For $X, Y \in \mathbb{R}_n$,

$$X = [x_1, x_2, \ldots, x_n] \qquad Y = [y_1, y_2, \ldots, y_n]$$

we define the *distance* between X and Y by

$$d(X, Y) = [(x_1 - y_1)^2 + (x_2 - y_2)^2 + \cdots + (x_n - y_n)^2]^{1/2} \qquad (2.2.4)$$

Then $d(X, Y)$ defines a metric (or metric topology) on $\mathbb{R}_n$. Under the topology on $\mathbb{R}_n$ induced by the metric of (2.2.4), $\mathbb{P}_n$ and $\mathbb{Q}^n$ are compact. Thus, *every continuous real-valued function on $\mathbb{P}_n$ is bounded and assumes both a maximum and a minimum value.* Similarly, *every continuous real-valued function on $\mathbb{Q}^n$ is bounded and assumes both a maximum and minimum value.*

---

[1] This set of vectors will appear again in Chap. 3 but with a different name.

[2] When studying the geometry of these spaces no distinction need be made between the vector space of n-dimensional row vectors $\mathbb{R}_n$ and the vector space of n-dimensional column vectors $\mathbb{R}_n$ or, in fact, between any of the vector spaces $\mathbb{R}_q^p$ provided only that $pq = n$.

Lemma 2.2.5 Let $Z = [z_i] \in \mathbb{R}^m$. Then[1]

1. $\displaystyle \min_{P \in \mathbb{P}_m} PZ = \min_{P \in \{I_{(1)}, I_{(2)}, \ldots, I_{(m)}\}} PZ = \min_{i=1, 2, \ldots, m} z_i$

2. $\displaystyle \max_{P \in \mathbb{P}_m} PZ = \max_{P \in \{I_{(1)}, I_{(2)}, \ldots, I_{(m)}\}} PZ = \max_{i=1, 2, \ldots, m} z_i$

PROOF It is sufficient to present the proof of part 1 since the proof of part 2 is essentially identical. Suppose that the rth entry of Z is the smallest, i.e.,

$$z_r = \min_{i=1, 2, \ldots, m} z_i$$

Let $P = [p_i]$ be any element of $\mathbb{P}_m$. Then

$$PZ = p_1 z_1 + p_2 z_2 + \cdots + p_m z_m$$
$$\geq p_1 z_r + p_2 z_r + \cdots + p_m z_r$$
$$= (p_1 + p_2 + \cdots + p_m) z_r = z_r$$

Hence $$PZ \geq z_r$$

for all $P \in \mathbb{P}_m$. On the other hand,

$$I_{(r)} Z = z_r$$

where $I_{(r)}$ is the rth unit vector of $\mathbb{R}_m$. Since $I_{(r)} \in \mathbb{P}_m$, it follows that

$$\min_{P \in \mathbb{P}_m} PZ = \min_{P \in \{I_{(1)}, I_{(2)}, \ldots, I_{(m)}\}} PZ = \min_{i=1, 2, \ldots, m} z_i = z_j \qquad ////$$

Lemma 2.2.5 was tacitly assumed throughout Sec. 2.1. In Example 2.1.4, to find an optimal strategy for player II, we considered all his possible strategies

$$Q = \begin{bmatrix} q_1 \\ q_2 \\ q_3 \end{bmatrix}$$

Then we computed player I's expectation when he plays each of his two pure strategies

$$E(H, Q) = I_{(1)} \begin{bmatrix} 1 & 2 & 4 \\ 3 & 1 & 0 \end{bmatrix} Q = q_1 + 2q_2 + 4q_3 = 4 - 3q_1 - 2q_2$$

$$E(T, Q) = I_{(2)} \begin{bmatrix} 1 & 2 & 4 \\ 3 & 1 & 0 \end{bmatrix} Q = 3q_1 + q_2$$

---

[1] It is also possible to prove that $\min_{P \in \mathbb{P}_m} PZ$ and $\max_{P \in \mathbb{P}_m} PZ$ exist by showing that $f(P) = PZ$ is a continuous real-valued function of P and then applying the results of the last paragraph.

We did not investigate what would happen should player I use a mixed strategy, say $P = [p \quad 1-p]$ against the strategy Q. In this case his expectation would be

$$E(P, Q) = P \begin{bmatrix} 1 & 2 & 4 \\ 3 & 1 & 0 \end{bmatrix} Q$$

Part 2 of Lemma 2.2.5 (with $Z = \begin{bmatrix} 1 & 2 & 4 \\ 3 & 1 & 0 \end{bmatrix} Q$) justifies this procedure. For each (fixed) strategy Q for player II, the maximum expectation for player I using all his possible strategies is the same as his maximum expectation using only his pure strategies. Thus, in Example 2.1.4, player II was able to compute the expectation against the two pure strategies of player I and then obtain his optimal strategy directly from Fig. 2.1.6.

There is one further difficulty concerning the graphical methods used to solve Example 2.1.4. We stated, after examining Fig. 2.1.9, that $n(q_1, q_2)$ assumes a minimum value (which it achieves at the point T). However, no attempt was made to show that the minimum value actually exists. To this end, we prove:

Theorem 2.2.6 Let A be an $m \times n$ matrix. There exist $\tilde{P} \in \mathbb{P}_m$ and $\tilde{Q} \in \mathbb{Q}^n$ such that

$$\min_{Q \in \mathbb{Q}^n} \left(\max_{P \in \mathbb{P}_m} PAQ \right) = \max_{P \in \mathbb{P}_m} PA\tilde{Q} = \tilde{P}A\tilde{Q}$$

PROOF For each i, $i = 1, 2, \ldots, m$, the function $E(I_{(i)}, Q)$ defined by

$$E(I_{(i)}, Q) = I_{(i)} AQ$$

is a continuous (in fact, linear) real-valued function whose domain is $\mathbb{Q}^n$. It follows that the function $n(Q)$ defined by

$$n(Q) = \max_{P \in \{I_{(1)}, I_{(2)}, \ldots, I_{(m)}\}} PAQ$$

being the maximum of only finitely many continuous functions, is also a real-valued continuous function whose domain is $\mathbb{Q}^n$. Moreover, by part 2 of Lemma 2.2.5

$$n(Q) = \max_{P \in \mathbb{P}_m} PAQ$$

Since $\mathbb{Q}^n$ is compact, $n(Q)$ assumes a minimum value at some point, say $\tilde{Q}$, of $\mathbb{Q}^n$. Thus

$$\min_{Q \in \mathbb{Q}^n} \left(\max_{P \in \mathbb{P}_m} PAQ \right) = \min_{Q \in \mathbb{Q}^n} n(Q) = n(\tilde{Q}) = \max_{P \in \mathbb{P}_m} PA\tilde{Q}$$

Similarly, with $\tilde{Q}$ held fixed,

$$m(P) = PA\tilde{Q}$$

defines a continuous real-valued function whose domain is the compact set $\mathbb{P}_m$. There exists a vector $\tilde{P} \in \mathbb{P}_m$ such that

$$m(\tilde{P}) = \max_{P \in \mathbb{P}_m} PA\tilde{Q} = \tilde{P}A\tilde{Q} \qquad ////$$

Theorem 2.2.6 describes the process used to locate the point T in Fig. 2.1.9 and the optimal strategy

$$\bar{Q} = \begin{bmatrix} \frac{1}{3} \\ \frac{2}{3} \\ 0 \end{bmatrix}$$

for player II. The corresponding process used to find the optimal strategies for player I depends on:

Theorem 2.2.7 Let A be an $m \times n$ matrix. There exist $\tilde{\tilde{P}} \in \mathbb{P}_m$ and $\tilde{\tilde{Q}} \in \mathbb{Q}^n$ such that

$$\max_{P \in \mathbb{P}_m} \left(\min_{Q \in \mathbb{Q}^n} PAQ \right) = \min_{Q \in \mathbb{Q}^n} \tilde{\tilde{P}}AQ = \tilde{\tilde{P}}A\tilde{\tilde{Q}}$$

For any choice of $Q \in \mathbb{Q}^n$, player I's best expectation is given by

$$n(Q) = \max_{P \in \mathbb{P}_m} PAQ$$

Thus, player II has it in his power, by selecting Q properly, to limit player I's expectation to winning no more than

$$\min_{Q \in \mathbb{Q}^n} n(Q) = \min_{Q \in \mathbb{Q}^n} \left(\max_{P \in \mathbb{P}_m} PAQ \right)$$

and, moreover, no better, i.e., lower, limitation to player I's expectation exists. Hence, if $\tilde{P} \in \mathbb{P}_m$, $\tilde{Q} \in \mathbb{Q}^n$ satisfy Theorem 2.2.6, it follows that $\tilde{Q}$ is an optimal strategy for player II in the sense that it limits player I's expectation as effectively as possible. Similarly, for any choice of $P \in \mathbb{P}_m$, player II's best expectation is to lose no more than

$$m(P) = \min_{Q \in \mathbb{Q}^n} PAQ$$

Thus, by selecting P properly, player I can obtain an expectation for himself of

$$\max_{P \in \mathbb{P}_m} m(P) = \max_{P \in \mathbb{P}_m} \left(\min_{Q \in \mathbb{Q}^n} PAQ \right)$$

and this is, in fact, the best expectation he is able to guarantee for himself. As before, if $\tilde{\tilde{P}} \in \mathbb{P}_m$, $\tilde{\tilde{Q}} \in \mathbb{Q}^n$ satisfy Theorem 2.2.7, then $\tilde{\tilde{P}}$ is an optimal strategy for player I. (Note that $\tilde{P}$ and $\tilde{\tilde{Q}}$ may or may not be optimal strategies for players I and II respectively.)

On the basis of the discussion above, it is reasonable to expect:

Theorem 2.2.8 For every $m \times n$ matrix A,

$$\min_{Q \in \mathbb{Q}^n} \left(\max_{P \in \mathbb{P}_m} PAQ \right) \geq \max_{P \in \mathbb{P}_m} \left(\min_{Q \in \mathbb{Q}^n} PAQ \right)$$

Theorem 2.2.8 is a special case of the more general Theorem 2.2.9 which makes no use of the special structure of $\mathbb{P}_m$, $\mathbb{Q}^n$, or A.

Theorem 2.2.9 Let $\mathbb{S}$, $\mathbb{T}$ be nonempty sets and let $F(s, t)$ be a real-valued function defined for all $s \in \mathbb{S}$, $t \in \mathbb{T}$. If $\max_{s \in \mathbb{S}} F(s, t_1)$ and $\min_{t \in \mathbb{T}} F(s_1, t)$ exist for all $s_1 \in \mathbb{S}$, $t_1 \in \mathbb{T}$, and if $\min_{t \in \mathbb{T}}(\max_{s \in \mathbb{S}} F(s, t))$ and $\max_{s \in \mathbb{S}}(\min_{t \in \mathbb{T}} F(s, t))$ both exist, then

$$\min_{t \in \mathbb{T}} \left(\max_{s \in \mathbb{S}} F(s, t) \right) \geq \max_{s \in \mathbb{S}} \left(\min_{t \in \mathbb{T}} F(s, t) \right)$$

PROOF. For each (fixed) $t_1 \in \mathbb{T}$, $F(s, t_1)$ and $\min_{t \in \mathbb{T}} F(s, t)$ are both real-valued functions whose domain is $\mathbb{S}$, and for each $s \in \mathbb{S}$ we have

$$F(s, t_1) \geq \min_{t \in \mathbb{T}} F(s, t)$$

Then, for every $t_1 \in \mathbb{T}$,

$$\max_{s \in \mathbb{S}} F(s, t_1) \geq \max_{s \in \mathbb{S}} \left(\min_{t \in \mathbb{T}} F(s, t) \right) \qquad (2.2.10)$$

The right-hand side of (2.2.10) is a fixed real number, while the left-hand side depends on the choice of t_1. Since (2.2.10) is satisfied for every $t_1 \in \mathbb{T}$, it follows that the minimum value which the left-hand side assumes also satisfies this inequality, i.e.,

$$\min_{t_1 \in \mathbb{T}} \left(\max_{s \in \mathbb{S}} F(s, t_1) \right) \geq \max_{s \in \mathbb{S}} \left(\min_{t \in \mathbb{T}} F(s, t) \right)$$

Replacing t_1 by t throughout the left-hand side of this inequality completes the proof. ////

In each of the examples solved in Sec. 2.1 we found that

$$\min_{Q \in \mathbb{Q}^n} \left(\max_{P \in \mathbb{P}_m} PAQ \right) = \max_{P \in \mathbb{P}_m} \left(\min_{Q \in \mathbb{Q}^n} PAQ \right) \qquad (2.2.11)$$

It was this which permitted us to say in each case that the game had a value and that the optimal strategies for each player that we found could not be improved upon (in the sense that no better guaranteed expectation exists). On the other hand, it is clear that if the choices of P and Q are restricted (e.g., to being unit vectors), then we need not have equality in (2.2.11). (See Exercises 1 and 2.)

The statement that equation (2.2.11) holds for every matrix A is called the *minimax theorem* and will be proved in Sec. 2.5. It is the fundamental theorem for finite zero-sum two-person games and establishes that every game has a well-defined solution.

EXERCISES

1 In the games of Examples 2.1.1, 2.1.2, 2.1.4, and 2.1.10, find

$$\max_{i=1,2,\ldots,m} \left(\min_{j=1,2,\ldots,n} I_{(i)} A I^{(j)} \right)$$

and

$$\min_{j=1,2,\ldots,n} \left(\max_{i=1,2,\ldots,m} I_{(i)} A I^{(j)} \right)$$

2 Find

$$\max_{i=1,2,3,4} \left(\min_{j=1,2,3,4,5} I_{(i)} A I^{(j)} \right)$$

and

$$\min_{j=1,2,3,4,5} \left(\max_{i=1,2,3,4} I_{(i)} A I^{(j)} \right)$$

when

(a) $A = \begin{bmatrix} 1 & 2 & -1 & 0 & 0 \\ 1 & 1 & -1 & -2 & 0 \\ 2 & 4 & -8 & -6 & 1 \\ 1 & 7 & 3 & 2 & 5 \end{bmatrix}$ (b) $A = \begin{bmatrix} -3 & 0 & 1 & 7 & 1 \\ 6 & -2 & 2 & 1 & -1 \\ -2 & 0 & -3 & 0 & 0 \\ 8 & -1 & 2 & 4 & -6 \end{bmatrix}$

3 Let $F(x, y) = x^2 - 4x - 4xy + 2y^2$. Show that $\max_{x \in \mathbb{R}} (\min_{y \in \mathbb{R}} F(x, y))$ exists and evaluate it. [*Hint:* Find $n(x) = \min_{y \in \mathbb{R}} F(x, y)$ by holding x fixed.] Show that $\min_{y \in \mathbb{R}} (\max_{x \in \mathbb{R}} x^2 - 4x - 4xy + 2y^2)$ does not exist.

4 Which of the following sets in the plane are convex?
 (a) $\{(x, y) | y = x^2\}$
 (b) $\{(x, y) | y \geq x^2\}$
 (c) $\{(x, y) | y \leq x^2)\}$
 (d) $\{(x, y) | y > x^2\}$
 (e) $\{(x, y) | x^2 + y^2 = 1\}$
 (f) $\{(x, y) | x^2 + y^2 \geq 1\}$
 (g) $\{(x, y) | x^2 + y^2 \leq 1\}$
 (h) $\{(x, y) | x^2 + y^2 > 1\}$

5 Show that $\mathbb{P}_m$ and $\mathbb{Q}^n$ are convex.

6 Show that $\mathbb{S}$ is convex if $\mathbb{S}$ contains all convex combinations of pairs of elements of $\mathbb{S}$.

7   Let $\mathbb{S}$ be the collection of all convex combinations of the vectors $X_1, X_2, \ldots, X_k$. Show that $\mathbb{S}$ is convex.
8   Prove that the intersection of convex sets is convex.
9   Let $Z \in \mathbb{R}^m$ and define $\mathbb{O}(Z)$ to be

$$\mathbb{O}(Z) = \{X \mid XZ \geq 0, \quad X \in \mathbb{R}_m\}$$

Show that $\mathbb{O}(Z)$ is convex. Show that $\mathbb{O}(Z)$ contains m linearly independent vectors but $\mathbb{O}(Z) \neq \mathbb{R}_m$ unless $Z = 0$.

10 A point X of a convex set $\mathbb{S}$ is an *extreme point* if it cannot be expressed as

$$X = \tfrac{1}{2}Y + \tfrac{1}{2}Z$$

for some $Y, Z \in \mathbb{S}$, $Y \neq Z$. Show that the unit vectors are the extreme points of $\mathbb{P}_m$ and $\mathbb{Q}^n$. Find the extreme points of those sets in Exercise 4 which are convex. Show that a nonzero vector space has no extreme points.

11 Let X, Y, Z be elements of the convex set $\mathbb{S}$, $Y \neq Z$. Show that if X can be expressed as

$$X = \lambda Y + (1 - \lambda)Z$$

for some real number λ, $0 < \lambda < 1$, then X is not an extreme point of $\mathbb{S}$. (This leads to an alternate definition of extreme point.)

12 Let $\mathbb{S}$ be a convex subset of $\mathbb{R}^n$ and let $X = [x_i]$ be a point of $\mathbb{S}$ whose distance from 0 is maximal; that is, X belongs to $\mathbb{S}$ and

$$\sum_{i=1}^n x_i^2 \geq \sum_{i=1}^n p_i^2$$

for every $P = [P_i]$ belonging to $\mathbb{S}$. Prove that X is an extreme point of $\mathbb{S}$. *Hint:* Suppose that $X = \tfrac{1}{2}Y + \tfrac{1}{2}Z$, where $Y \neq Z$ and $Y, Z \in \mathbb{S}$. Let $Y = [y_i]$, $Z = [z_i]$ and set $W = Y - Z$, $W = [w_i]$. Show that

$$\sum_{i=1}^n y_i^2 + \sum_{i=1}^n z_i^2 = \sum_{i=1}^n w_i^2 + 2\sum_{i=1}^n y_i z_i > 2\sum_{i=1}^n y_i z_i$$

13 Let $\mathbb{S}$ be a convex subset of $\mathbb{R}^n$ and let A be an $m \times n$ matrix. Show that $A\mathbb{S}$ is convex.

2.3 A MATHEMATICAL INTRODUCTION TO THE THEORY OF GAMES

In the previous sections we began our study of the theory of games with a discussion relying heavily on specific examples and intuitive arguments. To put the subject on a firmer foundation and also to develop more sophisticated

[2.3.4]   THE THEORY OF GAMES   197

methods for obtaining information about games we now develop the subject again, this time in a more formal way, but preserving as much of the language and notation of the previous sections as possible. In order to produce a self-contained and logical development some repetition of the material of the first two sections is necessary.

We will be concerned with a game G which can be represented by an $m \times n$ matrix $A = [a_{ij}]$. Henceforth we will not be concerned with specific games, as in Sec. 2.1. The rules of a game determine its matrix. However, once the matrix is determined, the rules of the game no longer interest us. Thus we will consider any two games which yield the same matrix as being the same game.

Definition 2.3.1 A *strategy* for player I is an element of $\mathbb{P}_m$. A *strategy* for player II is an element of $\mathbb{Q}^n$. The unit vectors of $\mathbb{P}_m$ are the *pure strategies* for player I, and the unit vectors of $\mathbb{Q}^n$ are the *pure strategies* for player II.

Let P, Q be strategies for players I and II respectively in a game G whose matrix is A. Then the *expected value* of G for player I with respect to P and Q is defined to be PAQ and is denoted by $E(P, Q)$.

Definition 2.3.2 Let G be a game with matrix A. If there exist strategies $\bar{P} \in \mathbb{P}_m$ and $\bar{Q} \in \mathbb{Q}^n$ and a real number v such that

$$E(\bar{P}, Q) \geq v \geq E(P, \bar{Q}) \qquad (2.3.3)$$

for all $P \in \mathbb{P}_m$, $Q \in \mathbb{Q}^n$, then the triple $(\bar{P}, \bar{Q}, v)$ is a *solution* of G. We say that v is a *value* of G and that $\bar{P}, \bar{Q}$ are *optimal strategies* (with respect to v) for players I and II, respectively.

Suppose $(\bar{P}, \bar{Q}, v)$ is a solution of some game. Since $\bar{Q} \in \mathbb{Q}^n$, we may set $Q = \bar{Q}$ in the first inequality of (2.3.3). Thus

$$E(\bar{P}, \bar{Q}) \geq v$$

Similarly, setting $P = \bar{P}$ in the second inequality of (2.3.3) yields

$$v \geq E(\bar{P}, \bar{Q})$$

Hence
$$E(\bar{P}, \bar{Q}) = v \qquad (2.3.4)$$

According to Definition 2.3.2, a game may have no values, one value, or more than one value. Also optimal strategies are defined with respect to a particular value of the game. This ambiguity is removed by:

Lemma 2.3.5 Let $(\bar{P}_1, \bar{Q}_1, v_1)$ and $(\bar{P}_2, \bar{Q}_2, v_2)$ be solutions of the same game. Then
$$v_1 = v_2 \quad (=v)$$
and $(\bar{P}_1, \bar{Q}_2, v)$ and $(\bar{P}_2, \bar{Q}_1, v)$ are also solutions.

PROOF Since $(\bar{P}_1, \bar{Q}_1, v_1)$ and $(\bar{P}_2, \bar{Q}_2, v_2)$ are both solutions of the game,
$$E(\bar{P}_1, Q) \geq v_1 \geq E(P, \bar{Q}_1) \quad (2.3.6)$$
$$E(\bar{P}_2, Q) \geq v_2 \geq E(P, \bar{Q}_2) \quad (2.3.7)$$
for all $P \in \mathbb{P}_m$, $Q \in \mathbb{Q}^n$. Substituting $Q = \bar{Q}_2$ in the first part of (2.3.6) and $P = \bar{P}_1$ in the second part of (2.3.7) yields
$$v_2 \geq E(\bar{P}_1, \bar{Q}_2) \geq v_1$$
Substituting $P = \bar{P}_2$ in the second part of (2.3.6) and $Q = \bar{Q}_1$ in the first part of (2.3.7) yields
$$v_1 \geq E(\bar{P}_2, \bar{Q}_1) \geq v_2$$
Hence
$$v_1 = v_2$$
The remainder of the lemma is immediate, for clearly
$$E(\bar{P}_1, Q) \geq v_1 = v_2 \geq E(P, \bar{Q}_2)$$
$$E(\bar{P}_2, Q) \geq v_2 = v_1 \geq E(P, \bar{Q}_1)$$
for all $P \in \mathbb{P}_m$, $Q \in \mathbb{Q}^n$. ////

We have proved that a game has at most one value (which we will denote by v or v_G or v_A). It is not until Sec. 2.5 that we will be able to prove that every game has a value and that each player has at least one optimal strategy. Actually finding a solution of a game is another matter entirely. Some elementary methods for obtaining solutions were developed in Sec. 2.1, and more will be added in Sec. 2.6.[1]

In Sec. 2.2 we discussed the best expectations that each of the players is able to guarantee and showed that these expectations can be expressed in terms of

$$\min_{Q \in \mathbb{Q}^n} \left(\max_{P \in \mathbb{P}_m} E(P, Q) \right) \quad \text{and} \quad \max_{P \in \mathbb{P}_m} \left(\min_{Q \in \mathbb{Q}^n} E(P, Q) \right)$$

[1] There are still other schemes for finding solutions of a given game. Undoubtedly the most important of these is the simplex method. Here a game is first transformed into a linear-programming problem. (See Owen [7] and Sec. 2.8.) The simplex method is particularly adaptable for use with electronic digital computers.

The existence of a solution for a game G is equivalent to these best expectations being equal; that is:

Theorem 2.3.8 The game G has a solution $(\bar{P}, \bar{Q}, v)$ if and only if

$$\max_{P \in \mathbb{P}_m} \left(\min_{Q \in \mathbb{Q}^n} E(P, Q) \right) = \min_{Q \in \mathbb{Q}^n} \left(\max_{P \in \mathbb{P}_m} E(P, Q) \right)$$

PROOF First, suppose

$$\max_{P \in \mathbb{P}_m} \left(\min_{Q \in \mathbb{Q}^n} E(P, Q) \right) = v = \min_{Q \in \mathbb{Q}^n} \left(\max_{P \in \mathbb{P}_m} E(P, Q) \right)$$

By Theorem 2.2.6 there exists $\bar{Q} \in \mathbb{Q}^n$ such that

$$v = \min_{Q \in \mathbb{Q}^n} \left(\max_{P \in \mathbb{P}_m} E(P, Q) \right) = \max_{P \in \mathbb{P}_m} E(P, \bar{Q})$$

Similarly, by Theorem 2.2.7 there exists $\bar{P} \in \mathbb{P}_n$ such that

$$\max_{P \in \mathbb{P}_n} \left(\min_{Q \in \mathbb{Q}^n} E(P, Q) \right) = \min_{Q \in \mathbb{Q}^n} E(\bar{P}, Q) = v$$

Hence, for all $Q \in \mathbb{Q}^n$, $P \in \mathbb{P}_m$, we have

$$E(\bar{P}, Q) \geq \min_{Q \in \mathbb{Q}^n} E(\bar{P}, Q) = v = \max_{P \in \mathbb{P}_m} E(P, \bar{Q}) \geq E(P, \bar{Q})$$

and it follows from Definition 2.3.2 that $(\bar{P}, \bar{Q}, v)$ is a solution of G.
  Conversely, let $(\bar{P}, \bar{Q}, v)$ be a solution of G. We wish to show that

$$\max_{P \in \mathbb{P}_m} \left(\min_{Q \in \mathbb{Q}^n} E(P, Q) \right) = v = \min_{Q \in \mathbb{Q}^n} \left(\max_{P \in \mathbb{P}_m} E(P, Q) \right)$$

In view of Theorem 2.2.8 it is only necessary to show that

$$\max_{P \in \mathbb{P}_m} \left(\min_{Q \in \mathbb{Q}^n} E(P, Q) \right) \geq \min_{Q \in \mathbb{Q}^n} \left(\max_{P \in \mathbb{P}_m} E(P, Q) \right)$$

Since $(\bar{P}, \bar{Q}, v)$ is a solution of G, we have

$$E(\bar{P}, Q) \geq v \geq E(P, \bar{Q})$$

for all $P \in \mathbb{P}_m$, $Q \in \mathbb{Q}^n$. From the first of these inequalities it follows that

$$\min_{Q \in \mathbb{Q}^n} E(\bar{P}, Q) \geq v$$

and from the second of these inequalities, it follows that

$$v \geq \max_{P \in \mathbb{P}_m} E(P, \bar{Q})$$

But then

$$\max_{P \in \mathbb{P}_m} \left(\min_{Q \in \mathbb{Q}^n} E(P, Q) \right) \geq \min_{Q \in \mathbb{Q}^n} E(\bar{P}, Q) \geq v \geq$$

$$\max_{P \in \mathbb{P}_m} E(P, \bar{Q}) \geq \min_{Q \in \mathbb{Q}^n} \left(\max_{P \in \mathbb{P}_m} E(P, Q) \right)$$

as required. ////

EXERCISES

1 Verify that

$$\left([\tfrac{2}{3} \ \tfrac{1}{3}], \begin{bmatrix} \tfrac{1}{3} \\ \tfrac{2}{3} \\ 0 \end{bmatrix}, \tfrac{5}{3} \right)$$

is a solution for the game of Example 2.1.4 using Definition 2.3.2.

2 Prove that the set of optimal strategies for each player is convex.

3 Let A be the matrix of a game whose value is v. Prove that a necessary and sufficient condition that $\bar{P} \in \mathbb{P}_m$ be an optimal strategy for player I is that every entry of $\bar{P}A$ is greater than or equal to v. State and prove the corresponding condition for $\bar{Q} \in \mathbb{Q}^n$ to be an optimal strategy for player II.

4 Show that $(I_{(r)}, I^{(s)}, a_{rs})$ is a solution of a game with matrix A if and only if

$$a_{rj} \geq a_{rs} \geq a_{is} \qquad \begin{matrix} i = 1, 2, \ldots, m \\ j = 1, 2, \ldots, n \end{matrix} \qquad (2.3.9)$$

An entry a_{rs} of A which satisfies (2.3.9) is called a *saddle point* of A.

5 Show that A has a saddle point if and only if

$$\max_{i=1,2,\ldots,m} \left(\min_{j=1,2,\ldots,n} E(I_{(i)}, I^{(j)}) \right) = \min_{j=1,2,\ldots,n} \left(\max_{i=1,2,\ldots,m} E(I_{(i)}, I^{(j)}) \right)$$

(See Exercises 1 and 2 of Sec. 2.2.) *Hint:* Show

$$\max_{i=1,2,\ldots,m} \left(\min_{j=1,2,\ldots,n} a_{ij} \right) = \min_{j=1,2,\ldots,n} a_{rj} = a_{rs}$$

6 Solve the following games by finding all the saddle points

(a) $A = \begin{bmatrix} 4 & 2 & 1 \\ -9 & 7 & -1 \end{bmatrix}$
(b) $A = \begin{bmatrix} -7 & 15 & 0 & 8 \\ 12 & -6 & -1 & -8 \\ 3 & 4 & 3 & 4 \\ 5 & -1 & -3 & 10 \end{bmatrix}$

(c) $A = \begin{bmatrix} 1 & -2 & -1 & 1 \\ -4 & -1 & -1 & 3 \\ 1 & -1 & -3 & -1 \\ 3 & -1 & -1 & 2 \end{bmatrix}$ (d) $A = \begin{bmatrix} -5 & -2 & 2 & 2 & 0 \\ 2 & -5 & -4 & 0 & -7 \\ 2 & 3 & 4 & 2 & 5 \\ -1 & 4 & 2 & -4 & 2 \end{bmatrix}$

7 Examine the games of Examples 2.1.1, 2.1.2, and 2.1.10 for saddle points.
8 Let a_{rs} be a saddle point of the matrix A for which

$$a_{rj} > a_{rs} > a_{is} \quad \begin{matrix} i \neq r \\ j \neq s \end{matrix}$$

Show that $(I_{(r)}, I^{(s)}, a_{rs})$ is the only solution of the game. (*Hint:* Use Exercise 3.)

9 Let a_{rs} and a_{tn} be saddle points of A. Show that a_{rn} and a_{ts} are saddle points of A and that $a_{rs} = a_{tn} = a_{rn} = a_{ts}$.
10 Let the matrix A' be formed from the matrix A by deleting one of the columns of A. Show that $v_{A'} \geq v_A$.
11 Let E be the $m \times n$ matrix all of whose entries are 1. Let G be a game whose matrix is the $m \times n$ matrix A and let G' be the game whose matrix is $A + \alpha E$. Show that $(\bar{P}, \bar{Q}, v)$ is a solution of G if and only if $(\bar{P}, \bar{Q}, v + \alpha)$ is a solution of G'.
12 Let $(\bar{P}_1, \bar{Q}_1, v_1)$ and $(\bar{P}_2, \bar{Q}_2, v_2)$ be solutions of games whose matrices are A_1 and A_2 respectively. Set

$$A = \begin{bmatrix} A_1 & 0 \\ 0 & A_2 \end{bmatrix}$$

Show that
(a) If $v_1 > 0$, $v_2 > 0$, then

$$\left(\alpha_1 [\bar{P}_1 \ 0] + \alpha_2 [0 \ \bar{P}_2], \alpha_1 \begin{bmatrix} \bar{Q}_1 \\ 0 \end{bmatrix} + \alpha_2 \begin{bmatrix} 0 \\ \bar{Q}_2 \end{bmatrix}, \frac{v_1 v_2}{v_1 + v_2} \right)$$

where $\alpha_1 = \dfrac{v_2}{v_1 + v_2}$ $\alpha_2 = \dfrac{v_1}{v_1 + v_2}$

is a solution of A.

(b) If $v_1 > 0$, $v_2 < 0$, then $\left([\bar{P}_1 \ 0], \begin{bmatrix} 0 \\ \bar{Q}_2 \end{bmatrix}, 0 \right)$ is a solution of A.

13 Show that the following game has a saddle point. Player I is dealt a card from a deck consisting of the two cards V and X. After examining the card, player I calls out either

"I have V"   or   "I have X"

Player II then calls out either

"I believe you"   or   "I do not believe you"

(a) If player I has told the truth and player II believes him, then player II wins 5 when player I holds V and 10 when player I holds X.

(b) If player I has told the truth and player II does not believe him, then player I wins 5 when he holds V and 10 when he holds X.

(c) On the other hand, if player I has lied and player II believes him, then player I wins 10 if he holds V and 20 if he holds X.

(d) Finally, if player I lied and player II catches the lie, then player II wins 10 if Player I holds V and 20 if player I holds X.

14 Suppose the payoff in the game of Exercise 13 is made to depend upon the card player I calls (rather than the card he holds). Does the resulting game have a saddle point?

2.4 LINEAR INEQUALITIES

In Sec. 2.5 we will prove that every (finite zero-sum two-person) game has a solution.[1] The main tool we will need for the proof presented here is a theorem concerning the existence of nonnegative solutions of a system of simultaneous linear inequalities. In Chap. 1 we saw that the behavior of a system of simultaneous linear equations is closely connected with the theory of vector spaces (see, for example, Exercises 23 and 24 of Sec. 1.3, Exercises 14 and 15 of Sec. 1.4, and Exercise 9 of Sec. 1.5). In the same way, problems dealing with the existence of nonnegative solutions of either simultaneous equations or simultaneous inequalities (or a mixture of both) lead to the theory of cones.

First we need some definitions.

Definition 2.4.1 A matrix A is said to be:

1 *Positive* if every entry of A is positive. We express this by writing $A > 0$.
2 *Semipositive* if every entry of A is nonnegative and $A \neq 0$. We express this by writing $A \geq 0$.
3 *Nonnegative* if every entry of A is nonnegative. In this case we write $A \geqq 0$.

---

[1] The first proof of this theorem was given by von Neumann [6] in 1928. Since then several other proofs have appeared by different authors. Most of these, though they require a knowledge either of topology (in particular, of fixed-point theorems) or of the geometry of convex sets, have the virtue of being intuitively clear—probably clearer than the proof presented here. (See Karlin [4] and McKinsey [5].) The main advantage of the proof we have chosen is that it requires few preliminaries, all of which can be presented in algebraic guise. For another proof using matrix methods see Dresher [1].

For two $m \times n$ matrices A and B we write

$$A > B \quad \text{if} \quad A - B > 0$$
$$A \geq B \quad \text{if} \quad A - B \geq 0$$
$$A \geqq B \quad \text{if} \quad A - B \geqq 0$$

All the vectors belonging to $\mathbb{P}_m$ and $\mathbb{Q}^n$ are semipositive. Further, it is not difficult to see that a vector of $\mathbb{R}_m$ (or $\mathbb{R}^n$) is semipositive if and only if it can be expressed as a positive multiple of some vector which belongs to $\mathbb{P}_m$ (or $\mathbb{Q}^n$).

Definition 2.4.2 Let $\mathbb{K}$ be a nonempty set of $m \times n$ matrices. $\mathbb{K}$ is called a *cone* if

1 $X + Y$ belongs to $\mathbb{K}$ whenever X and Y belong to $\mathbb{K}$.
2 αX belongs to $\mathbb{K}$ whenever X belongs to $\mathbb{K}$ and $\alpha \geqq 0$.

The first part of Definition 2.4.2 is identical to the first part of Definition 1.3.1. The second parts of the two definitions differ only in that a cone is required to be closed under multiplication by a nonnegative scalar. Consequently every vector space is a cone. However, the set of all $m \times n$ matrices which have nonnegative entries is a cone which is not a vector space.

Definition 2.4.3 Let $A_1, A_2, \ldots, A_k, B$ be elements of a vector space $\mathbb{V}$. We say that B is a nonnegative linear combination of $A_1, A_2, \ldots, A_k$ if there exist nonnegative numbers $\alpha_1, \alpha_2, \ldots, \alpha_k$ such that

$$B = \alpha_1 A_1 + \alpha_2 A_2 + \cdots + \alpha_k A_k$$

Corollary 2.4.4. Let $\mathbb{V}$ be a vector space and let $A_1, A_2, \ldots, A_k$ be elements of $\mathbb{V}$. The set $\mathbb{K}$ of all nonnegative linear combinations of $A_1, A_2, \ldots, A_k$ is a cone.

PROOF The proof is left as an exercise. ////

We say that the cone $\mathbb{K}$ is spanned by $A_1, A_2, \ldots, A_k$.

Definition 2.4.5 Let A be an $m \times n$ matrix. The cone spanned by the columns of A is called the *column cone* of A and is denoted by $\mathbb{K}(A)$. The cone spanned by the rows of A is called the *row cone* of A and is denoted by $\mathbb{K}_R(A)$.

Corollary 2.4.6 Let A be an $m \times n$ matrix and let $Z \in \mathbb{R}^m$. Then Z belongs to $\mathbb{K}(A)$ if and only if there exists $X \in \mathbb{R}^n$ such that

$$Z = AX \qquad X \geqq 0$$

PROOF The proof of the corollary, as well as the statement and proof of the corresponding result for $\mathbb{K}_R(A)$, is left as an exercise. ////

Corollary 2.4.6 is the analog of Theorem 1.4.4(b).

Definition 2.4.7 Let $\mathbb{S}$ be any subset of $\mathbb{R}_n^m$. The set of vectors $Y \in \mathbb{R}_m$ such that

$$YX \geqq 0$$

for all $X \in \mathbb{S}$ is a cone called the *nonnegative row complement* of $\mathbb{S}$ (or simply the *complement* of $\mathbb{S}$) and is denoted by $\mathbb{O}(\mathbb{S})$.

When $\mathbb{S}$ consists of a single matrix A, we write $\mathbb{O}(A)$ instead of $\mathbb{O}(\mathbb{S})$. For $A = [a_i] \in \mathbb{R}^m$ it is clear that $A^T \in \mathbb{O}(A)$ since

$$A^T A = a_1^2 + a_2^2 + \cdots + a_m^2 \geqq 0$$

Lemma 2.4.8 Let A be an $m \times n$ matrix and let Z be an element of $\mathbb{R}^m$. If

$$\mathbb{O}(A) \subseteq \mathbb{O}(Z)$$

then $Z \in \mathbb{C}(A)$; that is, the equation

$$Z = AX$$

has a solution.

PROOF It is sufficient to show that the row null space of A is contained in the row null space of Z, for then the lemma follows from Theorem 1.7.16. Let Y be a vector in the row null space of A. Clearly, the row null space of A is a vector space contained in $\mathbb{O}(A)$. Hence $-Y$ also belongs to $\mathbb{O}(A)$, and we have

$$YZ \geqq 0 \qquad (-Y)Z \geqq 0$$

Thus $YZ = 0$, and consequently Y belongs to the row null space of Z. ////

Lemma 2.4.8 can be strengthened considerably. We will in fact show that if $\mathbb{O}(A) \subseteq \mathbb{O}(Z)$, then $Z \in \mathbb{K}(A)$; that is, the system

$$Z = AX \qquad X \geqq 0$$

has a solution. This result is a part of Theorem 2.4.10, which states that $\mathbb{K}(Z) \subseteq \mathbb{K}(A)$ if and only if $\mathbb{O}(A) \subseteq \mathbb{O}(Z)$. Theorem 2.4.10 is the analog for cones of Theorem 1.7.16. First, however, we need a lemma.

Lemma 2.4.9 Let S, T be elements of $\mathbb{R}_m$, $\mathbb{R}^m$ respectively such that $ST \neq 0$ and let $B = TS$. Then the characteristic roots of B are 0, with geometric and algebraic multiplicity $m - 1$, and ST, with geometric and algebraic multiplicity 1. The characteristic vectors of B corresponding to the characteristic root 0 are the nonzero elements of $\mathbb{N}(S)$, and the unique (up to nonzero multiples) characteristic vector of B corresponding to ST is T.[1]

PROOF For $X \in \mathbb{N}(S)$, we have

$$BX = TSX = T0 = 0$$

Thus the nonzero elements of the $(m - 1)$-dimensional subspace $\mathbb{N}(S)$ of $\mathbb{R}^m$ are all characteristic vectors of B corresponding to the characteristic root 0. Consequently, 0 is a characteristic root of A of geometric multiplicity $m - 1$. (See Exercise 24 of Sec. 1.10). Also,

$$[B - (ST)I]T = (TS)T - (ST)T = (TS)T - T(ST) = 0$$

since ST is of order 1 and hence commutes with T. Thus T is a characteristic vector of B corresponding to the characteristic root ST. Since $ST \neq 0$, we have now determined n characteristic roots for B, counting each root according to its geometric multiplicity. Thus, in each case, the algebraic multiplicity is equal to the geometric multiplicity, and B has no other characteristic roots. Consequently, the geometric multiplicity of ST is 1. Hence T is the unique (up to nonzero multiples) characteristic vector of B corresponding to the characteristic root ST.   ////

Now we are able to prove:

Theorem 2.4.10 (Farkas, 1902) Let A be an $m \times n$ matrix and let $Z \in \mathbb{R}^m$. Then

$$\mathbb{K}(Z) \subseteq \mathbb{K}(A)$$

if and only if

$$\mathbb{O}(A) \subseteq \mathbb{O}(Z)$$

---

[1] Note that we are purposely confusing the 1×1 matrix ST with the entry of this matrix.

PROOF First suppose that $\mathbb{K}(Z) \subseteq \mathbb{K}(A)$; that is, suppose there exists a vector $X \in \mathbb{R}^n$ such that

$$Z = AX \qquad X \geqq 0$$

For $Y \in \mathbb{O}(A)$ we have

$$YA \geqq 0$$

and hence

$$YZ = YAX \geqq 0$$

Thus $Y \in \mathbb{O}(Z)$.

Conversely, suppose $\mathbb{O}(A) \subseteq \mathbb{O}(Z)$. By Lemma 2.4.8, Z belongs to $\mathbb{C}(A)$, and hence there exists $X \in \mathbb{R}^n$ such that $Z = AX$. It remains to show that a nonnegative vector X exists. The proof is by an induction on n. For $n = 1$, $X = [x]$, $x \in \mathbb{R}$, and Z is a scalar multiple of A. Since $A^T \in \mathbb{O}(A)$, it follows that $A^T \in \mathbb{O}(Z)$ and we have

$$A^T Z = A^T A X = (a_1^2 + a_2^2 + \cdots + a_m^2) x \geqq 0$$

If $A = 0$, then x is completely arbitrary and may be chosen to be nonnegative. If $A \neq 0$, then x is uniquely defined and is nonnegative.

Now assume that the theorem has been proved for all $m \times (n-1)$ matrices and let A be an $m \times n$ matrix. Denote by $\tilde{A}$ the $m \times (n-1)$ matrix formed from A by deleting the last column of A. Clearly $\mathbb{O}(A) \subseteq \mathbb{O}(\tilde{A})$. Here two cases arise. If $\mathbb{O}(\tilde{A}) \subseteq \mathbb{O}(Z)$, then, by the induction hypothesis, there is a nonnegative vector $\tilde{X}$ belonging to $\mathbb{R}^{n-1}$ such that $Z = \tilde{A}\tilde{X}$. Set

$$X = \begin{bmatrix} \tilde{X} \\ 0 \end{bmatrix}$$

Then X is nonnegative and

$$AX = [\tilde{A} \quad A^{(n)}] \begin{bmatrix} \tilde{X} \\ 0 \end{bmatrix} = \tilde{A}\tilde{X} = Z$$

proving the theorem.

Finally, let us assume $\mathbb{O}(\tilde{A}) \nsubseteq \mathbb{O}(Z)$; that is, there exists a vector $\tilde{Y} \in \mathbb{O}(\tilde{A})$ such that $\tilde{Y}Z < 0$. Then the induction hypothesis cannot be applied to $\tilde{A}$ and Z. Since $\tilde{Y} \notin \mathbb{O}(Z)$, it follows that $\tilde{Y} \notin \mathbb{O}(A)$; that is,

$$\tilde{Y}A = \tilde{Y}[\tilde{A} \quad A^{(n)}] = [\tilde{Y}\tilde{A} \quad \tilde{Y}A^{(n)}] \ngeqq 0$$

But $\tilde{Y} \in \mathbb{O}(\tilde{A})$, and hence $\tilde{Y}\tilde{A} \geqq 0$. Thus, $\tilde{Y}A^{(n)} < 0$.

Set

$$B = A^{(n)}\tilde{Y}$$
$$C = A^{(n)}\tilde{Y} - (\tilde{Y}A^{(n)})I$$

B and C are square matrices of order m. By Lemma 2.4.9, $A^{(n)}$ is a characteristic vector of B corresponding to the characteristic root $\tilde{Y}A^{(n)}$. Hence

$$CA^{(n)} = [B - (\tilde{Y}A^{(n)})I]A^{(n)} = 0 \qquad (2.4.11)$$

Set $\bar{A} = C\tilde{A}$ and $\bar{Z} = CZ$. We wish to show that the induction hypothesis can be applied to $\bar{A}$ and $\bar{Z}$; that is, we wish to show that $\mathbb{O}(\bar{A}) \subseteq \mathbb{O}(\bar{Z})$. Let $Y_0 \in \mathbb{O}(\bar{A})$ and set $\bar{Y} = Y_0 C$. Then, by equation (2.4.11),

$$\begin{aligned}\bar{Y}A = \bar{Y}[\tilde{A} \quad A^{(n)}] &= [\bar{Y}\tilde{A} \quad \bar{Y}A^{(n)}] \\ &= [Y_0 C\tilde{A} \quad Y_0 CA^{(n)}] \qquad (2.4.12) \\ &= [Y_0 \bar{A} \quad 0] \geq 0\end{aligned}$$

Equation (2.4.12) states that $\bar{Y} \in \mathbb{O}(A)$. By hypothesis $\mathbb{O}(A) \subseteq \mathbb{O}(Z)$, and hence

$$\bar{Y}Z = Y_0 CZ = Y_0 \bar{Z} \geq 0$$

Thus $Y_0 \in \mathbb{O}(\bar{Z})$. We may now apply the induction hypothesis to $\bar{A}$ and $\bar{Z}$. There exists a nonnegative vector $\bar{X} \in \mathbb{R}^{n-1}$ such that

$$\bar{Z} = \bar{A}\bar{X}$$

Then $CZ = C\tilde{A}\bar{X}$ which can be written

$$0 = C(Z - \tilde{A}\bar{X}) = [B - (\tilde{Y}A^{(n)})I](Z - \tilde{A}\bar{X})$$

Thus $Z - \tilde{A}\bar{X}$ is zero or is a characteristic vector of B corresponding to the characteristic root $\tilde{Y}A^{(n)}$. But by Lemma 2.4.9 $A^{(n)}$ is the only characteristic vector of B corresponding to the root $\tilde{Y}A^{(n)}$ (up to nonzero multiples). Hence

$$Z - \tilde{A}\bar{X} = \alpha A^{(n)} \qquad (2.4.13)$$

for some real number α. Equation (2.4.13) can be written

$$Z = [\tilde{A} \quad A^{(n)}]\begin{bmatrix}\bar{X} \\ \alpha\end{bmatrix} = A\begin{bmatrix}\bar{X} \\ \alpha\end{bmatrix}$$

and in order to complete the proof it remains only to show that $\alpha \geq 0$. Multiply equation (2.4.13) on the left by $\tilde{Y}$. Then

$$\tilde{Y}Z - \tilde{Y}\tilde{A}\bar{X} = \tilde{Y}\alpha A^{(n)} \qquad (2.4.14)$$

However, $\tilde{Y}$ was specifically defined as a vector satisfying the two conditions

$$\tilde{Y}Z < 0 \qquad \tilde{Y}\tilde{A} \geq 0$$

Since we also have $\bar{X} \geq 0$, the left-hand side of equation (2.4.14) is negative. But we proved earlier that $\tilde{Y}A^{(n)} < 0$, and hence it follows that $\alpha > 0$.

////

Theorem 2.4.15 Let A be an $m \times n$ matrix and let $Z \in \mathbb{R}^m$. Then the system

$$Z = AX \qquad X \geq 0$$

has a solution if and only if

$$\mathbb{O}(A) \subseteq \mathbb{O}(Z)$$

By means of a simple device we can also use Farkas' theorem to study the existence of nonnegative solutions of simultaneous linear inequalities. Let A be an $m \times n$ matrix and let $Z \in \mathbb{R}^m$. We wish to determine when the system of simultaneous linear inequalities

$$Z \geq AX \qquad X \geq 0 \qquad (2.4.16)$$

has a solution. We can change (2.4.16) into a system of m simultaneous linear equations in $m + n$ unknowns requiring a nonnegative solution by introducing the *slack vector* W. Set

$$W = Z - AX$$

Clearly (2.4.16) is equivalent to finding a vector $\begin{bmatrix} X \\ W \end{bmatrix}$ belonging to $\mathbb{R}^{m+n}$ such that

$$[A \quad I]\begin{bmatrix} X \\ W \end{bmatrix} = Z \qquad \begin{bmatrix} X \\ W \end{bmatrix} \geq 0 \qquad (2.4.17)$$

Conditions for the existence of such a vector are given by Theorem 2.4.15. We have:

Corollary 2.4.18 Let A be an $m \times n$ matrix and let $Z \in \mathbb{R}^m$. A necessary and sufficient condition that there exist $X \in \mathbb{R}^n$ satisfying

$$Z \geq AX \qquad X \geq 0$$

is that

$$YZ \geq 0$$

for all nonnegative vectors Y belonging to $\mathbb{O}(A)$; that is, every nonnegative element of $\mathbb{O}(A)$ belongs to $\mathbb{O}(Z)$.

PROOF   According to Theorem 2.4.15, the system (2.4.17) has a solution if and only if $\mathbb{O}([A \quad I]) \subseteq \mathbb{O}(Z)$. However, Y is contained in $\mathbb{O}([A \quad I])$ if and only if

$$YA \geqq 0 \qquad Y \geqq 0 \qquad\qquad ////$$

In the particular situation where we use Corollary 2.4.18 we will be concerned only with nonzero vectors X which satisfy the system of homogeneous inequalities

$$0 \geqq AX \qquad X \geqq 0 \qquad (2.4.19)$$

A nonnegative vector X is a zero vector if and only if the sum of its entries is zero. Thus in order to avoid the trivial solution, $X = 0$, for (2.4.19) we add the condition $E_{(n)} X > 0$ [where $E_{(n)}$ is the n-dimensional row vector all of whose entries are 1]. Clearly if X satisfies (2.4.19), then αX also satisfies (2.4.19) for every nonnegative number α. Hence we may replace the condition $E_{(n)} X > 0$ by the more easily handled condition

$$E_{(n)} X \geqq 1$$

In order to make this condition compatible with the form of Corollary 2.4.18 we write it as

$$-1 \geqq (-E_{(n)})X$$

We have now shown that (2.4.19) has a semipositive (i.e., nonzero nonnegative) solution if and only if the system of inequalities

$$\begin{bmatrix} 0 \\ -1 \end{bmatrix} \geqq \begin{bmatrix} A \\ -E_{(n)} \end{bmatrix} X \qquad X \geqq 0 \qquad (2.4.20)$$

has a solution.

Theorem 2.4.21   Let A be an $m \times n$ matrix. Exactly one of the systems

$$0 \geqq AX \qquad X \geqq 0 \qquad (2.4.22)$$

and

$$YA > 0 \qquad Y \geqq 0 \qquad (2.4.23)$$

has a solution.

PROOF   Let us assume that both (2.4.22) and (2.4.23) have solutions. From the pair of inequalities

$$0 \geqq AX \qquad Y \geqq 0$$

it follows that

$$0 \geqq YAX$$

On the other hand, from the inequalities

$$YA > 0 \quad X \geq 0$$

it follows that

$$YAX > 0$$

since some entry of X is positive and every entry of YA is positive. Thus we have

$$0 \geqq YAX > 0$$

which is a contradiction. Therefore (2.4.22) and (2.4.23) cannot both have solutions.

It remains to show that if (2.4.22) does not have a solution, then (2.4.23) does have a solution. According to the discussion preceding the theorem, the existence of a solution of (2.4.22) is equivalent to the existence of a solution of (2.4.20). Thus let us suppose (2.4.20) does not have a solution. By Corollary 2.4.18 [applied to (2.4.20)] there exists a nonnegative vector $Y_0 \in \mathbb{R}_{m+1}$ such that

$$Y_0 \begin{bmatrix} 0 \\ -1 \end{bmatrix} \not\geqq 0 \quad Y_0 \in \mathbb{O}\left(\begin{bmatrix} A \\ -E_{(n)} \end{bmatrix}\right) \quad (2.4.24)$$

Let

$$Y_0 = [Y, y_{m+1}]$$

where $Y \in \mathbb{R}_m$, $y_{m+1} \in \mathbb{R}$. Then the two parts of (2.4.24) become

$$-y_{m+1} \not\geqq 0 \quad YA - y_{m+1}E_{(n)} \geqq 0$$

which in turn can be written

$$y_{m+1} > 0 \quad YA \geqq y_{m+1}E_{(n)}$$

Hence Y satisfies (2.4.23). ////

Corollary 2.4.25 If A is a skew-symmetric matrix, then

$$0 \geqq AX \quad X \geq 0 \quad (2.4.26)$$

has a solution.

PROOF Suppose no solution of (2.4.26) exists. By Theorem 2.4.21, the system of inequalities

$$YA > 0 \quad Y \geq 0$$

has a solution. Taking the transpose of the first of these inequalities yields

$$A^T Y^T > 0$$

Since $A^T = -A$, we now have

$$AY^T < 0 \qquad Y^T \geq 0$$

Thus Y^T satisfies (2.4.26). ////

If we interpret the matrix A of Corollary 2.4.25 as the matrix of a symmetric game, the corollary states that there is an optimal strategy X (up to multiplication by some positive number) for player II.

EXERCISES

1 Which of the following subsets of $\mathbb{R}^3$ are cones?

(a) $\left\{ \begin{bmatrix} x \\ y \\ z \end{bmatrix} \middle| z \geq x^2 + y^2 \right\}$
(b) $\left\{ \begin{bmatrix} x \\ y \\ z \end{bmatrix} \middle| z^2 \geq x^2 + y^2, z \geq 0 \right\}$

(c) $\left\{ \begin{bmatrix} x \\ y \\ z \end{bmatrix} \middle| z = x^2 + y^2 \right\}$
(d) $\left\{ \begin{bmatrix} x \\ y \\ z \end{bmatrix} \middle| z^2 > x^2 + y^2, z \geq 0 \right\}$

(e) $\left\{ \begin{bmatrix} x \\ y \\ z \end{bmatrix} \middle| z^2 > x^2 + y^2 \right\}$
(f) $\left\{ \begin{bmatrix} x \\ y \\ z \end{bmatrix} \middle| z^2 \geq x^2 + y^2, z \leq 0 \right\}$

(g) $\left\{ \begin{bmatrix} x \\ y \\ z \end{bmatrix} \middle| x \geq 0, y > 0 \right\}$
(h) $\left\{ \begin{bmatrix} x \\ y \\ z \end{bmatrix} \middle| x \geq 0, y \geq 0, z = 0 \right\}$

2 Let

$$A = \begin{bmatrix} 1 & -1 \\ 2 & 1 \end{bmatrix} \qquad B = \begin{bmatrix} 0 & -1 \\ 1 & 4 \end{bmatrix}$$

Show that $\mathbb{K}(B) \subseteq \mathbb{K}(A)$:
(a) By expressing each column of B as a nonnegative linear combination of the columns of A.
(b) By sketching $\mathbb{K}(A)$ and $\mathbb{K}(B)$.
Show that $\mathbb{K}(B) \neq \mathbb{K}(A)$. Find all 2×2 matrices C such that $\mathbb{K}(A) = \mathbb{K}(C)$.

3 Let
$$A = \begin{bmatrix} 1 & -1 \\ 2 & 1 \end{bmatrix} \quad B = \begin{bmatrix} 0 & -1 \\ 1 & 4 \end{bmatrix}$$
Show that $\mathbb{O}(A) \subseteq \mathbb{O}(B)$:
(a) By using Exercise 2 and Farkas' theorem.
(b) By sketching $\mathbb{O}(A)$ and $\mathbb{O}(B)$.

4 Let
$$A = \begin{bmatrix} 2 & 3 \\ -1 & 1 \end{bmatrix} \quad Z = \begin{bmatrix} 1 \\ -2 \end{bmatrix}$$
Show that there does not exist a nonnegative vector X such that $Z \geqq AX$:
(a) By working directly with the inequalities.
(b) By showing that the condition of Corollary 2.4.18 is contradicted.
(c) By sketching Z and $\mathbb{K}(A)$.
Find a (necessarily not nonnegative) vector X such that $Z \geqq AX$. Find the set of all vectors Z for which there does exist a nonnegative vector X such that $Z \geqq AX$.

5 Let
$$A = \begin{bmatrix} 2 & 7 \\ 1 & -1 \end{bmatrix} \quad Z = \begin{bmatrix} 1 \\ 3 \end{bmatrix}$$
Show that there does not exist a nonnegative vector X such that $Z = AX$:
(a) By working directly with the simultaneous equations.
(b) By using Farkas' theorem.
(c) By sketching Z and $\mathbb{K}(A)$.
Show that there exists a semipositive vector X such that $Z \geqq AX$.

6 Let
$$A = \begin{bmatrix} -3 & 1 & -5 \\ 2 & -1 & 3 \end{bmatrix}$$
Show that
$$Z \geqq AX \quad X \geqq 0$$
has a solution for every $Z \in \mathbb{R}^2$. Find $Z \in \mathbb{R}^2$ such that
$$Z = AX \quad X \geqq 0$$
does not have a solution.

7 Prove Corollary 2.4.4.
8 Prove Corollary 2.4.6.
9 Prove that $\mathbb{O}(A)$ is a cone for every matrix A.

10 Prove that the intersection of two cones is a cone.
11 Prove that $\mathbb{O}(A)$ is a vector space if and only if $\mathbb{C}(A) = \mathbb{K}(A)$.
12 Let $A \in \mathbb{R}^n$, $A \neq 0$. Show that every element of $\mathbb{O}(A)$ can be expressed in the form

$$\alpha A^T + X$$

where $\alpha \geq 0$ and X belongs to the row null space of A.

13 Let A be an $m \times n$ matrix. Show that

$$AX \geqq Z \qquad X \geqq 0$$

has a solution if and only if

$$YZ \geqq 0$$

for all $Y \in \mathbb{O}(A)$ such that $Y \leqq 0$.

14 Let A be an $m \times n$ matrix. Prove that exactly one of the systems

$$0 = AX \qquad X \geq 0 \quad \text{and} \quad YA > 0$$

has a solution.

2.5 THE FUNDAMENTAL THEOREM

In Sec. 2.1 we defined a game to be symmetric if the rules of play and the payoff are the same for each player. Clearly such a game has a skew-symmetric matrix; that is, $A^T = -A$. Two-fingered morra (both as described in Example 2.1.10 and in the variant of Exercise 4) and the duels of Example 2.1.13 (provided the accuracy of the duelists are equal, as in Exercises 9 and 10 of Sec. 2.1) are examples of symmetric games. However, it is not difficult to devise a game in which the rules of play for the two players are completely different but which has the same matrix as one of the symmetric games we have discussed. Clearly the theory of such a game would be identical to the theory of the symmetric game with the same matrix. For this reason we now make the following definition.

Definition 2.5.1 A game is *symmetric* if its matrix is skew-symmetric.

In discussing symmetric games in the past, we have said that certain strategies were optimal for both players. The matrix of a symmetric game is square, and hence the strategies for each player have the same number of entries. Thus P is a strategy vector for player I (that is, $P \in \mathbb{P}_n$) if and only if P^T is a strategy vector for player II (that is, $P^T \in \mathbb{Q}^n$). At this point it is inconvenient that the strategies for player I and for player II have been written as row and column vectors respectively. In discussing symmetric games, we will identify the two strategies $P \in \mathbb{P}_n$ and $P^T \in \mathbb{Q}^n$ for players I and II and speak of them as being the same strategy.

Lemma 2.5.2 The value of a symmetric game (if it exists) is zero. Moreover, a strategy is an optimal strategy for player I if and only if it is an optimal strategy for player II, that is, P is optimal for player I if and only if P^T is optimal for player II.

PROOF Let G be a symmetric game with (skew-symmetric) matrix A and let $(\bar{P}, \bar{Q}, v)$ be a solution of G. Then

$$E(\bar{P}, Q) \geqq v \geqq E(P, \bar{Q}) \qquad (2.5.3)$$

for all $P \in \mathbb{P}_n$, $Q \in \mathbb{Q}^n$. Since A is skew-symmetric, it follows [from part (c) of Exercise 20 in Sec. 1.7] that

$$E(\bar{P}, Q) = \bar{P}AQ = (\bar{P}AQ)^T = Q^T A^T \bar{P}^T = -Q^T A \bar{P}^T = -E(Q^T, \bar{P}^T)$$
$$E(P, \bar{Q}) = PA\bar{Q} = (PA\bar{Q})^T = \bar{Q}^T A^T P^T = -\bar{Q}^T A P^T = -E(\bar{Q}^T, P^T)$$

Substituting these equations into (2.5.3) yields

$$E(\bar{Q}^T, P^T) \geqq -v \geqq E(Q^T, \bar{P}^T)$$

for all $Q^T \in \mathbb{P}_n$, $P^T \in \mathbb{Q}^n$. Thus $(\bar{Q}^T, \bar{P}^T, -v)$ is a solution of G. Since the value of a game is unique, we have

$$v = -v$$

and hence $v = 0$. Moreover $\bar{Q}^T$ and $\bar{P}^T$ are optimal strategies for players I and II respectively. ////

Corollary 2.5.4 Let G be a symmetric game with matrix A and let $Q \in \mathbb{Q}^n$. Then Q is an optimal strategy (for either player) in G if and only if

$$0 \geqq AQ$$

Theorem 2.5.5 Every symmetric game has a solution.

PROOF By Corollary 2.4.25 the system of inequalities

$$0 \geqq AX \qquad X \geqq 0$$

has a solution. Since X is semipositive, some positive multiple of X is a strategy vector, say

$$Q = \alpha X$$

Then

$$0 \geqq AQ \qquad ////$$

We now have all the machinery needed to prove the fundamental theorem of game theory.

Theorem 2.5.6 Every game has a solution.

PROOF Let A be an $m \times n$ matrix. Since $(\bar{P}, \bar{Q}, v)$ is a solution of the game with matrix A if and only if $(\bar{P}, \bar{Q}, v + \alpha)$ is a solution of the game with matrix $A + \alpha E$ (see Exercise 11 of Sec. 2.3), it is sufficient to prove the theorem for games with positive matrices. Hence we may assume that $A > 0$.

Define B to be the matrix

$$B = \begin{bmatrix} 0 & A & -E^{(m)} \\ -A^T & 0 & E^{(n)} \\ E_{(m)} & -E_{(n)} & 0 \end{bmatrix}$$

where $E_{(k)}$ and $E^{(k)}$ denote the k-dimensional row and column vectors respectively, all of whose entries are 1. B is a skew-symmetric matrix of order $m + n + 1$. Let G' be the game whose matrix is B. G' is symmetric, and, according to Theorem 2.5.5, G' has a solution. Let X be an optimal strategy for the first player in G' and partition X as

$$X = [U \quad V \quad w]$$

where $U \in \mathbb{R}_m$, $V \in \mathbb{R}_n$, $w \in \mathbb{R}$. Since the value of G' is zero and X is an optimal strategy for the first player, we have

$$XB \geq 0$$

Thus
$$wE_{(m)} \geq VA^T \geq 0 \qquad (2.5.7)$$
$$UA \geq wE_{(n)} \geq 0 \qquad (2.5.8)$$
$$VE^{(n)} \geq UE^{(m)} \geq 0 \qquad (2.5.9)$$

We wish to show that U, V, and w are all nonzero. Suppose

$$V = 0$$

Then $VE^{(n)} = 0$, and it follows from (2.5.9) that $UE^{(m)} = 0$. Since U is nonnegative, we have

$$U = 0$$

Then (2.5.8) yields

$$w = 0$$

and consequently $X = 0$. This is a contradiction. Hence

$$V \geq 0$$

Clearly
$$VA^T > 0$$
since A is positive and it now follows from (2.5.7) that
$$w > 0$$
Finally, from (2.5.8) we obtain
$$U \geq 0$$
Let us rewrite (2.5.7) as
$$wE^{(m)} \geq AV^T \qquad (2.5.10)$$
Multiply (2.5.8) and (2.5.10) on the right by V^T and on the left by U respectively. Then
$$wUE^{(m)} \geq UAV^T \geq wE_{(n)}V^T$$
Combining this with (2.5.9), we have
$$UE^{(m)} = VE^{(n)} > 0$$
Denote this common value by λ and set
$$\bar{P} = \frac{1}{\lambda} U \qquad \bar{Q} = \frac{1}{\lambda} V^T$$
Clearly $\bar{P} \in \mathbb{P}_m$, $\bar{Q} \in \mathbb{Q}^n$. Moreover,
$$\bar{P}A = \frac{1}{\lambda} UA \geq \frac{w}{\lambda} E_{(n)}$$
$$A\bar{Q} = \frac{1}{\lambda} AV^T \leq \frac{w}{\lambda} E^{(m)}$$

Thus $(\bar{P}, \bar{Q}, w/\lambda)$ is a solution of the game with matrix A. (See Exercise 3 of Sec. 2.3.) ////

Although Theorem 2.5.6 guarantees the existence of a solution of a game, it does not provide us with a means for finding it. In the next two sections we will turn our attention to the problem of actually obtaining solutions or, at least, to the task of getting useful information about solutions.

EXERCISES

1 Let $(\bar{P}, \bar{Q}, v)$, $v \geq 0$, be a solution of the game with matrix A. Show that
$$X = \left[\frac{1}{2+v} \bar{P} \quad \frac{1}{2+v} \bar{Q} \quad \frac{v}{2+v} \right]$$

is an optimal strategy in the game with matrix

$$B = \begin{bmatrix} 0 & A & -E^{(m)} \\ -A^T & 0 & E^{(n)} \\ E_{(m)} & -E_{(n)} & 0 \end{bmatrix}$$

2 Solve the game whose matrix is

$$\begin{bmatrix} 0 & 0 & 4 & 2 & -1 \\ 0 & 0 & 3 & 5 & -1 \\ -4 & -3 & 0 & 0 & 1 \\ -2 & -5 & 0 & 0 & 1 \\ 1 & 1 & -1 & -1 & 0 \end{bmatrix}$$

Exercises 3 to 7 describe the successive parts of a proof of Theorem 2.5.6 using Theorem 2.5.5 and the symmetrization described in Example 2.1.14. Let G be a game with $m \times n$ matrix A and let G' be the symmetrization of G described in Sec. 2.1.

3 Show that if S_i and T_j are optimal pure strategies for players I and II respectively in G, then

$$S'_{ij} = \begin{cases} \text{when playing as player I, play } S_i \\ \text{when playing as player II, play } T_j \end{cases}$$

is an optimal strategy in G'.

4 Let S'_{ij} of Exercise 3 be an optimal pure strategy in G'. Show that S_i, T_j are optimal strategies for players I and II respectively in G.

5 Let

$$P = [p_i] \in \mathbb{P}_m \qquad Q = [q_j] \in \mathbb{Q}^n$$

and let X be the strategy in G' in which S'_{ij} is played with probability $p_i q_j$. We say that X is the *composition* of P and Q and write

$$X = P \times Q$$

Verify that X is a strategy vector in G'. Show that for $P_1, P_2 \in \mathbb{P}_m$, $Q_1, Q_2 \in \mathbb{Q}^n$, we have

$$E(P_1 \times Q_1, P_2 \times Q_2) = \tfrac{1}{2}[E(P_1, Q_2) - E(P_2, Q_1)]$$

6 Let X be the strategy in G' in which S'_{ij} is played with probability x_{ij}. Define $P = [p_i]$ and $Q = [q_j]$ by

$$p_i = \sum_{j=1}^n x_{ij} \qquad i = 1, 2, \ldots, m$$

$$q_j = \sum_{i=1}^m x_{ij} \qquad j = 1, 2, \ldots, n$$

We say that X has been *decomposed* into P and Q. Show that
$$E(X,Y) = E(P \times Q, Y)$$
for every strategy Y for player II'. (Note that if X decomposes into P and Q, it does not follow that $X = P \times Q$.)

7 Prove:
  (a) If $(\bar{P}, \bar{Q}, v)$ is a solution of G, then $\bar{P} \times \bar{Q}$ is an optimal strategy in G'.
  (b) If $\bar{X}$ is an optimal strategy in G' and $\bar{X}$ decomposes into $\bar{P}$ and $\bar{Q}$, then $(\bar{P}, \bar{Q}, E(\bar{P}, \bar{Q}))$ is a solution of G.
  (c) Every game has a solution.

8 Show that the matrix
$$B = \begin{bmatrix} 0 & -3 & 2 & -1 \\ 3 & 0 & 0 & -3 \\ -2 & 0 & 0 & 2 \\ 1 & 3 & -2 & 0 \end{bmatrix}$$
is the symmetrization of a game with a 2×2 matrix. Use this fact to find a solution of the game whose matrix is B. Show that $[\tfrac{1}{5} \ \tfrac{1}{5} \ \tfrac{2}{5} \ \tfrac{1}{5}]$ is a solution which is not of the form $\bar{P} \times \bar{Q}$.

9 Show that the variant of two-fingered morra in Exercise 4 of Sec. 2.1 is not the symmetrization of any game having a 2×2 matrix (with the order of the strategies in G' given by $S'_{11}, S'_{12}, S'_{21}, S'_{22}$):
  (a) By working with the entries of the matrix directly.
  (b) By decomposing the solution given in Exercise 4 of Sec. 2.1 into strategies P and Q and then finding $P \times Q$.

10 Is the variant of two-fingered morra of Example 2.1.10 the symmetrization of a game having a 2×2 matrix?

11 Show that if G' is the symmetrization of the game with matrix A, then G' is the symmetrization of the game with matrix $A + \beta E$ for every β.

2.6 RELATIONS AMONG GAMES

Now that the existence of solutions of a game has been established, we turn our attention to the problem of obtaining information about the sets of optimal strategies for each player. It is often possible to find solutions for a given game G by first considering some simpler game G' which has the same value as G and in which the optimal strategies for each player yield optimal strategies for the corresponding player in G. The game G' is formed from G by eliminating those pure strategies in G which the players will never use. The dimensions of the resulting matrix A' (of G') are then smaller than the dimensions of the matrix A (of G). In several of the examples of Sec. 2.1, e.g., bluffing, three-card poker, duels, we decided (without any real justification) to ignore certain pure strategies which seemed obviously bad, and in this way we reduced the dimensions of the matrix. One way of finding pure strategies which are never used is given by:

Lemma 2.6.1 Let $\bar{P}$ be an optimal strategy for player I in a game with matrix A and value v. If, for some l, the lth entry of $\bar{P}A$ is greater than v, that is, if

$$\bar{P}A^{(l)} > v$$

then the lth entry of every optimal strategy for player II is zero.

Similarly, if $\bar{Q}$ is an optimal strategy for player II, and if for some k the kth entry of $A\bar{Q}$ is less than v, that is, if

$$A_{(k)}\bar{Q} < v$$

then the kth entry of every optimal strategy for player I is zero.

PROOF It is sufficient to prove the first half of the lemma. Set

$$\bar{P}A = [w_1 \quad w_2 \quad \ldots \quad w_n]$$

and let $\bar{Q} = [q_j]$ be an optimal strategy for player II. Since $(\bar{P}, \bar{Q}, v)$ is a solution of the game, we have

$$w_j = \bar{P}A^{(j)} = \bar{P}AI^{(j)} = E(\bar{P}, I^{(j)}) \geq v \quad j = 1, 2, \ldots, n$$
$$w_l = \bar{P}A^{(l)} > v$$

and
$$v = E(\bar{P}, \bar{Q}) = \bar{P}A\bar{Q} = \sum_{j=1}^{n} w_j q_j$$

Set
$$e_j = w_j - v \quad j = 1, 2, \ldots, n$$

Then $\sum_{j=1}^{n} e_j q_j = \sum_{j=1}^{n} (w_j - v)q_j = \sum_{j=1}^{n} w_j q_j - v\sum_{j=1}^{n} q_j = v - v = 0$ \quad (2.6.2)

Since $e_j \geq 0$ and $q_j \geq 0$, it follows from equation (2.6.2) that

$$e_1 q_1 = e_2 q_2 = \cdots = e_n q_n = 0$$

Clearly, if $e_l > 0$, then $q_l = 0$. ////

Definition 2.6.3 A pure strategy is said to be *essential* if it is played with positive probability in some optimal strategy. A pure strategy which is never played, i.e., played with probability zero in every optimal strategy, is said to be *inessential*.

The pure strategy $I^{(l)}$ for player II is essential when player II has at least one optimal strategy whose lth component is not zero. Lemma 2.6.1 states that if $\bar{P}A^{(l)} > v$ for some optimal strategy $\bar{P}$ for player I, then $I^{(l)}$ is inessential. We wish to prove the converse of Lemma 2.6.1. First it is necessary to transform our game into one which has the same sets of optimal strategies and which, in addition, has the value zero. If the value of the game with matrix A is v, the value of the game with matrix $A - vE$ is zero. Moreover the two games have the same sets of optimal strategies (see Exercise 11 of Sec. 2.3).

Theorem 2.6.4 If $\bar{P}A^{(l)} = v$ for every optimal strategy $\bar{P}$ for player I, then $I^{(l)}$ is essential for player II. Similarly, if $A_{(k)}\bar{Q} = v$ for every optimal strategy $\bar{Q}$ for player II, then $I_{(k)}$ is essential for player I.

PROOF It will be sufficient to prove the first assertion of the theorem. Suppose $\bar{P}A^{(l)} = v$ for every optimal strategy $\bar{P}$ According to the discussion above, we may assume that $v = 0$.

Let us assume that $I^{(l)}$ is not an essential strategy for player II. This assumption is equivalent to stating that there does not exist a vector $X \in \mathbb{R}^n$ such that

$$0 \geq AX \qquad I_{(l)}X \geq 1 \qquad X \geq 0 \qquad (2.6.5)$$

because some positive multiple of a vector X satisfying (2.6.5) would be an optimal strategy for player II with a nonzero lth entry. The inequalities of (2.6.5) can be expressed as

$$\begin{bmatrix} 0 \\ -1 \end{bmatrix} \geq \begin{bmatrix} A \\ -I_{(l)} \end{bmatrix} X \qquad X \geq 0 \qquad (2.6.6)$$

Since $I^{(l)}$ is assumed to be inessential, we require that inequalities (2.6.6) do not have a solution. By Corollary 2.4.18, there exists a nonnegative vector $Y \in \mathbb{R}_{m+1}$ such that

$$Y \in \mathbb{O}\left(\begin{bmatrix} A \\ -I_{(l)} \end{bmatrix}\right) \qquad Y \begin{bmatrix} 0 \\ -1 \end{bmatrix} < 0 \qquad (2.6.7)$$

Let $Y = [\tilde{Y} \quad y_{m+1}]$, where $\tilde{Y} \in \mathbb{R}_m$, $y_{m+1} \in \mathbb{R}$. Then (2.6.7) becomes

$$\tilde{Y}A - y_{m+1}I_{(l)} \geq 0 \qquad -y_{m+1} < 0$$

Hence $\qquad\qquad\qquad \tilde{Y}A \geq y_{m+1}I_{(l)} \qquad y_{m+1} > 0 \qquad (2.6.8)$

Some positive multiple of $\tilde{Y}$ is a strategy vector, say $\bar{P} = \alpha\tilde{Y}$. By (2.6.8)

$$\bar{P}A \geq \alpha y_{m+1}I_{(l)} \geq 0$$

and hence $\bar{P}$ is an optimal strategy for player I. Moreover,

$$\bar{P}A^{(l)} \geqq \alpha y_{m+1} I_{(l)} I^{(l)} = \alpha y_{m+1} > 0 = v$$

This contradicts the hypothesis of the theorem. Consequently the assumption that $I^{(l)}$ is not essential is false. ////

For convenience, we will say that the row $A_{(k)}$ (or the column $A^{(l)}$) of A is *essential* or *inessential* as the pure strategy $I_{(k)}$ (or $I^{(l)}$) is essential or inessential.

Definition 2.6.9 Let A be an $m \times n$ matrix. The matrix which results from deleting all the inessential rows and columns of A is called the *essential submatrix* of A and denoted by A^*. We will denote the dimensions of A^* by m^* and n^*. The game G^* whose matrix is A^* is called the *essential subgame* of G.

Let P be a strategy for player I in G. If the kth entry of P is deleted for all values of k for which $I_{(k)}$ is inessential, then the resulting vector, which we denote by P^*, is an m^*-dimensional strategy vector since all the deleted entries are zeros. Similarly, for a strategy Q for player II, the vector formed by deleting the lth entry whenever $I^{(l)}$ is inessential is an n^*-dimensional strategy vector and is denoted by Q^*. Clearly:

Theorem 2.6.10 If $(\bar{P}, \bar{Q}, v)$ is a solution of G, then $(\bar{P}^*, \bar{Q}^*, v)$ is a solution of G^*.

PROOF From $\bar{P}A \geqq vE_{(n)}$ it follows that $\bar{P}^*A^* = vE_{(n^*)}$. Similarly, from $A\bar{Q} \leqq vE^{(m)}$ it follows that $\bar{A}^*\bar{Q}^* = vE^{(m^*)}$. ////

Theorem 2.6.10, in addition to reducing the size of the matrix under consideration, also replaces a system of linear inequalities by a system of linear equations. There still remains the difficult task of finding the essential submatrix of a given matrix.

EXAMPLE 2.6.11 Find the essential submatrix of

$$A = \begin{bmatrix} 6 & 8 & 9 & 10 \\ 8 & 4 & 3 & 0 \end{bmatrix}$$

We can use the graphical method to find all of player I's optimal strategies. Let

$$P = [p \quad 1-p] \qquad 0 \leqq p \leqq 1$$

A simple computation shows that

$$E(P, I^{(1)}) = 8 - 2p$$
$$E(P, I^{(2)}) = 4 + 4p$$
$$E(P, I^{(3)}) = 3 + 6p$$
$$E(P, I^{(4)}) = 10p$$

Sketching these equations (in Fig. 2.6.12), we see that player I has the unique optimal strategy

$$\bar{P} = [\tfrac{2}{3} \quad \tfrac{1}{3}] \qquad v = \tfrac{20}{3}$$

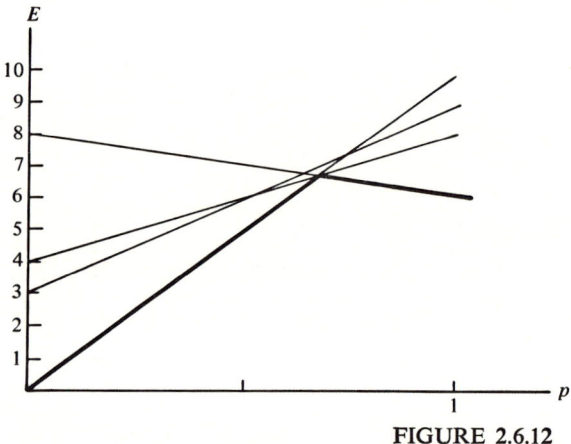

FIGURE 2.6.12

Both rows of A are essential. It follows from the equation

$$\bar{P}A = [\tfrac{20}{3} \quad \tfrac{20}{3} \quad \tfrac{21}{3} \quad \tfrac{20}{3}]$$

that the third column of A is inessential. Moreover, by the first part of Theorem 2.6.4 the other three columns of A are essential. Thus

$$A^* = \begin{bmatrix} 6 & 8 & 10 \\ 8 & 4 & 0 \end{bmatrix}$$

To obtain the optimal strategies for player II (or, rather, for player II*) we solve the equation

$$A^*\bar{Q}^* = \begin{bmatrix} \tfrac{20}{3} \\ \tfrac{20}{3} \end{bmatrix}$$

This yields

$$\bar{Q}^* = \begin{bmatrix} q_1 \\ \tfrac{5}{3} - 2q_1 \\ q_1 - \tfrac{2}{3} \end{bmatrix} \qquad \tfrac{2}{3} \leqq q_1 \leqq \tfrac{5}{6}$$

and a computation shows that every vector

$$\bar{Q} = \begin{bmatrix} q_1 \\ \tfrac{5}{3} - 2q_1 \\ 0 \\ q_1 - \tfrac{2}{3} \end{bmatrix} \qquad \tfrac{2}{3} \leqq q_1 \leqq \tfrac{5}{6}$$

is an optimal strategy for player II. ////

Unfortunately, the converse of Theorem 2.6.10 is not generally true (see Example 2.6.21). It is possible that A^* will have optimal strategies which do not arise from optimal strategies of A. However, we will show in Theorem 2.6.16 that every optimal strategy in G^* can be expressed as a linear combination of optimal strategies in G (after deleting zeros corresponding to inessential strategies). First we need:

Lemma 2.6.13 There exists an optimal strategy $\bar{P}_0$ for player I such that

1 $\bar{P}_0^* > 0$ and
2 $\bar{P}_0 A^{(l)} > v$

whenever $I^{(l)}$ is inessential.
There exists an optimal strategy $\bar{Q}_0$ for player II such that

1 $\bar{Q}_0^* > 0$ and
2 $A_{(k)} \bar{Q}_0 < v$

whenever $I_{(k)}$ is inessential.

PROOF It is sufficient to prove the first part of the lemma. By Definition 2.6.3, for each of the m^* essential strategies $I_{(i)}$ for player I, there exists an optimal strategy $\bar{P}_i$ whose ith entry is positive. Also, by Theorem 2.6.4, for each of the $n - n^*$ inessential strategies $I^{(j)}$ for player II, there exists an optimal strategy $\bar{P}_j$ for player I such that $\bar{P}_j A^{(j)} > v$. We will define $\bar{P}_0$ as a convex combination of these optimal strategies with positive coefficients, say

$$\bar{P}_0 = \frac{1}{2m^*} \sum_{i=1}^{m^*} \bar{P}_i + \frac{1}{2(n-n^*)} \sum_{j=1}^{n-n^*} \bar{P}_j$$

Clearly $\bar{P}_0$ is an optimal strategy for player I, $\bar{P}_0^* > 0$, and $\bar{P}_0 A^{(l)} > v$ for each l for which $I^{(l)}$ is inessential. ////

We call the strategies $\bar{P}_0$ and $\bar{Q}_0$ satisfying the conditions of Lemma 2.6.13 *central optimal strategies* for players I and II respectively. Because the positive vectors $\bar{P}_0^*$ and $\bar{Q}_0^*$ are optimal strategies in G^*, it follows that G^* has no inessential strategies and therefore the essential submatrix of A^* is A^*. Hence

$$\bar{P}A^* = vE_{(n^*)} \qquad A^*\bar{Q} = vE^{(m^*)} \qquad (2.6.14)$$

for all optimal strategies $\bar{P}$, $\bar{Q}$ of G^*, whether or not they arose from optimal strategies of A.

After all the solutions of the essential subgame of a given game have been found [by solving the equations of (2.6.14) for $\bar{P}$ and $\bar{Q}$], it is still necessary to determine the solutions of the original game. In order to do this, we let $\mathbb{U}$ be the subspace of $\mathbb{R}^{n^*}$ spanned by the optimal strategies for player II* in G^* and $\mathbb{V}$ be the subspace of $\mathbb{R}^n$ spanned by the optimal strategies for player II in G. For each inessential strategy $I^{(l)}$ of G the lth entry of every element of $\mathbb{V}$ is zero. Thus we may set

$$\mathbb{V}^* = \{Y^* \mid Y \in \mathbb{V}\}$$

Since optimal strategies of G yield optimal strategies of G^*, we have

$$\mathbb{V}^* \subseteq \mathbb{U}$$

Moreover, we can prove

Theorem 2.6.15 $\qquad\qquad\qquad \mathbb{V}^* = \mathbb{U}$

PROOF We may assume that $v = 0$. Then every optimal strategy $\bar{Q}$ for player II* in G^* satisfies

$$A^*\bar{Q} = 0$$

that is, $\bar{Q} \in \mathbb{N}(A^*)$. Hence we have

$$\mathbb{V}^* \subseteq \mathbb{U} \subseteq \mathbb{N}(A^*) \subseteq \mathbb{R}^{n^*}$$

In order to complete the proof it is sufficient to show that $\mathbb{N}(A^*) \subseteq \mathbb{V}^*$. Let $\mathbb{W}$ be the subspace of $\mathbb{R}^n$ spanned by the essential strategies for player II in G. Clearly

$$\mathbb{V} \subseteq \mathbb{W}$$

and $\qquad\qquad\qquad\qquad\qquad \mathbb{W}^* = \mathbb{R}^{n^*}$

Thus, for $X \in \mathbb{N}(A^*)$, there exists $Z \in \mathbb{W}$ such that
$$X = Z^*$$
It remains only to show that Z belongs to $\mathbb{V}$ (from which it follows that Z^* belongs to $\mathbb{V}^*$). Let $\bar{Q}_0$ be a central optimal strategy for player II and set
$$Y = Z + \alpha \bar{Q}_0 \quad (2.6.16)$$
where α is some (as yet undetermined) positive number. Clearly $Y \in \mathbb{W}$. Also, for sufficiently large values of α,
$$Y^* > 0 \quad (2.6.17)$$
since $\bar{Q}_0$ has a positive entry in each of player II's essential positions. When $I_{(k)}$ is an essential strategy for player I, we have
$$A_{(k)} Z = (A_{(k)})^* Z^* = (A_{(k)})^* X = 0$$
$$A_{(k)} \bar{Q}_0 = v = 0$$
since $(A_{(k)})^*$ is one of the rows of A^* (though not necessarily the kth row). Consequently
$$A_{(k)} Y = 0 \quad (2.6.18)$$
When $I_{(k)}$ is an inessential strategy for player I,
$$A_{(k)} \bar{Q}_0 < v = 0$$
and hence, for sufficiently large values of α, we have
$$A_{(k)} Y < 0 \quad (2.6.19)$$
We now fix α to be some number large enough so that (2.6.17) and all the inequalities of (2.6.19) are satisfied. Then we can combine (2.6.17), (2.6.18), and (2.6.19) into
$$AY \leqq 0 = vE^{(m)} \quad Y \geq 0$$
and some positive multiple of Y is a (central) optimal strategy for player II, say
$$\beta Y = \bar{Q}_1$$
Substituting for Y in equation (2.6.16) and solving for Z yields
$$Z = \frac{1}{\beta} \bar{Q}_1 - \alpha \bar{Q}_0 \in \mathbb{V}$$
Consequently
$$X = Z^* \in \mathbb{V}^* \quad ////$$

In Theorem 2.6.15 we proved that when $v = 0$, we have

$$\mathbb{V}^* = \mathbb{U} = \mathbb{N}(A^*)$$

Similarly, if $\mathbb{Z}$ is the subspace of $\mathbb{R}_m$ spanned by the optimal strategies for player I, then $\mathbb{Z}^*$ is the row null space of A^*. We can now conclude that

Theorem 2.6.20 For every game

$$n^* - \dim \mathbb{V} = m^* - \dim \mathbb{Z}$$

Moreover, when $v = 0$,

$$n^* - \dim \mathbb{V} = r(A^*) = m^* - \dim \mathbb{Z}$$

PROOF The first part of the theorem follows from the second. Hence, assume $v = 0$. Since only zero entries are deleted in passing from $\mathbb{V}$ to $\mathbb{V}^*$ it follows that

$$\dim \mathbb{V} = \dim \mathbb{V}^* = \dim \mathbb{N}(A^*) = v(A^*)$$

But by Theorem 1.6.1

$$r(A^*) = n^* - v(A^*) = n^* - \dim \mathbb{V}$$

Similarly, $\qquad\qquad\qquad r(A^*) = m^* - \dim \mathbb{Z} \qquad\qquad ////$

Note that Theorem 2.6.20 does not tell us when we have found all the optimal strategies of a game. Even when m^*, n^*, $\dim \mathbb{V}$, and $\dim \mathbb{Z}$ are known, Theorem 2.6.20 indicates only the number of linearly independent optimal strategies we can expect for each player. In the next section a method is given for finding all optimal strategies.

EXAMPLE 2.6.21 Find all solutions of the game with matrix

$$A = \begin{bmatrix} 1 & 1 \\ 4 & -3 \end{bmatrix}$$

Using the graphical method, we can show that the value of the game is 1, player I has the unique optimal strategy

$$\bar{P} = \begin{bmatrix} 1 & 0 \end{bmatrix}$$

and every vector

$$\bar{Q} = \begin{bmatrix} q \\ 1 - q \end{bmatrix} \qquad 0 \leq q \leq \tfrac{4}{7}$$

is an optimal strategy for player II. The essential submatrix of A is
$$A^* = [1 \quad 1]$$
Hence, $\qquad m = n = 2 \qquad m^* = 1 \qquad n^* = 2$

Note that every strategy for player II* is optimal in G^* but not every optimal strategy for player II* is an optimal strategy for player II. On the other hand (using the notation of Theorems 2.6.15 and 2.6.20),
$$\mathbb{U} = \mathbb{V} = \mathbb{V}^* = \mathbb{R}^2 \qquad \mathbb{Z}^* = \mathbb{R}$$
$\mathbb{Z}$ is the one-dimensional subspace of $\mathbb{R}_2$ which has $\bar{P} = [1 \quad 0]$ as a basis. Thus
$$n^* - \dim \mathbb{V} = m^* - \dim \mathbb{Z} = 0$$
Clearly every optimal strategy for player II can be expressed as a convex combination of the two optimal strategies
$$\bar{Q}_1 = \begin{bmatrix} \frac{4}{7} \\ \frac{3}{7} \end{bmatrix} \qquad \bar{Q}_2 = \begin{bmatrix} 0 \\ 1 \end{bmatrix} \qquad ////$$

EXERCISES

1. Find the essential submatrices of the matrices in Examples 2.1.1, 2.1.2, and 2.1.4.
2. Examine Example 2.1.10. Show that the strategies which were eliminated (S_3 and S_4) are actually inessential. Are any of the other strategies inessential?
3. Change the rules in Example 2.1.10 so that the winner wins an amount equal to the total number of fingers put out. (See Exercise 4 of Sec. 2.1.) Find the essential strategies.
4. Find all solutions of

    (a) $\begin{bmatrix} 2 & 4 \\ 2 & -1 \end{bmatrix}$ \qquad (b) $\begin{bmatrix} 2 & -1 & -4 \\ -3 & -1 & 5 \end{bmatrix}$

    Find all solutions of the essential subgame and verify Theorems 2.6.15 and 2.6.20.
5. Let
$$A = \begin{bmatrix} 0 & 1 & 2 & 3 & 4 \\ 1 & 0 & 1 & 2 & 3 \\ 2 & 1 & 0 & 1 & 2 \\ 3 & 2 & 1 & 0 & 1 \\ 4 & 3 & 2 & 1 & 0 \end{bmatrix}$$

Given that $[\tfrac{1}{2}\ 0\ 0\ 0\ \tfrac{1}{2}]$ is an optimal strategy for player I, find the essential submatrix and a maximal set of linearly independent optimal strategies for each player.

6 Prove:
   (a) The essential submatrix of a symmetric game is skew-symmetric.
   (b) $(A + vE)^* = A^* + vE$.

7 Let A be a matrix in which every column is essential. Show that if Q is a strategy vector for which

$$AQ = wE^{(m)}$$

then $v_A = w$ and Q is an optimal strategy for player II. Show that this result does not hold if one drops the hypothesis that every column of A is essential.

For $X, Y \in \mathbb{R}^m$ (or $\mathbb{R}_n$) we say that X dominates Y if $X \geq Y$ and X strictly dominates Y if $X > Y$.

8 Show that if $A_{(k)}$ strictly dominates $A_{(i)}$, then $I_{(i)}$ is an inessential strategy for player I. Formulate and prove the corresponding theorem for player II. (*Warning:* See Exercise 13.)

9 Show that if $A_{(i)}$ is strictly dominated by a convex combination of the other rows of A, then $I_{(i)}$ is inessential.

10 Show that if the ith row of A strictly dominates a convex combination of the other rows of A, then at least one of these rows is inessential.

11 Let A be a matrix in which $A_{(k)}$ dominates $A_{(i)}$, Let $\tilde{A}$ be the matrix formed from A by deleting $A_{(i)}$, and let $\tilde{G}$ be the game with matrix $\tilde{A}$. Prove:
   (a) $v_G = v_{\tilde{G}}$.
   (b) The optimal strategies for player II in G are identical with the optimal strategies for player $\tilde{\text{II}}$ in $\tilde{G}$.
   (c) Every optimal strategy for player $\tilde{\text{I}}$ in $\tilde{G}$ can be transformed into an optimal strategy for player I in G by inserting a zero entry at the ith position.

Formulate and prove the corresponding theorem when $A^{(l)}$ dominates $A^{(j)}$.

12 Find a solution of the games

(a) $\begin{bmatrix} 0 & 2 & -1 \\ 1 & 2 & 1 \\ -1 & 4 & 0 \end{bmatrix}$

(b) $\begin{bmatrix} -1 & -1 & -2 \\ -2 & -1 & -1 \\ -4 & 0 & 1 \\ 13 & 0 & -3 \end{bmatrix}$

$$(c) \begin{bmatrix} -1 & 0 & -3 & -1 \\ 1 & -2 & 3 & -2 \\ -1 & 0 & 1 & -4 \\ -2 & 2 & 0 & 1 \end{bmatrix}$$

using the concept of dominance.

13 Construct a 2×2 matrix in which $A_{(2)}$ dominates $A_{(1)}$ and $I_{(1)}$ is essential.

2.7 EXTREME OPTIMAL STRATEGIES

In Sec. 2.6 we showed that if $(\bar{P}, \bar{Q}, v)$ is any solution of the game whose matrix is A, then

$$\bar{P}^*A^* = vE_{(n^*)} \qquad A^*\bar{Q}^* = vE^{(m^*)} \qquad (2.7.1)$$

When the essential submatrix A^* of A is nonsingular (of order n^*), equations (2.7.1) can be solved uniquely for $\bar{P}^*$ and $\bar{Q}^*$,

$$\bar{P}^* = vE_{(n^*)}(A^*)^{-1} \qquad \bar{Q}^* = v(A^*)^{-1}E^{(n^*)} \qquad (2.7.2)$$

It then follows from equations (2.7.2) that G^* (and hence G) has only one solution. Further, it follows from equations (2.7.2) that

$$v \neq 0$$

and that

$$1 = \bar{P}^*E^{(n^*)} = vE_{(n^*)}(A^*)^{-1}E^{(n^*)}$$

Hence we have explicit formulas for v, $\bar{P}^*$, $\bar{Q}^*$, namely,

$$v = \frac{1}{E_{(n^*)}(A^*)^{-1}E^{(n^*)}}$$

$$\bar{P}^* = \frac{E_{(n^*)}(A^*)^{-1}}{E_{(n^*)}(A^*)^{-1}E^{(n^*)}} \qquad \bar{Q}^* = \frac{(A^*)^{-1}E^{(n^*)}}{E_{(n^*)}(A^*)^{-1}E^{(n^*)}} \qquad (2.7.3)$$

When A^* is singular, equations (2.7.1) cannot be solved explicitly for $\bar{P}^*$ and $\bar{Q}^*$ and the results in the paragraph above are no longer valid. We will, however, prove that when $v \neq 0$, relations such as (2.7.3) exist for some (special) optimal strategies. Moreover the remaining optimal strategies can easily be obtained from these.

Definition 2.7.4 An optimal strategy, which is an extreme point of the set of optimal strategies, i.e., which cannot be written as a convex combination of other optimal strategies, is an *extreme optimal strategy*.

When a player has a unique optimal strategy, it is clearly an extreme optimal strategy. It can be shown (see Theorem 2.7.23) that extreme optimal strategies always exist.

Theorem 2.7.5 Let $(\bar{P}, \bar{Q}, v)$, $v \neq 0$, be a solution of the game with $m \times n$ matrix A. Further, let $\bar{P}$ and $\bar{Q}$ be extreme optimal strategies. Then there exists a nonsingular submatrix $A^{\cdot}$ (having order $n^{\cdot}$) of A formed by deleting certain of the rows and columns of A such that

$$v = \frac{1}{E_{(n^{\cdot})}(A^{\cdot})^{-1}E^{(n^{\cdot})}}$$

and $\bar{P}$ and $\bar{Q}$ are given (up to zero entries in the deleted rows and columns) by

$$\bar{P}^{\cdot} = \frac{E_{(n^{\cdot})}(A^{\cdot})^{-1}}{E_{(n^{\cdot})}(A^{\cdot})^{-1}E^{(n^{\cdot})}} \qquad \bar{Q}^{\cdot} = \frac{(A^{\cdot})^{-1}E^{(n^{\cdot})}}{E_{(n^{\cdot})}(A^{\cdot})^{-1}E^{(n^{\cdot})}}$$

PROOF It will simplify the proof greatly to break it up into a series of lemmas. First let us reorder the rows and columns of A (and hence of $\bar{P}$ and $\bar{Q}$) so that

$$\bar{P} = [p_1 \quad p_2 \quad \cdots \quad p_r \quad 0 \quad 0 \quad \cdots \quad 0] \qquad \begin{matrix} p_i > 0 \\ i = 1, 2, \ldots, r \end{matrix}$$

$$\bar{Q} = \begin{bmatrix} q_1 \\ q_2 \\ \vdots \\ q_s \\ 0 \\ 0 \\ \vdots \\ 0 \end{bmatrix} \qquad \begin{matrix} q_j > 0 \\ j = 1, 2, \ldots, s \end{matrix}$$

Set

$$\bar{P}A = Y = [y_i] \geq vE_{(n)} \qquad A\bar{Q} = Z = [z_j] \leq vE^{(m)}$$

Since the first r pure strategies for player I (after the reordering) are all essential, it follows that

$$z_1 = z_2 = \cdots = z_r = v \qquad z_{r+1} \leq v, z_{r+2} \leq v, \ldots, z_m \leq v$$

Similarly

$$y_1 = y_2 = \cdots = y_s = v \qquad y_{s+1} \geq v, y_{s+2} \geq v, \ldots, y_n \geq v$$

We now reorder the last $n - r$ rows of A (leaving the first r rows of A fixed) so that

$$z_1 = z_2 = \cdots = z_r = z_{r+1} = \cdots = z_{r+\bar{r}} = v$$

and

$$z_{r+\bar{r}+1} < v, \quad z_{r+\bar{r}+2} < v, \ldots, z_m < v$$

Similarly we reorder the last $n - s$ columns of A (leaving the first s columns of A fixed) so that

$$y_1 = y_2 = \cdots = y_s = y_{s+1} = \cdots = y_{s+\bar{s}} = v$$

and

$$y_{s+\bar{s}+1} > v, \quad y_{s+\bar{s}+2} > v, \quad \ldots, \quad y_n > v$$

Note that the last $m - r - \bar{r}$ rows and $n - s - \bar{s}$ columns of A are all inessential.

We now partition $\bar{P}$, $\bar{Q}$, Y, Z, and A as

$$A = \begin{bmatrix} A_{11} & A_{12} & A_{13} \\ A_{21} & A_{22} & A_{23} \\ A_{31} & A_{32} & A_{33} \end{bmatrix} \begin{matrix} r \\ \bar{r} \\ m-r-\bar{r} \end{matrix} \quad \bar{Q} = \begin{bmatrix} \bar{Q}_1 \\ 0 \\ 0 \end{bmatrix} \quad Z = \begin{bmatrix} Z_1 \\ Z_2 \\ Z_3 \end{bmatrix} \begin{matrix} r \\ \bar{r} \\ m-r-\bar{r} \end{matrix}$$

with column labels s, $\bar{s}$, $n-s-\bar{s}$ for A and 1 for $\bar{Q}$ and Z.

$$\bar{P} = [\bar{P}_1 \quad 0 \quad 0]_1$$
$$Y = [Y_1 \quad Y_2 \quad Y_3]_1$$

where A_{11} is an $r \times s$ matrix, A_{12} is an $r \times \bar{s}$ matrix, A_{21} is an $\bar{r} \times s$ matrix, ..., $\bar{Q}_1$ and Z_1 are $r \times 1$ matrices, ..., and Y_3 is a $1 \times (n - s - \bar{s})$ matrix.[1] Set

$$\tilde{A} = \begin{bmatrix} A_{11} & A_{12} \\ A_{21} & A_{22} \end{bmatrix}$$

and note that

$$[\bar{P}_1 \quad 0]\tilde{A} = [Y_1 \quad Y_2] = vE_{(s+\bar{s})} \qquad (2.7.6)$$

$$\tilde{A}\begin{bmatrix} \bar{Q}_1 \\ 0 \end{bmatrix} = \begin{bmatrix} Z_1 \\ Z_2 \end{bmatrix} = vE^{(r+\bar{r})} \qquad (2.7.7)$$

$$Y_3 > vE_{(n-s-\bar{s})} \qquad Z_3 < vE^{(m-r-\bar{r})}$$

If $\tilde{A}$ is nonsingular, then it will serve for $A^{\cdot}$ and equations (2.7.6) and (2.7.7) can be solved for

$$\bar{P}^{\cdot} = [\bar{P}_1 \quad 0] \qquad \bar{Q}^{\cdot} = \begin{bmatrix} \bar{Q}_1 \\ 0 \end{bmatrix}$$

---

[1] Some of the numbers $\bar{r}$, $\bar{s}$, $m - r - \bar{r}$, $n - s - \bar{s}$ may be zero, in which case the corresponding matrices do not actually appear.

respectively. On the other hand, if $\tilde{A}$ is singular, we will delete some of its rows and columns until the resulting matrix is nonsingular and then apply the same process. In doing this it will be important not to delete any of the first r rows or s columns of $\tilde{A}$, that is, rows or columns through A_{11}. To show that there is a nonsingular submatrix of $\tilde{A}$ which contains A_{11} we first prove:

Lemma 2.7.8 The rows of the $r \times (s + \bar{s})$ submatrix

$$B = [A_{11} \quad A_{12}]$$

of $\tilde{A}$ are linearly independent.

PROOF Let $X = [x_i]$ be any element of $\mathbb{R}_r$ which satisfies

$$XB = 0 \qquad (2.7.9)$$

We wish to show that $X = 0$. Multiply equation (2.7.9) on the right by the $(s + \bar{s})$-dimensional column vector

$$\begin{bmatrix} \bar{Q}_1 \\ 0 \end{bmatrix}$$

Combining equations (2.7.9) and (2.7.7) yields

$$0 = XB \begin{bmatrix} \bar{Q}_1 \\ 0 \end{bmatrix} = vXE^{(r)}$$

Since $v \neq 0$, it follows that

$$XE^{(r)} = 0$$

i.e., the sum of the entries of X is 0. Let W be the m-dimensional row vector

$$W = [X \quad 0 \quad 0]$$

(with the same partition as above). For every $\varepsilon > 0$ the sum of the entries of

$$\bar{P} \pm \varepsilon W = [\bar{P}_1 \pm \varepsilon X \quad 0 \quad 0]$$

is 1. Moreover, since $\bar{P}_1 > 0$, both $\bar{P} + \varepsilon W$ and $\bar{P} - \varepsilon W$ are strategy vectors when ε is sufficiently small. In particular, when

$$0 < \varepsilon \leq \alpha = \min_{i = 1, 2, \ldots, r} \frac{p_i}{1 + |x_i|}$$

we have $p_i \pm \varepsilon x_i \geq 0$. (Note that $\alpha > 0$ since $\bar{P}_1 > 0$.) In fact a positive ε can be chosen so that $\bar{P} + \varepsilon W$ and $\bar{P} - \varepsilon W$ are both optimal strategies for player I. For, by equation (2.7.9)

$$(\bar{P} \pm \varepsilon W)A = \bar{P}A \pm \varepsilon WA = [Y_1 \quad Y_2 \quad Y_3 \pm \varepsilon X A_{13}]$$

According to equation (2.7.6),

$$[Y_1 \quad Y_2] = vE_{(s+\bar{s})}$$

Also, each entry of Y_3 is strictly greater than v, and thus

$$Y_3 \pm \varepsilon X A_{13} \geq vE_{(n-s-\bar{s})}$$

when ε is sufficiently small (but positive). Clearly

$$\bar{P} = \tfrac{1}{2}(\bar{P} - \varepsilon W) + \tfrac{1}{2}(\bar{P} + \varepsilon W) \qquad (2.7.10)$$

However by the hypotheses of Theorem 2.7.5, $\bar{P}$ is an extreme optimal strategy for player I. This contradicts equation (2.7.10) unless

$$\bar{P} = \bar{P} + \varepsilon W = \bar{P} - \varepsilon W$$

Thus $W = 0$. Consequently $X = 0$, and the rows of B are linearly independent, completing the proof of the lemma. ////

The same argument can be used to prove that the columns of the submatrix

$$C = \begin{bmatrix} A_{11} \\ A_{21} \end{bmatrix}$$

of $\tilde{A}$ are linearly independent.

Recall that we wish to delete some of the last $\bar{r}$ rows and $\bar{s}$ columns of $\tilde{A}$ so that the resulting matrix is nonsingular. Of course, in the resulting matrix the first s columns and the first r rows must still be linearly independent. In particular, we wish to delete as many columns of $\tilde{A}$ as possible subject to the condition that the first r rows remain linearly independent. That is, form

$$\tilde{B} = [A_{11} \quad A'_{12}]$$

from B by deleting columns of A_{12} so that:

1 The rows of $\tilde{B}$ are linearly independent.
2 Either A'_{12} does not appear, or when any of the columns of A'_{12} are deleted from $\tilde{B}$, the rows of the resulting matrix are linearly dependent.

Similarly form

$$\tilde{C} = \begin{bmatrix} A_{11} \\ A'_{21} \end{bmatrix}$$

from C by deleting rows of A_{21} so that:

1. The columns of $\tilde{C}$ are linearly independent.
2. Either A'_{21} does not appear, or when any of the rows of A'_{21} are deleted from $\tilde{C}$, the columns of the resulting matrix are linearly dependent.

Finally, set

$$A^{\cdot} = \begin{bmatrix} A_{11} & A'_{12} \\ A'_{21} & A'_{22} \end{bmatrix} \begin{matrix} r \\ r' \end{matrix}$$

with column labels s and s', where A'_{22} is formed from A_{22} by deleting the same rows and columns as were deleted to form A'_{21} and A'_{12} respectively.

It remains now to prove:

Lemma 2.7.11 $A^{\cdot}$ is nonsingular.

PROOF It is sufficient to show that the rows of $A^{\cdot}$ are linearly independent. If $r' = 0$, that is, A'_{21} and A'_{22} do not actually appear, then $A^{\cdot} = \tilde{B}$ and the result is immediate.

Now let us assume that $r' > 0$ and let

$$Z = [z_i] = [Z_1 \quad Z_2] \quad 1 \quad (2.7.12)$$

with column labels r and r', satisfy

$$ZA^{\cdot} = 0 \quad (2.7.13)$$

Let k be a (fixed) integer, $1 \leq k \leq r'$, and let A''_{12} be formed by deleting the kth row of A'_{21}. By the construction above there exists a nonzero vector Y such that

$$\begin{bmatrix} A_{11} \\ A''_{21} \end{bmatrix} Y = 0 \qquad W = [w_j] = \begin{bmatrix} A_{11} \\ A'_{21} \end{bmatrix} Y = \tilde{C}Y \neq 0 \quad (2.7.14)$$

Hence $w_{r+k} \neq 0$, and every other entry of W is zero. Combining equations (2.7.13) and (2.7.14) yields

$$0 = (ZA^{\cdot})\begin{bmatrix} Y \\ 0 \end{bmatrix} = Z\left(A^{\cdot}\begin{bmatrix} Y \\ 0 \end{bmatrix}\right) = ZW = z_{r+k} w_{r+k}$$

[2.7.16] THE THEORY OF GAMES 235

Thus we have $z_{r+k} = 0$. Therefore the matrix Z_2 of (2.7.12) is a zero matrix, and equation (2.7.13) becomes

$$0 = [Z_1 \; 0]A^{\cdot} = Z_1 \tilde{B}$$

Since the rows of $\tilde{B}$ are linearly independent, it follows that $Z_1 = 0$. Hence $Z = 0$, and the rows of $A^{\cdot}$ are linearly independent. ////

The proof of Theorem 2.7.5 is now immediate. $A^{\cdot}$ has been obtained by deleting some of the rows and columns of A. We now form $\bar{P}^{\cdot}$ by deleting the entries of $\bar{P}$ corresponding to those rows of A which were deleted to form $A^{\cdot}$. Since none of the rows of A through A_{11} have been removed, it follows that $\bar{P}^{\cdot}$ contains all the nonzero entries of $\bar{P}$. Similarly $\bar{Q}^{\cdot}$ is formed by deleting the entries of $\bar{Q}$ corresponding to those columns of A which have been deleted, and $\bar{Q}^{\cdot}$ contains all the nonzero entries of $\bar{Q}$. Clearly

$$\bar{P}^{\cdot} A^{\cdot} = v E_{(n \cdot)} \qquad A^{\cdot} \bar{Q}^{\cdot} = v E^{(n \cdot)}$$

Since $A^{\cdot}$ is nonsingular, we have

$$v = \frac{1}{E_{(n \cdot)}(A^{\cdot})^{-1} E^{(n \cdot)}}$$

$$\bar{P}^{\cdot} = \frac{E_{(n \cdot)}(A^{\cdot})^{-1}}{E_{(n \cdot)}(A^{\cdot})^{-1} E^{(n \cdot)}} \qquad \bar{Q}^{\cdot} = \frac{(A^{\cdot})^{-1} E^{(n \cdot)}}{E_{(n \cdot)}(A^{\cdot})^{-1} E^{(n \cdot)}} \qquad (2.7.15)$$

////

EXAMPLE 2.7.16 Let

$$A = \begin{bmatrix} 6 & 8 & 9 & 10 \\ 8 & 4 & 3 & 0 \end{bmatrix}$$

Clearly A contains six 2×2 submatrices and eight 1×1 submatrices. We can consider each of them in turn as a candidate for $A^{\cdot}$. (It is clear, even if we did not have the results of Example 2.6.11, that $v > 0$.) First let us assume

$$A^{\cdot} = \begin{bmatrix} 6 & 8 \\ 8 & 4 \end{bmatrix}$$

Then

$$(A^{\cdot})^{-1} = \begin{bmatrix} -\frac{2}{20} & \frac{4}{20} \\ \frac{4}{20} & -\frac{3}{20} \end{bmatrix}$$

and equations (2.7.15) yield

$$v = \frac{1}{\frac{3}{20}} = \frac{20}{3}$$

$$\bar{P}^{\cdot} = \tfrac{20}{3}[\tfrac{2}{20} \; \tfrac{1}{20}] = [\tfrac{2}{3} \; \tfrac{1}{3}] \qquad \bar{Q}^{\cdot} = \frac{20}{3}\begin{bmatrix} \frac{2}{20} \\ \frac{1}{20} \end{bmatrix} = \begin{bmatrix} \frac{2}{3} \\ \frac{1}{3} \end{bmatrix}$$

Thus, we have actually obtained a solution. Note that

$$\bar{P} = [\tfrac{2}{3} \ \tfrac{1}{3}] \quad \bar{Q} = \begin{bmatrix} \tfrac{2}{3} \\ \tfrac{1}{3} \\ 0 \\ 0 \end{bmatrix}$$

are extreme optimal strategies for players I and II, respectively.

Next, let us assume

$$\dot{A} = \begin{bmatrix} 6 & 10 \\ 8 & 0 \end{bmatrix}$$

Then

$$(A^{\cdot})^{-1} = \begin{bmatrix} 0 & \tfrac{5}{40} \\ \tfrac{4}{40} & -\tfrac{3}{40} \end{bmatrix}$$

and equations (2.7.15) yield

$$v = \frac{1}{\tfrac{6}{40}} = \frac{20}{3}$$

$$\bar{P}^{\cdot} = \tfrac{20}{3}[\tfrac{4}{40} \ \tfrac{2}{40}] = [\tfrac{2}{3} \ \tfrac{1}{3}] \quad \bar{Q}^{\cdot} = \tfrac{20}{3}\begin{bmatrix} \tfrac{5}{40} \\ \tfrac{1}{40} \end{bmatrix} = \begin{bmatrix} \tfrac{5}{6} \\ \tfrac{1}{6} \end{bmatrix}$$

Again we have found extreme optimal strategies

$$\bar{P} = [\tfrac{2}{3} \ \tfrac{1}{3}] \quad \bar{Q} = \begin{bmatrix} \tfrac{5}{6} \\ 0 \\ 0 \\ \tfrac{1}{6} \end{bmatrix}$$

for players I and II, respectively. Note that the two choices for $A^{\cdot}$ yield the same $\bar{P}$.

Now, assume

$$A^{\cdot} = \begin{bmatrix} 9 & 10 \\ 3 & 0 \end{bmatrix}$$

Then

$$(A^{\cdot})^{-1} = \begin{bmatrix} 0 & \tfrac{10}{30} \\ \tfrac{3}{30} & -\tfrac{9}{30} \end{bmatrix}$$

and equations (2.7.15) yield

$$v = \frac{1}{\tfrac{4}{30}} = \frac{15}{2}$$

$$\bar{P}^{\cdot} = \tfrac{15}{2}[\tfrac{3}{30} \ \tfrac{1}{30}] = [\tfrac{3}{4} \ \tfrac{1}{4}] \quad \bar{Q}^{\cdot} = \tfrac{15}{2}\begin{bmatrix} \tfrac{10}{30} \\ -\tfrac{6}{30} \end{bmatrix} = \begin{bmatrix} \tfrac{5}{2} \\ -\tfrac{3}{2} \end{bmatrix}$$

which is clearly not a solution. We leave it as an exercise to test the remaining 11 square submatrices of A. ////

In order to apply Theorem 2.7.5 we must be certain that $v \neq 0$, so that $A^{\cdot}$ is nonsingular and $(A^{\cdot})^{-1}$ exists. However, in Theorem 1.9.32 an expression is given for the inverse of a nonsingular matrix in terms of its adjoint,

$$A^{-1} = \frac{1}{\det A} \text{ adj } A$$

Substituting this expression into equations (2.7.15) yields

$$v = \frac{\det A^{\cdot}}{E_{(n\cdot)}(\text{adj } A^{\cdot})E^{(n\cdot)}}$$

$$\bar{P}^{\cdot} = \frac{E_{(n\cdot)}(\text{adj } A^{\cdot})}{E_{(n\cdot)}(\text{adj } A^{\cdot})E^{(n\cdot)}} \qquad \bar{Q}^{\cdot} = \frac{(\text{adj } A^{\cdot})E^{(n\cdot)}}{E_{(n\cdot)}(\text{adj } A^{\cdot})E^{(n\cdot)}} \qquad (2.7.17)$$

The restriction that $v \neq 0$ will, of course, carry over from Theorem 2.7.5 to equations (2.7.17). We can prove, in fact, that this restriction is now unnecessary. First we need the following matrix identities.

Lemma 2.7.18 Let A be a square matrix of order n. Then

$$E_{(n)} \text{ adj}(A + xE) = E_{(n)} \text{ adj } A \qquad (2.7.19a)$$
$$[\text{adj}(A + xE)]E^{(n)} = (\text{adj } A)E^{(n)} \qquad (2.7.19b)$$
$$E_{(n)}[\text{adj}(A + xE)]E^{(n)} = E_{(n)}(\text{adj } A)E^{(n)} \qquad (2.7.19c)$$

for every number x.

PROOF It is sufficient to prove (2.7.19a). Set

$$C = A + xE$$

and denote the entry in the (i,j) position of adj C and adj A by C_{ji} and A_{ji} respectively. The ith entries of the n-dimensional row vectors $E_{(n)}$ adj A and $E_{(n)}$ adj C are

$$\sum_{p=1}^{n} A_{ip} \quad \text{and} \quad \sum_{p=1}^{n} C_{ip}$$

respectively.
Set

$$B = \begin{bmatrix} a_{11} + x & a_{12} + x & \cdots & a_{1n} + x \\ \cdots\cdots\cdots\cdots\cdots\cdots\cdots\cdots\cdots\cdots\cdots \\ a_{i-1,1} + x & a_{i-1,2} + x & \cdots & a_{i-1,n} + x \\ 1 & 1 & \cdots & 1 \\ a_{i+1,1} + x & a_{i+1,2} + x & \cdots & a_{i+1,n} + x \\ \cdots\cdots\cdots\cdots\cdots\cdots\cdots\cdots\cdots\cdots\cdots \\ a_{n1} + x & a_{n2} + x & \cdots & a_{nn} + x \end{bmatrix}$$

Since B is obtained from C by replacing the entries in the ith row by 1s, the ith row expansion for det B yields

$$\det B = \sum_{p=1}^{n} C_{ip}$$

However, det B is independent of x; for when the ith row multiplied by x is subtracted from every other row, the determinant is unchanged. When $x = 0$, this expansion becomes

$$\det B = \sum_{p=1}^{n} A_{ip}$$

and hence $\quad \sum_{p=1}^{n} A_{ip} = \sum_{p=1}^{n} C_{ip} \quad i = 1, 2, \ldots, n \quad$ ////

Corollary 2.7.20

$$\det(A + xE) = \det A + xE_{(n)}(\text{adj } A)E^{(n)}$$

PROOF Multiplying (2.7.19a) on the right by $A + xE$ and applying Theorem 1.9.32 yields

$$E_{(n)}[(\det(A + xE))I] = E_{(n)}[(\det A)I] + xE_{(n)}(\text{adj } A)E$$

Since $E = E^{(n)}E_{(n)}$, we have

$$[\det(A + xE)]E_{(n)} = (\det A)E_{(n)} + xE_{(n)}(\text{adj } A)(E^{(n)}E_{(n)})$$
$$= [(\det A + xE_{(n)}(\text{adj } A)E^{(n)}]E_{(n)} \quad ////$$

We can now establish equations (2.7.17) without any restrictions on v.

Theorem 2.7.21 Let $(\bar{P}, \bar{Q}, v)$ be a solution of the game with matrix A and let $\bar{P}$ and $\bar{Q}$ be extreme optimal strategies for players I and II respectively. Then there exists a submatrix $A^{\cdot}$ (having order $n^{\cdot}$) of A such that

$$v = \frac{\det A^{\cdot}}{E_{(n^{\cdot})}(\text{adj } A^{\cdot})E^{(n^{\cdot})}}$$

and $\bar{P}$ and $\bar{Q}$ are given (up to zero entries) by

$$\bar{P}^{\cdot} = \frac{E_{(n^{\cdot})}(\text{adj } A^{\cdot})}{E_{(n^{\cdot})}(\text{adj } A^{\cdot})E^{(n^{\cdot})}} \qquad \bar{Q}^{\cdot} = \frac{(\text{adj } A^{\cdot})E^{(n^{\cdot})}}{E_{(n^{\cdot})}(\text{adj } A^{\cdot})E^{(n^{\cdot})}}$$

PROOF If $v \neq 0$, the theorem follows from Theorem 2.7.5 [as in (2.7.17)].
Now assume $v = 0$. For any real number x, $(\bar{P}, \bar{Q}, x)$ is a solution of the game with matrix $A + xE$, and $\bar{P}$ and $\bar{Q}$ are extreme optimal

strategies. Let $x \neq 0$. According to (2.7.17), there exists a submatrix $A^{\cdot} + xE$ of $A + xE$ such that

$$x = \frac{\det(A^{\cdot} + xE)}{E_{(n\cdot)}[\mathrm{adj}(A^{\cdot} + xE)]E^{(n\cdot)}}$$

$$\bar{P}^{\cdot} = \frac{E_{(n\cdot)}\,\mathrm{adj}(A^{\cdot} + xE)}{E_{(n\cdot)}[\mathrm{adj}(A^{\cdot} + xE)]E^{(n\cdot)}} \qquad \bar{Q}^{\cdot} = \frac{[\mathrm{adj}(A^{\cdot} + xE)]E^{(n\cdot)}}{E_{(n\cdot)}[\mathrm{adj}(A^{\cdot} + xE)]E^{(n\cdot)}} \qquad (2.7.22)$$

Using Lemma 2.7.18 and Corollary 2.7.20, equations (2.7.22) become

$$0 = \frac{\det A^{\cdot}}{E_{(n\cdot)}(\mathrm{adj}\ A^{\cdot})E^{(n\cdot)}}$$

$$\bar{P}^{\cdot} = \frac{E_{(n\cdot)}(\mathrm{adj}\ A^{\cdot})}{E_{(n\cdot)}(\mathrm{adj}\ A^{\cdot})E^{(n\cdot)}} \qquad \bar{Q}^{\cdot} = \frac{(\mathrm{adj}\ A^{\cdot})E^{(n\cdot)}}{E_{(n\cdot)}(\mathrm{adj}\ A^{\cdot})E^{(n\cdot)}} \qquad ////$$

Theorem 2.7.21 says nothing about the existence of extreme optimal strategies (except that there can be only a finite number of them). It can, however, be shown that:

Theorem 2.7.23 Every optimal strategy can be expressed as a convex combination of extreme optimal strategies.[1]

Theorem 2.7.23 follows from the fact that for each player the set of optimal strategies is convex and compact. The proof that in $\mathbb{R}^n$ every element of a convex compact set can be expressed as a convex combination of extreme points requires a knowledge of the geometry of convex sets that is beyond the scope of this book. Theorems 2.7.21 and 2.7.23, taken together, provide a (theoretical, at least) method for obtaining all solutions of a game.

EXERCISES

1 Find all extreme optimal strategies of each of the games:

(a) $\begin{bmatrix} 2 & 4 \\ 2 & -1 \end{bmatrix}$ (b) $\begin{bmatrix} 2 & -1 & -4 \\ -3 & -1 & 5 \end{bmatrix}$

(c) $\begin{bmatrix} 3 & -2 & -3 \\ -2 & 4 & 5 \end{bmatrix}$ (d) $\begin{bmatrix} 2 & 2 & 2 \\ 4 & 5 & -6 \end{bmatrix}$

by means of Theorem 2.7.21. (See Exercise 4 of Sec. 2.6.)

2 Find all solutions of the game of Exercise 5 of Sec. 2.6.

---

[1] For a proof of this theorem see Karlin [4, vol. I, p. 401].

3 Solve:

$$(a) \begin{bmatrix} 4 & -2 & -1 \\ 3 & -1 & 1 \\ -3 & 1 & 1 \end{bmatrix} \qquad (b) \begin{bmatrix} 0 & 1 & -1 \\ -1 & -1 & 0 \\ 1 & -1 & 1 \end{bmatrix}$$

by means of Theorem 2.7.21.

4 Let G be a game in which each player has a positive extreme optimal strategy. Show that the matrix of the game is square and that these are the only optimal strategies for each player.

5 Find all the extreme optimal strategies in the game of Exercise 2 of Sec. 2.5.

6 Let $\bar{P}$ and $\bar{Q}$ be optimal strategies satisfying equations (2.7.15) for some matrix $A^{\cdot}$. Show that $\bar{P}$ and $\bar{Q}$ are extreme optimal strategies.

2.8 LINEAR PROGRAMMING

Linear programming, which we study in this section, and the theory of games both involve finding nonnegative solutions for a system of simultaneous linear inequalities. Although they were originally treated as completely separate subjects, we can use the machinery of game theory to develop the theory of linear programming. However, unlike a game, a linear-programming problem need not have a solution.

EXAMPLE 2.8.1 **Maximum-profit problem** A farmer has 400 acres of land which is suitable for the growing of alfalfa, corn, and wheat. He also has 640 man-hours of labor per week at his disposal. Suppose that each acre devoted to alfalfa requires 1 man-hour of labor per week and will yield a profit of $40, each acre devoted to corn requires 2 man-hours of labor per week and will yield a profit of $60, and each acre of wheat requires 4 man-hours of labor per week and will yield a profit of $80. How many acres each of alfalfa, corn, and wheat should the farmer plant in order to maximize his profit?

Let a, c, and w be the number of acres to be planted with alfalfa, corn, and wheat respectively. Then the limitations under which the farmer must operate are described by the system of inequalities

$$a \geq 0 \qquad c \geq 0 \qquad w \geq 0$$

Land: $\qquad a + c + w \leq 400 \qquad (2.8.2)$

Labor: $\qquad a + 2c + 4w \leq 640$

A set of numbers a, c, w which satisfy the system of inequalities (2.8.2) will yield a profit of

$$P = 40a + 60c + 80w$$

Thus the farmer wishes to find values of a, c, w satisfying inequalities (2.8.2) which maximize $40a + 60c + 80w$.

Because there are only three unknowns in this problem, we can use graphical means to find all points (a, c, w) satisfying the inequalities (2.8.2).

The solid with vertices $0, A, B, C, D, E$ in Fig. 2.8.3 consists of all points (a, c, w) satisfying inequalities (2.8.2). We call these points the *feasible points* of the problem. It remains to determine at which feasible points the value of $P = 40a + 60c + 80w$ is a maximum. This can be done by intersecting the set of planes

$$40a + 60c + 80w = k$$

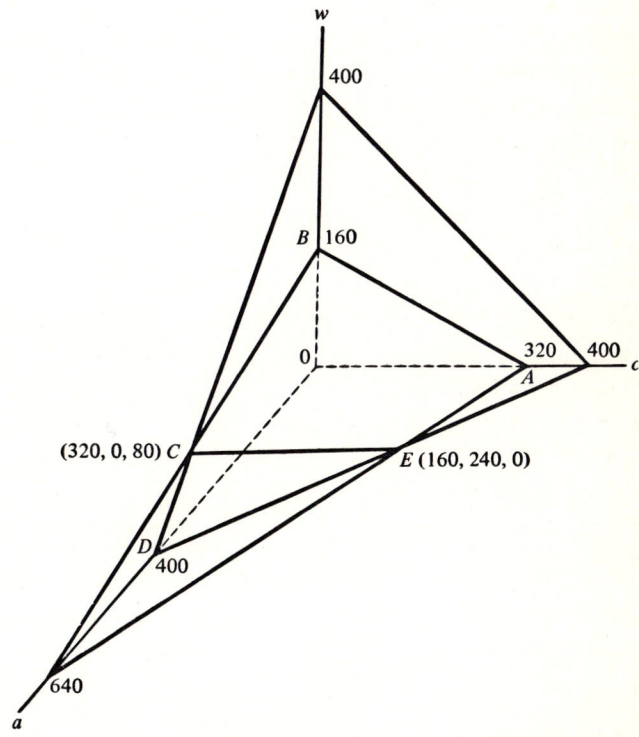

FIGURE 2.8.3

for various values of k with the set of feasible points. It is easily seen from Fig. 2.8.4 that the maximum value of P occurs at E and that the farmer is able to make a profit of $20,800 by planting 160 acres of alfalfa and 240 acres of corn (and no wheat). $E = (160, 240, 0)$ is called an *optimal point* or *solution* of the maximum (linear-programming) problem.

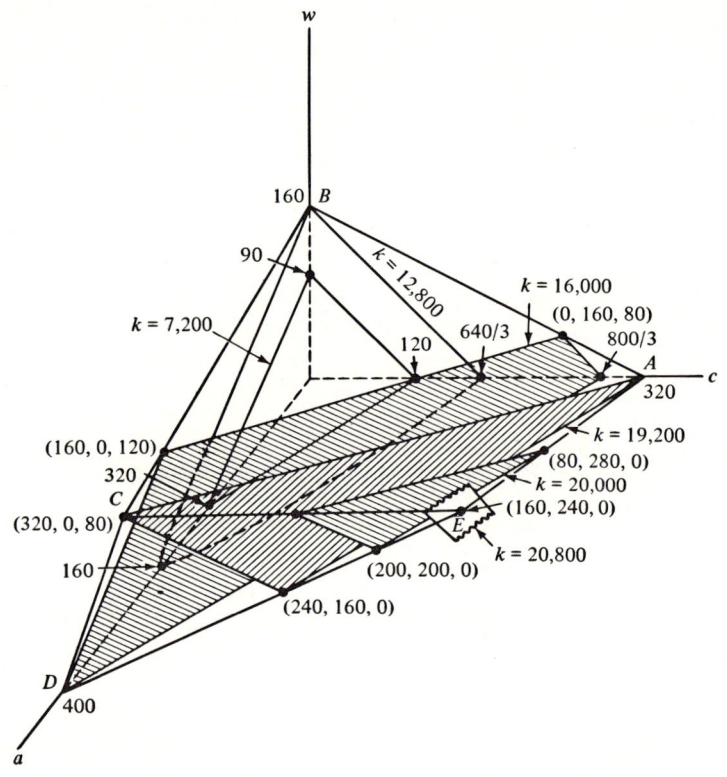

FIGURE 2.8.4

In order to apply the methods we have developed earlier in this chapter, we must restate Example 2.8.1 in the language of matrices. Thus we wish to find vectors

$$X = \begin{bmatrix} a \\ c \\ w \end{bmatrix}$$

belonging to $\mathbb{R}^3$ satisfying

$$\begin{bmatrix} 1 & 1 & 1 \\ 1 & 2 & 4 \end{bmatrix} X \leq \begin{bmatrix} 400 \\ 640 \end{bmatrix} \qquad X \geq 0 \qquad (2.8.5)$$

which maximize the expression

$$P = [40 \quad 60 \quad 80] X \qquad ////$$

EXAMPLE 2.8.6 **Minimum-cost problem** The same farmer is able to raise some livestock. Each animal will eat alfalfa and hay, and from these he must obtain at least 40 units of minerals, 60 units of protein, and 80 units of vitamins every week. Suppose that alfalfa and hay each have a mineral content of 1 unit per hundredweight. Suppose also that the protein contents per hundredweight of alfalfa and hay are 1 and 2 units respectively and that the vitamin contents per hundredweight of alfalfa and hay are 1 and 4 units respectively. In addition, the price of alfalfa is $4.00 per hundredweight, and the price of hay is $6.40 per hundredweight. What diet will provide the animals with adequate amounts of minerals, protein, and vitamins at the least cost?

Let a and h be the number of hundredweights of alfalfa and hay that the animals are to be fed. Then the system of inequalities

$$a \geq 0 \qquad h \geq 0$$

Minerals: $\qquad a + h \geq 40$

Protein: $\qquad a + 2h \geq 60 \qquad (2.8.7)$

$$a + 4h \geq 80$$

describes the dietary requirements of the animals. Thus the farmer wishes to find the values of a and h which satisfy inequalities (2.8.7) and which minimize the cost

$$C = 400a + 640h$$

Again we can solve the problem graphically. First we find all the feasible points of the problem (the shaded area of Fig. 2.8.8).

The set of feasible points for this problem is an unbounded portion of the plane. As before, we intersect the set of lines

$$400a + 640h = k$$

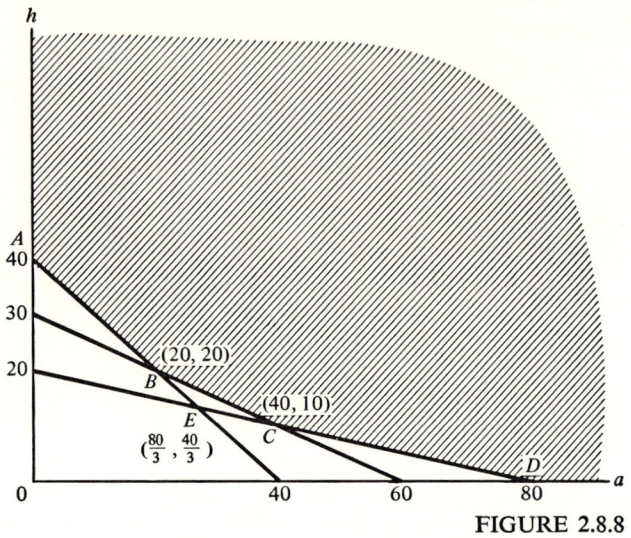

FIGURE 2.8.8

for various values of k, with the set of feasible points. From Fig. 2.8.9 it is clear that the minimum value of the cost occurs at $B = (20, 20)$. Thus the farmer is able to give his animals a satisfactory diet at a cost of $208.00 per week by using 20 hundredweight of alfalfa and 20 hundredweight of hay.

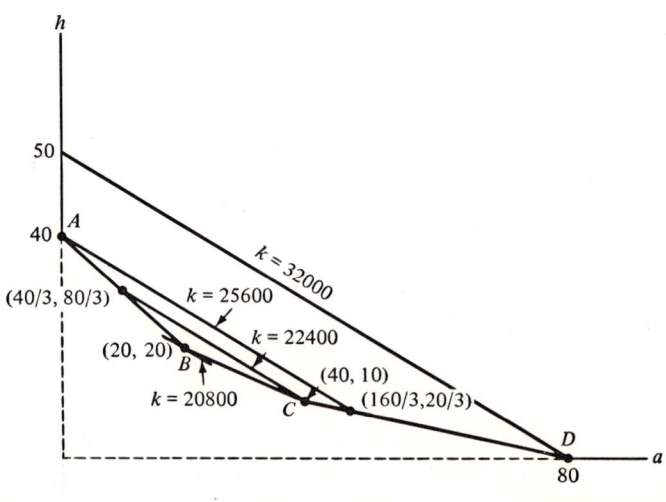

FIGURE 2.8.9

In matrix terminology, Example 2.8.6 requires us to find vectors

$$Y = [a \quad h]$$

belonging to $\mathbb{R}_2$ satisfying

$$Y \begin{bmatrix} 1 & 1 & 1 \\ 1 & 2 & 4 \end{bmatrix} \geq [40 \quad 60 \quad 80] \qquad Y \geq 0 \qquad (2.8.10)$$

which minimizes

$$C = Y \begin{bmatrix} 400 \\ 640 \end{bmatrix} \qquad ////$$

Let us now begin again and put the study of linear-programming problems on a more solid foundation.

Definition 2.8.11 Let A, Z_1, Z_2 be $m \times n$, $m \times 1$, $1 \times n$ matrices respectively. The two problems

    PROBLEM 1   MAXIMIZATION PROBLEM   Find a vector $X \in \mathbb{R}^n$ satisfying

$$AX \leq Z_1 \qquad X \geq 0 \qquad (2.8.12)$$

which maximizes $Z_2 X$.

    PROBLEM 2:   MINIMIZATION PROBLEM   Find a vector $Y \in \mathbb{R}_m$ satisfying

$$YA \geq Z_2 \qquad Y \geq 0 \qquad (2.8.13)$$

which minimizes YZ_1.

form a *dual pair of linear-programming problems*.

We denote this pair of problems by (A, Z_1, Z_2). Thus Examples 2.8.1 and 2.8.6 together constitute a dual pair of linear-programming problems (A, Z_1, Z_2) in which

$$A = \begin{bmatrix} 1 & 1 & 1 \\ 1 & 2 & 4 \end{bmatrix} \qquad Z_1 = \begin{bmatrix} 400 \\ 640 \end{bmatrix} \qquad Z_2 = [40 \quad 60 \quad 80]$$

Definition 2.8.14 A vector $X \in \mathbb{R}^n$ which satisfies the system of inequalities (2.8.12) is called a *feasible vector* of the maximization problem of the dual pair (A, Z_1, Z_2). A feasible vector X which maximizes the quantity $Z_2 X$ is called an *optimal vector* or *solution* of the maximization problem, and $Z_2 X$ is called the *value* of the problem.

A vector $Y \in \mathbb{R}_m$ which satisfies the system of inequalities (2.8.13) is called a *feasible vector* of the minimization problem of the dual pair (A, Z_1, Z_2). A feasible vector which minimizes the quantity YZ_1 is called an *optimal vector* or *solution* of the minimization problem, and YZ_1 is called the *value* of the problem.

It is possible that the inequalities of a linear-programming problem are inconsistent, i.e., that no feasible vectors exist. In this case, of course, the problem has no solution. We call such a problem an *inconsistent problem*. Another way in which a linear-programming problem might fail to have a solution is for $Z_2 X$ to be unbounded from above or for YZ_1 to be unbounded from below. Such a problem is said to be *unbounded*. In the exercises we shall give examples to show that both these situations can actually occur.

Maximization problems and minimization problems are essentially the same thing, for any problem can be expressed either way. The minimization problem

PROBLEM 2' Find a vector $X^T \in \mathbb{R}_n$ satisfying
$$X^T(-A^T) \geqq -Z_1^T \qquad X^T \geqq 0$$
which minimizes $X^T(-Z_2^T)$.

is identical with the maximization problem (Problem 1) of Definition 2.8.11.

Similarly, Problem 2 of Definition 2.8.11 can be expressed as a maximization problem.

PROBLEM 1' Find a vector $Y^T \in \mathbb{R}^m$ satisfying
$$(-A^T)Y^T \leqq -Z_2^T \qquad Y^T \geqq 0$$
which maximizes $(-Z_1^T)Y^T$.

Clearly Problems 1' and 2' also form a dual pair of linear-programming problems $(-A^T, -Z_2^T, -Z_1^T)$. Thus we need not make any distinction between maximization and minimization problems in a dual pair.

Lemma 2.8.15 If X, Y are feasible vectors of the dual pair of linear programming problems (A, Z_1, Z_2), then
$$Z_2 X \leqq YZ_1$$

PROOF Since X and Y are feasible vectors of the two problems of (A, Z_1, Z_2), we have

$$AX \leq Z_1 \quad X \geq 0 \quad YA \geq Z_2 \quad Y \geq 0$$

and hence
$$Z_2 X \leq YAX \leq YZ_1 \qquad ////$$

Corollary 2.8.16 If both problems of a dual pair have feasible vectors, then neither problem is unbounded.

According to Lemma 2.8.15, when both problems of a dual pair have solutions, then the value of the maximization problem is less than or equal to the value of the minimization problem. We shall show, in fact, in Theorem 2.8.19 that when both problems have solutions, their values are equal. (Note Examples 2.8.1 and 2.8.6). Also, it follows from Corollary 2.8.16 that the dual problem of an unbounded problem can have no feasible vectors and hence is inconsistent. Thus, neither problem will have a solution. Conversely we have:

Theorem 2.8.17 If a linear-programming problem is inconsistent, then its dual does not have a solution.

PROOF It is sufficient to treat the case in which the maximization problem is inconsistent. If the dual problem is also inconsistent, the theorem is immediate. Hence, we suppose that the maximization problem of (A, Z_1, Z_2) is inconsistent and that Y is a feasible vector of the minimization problem. Then

$$Y \geq 0 \quad YA \geq Z_2$$

and there does not exist a vector $X \in \mathbb{R}^n$ such that

$$X \geq 0 \quad AX \leq Z_1$$

By Corollary 2.4.18 there exists a vector $\tilde{Y} \in \mathbb{R}_m$ such that

$$\tilde{Y} \geq 0 \quad \tilde{Y}Z_1 < 0 \quad \tilde{Y}A \geq 0$$

Clearly $Y + \alpha \tilde{Y}$ is a feasible vector of the minimization problem for every nonnegative real number α. Since $\tilde{Y}Z_1 < 0$, it follows that $(Y + \alpha \tilde{Y})Z_1$ is unbounded from below. Hence the minimization problem is unbounded and so does not have a solution. ////

In the exercises we will give examples to show that it is possible for the dual of an inconsistent problem to be either inconsistent or unbounded.

Thus far we have not given any indication when optimal vectors exist. To this end we first prove Lemma 2.8.18, which gives an efficient method for recognizing optimal vectors.

Lemma 2.8.18 If $\bar{X}$, $\bar{Y}$ are feasible vectors of the dual pair of linear-programming problems (A, Z_1, Z_2) for which

$$\bar{Y}Z_1 \leq Z_2 \bar{X} \qquad (2.8.19)$$

then $\bar{X}$, $\bar{Y}$ are optimal vectors and $\bar{Y}Z_1 = Z_2 \bar{X}$.

PROOF For any feasible vectors X, Y of the dual problems (A, Z_1, Z_2) we have, from Lemma 2.8.15,

$$Z_2 X \leq \bar{Y}Z_1 \leq Z_2 \bar{X}$$
$$\bar{Y}Z_1 \leq Z_2 \bar{X} \leq YZ_1$$

Since $\bar{X}$, $\bar{Y}$ are themselves feasible vectors, it follows that

$$Z_2 \bar{X} = \max\{Z_2 X \mid X \text{ is a feasible vector of Problem 1}\}$$
$$\bar{Y}Z_1 = \min\{YZ_1 \mid Y \text{ is a feasible vector of Problem 2}\}$$

Thus $\bar{X}$ and $\bar{Y}$ are optimal vectors.

It is an immediate consequence of equation (2.8.19) and Lemma 2.8.15 that $Z_2 \bar{X} = \bar{Y}Z_1$. ////

Note that Lemma 2.8.18 describes a situation in which the values of the problems in a dual pair are equal. However, Lemma 2.8.18 does not give any indication that such a situation can ever occur, much less that it is the only way in which optimal vectors can exist. That this is, in fact, the case is contained in:

Theorem 2.8.20 If both linear-programming problems of a dual pair have feasible vectors, then both have optimal vectors and the values of the two problems are equal.

PROOF Let (A, Z_1, Z_2) be a dual pair of problems both of which have feasible vectors. We wish to show that there exist vectors $X \in \mathbb{R}^n$, $Y \in \mathbb{R}_m$ such that

$$AX \leq Z_1 \quad X \geq 0 \quad YA \geq Z_2 \quad Y \geq 0$$
$$Z_2 X - YZ_1 \geq 0 \qquad (2.8.21)$$

for, according to Lemma 2.8.18, a pair of vectors satisfying all these inequalities would be a pair of optimal vectors. It is more convenient to express the inequalities of (2.8.21) in the form

$$AX \leq Z_1 \quad X \geq 0 \quad (-A^T)Y^T \leq -Z_2^T \quad Y^T \geq 0$$
$$-Z_2 X + Z_1^T Y^T \leq 0$$

for these inequalities can then be combined into the pair of inequalities

$$\begin{bmatrix} 0 & A \\ -A^T & 0 \\ Z_1^T & -Z_2 \end{bmatrix} \begin{bmatrix} Y^T \\ X \end{bmatrix} \leq \begin{bmatrix} Z_1 \\ -Z_2^T \\ 0 \end{bmatrix} \qquad \begin{bmatrix} Y^T \\ X \end{bmatrix} \geq 0 \qquad (2.8.22)$$

Suppose the inequalities (2.8.22) do not have a solution. It follows from Corollary 2.4.18 that there exists a nonnegative vector $W \in \mathbb{R}_{m+n+1}$ such that

$$W \begin{bmatrix} 0 & A \\ -A^T & 0 \\ Z_1^T & -Z_2 \end{bmatrix} \geq 0 \qquad W \begin{bmatrix} Z_1 \\ -Z_2^T \\ 0 \end{bmatrix} < 0 \qquad (2.8.23)$$

Partition W as

$$W = [W_1 \quad W_2 \quad w_3] \qquad W_1 \in \mathbb{R}_m \quad W_2 \in \mathbb{R}_n \quad w_3 \in \mathbb{R}$$

Then inequalities (2.8.23) become

$$\begin{aligned} -W_2 A^T + w_3 Z_1^T \geq 0 & \qquad W_1 A - w_3 Z_2 \geq 0 \\ W_1 Z_1 - W_2 Z_2^T < 0 & \end{aligned} \qquad (2.8.24)$$

Now, two cases arise.

CASE 1 First let us suppose that $w_3 \neq 0$. We can rewrite (2.8.24) as

$$A\left(\frac{1}{w_3} W_2^T\right) \leq Z_1 \qquad \left(\frac{1}{w_3} W_1\right) A \geq Z_2$$

$$\left(\frac{1}{w_3} W_1\right) Z_1 - Z_2\left(\frac{1}{w_3} W_2^T\right) < 0$$

Consequently $\qquad \bar{X} = \frac{1}{w_3} W_2^T \qquad \bar{Y} = \frac{1}{w_3} W_1$

are feasible vectors for the dual pair of problems, and

$$\bar{Y} Z_1 - Z_2 \bar{X} < 0$$

which contradicts Lemma 2.8.15.

CASE 2 Next, let us suppose that $w_3 = 0$. Let X, Y be any pair of feasible vectors. The first two inequalities of (2.8.24) yield

$$W_1 Z_1 \geq W_1 A X \geq 0 \cdot X = 0 = Y \cdot 0 \geq Y A W_2^T \geq Z_2 W_2^T = W_2 Z_2^T$$

which contradicts the last inequality of (2.8.24).

Thus, for either supposition for w_3, the original supposition that no solutions of inequalities (2.8.22) exist has led to a contradiction. By Lemma 2.8.18, both problems have solutions, and their values are equal.   ////

Corollary 2.8.25 Either both linear-programming problems of a dual pair have optimal vectors or neither has optimal vectors.

PROOF According to Theorem 2.8.20, either both linear-programming problems of a dual pair have optimal vectors or at least one of the problems is inconsistent. By Theorem 2.8.17, if one of the problems is inconsistent, then its dual does not have optimal vectors.   ////

We conclude this chapter with a theorem which shows the close connection between linear-programming problems and the theory of games. It is possible to associate a symmetric game with a dual pair of linear-programming problems in such a way that all the optimal vectors (if any) of the linear-programming problems are obtainable from the optimal strategies of the game. We associate the game whose matrix is the $(m+n+1) \times (m+n+1)$ skew-symmetric matrix

$$B = \begin{bmatrix} 0 & A & -Z_1 \\ -A^T & 0 & Z_2^T \\ Z_1^T & -Z_2 & 0 \end{bmatrix}$$

with the dual pair (A, Z_1, Z_2).

Theorem 2.8.26 If X, Y are optimal vectors for the dual pair of linear-programming problems (A, Z_1, Z_2), then some positive multiple of

$$U = \begin{bmatrix} Y & X^T & 1 \end{bmatrix}$$

is an optimal strategy for each player in the game whose matrix is B. Conversely, if

$$P = \begin{bmatrix} W_1 & W_2 & w_3 \end{bmatrix} \qquad W_1 \in \mathbb{R}_m \qquad W_2 \in \mathbb{R}_n \qquad w_3 \in \mathbb{R}$$

is an optimal strategy in the game whose matrix is B, and if $w_3 \neq 0$, then

$$X = \frac{1}{w_3} W_2^T \qquad Y = \frac{1}{w_3} W_1$$

are optimal vectors of the dual pair of linear-programming problems (A, Z_1, Z_2).

Finally, the dual pair (A, Z_1, Z_2) have no optimal vectors if and only if $w_3 = 0$ for every optimal strategy of B.

PROOF Let X, Y be optimal vectors for the dual pair (A, Z_1, Z_2). By Theorem 2.8.20 we have

$$AX \leqq Z_1 \qquad YA \geqq Z_2 \qquad Z_2 X = YZ_1$$

and hence

$$UB = [Y \quad X^T \quad 1] \begin{bmatrix} 0 & A & -Z_1 \\ -A^T & 0 & Z_2^T \\ Z_1^T & -Z_2 & 0 \end{bmatrix} \geqq 0$$

Since U is semipositive and the value of the game with matrix B is zero, it follows that some positive multiple of U is an optimal strategy of B. It only remains to divide U by the sum of its entries (which is positive) in order to obtain a strategy vector.

Conversely, if $P = [W_1 \quad W_2 \quad w_3]$ is an optimal strategy of B, then $PB \geqq 0$ and hence

$$-W_2 A^T + w_3 Z_1^T \geqq 0 \qquad W_1 A - w_3 Z_2 \geqq 0 \qquad -W_1 Z_1 + W_2 Z_2^T \geqq 0 \tag{2.8.27}$$

As before, if $w_3 \neq 0$, then inequalities (2.8.27) can be written as

$$A\left(\frac{1}{w_3} W_2^T\right) \leqq Z_1 \qquad \left(\frac{1}{w_3} W_1\right) A \geqq Z_2 \qquad \left(\frac{1}{w_3} W_1\right) Z_1 \leqq \left(\frac{1}{w_3} W_2\right) Z_2^T$$

and, by Lemma 2.8.18,

$$\bar{X} = \left(\frac{1}{w_3} W_2^T\right) \qquad \bar{Y} = \left(\frac{1}{w_3} W_1\right)$$

are optimal vectors of (A, Z_1, Z_2).

It is now clear that $w_3 = 0$ for every optimal strategy P of B if and only if the dual pair of linear-programming problems do not have any optimal vectors, since every pair of optimal vectors leads to an optimal strategy of B in which $w_3 \neq 0$, and conversely. ////

EXERCISES

1 Prove that the system of inequalities

$$x + 2y + z - w \leqq -1$$
$$-x + y - 2z + 5w \leqq 4$$

has no nonnegative solutions (*a*) directly from the inequalities, and (*b*) by constructing an inconsistent linear-programming problem with these inequalities. (*Hint:* Try to maximize $x + y + z + w$. Show that the dual problem is unbounded.)

2 Show that the pair of linear-programming problems

$$\left(\begin{bmatrix} 2 & -5 \\ -7 & 10 \end{bmatrix}, \begin{bmatrix} \frac{1}{2} \\ -1 \end{bmatrix}, [2 \;\; -1]\right)$$

does not have any solutions (a) by the graphical methods of this section, and (b) by showing that the last pure strategy of the associated game is inessential. Determine whether each problem is inconsistent or unbounded.

3 Solve:

(a) $\left(\begin{bmatrix} 2 & 1 & -1 \\ 1 & -1 & -1 \end{bmatrix}, \begin{bmatrix} -2 \\ 1 \end{bmatrix}, [4 \;\; 1 \;\; 1]\right)$.

(b) $\left(\begin{bmatrix} 1 & 3 \\ 2 & 1 \end{bmatrix}, \begin{bmatrix} 2 \\ 3 \end{bmatrix}, [2 \;\; 5]\right)$.

(c) $\left(\begin{bmatrix} 1 & 4 & -2 \\ 1 & 1 & 1 \\ 1 & -2 & 0 \end{bmatrix}, \begin{bmatrix} 4 \\ 4 \\ 1 \end{bmatrix}, [3 \;\; 4 \;\; 2]\right)$.

4 A farmer has 7 acres of land and $2,400 capital. He can raise either livestock or vegetables. An acre of land is sufficient for either 10 animals or 10 tons of vegetables. Raising livestock requires an investment of $20 per animal, while raising vegetables requires a capital outlay of $60 per ton. In addition, the farmer must raise at least enough vegetables to feed his livestock, and each ton of vegetables feeds six animals. How can the farmer maximize his profit if he realizes a profit of $45 for each animal he sells and $30 for each ton of vegetables he sells (after feeding his livestock)? Can the farmer make more money if he is able to buy vegetables to feed his livestock at $30 per ton?

5 Show that the set of feasible vectors of a linear-programming problem is convex.

6 Show that the set of optimal vectors of a linear-programming problem is convex.

7 Show that if A, Z_1, and Z_2 are positive matrices, both problems of (A, Z_1, Z_2) have solutions.

8 Show that (A, Z_1, Z_2) does not have any solutions if and only if there exist vectors $W_1 \in \mathbb{R}_m$ and $W_2 \in \mathbb{R}^n$ such that

$$W_1 A \geq 0 \qquad W_1 \geq 0 \qquad A W_2 \leq 0 \qquad W_2 \geq 0$$
$$Z_2 W_2 > W_1 Z_1$$

9 An optimal vector is *extreme* if it cannot be expressed as a convex combination of other optimal vectors. Let X and Y be optimal vectors of (A, Z_1, Z_2) and let

$$P = \frac{1}{\alpha}[Y \;\; X^T \;\; 1]$$

where α is chosen so that P is a strategy vector. Show that P is an extreme optimal strategy of the associated game if and only if both X and Y are extreme optimal vectors.

REFERENCES

Although the theory of games is a relatively new branch of mathematics, there is already an extensive literature on the subject dealing not only with the finite zero-sum two-person games in this chapter but also with several more general situations. The fundamental theorem remains valid for a large and important class of infinite zero-sum two-person games, i.e., games in which there are infinitely many pure strategies. Many of the results of this chapter can be extended to these games. A number of interesting infinite games are described by Karlin [4] and Dresher [1]. The latter discusses in great detail duels in which each player can fire at any time. In addition to studying infinite games, Owen [7] treats non-zero-sum games, e.g., a duel in which both players lose if both are killed, and n-person games. In a finite non-zero-sum game there is a payoff matrix for each player. The players are often not fighting each other, and it is possible that both can benefit by cooperating with each other. The possibility of players forming coalitions is even more important in n-person games.

We have chosen to study linear-programming problems in terms of games. Hillier and Lieberman [3] show that linear programming can be treated independently and that a game can be solved by transforming it into a linear-programming problem. In addition they develop the fundamentals of nonlinear programming. The theory of linear programming is also discussed by Gass [2], who also presents both the theory and the computational aspects of the simplex method.

1 Dresher, M.: "Games of Strategy," Prentice-Hall, Inc., Englewood Cliffs, N.J., 1961.
2 Gass, S. I.: "Linear Programming," 2d ed., McGraw-Hill Book Company, New York, 1964.
3 Hillier, F. S., and G. J. Lieberman: "Introduction to Operations Research," Holden-Day, Inc., San Francisco, 1967.
4 Karlin, S.: "Mathematical Methods and Theory in Games, Programming and Economics," vols. I and II, Addison-Wesley Publishing Company, Inc., Reading, Mass., 1959
5 McKinsey, J. C. C.: "Introduction to the Theory of Games," McGraw-Hill Book Company, New York, 1952.
6 Neumann, J. von: Zur Theorie der Gesellschaftsspiele, *Math. Ann.*, *100*: 295–320 (1928).
7 Owen, G.: "Game Theory," W. B. Saunders Company, Philadelphia, 1968.

3

FINITE MARKOV CHAINS

3.1 EXAMPLES OF FINITE MARKOV CHAINS

A stochastic process is concerned with events that change in a random way with time. A typical stochastic process might predict the motion of an object which is constrained to be at any one time in exactly one of a number of possible states. It would then be a scheme for determining the probability of the object's being in a specific state at a specific time. This probability would generally depend on a large number of factors, e.g., (1) the state, (2) the time, (3) some or all of the previous states the object has been in, (4) the states other objects are in or have been in. For example, the object might be the reader, and the states might be the countries of the world. Clearly aspects of all four factors mentioned above will influence the probability that the reader will be in a specific state at a specific time. If the probability that an object in one state will move to another state depends only on the two states involved (and not on earlier states, or time, or other factors), then the stochastic process is called a *Markov process*.[1] If the number of states the object can be in is finite (in which case we denote them by $S_1, S_2, \ldots, S_n$), the Markov process is called a *finite Markov chain*.

[1] Named for the Russian mathematician A. A. Markov (1856–1922), who introduced it in 1907.

The probability that an object in state S_j at a given time will be in state S_i at the end of a specified time interval is called the *transition probability* and will be denoted by a_{ij}. Thus the stochastic process is a Markov chain if a_{ij} is a constant (depending only on i and j). Returning to the example above, a_{ij} might describe the probability that the reader, if he is in state S_j (say, Japan) on the first day of any month, will be in state S_i (say Mexico) on the first day of the next month. For this example to be a Markov chain the probability of being in Mexico a month after being in Japan must not depend on anything other than the two countries involved.

The $n \times n$ matrix $A = [a_{ij}]$ which has the transition probability a_{ij} in the (i, j) position is called the *transition matrix* of the Markov chain. For a Markov chain with n states, A is a real square semipositive matrix of order n. Since, for each j, a Markov chain moves an object from state S_j to some state, it follows that

$$a_{1j} + a_{2j} + \cdots + a_{nj} = 1$$

i.e., the sum of the entries in any column of A is 1.

To define a Markov chain completely it is necessary to specify two things:

1 The transition matrix
2 For each j, the probability $p_j^{(0)}$ that the object is initially in the state S_j

The vector

$$P^{(0)} = \begin{bmatrix} p_1^{(0)} \\ p_2^{(0)} \\ \vdots \\ p_n^{(0)} \end{bmatrix}$$

is called the *initial* or *0-step probability vector*. Clearly

$$\begin{aligned} p_j^{(0)} &\geqq 0 \quad j = 1, 2, \ldots, n \\ p_1^{(0)} + p_2^{(0)} + \cdots + p_n^{(0)} &= 1 \end{aligned} \quad (3.1.1)$$

More generally, for $k = 0, 1, 2, \ldots$, we define the *k-step probability vector* to be

$$P^{(k)} = \begin{bmatrix} p_1^{(k)} \\ p_2^{(k)} \\ \vdots \\ p_n^{(k)} \end{bmatrix}$$

where $p_j^{(k)}$ is the probability that the object is in state S_j after the Markov chain process has been applied k times, beginning with the initial probability vector $P^{(0)}$. Any column vector whose entries satisfy conditions (3.1.1) is called a *probability vector*.[1] Thus every column of a transition matrix is a probability vector. The collection of all *n*-dimensional probability vectors will be denoted by $\mathbb{P}^n$.

Let $A = [a_{ij}]$ be the transition matrix of a Markov chain process and let $P^{(0)} = [p_j^{(0)}]$ be the initial probability vector. We wish to find the probability $p_i^{(1)}$ that the object will be in state S_i after a single application of the Markov chain process. Let (S_i, S_j) denote the event "the object is initially in the state S_j and is in the state S_i after the Markov chain process has acted once." Thus, in order for the event (S_i, S_j) to occur the object must initially be in state S_j (and this happens with probability $p_j^{(0)}$) and must then move to state S_i (and this happens with probability a_{ij}). Hence, the probability of the event (S_i, S_j) is $a_{ij} p_j^{(0)}$. The probability that the object is in the state S_i after a single application of the Markov chain process is the sum of the probabilities of the events (S_i, S_j) for all j, that is,

$$p_i^{(1)} = \sum_{j=1}^{n} a_{ij} p_j^{(0)}$$

Thus:

Theorem 3.1.2 If A is the transition matrix of a Markov chain with initial probability vector $P^{(0)}$, then

$$P^{(1)} = AP^{(0)}$$

Corollary 3.1.3 For any nonnegative integers k, l

$$P^{(k+l)} = A^k P^{(l)} \qquad (3.1.4)$$

EXAMPLE 3.1.5 The men in a certain community are classified as either smokers or nonsmokers. It is found that among the sons of smokers, $\frac{3}{5}$ are smokers and $\frac{2}{5}$ are nonsmokers, while among the sons of nonsmokers, $\frac{4}{5}$ are nonsmokers and $\frac{1}{5}$ are smokers. If $\frac{2}{3}$ of the current population smokes, what portion of each of the next two generations will smoke?

A research organization wishes to study a sample population of this community. Is it possible to construct a sample such that the proportion of smokers in the next generation will be the same as the proportion of smokers in the current generation?

---

[1] In Chap. 2 we called such a vector a strategy, but the term probability vector is more appropriate here.

[3.1.5]

Note that the probability that a man will be a smoker depends only on the state, i.e., smoker or nonsmoker, of his father. We will classify the men of the present generation as either:

1 S, smokers
2 N, nonsmokers

and the men of the next generation as:

1 SS, smoking sons of smokers
2 SN, smoking sons of nonsmokers
3 NS, nonsmoking sons of smokers
4 NN, nonsmoking sons of nonsmokers

Since $\frac{2}{3}$ of the present generation smokes, it follows that $\frac{2}{3}$ of the next generation will be in either the SS or NS categories.[1] But, $\frac{3}{5}$ of these will be smokers (SS), and $\frac{2}{5}$ will be nonsmokers (NS). Hence the proportion of SS and NS in the next generation will be $\frac{2}{3} \cdot \frac{3}{5} = \frac{6}{15}$ and $\frac{2}{3} \cdot \frac{2}{5} = \frac{4}{15}$, respectively. In the same way, $\frac{1}{3}$ of the present generation are nonsmokers, and hence $\frac{1}{3}$ of the next generation will be in the SN and NN categories. Since $\frac{1}{5}$ of these will be smokers (SN) and $\frac{4}{5}$ will be nonsmokers (NN), the proportion of SN and NN in the next generation will be $\frac{1}{3} \cdot \frac{1}{5} = \frac{1}{15}$ and $\frac{1}{3} \cdot \frac{4}{5} = \frac{4}{15}$, respectively. Thus, in the next generation the smokers (SS and SN combined) will constitute $\frac{6}{15} + \frac{1}{15} = \frac{7}{15}$ of the population while the nonsmokers (NS and NN combined) will constitute $\frac{4}{15} + \frac{4}{15} = \frac{8}{15}$ of the population. We can find the proportion of smokers and nonsmokers in all succeeding generations in the same manner.

Let us denote the proportion of smokers and nonsmokers in the kth generation by $S^{(k)}$ and $N^{(k)}$ respectively.[2] Using the method of the last paragraph, it is clear that

$$S^{(k+1)} = \tfrac{3}{5}S^{(k)} + \tfrac{1}{5}N^{(k)}$$
$$N^{(k+1)} = \tfrac{2}{5}S^{(k)} + \tfrac{4}{5}N^{(k)}$$

Thus we have a Markov chain in which the transition matrix is

$$A = \begin{bmatrix} \tfrac{3}{5} & \tfrac{1}{5} \\ \tfrac{2}{5} & \tfrac{4}{5} \end{bmatrix}$$

and the initial probability vector is

$$P^{(0)} = \begin{bmatrix} S^{(0)} \\ N^{(0)} \end{bmatrix} = \begin{bmatrix} \tfrac{2}{3} \\ \tfrac{1}{3} \end{bmatrix}$$

---

[1] We are assuming that smoking and nonsmoking fathers have the same number of sons. In probabilistic terms, we are assuming that smoking and fertility are independent attributes.

[2] It will be convenient for us to refer to the present generation as the zeroth generation.

The requirement that the proportion of smokers in the next generation be the same as the proportion of smokers in the current generation can be expressed as

$$P^{(1)} = P^{(0)}$$

Since $P^{(1)} = AP^{(0)}$, we are looking for a solution $P^{(0)}$ of the equation

$$AP^{(0)} = P^{(0)} \qquad (3.1.6)$$

i.e., we are looking for a probability vector which is a characteristic vector of A corresponding to the characteristic root $+1$. A simple calculation will show that the characteristic roots of A are $+1$ and $\frac{2}{3}$ and that the characteristic vector corresponding to $+1$ is

$$\begin{bmatrix} \frac{1}{3} \\ \frac{2}{3} \end{bmatrix}$$

Thus if any sample of the community contains $\frac{1}{3}$ smokers and $\frac{2}{3}$ nonsmokers, the next generation (and, in fact, every succeeding generation) will also contain $\frac{1}{3}$ smokers and $\frac{2}{3}$ nonsmokers. We call a probability vector satisfying equation (3.1.6) a *stationary vector* of the Markov chain. ////

When dealing with a number $N > 1$ of objects all of which are subject to the same Markov chain process, we may replace the initial probability vector by the *initial population vector*

$$Q^{(0)} = \begin{bmatrix} q_1^{(0)} \\ q_2^{(0)} \\ \vdots \\ q_n^{(0)} \end{bmatrix}$$

where $q_j^{(0)}$ is the number of objects initially in state S_j. Similarly we replace the k-step probability vector $P^{(k)}$ by the *k-step population vector*

$$Q^{(k)} = \begin{bmatrix} q_1^{(k)} \\ q_2^{(k)} \\ \vdots \\ q_n^{(k)} \end{bmatrix}$$

where $q_j^{(k)}$ is the number of objects in state S_j after the Markov chain process has been applied k times, beginning with the initial population vector $Q^{(0)}$. Clearly

$$q_j^{(k)} = N p_j^{(k)} \qquad \begin{array}{l} j = 1, 2, \ldots, n \\ k = 0, 1, 2, \ldots \end{array} \qquad (3.1.7)$$

and equation (3.1.4) becomes

$$Q^{(k+l)} = A^k Q^{(l)} \qquad k, l = 0, 1, 2, \ldots \qquad (3.1.8)$$

EXAMPLE 3.1.9 The population of Ruritania is 1,800. There are three cities in Ruritania, S, T, and U, and every year the entire population of each city moves, half to each of the other two cities. If the populations of S, T, and U are currently 200, 600, and 1,000 respectively, what will the population of each city be next year and the year following? What are the long-range population expectations for each city?

Let $s^{(k)}$, $t^{(k)}$, $u^{(k)}$ be the population of S, T, U respectively after k years and set

$$Q^{(k)} = \begin{bmatrix} s^{(k)} \\ t^{(k)} \\ u^{(k)} \end{bmatrix} \qquad k = 0, 1, 2, \ldots$$

Clearly we have a Markov chain with three states. The transition matrix and the initial population vector are

$$A = \begin{bmatrix} 0 & \tfrac{1}{2} & \tfrac{1}{2} \\ \tfrac{1}{2} & 0 & \tfrac{1}{2} \\ \tfrac{1}{2} & \tfrac{1}{2} & 0 \end{bmatrix} \qquad Q^{(0)} = \begin{bmatrix} 200 \\ 600 \\ 1{,}000 \end{bmatrix}$$

respectively. It follows from Theorem 3.1.2 and equations (3.1.8) that

$$Q^{(1)} = AQ^{(0)} = \begin{bmatrix} 800 \\ 600 \\ 400 \end{bmatrix} \qquad Q^{(2)} = AQ^{(1)} = A^2 Q^{(0)} = \begin{bmatrix} 500 \\ 600 \\ 700 \end{bmatrix}$$

We can show by an induction[1] on k that

$$A^k = \begin{bmatrix} a_k & a_{k+1} & a_{k+1} \\ a_{k+1} & a_k & a_{k+1} \\ a_{k+1} & a_{k+1} & a_k \end{bmatrix} \qquad k = 0, 1, 2, \ldots \qquad (3.1.10)$$

where

$$a_k = \frac{1}{3}\left[1 + \frac{(-1)^k}{2^{k-1}}\right]$$

[The verification of equation (3.1.10) is left as an exercise.] Substituting these values into equation (3.1.8), we have

$$Q^{(k)} = \begin{bmatrix} 600 + \dfrac{400(-1)^{k+1}}{2^k} \\ 600 \\ 600 - \dfrac{400(-1)^{k+1}}{2^k} \end{bmatrix} \qquad k = 0, 1, 2, \ldots$$

---

[1] For a more explicit method for obtaining equation (3.1.10) see Exercises 10 to 12 of Sec. 1.13.

Thus, the population of T remains at 600. The population of S and U both approach 600 as a limit, and each is alternately greater than and less than 600 ////

EXAMPLE 3.1.11 Player I draws a card from a three-card deck consisting of two cards marked H and one card marked L. If player I draws L, he pays player II $1; while if he draws H, player II pays him $1. At the outset, player I has $1, and player II has $3. The game is over when one of the players has won all the other player's money. Find the probability of each player's emerging the winner. What is the probability that a tenth draw will be required, i.e., that after each of the first nine draws, both players have some money?

We will take player I to be the object and use the amount of money that he has to describe the various states. The most natural way to do this would be to define S_1 to be the state "player I has $0," S_2 to be the state "player I has $1," S_3 to be the state "player I has $2," etc. This will lead to the transition matrix

$$\begin{bmatrix} 1 & \frac{1}{3} & 0 & 0 & 0 \\ 0 & 0 & \frac{1}{3} & 0 & 0 \\ 0 & \frac{2}{3} & 0 & \frac{1}{3} & 0 \\ 0 & 0 & \frac{2}{3} & 0 & 0 \\ 0 & 0 & 0 & \frac{2}{3} & 1 \end{bmatrix}$$

However, it will be much more convenient to order the states, still in terms of player I's holdings, $1, $2, $3, $0, $4, for then the transition matrix becomes

$$A = \begin{bmatrix} 0 & \frac{1}{3} & 0 & 0 & 0 \\ \frac{2}{3} & 0 & \frac{1}{3} & 0 & 0 \\ 0 & \frac{2}{3} & 0 & 0 & 0 \\ \frac{1}{3} & 0 & 0 & 1 & 0 \\ 0 & 0 & \frac{2}{3} & 0 & 1 \end{bmatrix}$$

We now partition A into

$$A = \begin{bmatrix} B & 0 \\ C & I \end{bmatrix}$$

where

$$C = \begin{bmatrix} \frac{1}{3} & 0 & 0 \\ 0 & 0 & \frac{2}{3} \end{bmatrix} \quad B = \begin{bmatrix} 0 & \frac{1}{3} & 0 \\ \frac{2}{3} & 0 & \frac{1}{3} \\ 0 & \frac{2}{3} & 0 \end{bmatrix}$$

It can easily be verified, using an induction on k, that

$$A^k = \begin{bmatrix} B^k & 0 \\ C(I + B + B^2 + \cdots + B^{k-1}) & I \end{bmatrix} \quad k = 1, 2, \ldots \quad (3.1.12)$$

By a direct computation we can show that

$$B^3 = \tfrac{4}{9}B$$

and hence

$$B + B^3 + B^5 + \cdots + B^{2l+1} = [1 + \tfrac{4}{9} + (\tfrac{4}{9})^2 + \cdots + (\tfrac{4}{9})^l]B$$
$$B^2 + B^4 + B^6 + \cdots + B^{2l} = [1 + \tfrac{4}{9} + (\tfrac{4}{9})^2 + \cdots + (\tfrac{4}{9})^l]B^2$$
$$l = 1, 2, \ldots$$

Since the infinite geometric series $1 + \tfrac{4}{9} + (\tfrac{4}{9})^2 + (\tfrac{4}{9})^3 + \cdots$ converges to $\tfrac{9}{5}$, it is reasonable to say that the infinite series of matrices $I + B + B^2 + B^3 + \cdots$ converges to[1]

$$I + \tfrac{9}{5}B + \tfrac{9}{5}B^2 = \begin{bmatrix} \tfrac{7}{5} & \tfrac{3}{5} & \tfrac{1}{5} \\ \tfrac{6}{5} & \tfrac{9}{5} & \tfrac{3}{5} \\ \tfrac{4}{5} & \tfrac{6}{5} & \tfrac{7}{5} \end{bmatrix}$$

and that

$$\lim_{k \to \infty} C(I + B + B^2 + \cdots + B^{k-1}) =$$

$$\begin{bmatrix} \tfrac{1}{3} & 0 & 0 \\ 0 & 0 & \tfrac{2}{3} \end{bmatrix} \begin{bmatrix} \tfrac{7}{5} & \tfrac{3}{5} & \tfrac{1}{5} \\ \tfrac{6}{5} & \tfrac{9}{5} & \tfrac{3}{5} \\ \tfrac{4}{5} & \tfrac{6}{5} & \tfrac{7}{5} \end{bmatrix} = \begin{bmatrix} \tfrac{7}{15} & \tfrac{3}{15} & \tfrac{1}{15} \\ \tfrac{8}{15} & \tfrac{12}{15} & \tfrac{14}{15} \end{bmatrix}$$

Moreover, from $B^3 = \tfrac{4}{9}B$ it follows that

$$B^k = \begin{cases} \left(\tfrac{2}{3}\right)^k \begin{bmatrix} 0 & \tfrac{1}{2} & 0 \\ 1 & 0 & \tfrac{1}{2} \\ 0 & 1 & 0 \end{bmatrix} & k \text{ odd} \\ \left(\tfrac{2}{3}\right)^k \begin{bmatrix} \tfrac{1}{2} & 0 & \tfrac{1}{4} \\ 0 & 1 & 0 \\ 1 & 0 & \tfrac{1}{2} \end{bmatrix} & k \text{ even} \end{cases} \quad (3.1.13)$$

Thus,
$$\lim_{k \to \infty} B^k = 0$$

Combining these results, we have

$$\lim_{k \to \infty} A^k = \begin{bmatrix} 0 & 0 & 0 & 0 & 0 \\ 0 & 0 & 0 & 0 & 0 \\ 0 & 0 & 0 & 0 & 0 \\ \tfrac{7}{15} & \tfrac{3}{15} & \tfrac{1}{15} & 1 & 0 \\ \tfrac{8}{15} & \tfrac{12}{15} & \tfrac{14}{15} & 0 & 1 \end{bmatrix} \quad (3.1.14)$$

---

[1] All the material introduced here concerning infinite series of matrices, limits of matrix functions, etc., will be reintroduced and treated more rigorously in Sec. 3.2.

Up to this point we have made no mention of the initial probability vector. Since, by hypothesis, player I is initially in state S_1 (he has \$1), the initial probability vector for the Markov chain is

$$P^{(0)} = \begin{bmatrix} 1 \\ 0 \\ 0 \\ 0 \\ 0 \end{bmatrix}$$

Multiplying $P^{(0)}$ by the matrix of equation (3.1.14) shows us that the probability of player I's eventually emerging the loser, i.e., being in state S_4, is $\frac{7}{15}$ and the probability of player I's eventually emerging the winner, i.e., being in state S_5, is $\frac{8}{15}$. It also shows that the probability of the game's continuing endlessly is zero.

In the same way, it follows from equations (3.1.12) and (3.1.13) (with $k = 9$) that there is no possibility of player I's being in states S_1 or S_3 after nine draws and that the probability of his being in state S_2 is $(\frac{2}{3})^9$. Thus the probability that a tenth draw will be required is $(\frac{2}{3})^9$, which is approximately 0.026.   ////

EXAMPLE 3.1.15   A small mouse and an enormous cheese are placed in the maze shown (though not necessarily in the same room).   We make four assumptions:

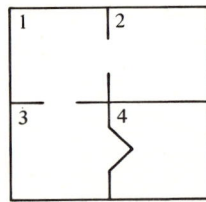

1 Rooms 1 and 3 have two exits, room 2 has one exit, and room 4 has no exits.   The mouse cannot leave room 4 once he has entered it.

2 The mouse changes rooms exactly once every minute (provided he is not in room 4).

3 The mouse will choose each of the exits of the room he is in with equal probability.

4 The mouse is given a piece of cheese every time he enters the room which has the cheese.

How many pieces of cheese can the mouse be expected to get if he is put into room j and the cheese is put in room i, provided this process continues indefinitely?

We can express the location of the mouse at any time by means of a Markov chain process with the four states

$$S_j = \text{the mouse is in room } j \quad j = 1, 2, 3, 4$$

The transition matrix is

$$A = [a_{ij}] = \begin{bmatrix} 0 & 1 & \frac{1}{2} & 0 \\ \frac{1}{2} & 0 & 0 & 0 \\ \frac{1}{2} & 0 & 0 & 0 \\ 0 & 0 & \frac{1}{2} & 1 \end{bmatrix}$$

and the initial probability vector is $I^{(j)}$. The mouse will get a piece of cheese at the 0-minute point, i.e., before he makes his first move, provided he is placed in the same room as the cheese, that is, $j = i$. The probability that the mouse will be in room i and hence get a piece of cheese at the 1-minute point, i.e., after his first move, is equal to the ith entry of

$$AI^{(j)}$$

Similarly, for $k = 1, 2, \ldots$, the probability that he will be in room i at the k-minute point is given by the ith entry of

$$A^k I^{(j)}$$

Since the mouse is given one piece of cheese every time he is fed, the total number of pieces of cheese he can expect is equal to the ith entry of

$$(I + A + A^2 + \cdots) I^{(j)}$$

Let us partition A into

$$A = \begin{bmatrix} B & 0 \\ C & I \end{bmatrix}$$

where
$$B = \begin{bmatrix} 0 & 1 & \frac{1}{2} \\ \frac{1}{2} & 0 & 0 \\ \frac{1}{2} & 0 & 0 \end{bmatrix} \quad C = [0 \ 0 \ \tfrac{1}{2}]$$

As in (3.1.12) of Example 3.1.11,

$$A^k = \begin{bmatrix} B^k & 0 \\ C(I + B + B^2 + \cdots + B^{k-1}) & I \end{bmatrix} \quad k = 1, 2, \ldots$$

and consequently

$I + A + A^2 + \cdots + A^k =$

$$\begin{bmatrix} I + B + B^2 + \cdots + B^k & 0 \\ C[kI + (k-1)B + (k-2)B^2 + \cdots + B^{k-1}] & (k+1)I \end{bmatrix}$$

A direct computation shows that
$$B^3 = \tfrac{3}{4}B$$
and hence
$$C[kI + (k-1)B + (k-2)B^2 + \cdots + B^{k-1}] = C(kI + \alpha B + \beta B^2)$$
where
$$\alpha = (k-1) + \tfrac{3}{4}(k-3) + (\tfrac{3}{4})^2(k-5) + \cdots$$
$$\beta = (k-2) + \tfrac{3}{4}(k-4) + (\tfrac{3}{4})^2(k-6) + \cdots$$

Moreover, since
$$1 + \tfrac{3}{4} + (\tfrac{3}{4})^2 + \cdots = 4$$
it follows that
$$I + B + B^2 + \cdots = I + 4B + 4B^2 = \begin{bmatrix} 4 & 4 & 2 \\ 2 & 3 & 1 \\ 2 & 2 & 2 \end{bmatrix}$$

Combining these results, we have
$$I + A + A^2 + \cdots = \begin{bmatrix} 4 & 4 & 2 & 0 \\ 2 & 3 & 1 & 0 \\ 2 & 2 & 2 & 0 \\ \infty & \infty & \infty & \infty \end{bmatrix} \qquad (3.1.16)$$

The mouse's expectation can be read directly from the matrix of (3.1.16). Thus, if the cheese is placed in room 4, the mouse will get infinitely many pieces of cheese irrespective of his initial position; if the cheese is placed in room 1 and the mouse is placed in either room 1 or room 2, he can expect four pieces of cheese, etc. ////

We will call a state which once it is entered can never be left, e.g., the states $0 and $4 in Example 3.1.11 and room 4 in Example 3.1.15, an *absorbing state*. The state S_q of a Markov chain process is absorbing if and only if the qth column of the transition matrix is $I^{(q)}$, that is
$$a_{qq} = 1 \qquad a_{pq} = 0 \qquad p \neq q$$
Let $a_{ij}^{(k)}$ denote the entry in the (i,j) position of A^k. Then
$$a_{ij}^{(k+1)} = \sum_{l=1}^{n} a_{il}^{(k)} a_{lj}^{(1)} = \sum_{m=1}^{n} a_{im}^{(1)} a_{mj}^{(k)} \qquad (3.1.17)$$
When S_j is an absorbing state, the first part of equation (3.1.17) becomes
$$a_{ij}^{(k+1)} = a_{ij}^{(k)} \qquad i = 1, 2, \ldots, n$$

and it follows that

$$a_{ij} = a_{ij}^{(1)} = a_{ij}^{(2)} = \cdots = 0 \qquad i \neq j$$
$$a_{jj} = a_{jj}^{(1)} = a_{jj}^{(2)} = \cdots = 1$$

Moreover, when S_i is an absorbing state, equation (3.1.17) yields

$$a_{ij}^{(k+1)} = \sum_{m=1}^{n} a_{im}^{(1)} a_{mj}^{(k)} = a_{ij}^{(k)} + \sum_{m \neq i} a_{im}^{(1)} a_{mj}^{(k)} \geq a_{ij}^{(k)} \qquad j = 1, 2, \ldots, n$$

That is, for each value of j, $a_{ij}^{(k)}$ is a nondecreasing function of k (which has, of course, an upper bound of 1). Thus when S_i is an absorbing state, we know that $\lim_{k \to \infty} a_{ij}^{(k)}$ exists. We will be able to make use of this in:

EXAMPLE 3.1.18 A mouse is placed in room j of the maze. Let us assume that:

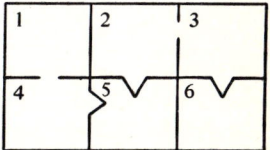

1 The mouse cannot leave room 5 or room 6 once he has entered it.
2 The mouse changes rooms exactly once every minute (provided he is not in room 5 or 6).
3 The mouse will choose each exit of the room he is in with equal probability.

Find the probability (which we denote by f_{ij}) that the mouse will ever enter room i.

Let us first consider the problem when room i corresponds to an absorbing state S_i. Note that the absorbing states of this Markov chain process are S_5 (room 5) and S_6 (room 6). Since the mouse may never leave either of these rooms, the probability that he will have entered one of them at or before the kth minute is the same as the probability that he will be in that room at the kth minute. Hence, for an absorbing state S_i, the probability that the mouse will have entered S_i (being originally in S_j) at or before the kth minute is given by

$$a_{ij}^{(k)}$$

and the probability that he will ever enter S_i is

$$f_{ij} = \lim_{k \to \infty} a_{ij}^{(k)} \qquad (3.1.19)$$

The transition matrix of this Markov chain process is

$$A = \begin{bmatrix} 0 & 0 & 0 & \frac{1}{2} & 0 & 0 \\ 0 & 0 & \frac{1}{2} & 0 & 0 & 0 \\ 0 & \frac{1}{2} & 0 & 0 & 0 & 0 \\ 1 & 0 & 0 & 0 & 0 & 0 \\ 0 & \frac{1}{2} & 0 & \frac{1}{2} & 1 & 0 \\ 0 & 0 & \frac{1}{2} & 0 & 0 & 1 \end{bmatrix}$$

We partition A as

$$A = \begin{bmatrix} B & 0 \\ C & I \end{bmatrix}$$

where $B = \begin{bmatrix} 0 & 0 & 0 & \frac{1}{2} \\ 0 & 0 & \frac{1}{2} & 0 \\ 0 & \frac{1}{2} & 0 & 0 \\ 1 & 0 & 0 & 0 \end{bmatrix}$ $C = \begin{bmatrix} 0 & \frac{1}{2} & 0 & \frac{1}{2} \\ 0 & 0 & \frac{1}{2} & 0 \end{bmatrix}$

As in (3.1.12) of Example 3.1.11,

$$A^k = \begin{bmatrix} B^k & 0 \\ C(I + B + B^2 + \cdots + B^{k-1}) & I \end{bmatrix}$$

Since

$$B^4 = \tfrac{3}{4}B^2 - \tfrac{1}{8}I$$

for each k, $k = 4, 5, 6, \ldots$ there exist scalars α_k, β_k, γ_k, δ_k such that

$$B^k = \alpha_k B^3 + \beta_k B^2 + \gamma_k B + \delta_k I \qquad (3.1.20)$$

The characteristic roots of B are $\pm\tfrac{1}{2}$, $\pm\sqrt{2}/2$, and it follows from equation (3.1.20) that

$$(\tfrac{1}{2})^k = \alpha_k(\tfrac{1}{2})^3 + \beta_k(\tfrac{1}{2})^2 + \gamma_k(\tfrac{1}{2}) + \delta_k$$

$$(-\tfrac{1}{2})^k = \alpha_k(-\tfrac{1}{2})^3 + \beta_k(-\tfrac{1}{2})^2 + \gamma_k(-\tfrac{1}{2}) + \delta_k$$

$$\left(\frac{\sqrt{2}}{2}\right)^k = \alpha_k\left(\frac{\sqrt{2}}{2}\right)^3 + \beta_k\left(\frac{\sqrt{2}}{2}\right)^2 + \gamma_k\left(\frac{\sqrt{2}}{2}\right) + \delta_k$$

$$\left(-\frac{\sqrt{2}}{2}\right)^k = \alpha_k\left(-\frac{\sqrt{2}}{2}\right)^3 + \beta_k\left(-\frac{\sqrt{2}}{2}\right)^2 + \gamma_k\left(-\frac{\sqrt{2}}{2}\right) + \delta_k$$

Solving these equations yields

$$\alpha_k = \begin{cases} 0 & \text{when } k \text{ is even} \\ \dfrac{2^l - 1}{2^{2(l-1)}} & \text{when } k \text{ is odd, } k = 2l + 1 \end{cases}$$

$$\beta_k = \begin{cases} \dfrac{2^l - 1}{2^{2(l-1)}} & \text{when } k \text{ is even, } k = 2l \\ 0 & \text{when } k \text{ is odd} \end{cases}$$

$$\gamma_k = \begin{cases} 0 & \text{when } k \text{ is even} \\ \dfrac{2 - 2^l}{2^{2l}} & \text{when } k \text{ is odd, } k = 2l + 1 \end{cases}$$

$$\delta_k = \begin{cases} \dfrac{2 - 2^l}{2^{2l}} & \text{when } k \text{ is even, } k = 2l \\ 0 & \text{when } k \text{ is odd} \end{cases}$$

Consequently
$$\lim_{k \to \infty} B^k = 0$$

$$\lim_{k \to \infty}(I + B + B^2 + \cdots + B^{k-1})$$

$$= I + B + B^2 + B^3 + I\left(\sum_{l=2}^{\infty} \frac{2 - 2^l}{2^{2l}}\right) + B\left(\sum_{l=2}^{\infty} \frac{2 - 2^l}{2^{2l}}\right)$$

$$+ B^2\left(\sum_{l=2}^{\infty} \frac{2^l - 1}{2^{2(l-1)}}\right) + B^3\left(\sum_{l=2}^{\infty} \frac{2^l - 1}{2^{2(l-1)}}\right)$$

$$= \tfrac{2}{3}I + \tfrac{2}{3}B + \tfrac{8}{3}B^2 + \tfrac{8}{3}B^3 = \begin{bmatrix} 2 & 0 & 0 & 1 \\ 0 & \tfrac{4}{3} & \tfrac{2}{3} & 0 \\ 0 & \tfrac{2}{3} & \tfrac{4}{3} & 0 \\ 2 & 0 & 0 & 2 \end{bmatrix}$$

and
$$\lim_{k \to \infty} C(I + B + B^2 + B^3 + \cdots + B^{k-1}) = \begin{bmatrix} 1 & \tfrac{2}{3} & \tfrac{1}{3} & 1 \\ 0 & \tfrac{1}{3} & \tfrac{2}{3} & 0 \end{bmatrix}$$

Thus we have

$$f_{51} = 1 \quad f_{52} = \tfrac{2}{3} \quad f_{53} = \tfrac{1}{3} \quad f_{54} = 1$$
$$f_{61} = 0 \quad f_{62} = \tfrac{1}{3} \quad f_{63} = \tfrac{2}{3} \quad f_{64} = 0$$

i.e., if the mouse is placed in room 2, the probability that he will ever enter (and hence forever remain in) room 5 is $\frac{2}{3}$, while the probability that he will ever enter room 6 is $\frac{1}{3}$, etc.

Now let us consider a case in which the state corresponding to room i is not absorbing. In particular we determine f_{1j}, for $j = 2, 3, \ldots, 6$. First, construct another maze, identical to the given one except that the mouse can never leave room 1 after he has entered it. If the mouse is not originally in

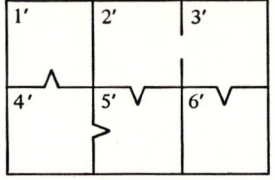

room 1, the probability that he will enter room 1 for the first time at the kth minute is the same for both mazes (though the mouse's subsequent movements are completely different). Consequently the probability that the mouse will ever enter room 1 is the same in both mazes. The transition matrix for the Markov chain process in which S_1 is absorbing is

$$A' = \begin{bmatrix} 1 & 0 & 0 & \frac{1}{2} & 0 & 0 \\ 0 & 0 & \frac{1}{2} & 0 & 0 & 0 \\ 0 & \frac{1}{2} & 0 & 0 & 0 & 0 \\ 0 & 0 & 0 & 0 & 0 & 0 \\ 0 & \frac{1}{2} & 0 & \frac{1}{2} & 1 & 0 \\ 0 & 0 & \frac{1}{2} & 0 & 0 & 1 \end{bmatrix}$$

which we partition as

$$A' = \begin{bmatrix} 1 & D' & 0 \\ 0 & B' & 0 \\ 0 & C' & I \end{bmatrix}$$

where $D' = [0 \ 0 \ \frac{1}{2}]$ $B' = \begin{bmatrix} 0 & \frac{1}{2} & 0 \\ \frac{1}{2} & 0 & 0 \\ 0 & 0 & 0 \end{bmatrix}$ $C' = \begin{bmatrix} \frac{1}{2} & 0 & \frac{1}{2} \\ 0 & \frac{1}{2} & 0 \end{bmatrix}$

Clearly

$$A'^k = \begin{bmatrix} 1 & D'(I + B' + B'^2 + \cdots + B'^{k-1}) & 0 \\ 0 & B'^k & 0 \\ 0 & C'(I + B' + B'^2 + \cdots + B'^{k-1}) & I \end{bmatrix}$$

Since $B'^3 = \frac{1}{4}B'$

and $I + B' + B'^2 + \cdots = I + \frac{4}{3}B' + \frac{4}{3}B'^2 = \begin{bmatrix} \frac{4}{3} & \frac{2}{3} & 0 \\ \frac{2}{3} & \frac{4}{3} & 0 \\ 0 & 0 & 1 \end{bmatrix}$

we find that

$$D'(I + B' + B'^2 + \cdots) = [0 \quad 0 \quad \tfrac{1}{2}]$$

$$C'(I + B' + B'^2 + \cdots) = \begin{bmatrix} \tfrac{2}{3} & \tfrac{1}{3} & \tfrac{1}{2} \\ \tfrac{1}{3} & \tfrac{2}{3} & 0 \end{bmatrix}$$

Thus
$$\lim_{k \to \infty} A'^k = \begin{bmatrix} 1 & 0 & 0 & \tfrac{1}{2} & 0 & 0 \\ 0 & 0 & 0 & 0 & 0 & 0 \\ 0 & 0 & 0 & 0 & 0 & 0 \\ 0 & 0 & 0 & 0 & 0 & 0 \\ 0 & \tfrac{2}{3} & \tfrac{1}{3} & \tfrac{1}{2} & 1 & 0 \\ 0 & \tfrac{1}{3} & \tfrac{2}{3} & 0 & 0 & 1 \end{bmatrix}$$

and we see that

$$f_{12} = f_{13} = 0 \quad f_{14} = \tfrac{1}{2} \quad f_{15} = f_{16} = 0$$

Finally, let us find the probability that the mouse if originally placed in room 1 will be in room 1 at any later time. (We will denote this probability by f_{11}.) At the first minute, the mouse must leave room 1. In this example his only choice is to enter room 4. We have already shown that after entering room 4 the probability that he will ever subsequently be in room 1 is $\tfrac{1}{2}$. Hence

$$f_{11} = \tfrac{1}{2}$$

We leave it to the reader to find f_{ij}, $i \neq 1, 5, 6$. ////

EXAMPLE 3.1.21 A mouse is placed in room j ($j \neq 3$) of the maze. What is

| 1 | 2 | 3 |

the probability that the mouse will ever enter room 3? If the mouse dies as soon as he enters room 3, what is his average life expectancy? We have just seen that the first problem can be solved by the methods of Example 3.1.18. Here we present an alternate way of solving the first problem which at the same time provides a solution for the second problem. Let us denote the probability that the mouse will enter room 3 for the first time at the kth minute by

$$f_{3j}^{(k)}, \quad k = 1, 2, \ldots$$

Then
$$f_{3j}^{(1)} + f_{3j}^{(2)} + \cdots + f_{3j}^{(k)}$$

is the probability that he will enter room 3 at least once during the first k minutes, and the probability that he will ever enter room 3 (which we have previously denoted by f_{3j}) is given by the infinite series

$$f_{3j} = f_{3j}^{(1)} + f_{3j}^{(2)} + f_{3j}^{(3)} + \cdots \qquad (3.1.22)$$

In order to evaluate $f_{3j}^{(k)}$ we transform the state corresponding to room 3

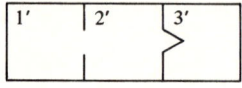

into an absorbing state. The transition matrix of the new maze is

$$A' = \begin{bmatrix} 0 & \tfrac{1}{2} & 0 \\ 1 & 0 & 0 \\ 0 & \tfrac{1}{2} & 1 \end{bmatrix}$$

which we partition as

$$A' = \begin{bmatrix} B' & 0 \\ C' & 1 \end{bmatrix}$$

where

$$B' = \begin{bmatrix} 0 & \tfrac{1}{2} \\ 1 & 0 \end{bmatrix} \qquad C' = [0 \ \tfrac{1}{2}]$$

$f_{3j}^{(k)}$ is also the probability that the mouse will enter room 3' of the second maze at the kth minute. Because he can never leave room 3' after he has entered it, the probability that the mouse will be in room 3' at the kth minute is the sum of the probabilities of entering room 3' for the first time in each of the first k minutes, i.e.,

$$a'^{(k)}_{3j} = f_{3j}^{(1)} + f_{3j}^{(2)} + \cdots + f_{3j}^{(k)} \qquad (3.1.23)$$

[Compare equation (3.1.23) with equation (3.1.19).] It follows from equation (3.1.23) that

$$a'^{(k)}_{3j} - a'^{(k-1)}_{3j} = f_{3j}^{(k)} \qquad k = 2, 3, \ldots$$

and consequently $f_{3j}^{(k)}$ is equal to the entry in the $(3, j)$ position of

$$(A')^k - (A')^{k-1}$$

Moreover, by equation (3.1.12),

$$(A')^k - (A')^{k-1} = \begin{bmatrix} (B')^k - (B')^{k-1} & 0 \\ C'(B')^{k-1} & 0 \end{bmatrix}$$

and $f_{3j}^{(k)}$ is the jth entry of the two-dimensional row vector $C'(B')^{k-1}$. A simple computation shows that

$$(B')^2 = \tfrac{1}{2} I$$

and consequently
$$(B')^{k-1} = \begin{cases} (\tfrac{1}{2})^{(k-1)/2} I & \text{when } k \text{ is odd} \\ (\tfrac{1}{2})^{(k-2)/2} B' & \text{when } k \text{ is even} \end{cases}$$

Thus
$$C'(B')^{k-1} = \begin{cases} [0 \quad (\tfrac{1}{2})^{(k+1)/2}] & \text{when } k \text{ is odd} \\ [(\tfrac{1}{2})^{k/2} \quad 0] & \text{when } k \text{ is even} \end{cases}$$

and
$$f_{31}{}^{(k)} = \begin{cases} 0 & \text{when } k \text{ is odd} \\ (\tfrac{1}{2})^{k/2} & \text{when } k \text{ is even} \end{cases}$$

$$f_{32}{}^{(k)} = \begin{cases} (\tfrac{1}{2})^{(k+1)/2} & \text{when } k \text{ is odd} \\ 0 & \text{when } k \text{ is even} \end{cases} \quad (3.1.24)$$

Substituting equations (3.1.24) into equation (3.1.22) yields

$$f_{31} = f_{32} = \tfrac{1}{2} + \tfrac{1}{4} + \tfrac{1}{8} + \cdots = 1 \quad (3.1.25)$$

Equation (3.1.25) states that the mouse is certain to die. The probability that he will live exactly k minutes (provided he is initially in room j) is

$$f_{3j}{}^{(k)}$$

The mouse's average life expectancy is given by

$$E_j = \sum_{k=1}^{\infty} k f_{3j}{}^{(k)} \quad (3.1.26)$$

For $j = 1$, substituting the first equation of (3.1.24) into equation (3.1.26) and setting $k = 2l$ yields[1]

$$E_1 = \sum_{l=1}^{\infty} 2l(\tfrac{1}{2})^l = 4 \text{ minutes}$$

For $j = 2$, substituting the second equation of (3.1.24) into equation (3.1.26) and setting $k = 2l - 1$ yields

$$E_2 = \sum_{l=1}^{\infty} (2l - 1)(\tfrac{1}{2})^l = 3 \text{ minutes} \quad ////$$

EXERCISES

1 Show that in Example 3.1.9 the same ultimate population will occur (namely, 600 people in each city) no matter where the 1,800 citizens of Ruritania are initially located.

---

[1] It can be shown by an induction on p that
$$1x + 2x^2 + 3x^3 + \cdots + px^p = \frac{px^{p+2} - (p+1)x^{p+1} + x}{(1-x)^2}$$
and hence, for $|x| < 1$,
$$1x + 2x^2 + 3x^3 + \cdots = \frac{x}{(1-x)^2}$$

2 Verify equation (3.1.10).
3 Show that the stationary vectors in the Markov chain of Example 3.1.11 are exactly those vectors in the set of all five-dimensional probability vectors whose first three entries are zero.
4 Repeat Example 3.1.11 with a deck consisting of two cards marked H and two cards marked L.
5 Find the probability that a mouse initially in room 2 of the maze of Example 3.1.15 will ever enter room 3. Find the probability that a mouse initially in room 3 will ever enter room 2.
6 In example 3.1.15 the maze is replaced by the maze shown. Find the number of pieces of cheese the mouse can be expected to get if he is placed in room j and the cheese is in room i.

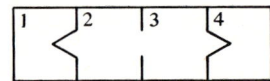

7 Show that a mouse initially in room 2 of the maze of Exercise 6 is certain to enter either room 1 or room 4. What is the probability that he will enter room 4? What is the probability that he will enter room 1?
8 If the mouse of Exercise 7 dies when he enters either room 1 or room 4, what is his life expectancy? (*Hint*: Show that the maze can be replaced by one with three rooms, one of which has two one-way doors leading into it.)
9 Repeat Example 3.1.18 with the maze replaced by the one sketched.

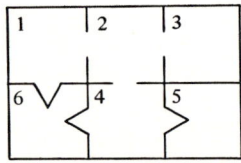

10 There are two coins, of which one when tossed shows heads and tails with equal probability and the other shows heads two-thirds of the time. If a coin is tossed and shows tails, the other coin is used for the next toss. If a coin is tossed and shows heads, the same coin is used for the next toss. The first coin to be tossed is picked at random. What is the probability that the biased coin will be used for the third toss? What is the probability that the third toss will be heads? (*Hint:* There are four states.)
11 The following problem describes a common model used to explain diffusion in gases. We begin with two baskets, each of which contains five balls. The balls in one basket are all white, and those in the other basket are all

black. Each second one ball is selected from each basket and moved to the other basket. Show that this is a Markov chain process and find the transition matrix and the initial vector. What is the probability that both baskets will have at least one white ball after 3 seconds? After 5 seconds?

12 In a certain community it is found that half the sons of college-educated fathers finish college, a quarter finish high school but not college, and a quarter do not finish high school. Among sons of fathers who finish high school but not college one-third finish college, one-third finish high school but not college, and one-third do not finish high school. Among sons of fathers who did not finish high school one-sixth finish college, one-half finish high school but not college, and one-third do not finish high school. Find a transition matrix which expresses the relation between one generation and the next. What proportion of each type of man will yield a stationary population?

13 Prove that the product of two transition matrices of the same order is a transition matrix.

14 Prove that the product of a transition matrix and a probability vector is a probability vector.

15 Let A be a square matrix of order n. Prove that if AX is a probability vector for every probability vector X, then A is a transition matrix.

16 Prove that the set of stationary vectors of a Markov chain is convex.

17 Let A be the transition matrix of a Markov chain and let $\tilde{A}$ be the transition matrix of the same Markov chain after the order of the states has been changed. Show that A and $\tilde{A}$ are similar.

18 A set $\mathbb{S}$ of states is *closed* if it is impossible to go from a state in $\mathbb{S}$ to a state not in $\mathbb{S}$. (Thus a state is absorbing if the set consisting of that state is closed.) Describe the transition matrix of a Markov chain with n states in which the set $\mathbb{S} = \{S_1, S_2, \ldots, S_k\}$ is closed. Describe the transition matrix if the set $\mathbb{S}' = \{S_{k+1}, S_{k+2}, \ldots, S_n\}$ is also closed.

3.2 INFINITE PROCESSES

In Sec. 3.1 we introduced several procedures involving limits of matrix expressions in order to solve problems that arose in Examples 3.1.11, 3.1.15, and 3.1.18. No attempt was made to justify what was done or even to define explicitly the terms introduced. This will be rectified now.

It is possible to impose a meaningful *distance*, or *metric*, function on the set $\mathbb{C}_n^m$ of all $m \times n$ matrices with complex entries. It is the existence of this distance function which makes it possible for us to define limits of certain matrix expressions.

Definition 3.2.1 Let $A = [a_{ij}] \in \mathbb{C}_n^m$. We define the *norm* of A, written $\|A\|$, by

$$\|A\| = \max_{\substack{i=1,2,\ldots,m \\ j=1,2,\ldots,n}} \{|a_{ij}|\}$$

When $m = n = 1$, the norm of A is simply the absolute value of the single entry of A.

Lemma 3.2.2 Let A and B be $m \times n$ matrices, C be an $n \times p$ matrix and α be a complex number. Then:

1. $\|A\|$ is a nonnegative real number and $\|A\| = 0$ if and only if $A = 0$.
2. $\|\alpha A\| = |\alpha|\,\|A\|$.
3. $\|A + B\| \leq \|A\| + \|B\|$ (the triangle law for matrices).
4. $\|AC\| \leq n\|A\|\,\|C\|$.

PROOF   Parts 1 and 2 are immediate. The remainder of the lemma follows from the triangle law for complex numbers.

(3) Let $A = [a_{ij}]$, $B = [b_{ij}]$. Then

$$\|A + B\| = \max_{\substack{i=1,2,\ldots,m \\ j=1,2,\ldots,n}} \{|a_{ij} + b_{ij}|\}$$

$$\leq \max_{\substack{i=1,2,\ldots,m \\ j=1,2,\ldots,n}} \{|a_{ij}| + |b_{ij}|\}$$

$$\leq \max_{\substack{i=1,2,\ldots,m \\ j=1,2,\ldots,n}} |a_{ij}| + \max_{\substack{i=1,2,\ldots,m \\ j=1,2,\ldots,n}} |b_{ij}|$$

$$= \|A\| + \|B\|$$

(4) Let $C = [c_{jk}]$ and set

$$AC = D = [d_{ik}] \quad \begin{array}{l} i = 1, 2, \ldots, m \\ j = 1, 2, \ldots, n \\ k = 1, 2, \ldots, p \end{array}$$

Then

$$\|AC\| = \|D\| = \max_{\substack{i=1,2,\ldots,m \\ k=1,2,\ldots,p}} \{|d_{ik}|\} = \max_{\substack{i=1,2,\ldots,m \\ k=1,2,\ldots,p}} \left\{\left|\sum_{j=1}^n a_{ij}c_{jk}\right|\right\}$$

$$\leq \max_{\substack{i=1,2,\ldots,m \\ k=1,2,\ldots,p}} \left\{\sum_{j=1}^n |a_{ij}c_{jk}|\right\} = \max_{\substack{i=1,2,\ldots,m \\ k=1,2,\ldots,p}} \left\{\sum_{j=1}^n |a_{ij}|\,|c_{jk}|\right\}$$

$$\leq \max_{\substack{i=1,2,\ldots,m \\ k=1,2,\ldots,p}} \left\{\sum_{j=1}^n \|A\|\,\|C\|\right\} = n\|A\| \cdot \|C\| \qquad \text{////}$$

The norm of a matrix generally behaves very much like the absolute value and will play the same role in matrix situations as the absolute value does when treating real- or complex-valued functions.

Definition 3.2.3 Let $L, A_0, A_1, A_2, \ldots$ be $m \times n$ matrices. We say that the sequence of matrices $A_0, A_1, A_2 \ldots$, *converges* to L and write

$$\lim_{k \to \infty} A_k = L$$

if for every $\varepsilon > 0$ there exists a positive integer K (depending on ε) such that

$$\|A_k - L\| < \varepsilon$$

for every integer $k > K$.

Most frequently we shall define A_k to be A^k, where A is a square matrix. The main result of this section (Theorem 3.2.9) gives a set of necessary and sufficient conditions for

$$\lim_{k \to \infty} A^k$$

to exist. First we must show that many of the familiar limit properties of functions of a real or complex variable can be proved for matrix functions as well, usually in the same way.

Lemma 3.2.4 Let B be an $m \times n$ matrix and let $A_0, A_1, A_2, \ldots$ be a sequence of $n \times p$ matrices. If $\lim_{k \to \infty} A_k$ exists, then $\lim_{k \to \infty} BA_k$ exists and

$$\lim_{k \to \infty} BA_k = B\left(\lim_{k \to \infty} A_k\right)$$

PROOF If $B = 0$, the result is clear. Now suppose that $B \neq 0$. Then $\|B\| \neq 0$. Set

$$\lim_{k \to \infty} A_k = L$$

and let $\varepsilon > 0$. There exists a positive integer K such that

$$\|A_k - L\| < \frac{\varepsilon}{n\|B\|}$$

whenever $k > K$. It follows from part 4 of Lemma 3.2.2 that

$$\|BA_k - BL\| = \|B(A_k - L)\| \leq n\|B\| \cdot \|A_k - L\| < \varepsilon$$

whenever $k > K$. Thus

$$\lim_{k \to \infty} BA_k = BL = B\left(\lim_{k \to \infty} A_k\right) \qquad ////$$

An important consequence of Lemma 3.2.4 is:

Corollary 3.2.5 If $\lim_{k \to \infty} A^k$ exists and P is nonsingular, then $\lim_{k \to \infty} (P^{-1}AP)^k$ exists and

$$P^{-1}\left(\lim_{k \to \infty} A^k\right)P = \lim_{k \to \infty} (P^{-1}AP)^k$$

PROOF   Clearly

$$(P^{-1}AP)^k = P^{-1}A^kP \qquad k = 1, 2, \ldots$$

By two applications of Lemma 3.2.4, one on the left and the other on the right, we have

$$P^{-1}\left(\lim_{k \to \infty} A^k\right)P = \lim_{k \to \infty} (P^{-1}A^kP) = \lim_{k \to \infty} (P^{-1}AP)^k \qquad ////$$

Note that Lemma 3.2.4 and Corollary 3.2.5 each state two things: (1) the limits exist, and (2) they are equal. Lemma 3.2.4 was tacitly assumed while solving Example 3.1.11 of Sec. 3.1. The probability of each player's ultimately emerging the winner in Example 3.1.11 is given by the entries of

$$\lim_{k \to \infty} P^{(k)} = \lim_{k \to \infty} (A^k P^{(0)})$$

However, we actually computed

$$\left(\lim_{k \to \infty} A^k\right) P^{(0)}$$

Lemma 3.2.4 shows that these two expressions (when they exist) are equal.

Corollary 3.2.5 was also tacitly assumed in Example 3.1.11 of Sec. 3.1 when we changed the order of (i.e., permuted) the states. The resulting transition matrix A is related to the original transition matrix $\tilde{A}$ by

$$\tilde{A} = P^{-1}AP$$

where P is the permutation matrix corresponding to the permutation of the states.

Lemma 3.2.6 For $i = 1, 2, \ldots, l$, let A_{ii} be a square matrix of order n_i and let $A = A_{11} \oplus A_{22} \oplus \cdots \oplus A_{ll}$. Then $\lim_{k \to \infty} A^k$ exists if and only if $\lim_{k \to \infty} A_{ii}^k$ exists for each i. In this event

$$\lim_{k \to \infty} A^k = \lim_{k \to \infty} A_{11}^k \oplus \lim_{k \to \infty} A_{22}^k \oplus \cdots \oplus \lim_{k \to \infty} A_{ll}^k$$

PROOF The proof is left as an exercise. ////

Let A be an arbitrary square matrix. We wish to determine whether or not $\lim_{k\to\infty} A^k$ exists. According to Corollary 3.2.5, A may be replaced by any matrix similar to it; i.e., instead of examining $\lim_{k\to\infty} A^k$, we may examine $\lim_{k\to\infty}(P^{-1}AP)^k$ for any nonsingular matrix P. In Chapter 1 we discussed various simple forms $P^{-1}AP$ can have. According to Theorem 1.13.21, we can construct a nonsingular matrix P such that $P^{-1}AP$ is the direct sum of the hypercompanion matrices of the elementary divisors of A. By virtue of Lemma 3.2.6, it is sufficient to determine under what conditions $\lim_{k\to\infty} C^k$ exists when C has the form of (1.13.23).

It is convenient to consider the field from which the entries of A are taken to be the complex field $\mathbb{C}$ (rather than the real field $\mathbb{R}$). The fundamental theorem of algebra[1] states that *in the set of polynomials with complex coefficients, the irreducible polynomials are exactly the polynomials of degree 1.* The companion matrix of

$$p(x) = x - \lambda$$

is the 1×1 matrix $[\lambda]$. Hence, A is similar to the direct sum of matrices of the form

$$C = \begin{bmatrix} \lambda & 0 & 0 & \cdots & 0 & 0 \\ 1 & \lambda & 0 & \cdots & 0 & 0 \\ 0 & 1 & \lambda & \cdots & 0 & 0 \\ \multicolumn{6}{c}{\dotfill} \\ 0 & 0 & 0 & \cdots & 1 & \lambda \end{bmatrix} \quad (1.13.23)$$

and it is sufficient to determine under what conditions $\lim_{k\to\infty} C^k$ exists when C is a matrix of the form (1.13.23). Set

$$N = \begin{bmatrix} 0 & 0 & 0 & \cdots & 0 & 0 \\ 1 & 0 & 0 & \cdots & 0 & 0 \\ 0 & 1 & 0 & \cdots & 0 & 0 \\ \multicolumn{6}{c}{\dotfill} \\ 0 & 0 & 0 & \cdots & 1 & 0 \end{bmatrix}$$

Clearly

$$C = \lambda I + N$$

---

[1] For a proof of this result see, for example, E. Hille, "Analytic Function Theory," vol. I, p. 207, Ginn and Co., Boston, 1959.

Moreover, if the order of C (and hence N) is r then $N^r = 0$. Hence, for $k \geq 0$ we have

$$C^k = \lambda^k I + \frac{k}{1} \lambda^{k-1} N + \frac{k(k-1)}{1 \cdot 2} \lambda^{k-2} N^2 + \cdots + \frac{k(k-1) \cdots (k-r+2)}{1 \cdot 2 \cdots (r-1)} \lambda^{k-r+1} N^{r-1} =$$

$$\begin{bmatrix} \lambda^k & 0 & 0 & \cdots & 0 & 0 & 0 \\ k\lambda^{k-1} & \lambda^k & 0 & \cdots & 0 & 0 & 0 \\ \frac{k(k-1)}{2} \lambda^{k-2} & k\lambda^{k-1} & \lambda^k & \cdots & 0 & 0 & 0 \\ & \frac{k(k-1)}{2} \lambda^{k-2} & k\lambda^{k-1} & & & & \\ & & \frac{k(k-1)}{2} \lambda^{k-2} & \cdots & & & \\ \vdots & & & & & & \\ \frac{k(k-1) \cdots (k-r+2)}{1 \cdot 2 \cdots (r-1)} \lambda^{k-r+1} & & & \cdots & \frac{k(k-1)}{2} \lambda^{k-2} & k\lambda^{k-1} & \lambda^k \end{bmatrix}$$

(3.2.7)

Lemma 3.2.8 Let C be a square matrix of the form of (1.13.23). Then $\lim_{k \to \infty} C^k$ exists if and only if

$$|\lambda| < 1$$

in which case $\lim_{k \to \infty} C^k = 0$, or

$$\lambda = +1 \quad \text{and} \quad r = 1$$

in which case $\lim_{k \to \infty} C^k = [1]$.

PROOF Since $\lim_{k \to \infty} \lambda^k$ does not exist unless $|\lambda| < 1$ or $\lambda = +1$, it follows from equation (3.2.7) that $\lim_{k \to \infty} C^k$ does not exist unless $|\lambda| < 1$ or $\lambda = +1$.

Conversely, if $|\lambda| < 1$, it can be shown, using L'Hospital's rule, that

$$\lim_{k \to \infty} \lambda^k = \lim_{k \to \infty} k\lambda^{k-1} = \lim_{k \to \infty} \frac{k(k-1)}{2} \lambda^{k-1} = \cdots$$

$$= \lim_{k \to \infty} \frac{k(k-1) \cdots (k-r+2)}{1 \cdot 2 \cdots (r-1)} \lambda^{k-r+1} = 0$$

Hence $\lim_{k \to \infty} C^k = 0$.

Finally, let us suppose $\lambda = +1$. If $r \geq 2$, then $\lim_{k \to \infty} C^k$ does not exist since $\lim_{k \to \infty} k \lambda^{k-1} = \lim_{k \to \infty} k$ does not exist. On the other hand, if $r = 1$, then $C = [1]$ and $\lim_{k \to \infty} C^k = [1]$. ////

Since the minimal polynomial of the $r \times r$ matrix of (1.13.23) is $(x - \lambda)^r$, it follows that, in the event that $\lambda = +1$, $\lim_{k \to \infty} C^k$ exists if and only if $(x - 1)^2$ does not divide the minimal polynomial of C.

Now, let A be an arbitrary $n \times n$ matrix with entries from $\mathbb{C}$. Each elementary divisor of A is of the form $(x - \lambda)^r$ and gives rise (by Theorem 1.13.21) to an $r \times r$ matrix of the form of (1.13.23), having $(x - \lambda)^r$ as its minimal polynomial. By Lemma 3.2.8 and the discussion preceding it, $\lim_{k \to \infty} A^k$ exists if and only if each elementary divisor of A is of the form

$$(x - \lambda)^r \quad \text{where} \quad |\lambda| < 1$$

or is of the form

$$x - 1$$

However, the minimal polynomial of A is the least common multiple of the elementary divisors of A. Thus we have proved the main theorem of this section.

Theorem 3.2.9 Let A be a square matrix with entries from $\mathbb{C}$. Then $\lim_{k \to \infty} A^k$ exists if and only if the following conditions are satisfied:

1 The absolute value of every characteristic root of A is at most 1.
2 The only characteristic root of A of absolute value 1 (if any) is $+1$.
3 $(x - 1)^2$ does not divide the minimal polynomial of A.

In Examples 3.1.11, 3.1.15, and 3.1.18 we were also faced with an infinite series of matrices. Just as when dealing with infinite series of numbers, we define convergence of an infinite series by means of partial sums. Thus, if

$$\alpha_0 I + \alpha_1 A + \alpha_2 A^2 + \cdots$$

is an infinite series in the square matrix A, we first define the *partial sums*

$$A_0 = \alpha_0 I$$
$$A_1 = \alpha_0 I + \alpha_1 A$$
$$A_2 = \alpha_0 I + \alpha_1 A + \alpha_2 A^2$$
$$\dotsb\dotsb\dotsb\dotsb\dotsb$$

and then say that the infinite series $a_0 I + a_1 A + a_2 A^2 + \cdots$ *converges* to B if the sequence $A_0, A_1, A_2, \ldots$ converges to B.

Theorem 3.2.10 For a square matrix A, the following conditions are equivalent:

1. $\lim_{k \to \infty} A^k = 0$.
2. The absolute value of every characteristic root of A is less than 1.
3. $I + A + A^2 + \cdots$ converges.

Moreover, when one (and hence all) of these conditions are satisfied, we also have

4. $I - A$ is nonsingular.
5. $I + A + A^2 + \cdots$ converges to $(I - A)^{-1}$.

PROOF   The equivalence of conditions 1 and 2 is contained in Lemma 3.2.8. Now, let A be a matrix which satisfies conditions 1 and 2. Clearly λ is a characteristic root of A if and only if $1 - \lambda$ is a characteristic root of $I - A$. Hence 0 is not a characteristic root of $I - A$ and it follows that $I - A$ is nonsingular. Then conditions 3 and 5 are immediate consequences of the identity

$$(I - A)^{-1} - (I + A + A^2 + \cdots + A^{k-1}) = (I - A)^{-1} A^k$$

Finally, let A be a matrix which satisfies condition 3. We may assume that A has the form of the direct sum of hypercompanion matrices. It is sufficient to consider the infinite series

$$I + C + C^2 + \cdots$$

where C has the form of (1.13.23). The diagonal entries of $I + C + C^2 + \cdots$ are

$$1 + \lambda + \lambda^2 + \cdots$$

which converges only when $|\lambda| < 1$. Hence A satisfies condition 2.   ////

EXERCISES

1 Find $\lim_{k \to \infty} A^k$ where A is

(a) $\begin{bmatrix} 1 & -\frac{1}{2} & \frac{1}{2} \\ 0 & \frac{3}{2} & -1 \\ 0 & 1 & -\frac{1}{2} \end{bmatrix}$ (b) $\begin{bmatrix} -\frac{1}{2} & \frac{3}{2} & -\frac{3}{2} \\ 1 & 0 & -\frac{1}{2} \\ 1 & -1 & \frac{1}{2} \end{bmatrix}$

(a) $\begin{bmatrix} 2 & -4 & -8 \\ -3 & 12 & 21 \\ 2 & -8 & -14 \end{bmatrix}$ (d) $\begin{bmatrix} -\frac{1}{2} & 1 & -3 \\ 1 & -\frac{1}{2} & 3 \\ \frac{1}{2} & -\frac{1}{2} & 2 \end{bmatrix}$

2 Let
$$A = \begin{bmatrix} 0 & \frac{1}{2} & 0 \\ \frac{1}{2} & 0 & \frac{3}{2} \\ \frac{1}{2} & -\frac{1}{2} & \frac{3}{2} \end{bmatrix}$$

Find $\lim_{k \to \infty} A^k$ by (a) using Corollary 3.2.5 and (b) proving and using the recurrence relation
$$A^{k+2} = \left(2 - \frac{1}{2^k}\right)A^2 - \left(1 - \frac{1}{2^k}\right)A \qquad k = 1, 2, \ldots$$

3 Let
$$A = \begin{bmatrix} 0 & -1 & 0 \\ 0 & \frac{1}{2} & 0 \\ 2 & 0 & 0 \end{bmatrix}$$

Find $I + A + A^2 + \cdots$ (a) by a direct calculation and (b) by means of Theorem 3.2.10.

4 Let
$$A = \begin{bmatrix} 1 & 0 & 1 \\ \frac{1}{2} & 0 & \frac{1}{2} \\ 0 & -\frac{1}{2} & 0 \end{bmatrix}$$

Show that $\lim_{k \to \infty} A^k = 0$ by (a) Theorem 3.2.10 and (b) proving and using the recurrence relation
$$A^{k+2} = \frac{k+1}{2^k} A^2 - \frac{k}{2^{k+1}} A \qquad k = 1, 2, 3, \ldots$$

Find $I + A + A^2 + \cdots$ by (a) using Theorem 3.2.10 and (b) means of the above recurrence relation. (*Hint:* $\sum_{k=1}^{\infty} \frac{k}{2^k} = 2$.)

5 Prove Lemma 3.2.6.
6 Verify equation (3.2.7).
7 Show that if $\lim_{k \to \infty} A^k$ exists, then:

(a) $A\left(\lim_{k \to \infty} A^k\right) = \left(\lim_{k \to \infty} A^k\right)A = \lim_{k \to \infty} A^k$.

(b) $\left(\lim_{k \to \infty} A^k\right)^2 = \lim_{k \to \infty} A^k$.

(c) $\left(\lim_{k \to \infty} A^k\right)^T = \lim_{k \to \infty} (A^T)^k$.

(d) $\lim_{k \to \infty} (\det A)^k = \det\left(\lim_{k \to \infty} A^k\right)$.

8 Show that if λ is a characteristic root of the square matrix A of order n, then
$$|\lambda| \leq n\|A\|$$

9 Let A and B be square matrices of the same order such that $AB = BA$. Show that if $\lim_{k \to \infty} A^k$ and $\lim_{k \to \infty} B^k$ both exist, then
$$\lim_{k \to \infty} (AB)^k = \left(\lim_{k \to \infty} A^k\right)\left(\lim_{k \to \infty} B^k\right)$$

10 Find a pair of 2×2 matrices, A and B, such that $\lim_{k \to \infty} A^k$ and $\lim_{k \to \infty} B^k$ both exist but $\lim_{k \to \infty}(AB)^k$ does not exist.

11 Prove that $\alpha_0 I + \alpha_1 C + \alpha_2 C^2 + \cdots$ converges to B if and only if
$$\alpha_0 I + \alpha_1 P^{-1}CP + \alpha_2 (P^{-1}CP)^2 + \cdots$$
converges to $P^{-1}BP$.

12 Show that the infinite series
$$e^A = I + \frac{1}{1!}A + \frac{1}{2!}A^2 + \frac{1}{3!}A^3 + \cdots$$
converges for every square matrix A. Find e^A when

(a) $A = I$ (b) $A = \begin{bmatrix} 2 & 0 \\ 1 & 2 \end{bmatrix}$ (c) $A = \begin{bmatrix} 3 & 1 \\ -1 & 1 \end{bmatrix}$

What are the characteristic roots of e^A?

13 Let A be the matrix of (1.13.23). Prove that
$$A^k = \begin{bmatrix} \frac{1}{0!}x^k|_{x=\lambda} & 0 & 0 & \cdots & 0 \\ \frac{1}{1!}Dx^k|_{x=\lambda} & \frac{1}{0!}x^k|_{x=\lambda} & 0 & \cdots & 0 \\ \frac{1}{2!}D^2x^k|_{x=\lambda} & \frac{1}{1!}Dx^k|_{x=\lambda} & \frac{1}{0!}x^k|_{x=\lambda} & \cdots & 0 \\ \cdots & \cdots & \cdots & \cdots & \cdots \\ \frac{1}{r!}D^rx^k|_{x=\lambda} & \frac{1}{(r-1)!}D^{r-1}x^k|_{x=\lambda} & \frac{1}{(r-2)!}D^{r-2}x^k|_{x=\lambda} & \cdots & \frac{1}{0!}x^k|_{x=\lambda} \end{bmatrix}$$
where $D^s x^k|_{x=\lambda}$ is the sth derivative of x^k evaluated at $x = \lambda$ and $0! = 1$.

14 Let A be a square matrix for which $L = \lim_{k \to \infty} A^k$ exists. Let $Y, AY, A^2Y, \ldots, A^{r-1}Y$ be a cyclic basis for an r-dimensional A-invariant cyclic subspace $\mathbb{U}$ of $\mathbb{N}[(A - \lambda I)^e]$. Show that if $|\lambda| < 1$, then $Y, AY, A^2Y, \ldots, A^{r-1}Y$ are characteristic vectors of L corresponding to the characteristic root 0. State and prove a corresponding result for $\lambda = 1$.

15 Prove that det A is a continuous function of A with respect to $\|\ \|$; that is, show that for any $\varepsilon > 0$ there exists $\delta > 0$ such that

$$|\det B - \det A| < \varepsilon$$

whenever $\|B - A\| < \delta$.

3.3 A MATHEMATICAL INTRODUCTION TO THE THEORY OF FINITE MARKOV CHAINS

In Sec. 3.1 we began our study of the theory of finite Markov chains with a heuristic discussion and then presented a number of numerical examples. Throughout we drew extensively from the elementary theory of probability. It is beyond the scope of this book to present rigorous proofs (or even rigorous definitions) when dealing with the basics of probability theory. Thus, we now develop the subject of finite Markov chains once again, from the beginning, obtaining all our results as theorems about matrices and then interpreting them in the language of probability theory. The entries of these matrices will be taken either from the real field $\mathbb{R}$ or the complex field $\mathbb{C}$.

Definition 3.3.1 A vector $P = [p_j]$ belonging to $\mathbb{R}^n$ is a *probability vector* if

$$P \geq 0 \qquad \sum_{j=1}^{n} p_j = 1$$

A matrix $A = [a_{ij}]$ belonging to $\mathbb{R}_n{}^n$ is a *transition matrix* if each of its columns is a probability vector.

It is equivalent to define a transition matrix to be a nonnegative real matrix of order n whose transpose has

$$E^{(n)} = \begin{bmatrix} 1 \\ 1 \\ \vdots \\ 1 \end{bmatrix}$$

as a characteristic vector corresponding to the characteristic root $+1$.

Definition 3.3.2 A *Markov chain* is a pair $(A, P^{(0)})$, where A is a transition matrix and $P^{(0)}$ is a probability vector, called the *initial vector*.

We will interpret a_{ij} as the probability that an object in state S_j at a given time will be in state S_i at the "next" time. Also we will interpret the ith entry of $P^{(0)}$ as the probability of the object's initially being in the state S_i, that is, before the process begins. Further, we define[1] the probability vector $P^{(k)}$ by

$$P^{(k)} = A^k P^{(0)} \qquad k = 0, 1, 2, \ldots \qquad (3.3.3)$$

and say that the ith entry of $P^{(k)}$ is the probability of the object's being in the state S_i at time k.

Definition 3.3.4 A probability vector P is a *stationary vector* of the transition matrix A if $AP = P$, that is, if P is a characteristic vector of A corresponding to the characteristic root $+1$.

Clearly, if $P^{(0)}$ is a stationary vector of A, then

$$P^{(0)} = P^{(1)} = P^{(2)} = \cdots$$

The main theorem of this section (Theorem 3.3.26) states that every transition matrix has at least one stationary vector. When $\lim_{k \to \infty} A^k$ exists, the theorem follows immediately from part (*a*) of Exercise 7 in Sec. 3.2 and each column of $\lim_{k \to \infty} A^k$ is a stationary vector of A. In Theorem 3.3.18 we will show that $\lim_{k \to \infty} A^k$ exists for a large class of transition matrices, and in Theorem 3.3.22 we show how the results of Theorem 3.3.18 can be used to prove the more general Theorem 3.3.25.

Before beginning the study of transition matrices we present a few lemmas which will be needed later. These lemmas provide us with upper bounds for the absolute value of the characteristic roots of a matrix in terms of the entries of the matrix. Let $A = [a_{ij}]$ be a square matrix of order n. We set

$$\rho_i(A) = \sum_{j=1}^{n} |a_{ij}| \qquad i = 1, 2, \ldots, n$$

$$\gamma_j(A) = \sum_{i=1}^{n} |a_{ij}| \qquad j = 1, 2, \ldots, n$$

$$\rho(A) = \max_{i=1, 2, \cdots, n} \{\rho_i(A)\} \qquad \gamma(A) = \max_{j=1, 2, \cdots, n} \{\gamma_j(A)\}$$

and call $\rho(A)$ and $\gamma(A)$ the *row sum* and the *column sum* of A respectively.

Lemma 3.3.5 If λ is a characteristic root of A, then

$$|\lambda| \leq \min \{\rho(A), \gamma(A)\}$$

---

[1] There are many texts in which the development of finite Markov chains is based on probability theory and which make virtually no use of matrix methods (see, for example, Chung [1] or Parzen [8]). In such a treatment the basic tool is the Chapman-Kolmogorov theorem, which is equivalent to equation (3.3.3). A proof of this theorem would lead us too far afield into probability theory, and so we choose to "define" our way around this difficulty.

PROOF Let $X = [x_i]$ be a characteristic vector of A corresponding to λ. Then X satisfies the matrix equation

$$AX = \lambda X$$

which can be written as a system of n simultaneous linear equations

$$\sum_{j=1}^{n} a_{ij} x_j = \lambda x_i \quad i = 1, 2, \ldots, n \quad (3.3.6)$$

Set

$$M = \max_{j=1,2,\ldots,n} |x_j|$$

Suppose this maximum takes place at $j = k$, that is,

$$M = |x_k| \geq |x_j| \quad j = 1, 2, \ldots, n$$

From the equation of (3.3.6) in which $i = k$ we have

$$|\lambda| M = |\lambda| |x_k| = |\lambda x_k| = \left| \sum_{j=1}^{n} a_{kj} x_j \right| \leq \sum_{j=1}^{n} |a_{kj} x_j|$$

$$= \sum_{j=1}^{n} |a_{kj}| |x_j| \leq \sum_{j=1}^{n} |a_{kj}| M = \rho_k(A) M \leq \rho(A) M \quad (3.3.7)$$

However, $M > 0$ since X, being a characteristic vector of A, is not the zero vector. Thus

$$|\lambda| \leq \rho(A)$$

Now let us repeat this process using A^T and a characteristic vector of A^T corresponding to the characteristic root λ. (Recall that A and A^T have the same characteristic roots.) Then

$$|\lambda| \leq \rho(A^T) = \gamma(A)$$

and consequently

$$|\lambda| \leq \min \{\rho(A), \gamma(A)\} \quad \text{////}$$

Since every column sum of a transition matrix is $+1$, we have:

Corollary 3.3.8 If λ is a characteristic root of a transition matrix, then

$$|\lambda| \leq +1$$

Every transition matrix A does in fact have $+1$ as a characteristic root since, by definition, $+1$ is a characteristic root of A^T and $E^{(n)}$ is a corresponding characteristic vector.

It will be important for us to know when there is equality throughout (3.3.7). An inequality can occur at three places. In order for the first inequality to be an equality, i.e., in order that

$$\left|\sum_{j=1}^{n} a_{kj} x_j\right| = \sum_{j=1}^{n} |a_{kj} x_j|,$$

it is both necessary and sufficient that all the terms

$$a_{kj} x_j \quad j = 1, 2, \ldots, n$$

be nonnegative real multiples of some fixed complex number, i.e., that there exist a complex number α and nonnegative real numbers $b_1, b_2, \ldots, b_n$ such that

$$a_{kj} x_j = b_j \alpha \quad j = 1, 2, \ldots, n \qquad (3.3.9)$$

For the second inequality to be an equality it is necessary and sufficient to have, for each j, either

$$a_{kj} = 0 \quad \text{or} \quad |x_j| = M \qquad (3.3.10)$$

Finally, the third inequality is an equality if and only if

$$\rho_k(A) = \rho(A)$$

We will apply these results in proving:

Lemma 3.3.11 Let A be a positive square matrix of order n and let λ be a characteristic root of A for which

$$|\lambda| = \rho(A)$$

Then

1. $\lambda = \rho(A)$.
2. The absolute value of every other characteristic root of A is less than $\rho(A)$.
3. $E^{(n)}$ is the only characteristic vector of A corresponding to λ.

PROOF Since $|\lambda| = \rho(A)$, the three inequalities of (3.3.7) are all equalities. Moreover, since every entry of A is positive, it follows from (3.3.10) that

$$|x_j| = M \quad j = 1, 2, \ldots, n$$

Finally, according to equations (3.3.9), there exist nonnegative real numbers $b_1, b_2, \ldots, b_n$ and a number α such that

$$a_{kj} x_j = b_j \alpha \quad j = 1, 2, \ldots, n$$

Then
$$x_j = \frac{b_j}{a_{kj}} \alpha \qquad j = 1, 2, \ldots, n$$

and hence $M = |x_j| = \left|\dfrac{b_j}{a_{kj}}\right| |\alpha| = \dfrac{b_j}{a_{kj}} |\alpha| \qquad j = 1, 2, \ldots, n$

since b_j and a_{kj} are real, b_j is nonnegative and a_{kj} is positive. Consequently
$$x_j = \frac{b_j}{a_{kj}} \alpha = \frac{M}{|\alpha|} \alpha \qquad j = 1, 2, \ldots, n$$

that is
$$x_1 = x_2 = \cdots = x_n = \frac{M}{|\alpha|} \alpha$$

However, a nonzero multiple of a characteristic vector corresponding to the characteristic root λ is again a characteristic vector corresponding to λ. Thus we may set
$$x_1 = x_2 = \cdots = x_n = 1$$
and hence $X = E^{(n)}$.

We have now proved that every characteristic root of A whose absolute value is $\rho(A)$ has $E^{(n)}$ as a characteristic vector, and it follows that there is only one such characteristic root. Combining this with Lemma 3.3.5, we may conclude that the absolute value of every other characteristic root of A is less than $\rho(A)$.

It remains only to show that $\lambda = \rho(A)$. Since
$$AE^{(n)} = \lambda E^{(n)} \qquad (3.3.12)$$
and the left-hand side of equation (3.3.12) is a positive element of $\mathbb{R}^n$, λ is a positive number [whose absolute value is $\rho(A)$]. Thus $\lambda = \rho(A)$. ////

Corollary 3.3.13 Let A be a positive transition matrix. The only characteristic root of A of absolute value $+1$ is $+1$. Moreover, the only characteristic vector of A^T corresponding to the root $+1$ is $E^{(n)}$.

In order to be able to conclude from Theorem 3.2.9 that $\lim_{k \to \infty} A^k$ exists when A is a positive transition matrix, it remains only to show that $(x-1)^2$ does not divide the minimal polynomial of A. Note that the first two conditions in Theorem 3.2.9 are contained in Corollaries 3.3.8 and 3.3.13. According to Theorem 1.12.26, if $(x-1)^e$ is the largest power of $x-1$ which divides the minimal polynomial of A (or, equivalently, the minimal polynomial of A^T), then there exists a vector Y such that
$$(A^T - I)^{e-1} Y \neq 0 \qquad (A^T - I)^e Y = 0$$

Thus $(A^T - I)^{e-1}Y$ is a characteristic vector of A^T corresponding to the characteristic root $+1$. By Corollary 3.3.13, $(A^T - I)^{e-1}Y$ is a nonzero multiple of $E^{(n)}$. Hence we have shown that if $e \geq 2$, there exists a vector $X = (A^T - I)^{e-2}Y$ such that $(A^T - I)X = E^{(n)}$. Consequently, in order to show that $e = 1$, it is sufficient to prove:

Lemma 3.3.14 Let A be a transition matrix. There does not exist a vector X such that $(A^T - I)X = E^{(n)}$.

PROOF Suppose there does exist a vector X such that

$$A^T X = X + E^{(n)}$$

Multiplying this equation on the left by $E_{(n)}$, we have

$$E_{(n)} A^T X = E_{(n)} X + E_{(n)} E^{(n)} = E_{(n)} X + n$$

We can show by an induction on k that

$$E_{(n)}(A^T)^k X = E_{(n)} X + kn \qquad k = 1, 2, \ldots \qquad (3.3.15)$$

For, if we suppose that equation (3.3.15) has been established for k, then

$$E_{(n)}(A^T)^{k+1} X = E_{(n)}(A^T)^k [A^T X] = E_{(n)}(A^T)^k [X + E^{(n)}]$$

$$= E_{(n)}(A^T)^k X + E_{(n)}(A^T)^k E^{(n)}$$

$$= (E_{(n)} X + kn) + n = E_{(n)} X + (k+1)n$$

since $(A^T)^k E^{(n)} = E^{(n)}$.

Rewrite equation (3.3.15) as

$$kn = E_{(n)}(A^T)^k X - E_{(n)} X \qquad k = 1, 2, \ldots$$

By Lemma 3.2.2 we have

$$kn = \|E_{(n)}(A^T)^k X - E_{(n)} X\| \leq \|E_{(n)}(A^T)^k X\| + \|E_{(n)} X\|$$

$$\leq n^2 \|E_{(n)}\| \, \|(A^T)^k\| \, \|X\| + n \|E_{(n)}\| \, \|X\|$$

$$\leq (n^2 + n) \|X\|$$

But n and $\|X\|$ are fixed while k is permitted to become arbitrarily large. Clearly this is impossible, and we have arrived at a contradiction. Hence, no such vector X exists. ////

Theorem 3.3.16 If A is a positive transition matrix, then $\lim_{k \to \infty} A^k$ exists. Moreover, $\lim_{k \to \infty} A^k > 0$.

PROOF According to the discussion above, $\lim_{k\to\infty} A^k$ exists. Since $A^k \geq 0$ for $k = 1, 2, \ldots$, it follows that $\lim_{k\to\infty} A^k \geq 0$. Moreover, by Lemma 3.2.4, with $B = E_{(n)}$, $\lim_{k\to\infty} A^k$ is a transition matrix. Applying part (*a*) of Exercise 7 in Sec. 3.2, we have $\lim_{k\to\infty} A^k > 0$. ////

Theorem 3.3.16 can be strengthened considerably. Instead of A being positive, we will show that it is sufficient for A^h to be positive for some positive integer h.

Definition 3.3.17 A transition matrix is *regular* if some power of it is positive. A Markov chain is *regular* if its transition matrix is regular.

The transition matrix of Example 3.1.5 is positive and consequently regular. The transition matrix of Example 3.1.9 is not positive but is regular since its square is positive. The Markov chains of the remaining examples of Sec. 3.1 are nonregular.

Theorem 3.3.18 If A is a regular transition matrix, then:

1 the only characteristic root of A of absolute value $+1$ is $+1$.
2 the only characteristic vector of A^T corresponding to the root $+1$ is $E^{(n)}$.
3 $\lim_{k\to\infty} A^k$ exists and is positive.
4 The columns of the matrix $L = \lim_{k\to\infty} A^k$ are identical, and each is the unique stationary vector of A.

PROOF Let h be a positive integer for which

$$H = A^h > 0$$

H is also a transition matrix, and we may apply Corollary 3.3.13 to H. Let X be a characteristic vector of A^T corresponding to the characteristic root λ and let $|\lambda| = +1$. Then X is also a characteristic vector of H^T corresponding to the characteristic root λ^h, and $|\lambda^h| = |\lambda|^h = +1$. By Corollary 3.3.13

$$X = E^{(n)}$$

However, $E^{(n)}$ is a characteristic vector of A corresponding to the characteristic root $+1$. Thus we have proved that $+1$ is the only characteristic root of A of absolute value $+1$ and that the only characteristic vector of A^T corresponding to the root $+1$ is $E^{(n)}$.

If λ is a characteristic root of A and $\lambda \neq +1$, then $|\lambda| \neq +1$ and, by Corollary 3.3.8,

$$|\lambda| < 1$$

Furthermore, according to Lemma 3.3.14, $(x-1)^2$ does not divide the minimal polynomial of A. It now follows from Theorem 3.2.9 that $\lim_{k \to \infty} A^k$ exists. Let

$$L = \lim_{k \to \infty} A^k$$

By Lemma 3.2.4 and part (a) of Exercise 7 in sec 3.2, L is a positive transition matrix.

In order to determine the rank of L it is necessary to examine the rational canonical form of A^T. Since $(x-1)^2$ does not divide the minimal polynomial of A^T, every $(x-1)$-elementary divisor of A^T gives rise to a characteristic vector corresponding to $+1$. But we have just shown that $E^{(n)}$ is the only such characteristic vector, and thus $x-1$ appears only once among the elementary divisors of A^T. The other elementary divisors of A^T, and hence of A, are of the form $(x - \lambda)^e$, where $|\lambda| < 1$. By Theorem 1.13.8 there exists a nonsingular matrix P such that

$$A = P \begin{bmatrix} 1 & 0 \\ 0 & \tilde{A} \end{bmatrix} P^{-1}$$

where $\tilde{A}$ has order $n-1$ and the absolute value of each of its characteristic roots is less than 1. By Lemma 3.2.8

$$\lim_{k \to \infty} \tilde{A}^k = 0$$

and hence

$$L = \lim_{k \to \infty} A^k = \lim_{k \to \infty} P \begin{bmatrix} 1 & 0 \\ 0 & \tilde{A} \end{bmatrix}^k P^{-1} = P \left(\lim_{k \to \infty} \begin{bmatrix} 1 & 0 \\ 0 & \tilde{A}^k \end{bmatrix} \right) P^{-1}$$

$$= P \begin{bmatrix} 1 & 0 \\ 0 & 0 \end{bmatrix} P^{-1} \tag{3.3.19}$$

The rank of L is 1, and each column of L is a scalar multiple of the first column. But the sum of the entries in each column of L is $+1$. Thus the columns of L are identical. Also,

$$AL = A \lim_{k \to \infty} A^k = \lim_{k \to \infty} A^{k+1} = L$$

and consequently each column of L is a stationary vector of A.

Since the rank of $A^T - I$ (and therefore of $A - I$) is $n - 1$, the stationary vector of A is unique. ////

If S_j is a fixed state of a Markov chain with transition matrix A, then

$$\mathbb{U}_1 = \{S_i | a_{ij} \neq 0\}$$

is the set of states to which it is *possible* to go (i.e., the probability is positive) from S_j in exactly one step. Clearly a transition matrix is positive if and only if it is possible to go from any state to any state in one step. Similarly,

$$\mathbb{U}_2 = \{S_i | a_{ik} a_{kj} \neq 0 \text{ for some } k\}$$
$$= \{S_i | a_{ij}^{(2)} \neq 0\}$$

is the set of states to which it is possible to go from S_j in exactly two steps. More generally

$$\mathbb{U} = \{S_i | a_{ik_1} a_{k_1 k_2} \cdots a_{k_{l-1} j} \neq 0 \text{ for some } k_1, k_2, \ldots, k_{l-1}\}$$
$$= \{S_i | a_{ij}^{(l)} \neq 0\}$$

is the set of states to which it is possible to go from S_j in exactly l steps. A transition matrix is regular if and only if there is a positive number h such that it is possible to go from any state to any state in exactly h moves.

The main theorem of this section (Theorem 3.3.30) states that every Markov chain has a stationary vector. The existence of a unique stationary vector for a regular Markov chain is guaranteed by Theorem 3.3.18. In order to prove Theorem 3.3.30 we must first be able to transform an arbitrary transition matrix into one with a simple form. This is accomplished by a process which is equivalent to reordering the states of the Markov chain. (Actually the same results can be obtained without this reordering, but the notation would be terribly cumbersome and make even the simplest proof seem opaque.) We have seen in Example 3.1.11 that when the states are reordered, both the rows and the columns of the transition matrix must undergo the same reordering as was performed on the states. A reordering is, of course, a permutation, and the operation or performing the same permutation simultaneously on the rows and columns of a matrix is a similarity transformation using the corresponding permutation matrix. Similarity transformations leave unchanged the properties of transition matrices that we are primarily interested in, namely, the characteristic roots, the existence of characteristic vectors (though not the characteristic vectors themselves), the existence of the limit of powers, etc.

Definition 3.3.20 An $n \times n$ matrix is *reducible* if it can be brought into the form

$$\begin{bmatrix} C & 0 \\ B & D \end{bmatrix} \quad (3.3.21)$$

where B, C, and D are $(n-l) \times l$, $l \times l$ and $(n-l) \times (n-l)$ matrices, respectively, by applying the same permutation to its rows and columns. A matrix which is not reducible is *irreducible*.

A Markov chain is *reducible* or *irreducible* as its transition matrix is reducible or irreducible.

If A is reducible, then every power of A is also reducible, for if

$$P^{-1}AP = \begin{bmatrix} C & 0 \\ B & D \end{bmatrix}$$

then $P^{-1}A^kP$ has the form

$$P^{-1}A^kP = \begin{bmatrix} C^k & 0 \\ B_k & D^k \end{bmatrix} \quad k = 1, 2, \ldots$$

Consequently, a regular transition matrix is irreducible.

Since the transition matrices of Examples 3.1.5 and 3.1.9 are regular, they are also irreducible. On the other hand, Example 3.1.21 furnishes an example of an irreducible Markov chain which is not regular. The Markov chain of the remaining examples of Sec. 3.1 are reducible.

Note that if the matrix of (3.3.21) is a transition matrix, D is also a transition matrix, acting on the states $S_{l+1}, S_{l+2}, \ldots, S_n$. Once an object is in any of the states $S_{l+1}, S_{l+2}, \ldots, S_n$, it can *never* (i.e., with probability 0) afterwards appear in any of the states $S_1, S_2, \ldots, S_l$. We call such a set of states a *closed* set, i.e., a set of states with the property that whenever an object enters the set, it can never leave. Clearly the intersection and union of closed sets of states are closed.

A reducible Markov chain contains a proper closed set of states. Conversely, if there is a proper closed set of states in a Markov chain, we may reorder the states in such a way that the resulting transition matrix has the form of (3.3.20). Hence, *a transition matrix is irreducible if and only if the only closed set of states is the set of all states*.[1] A minimal closed set of states, i.e., a closed set of states which does not properly contain another closed set of states, is called an *ergodic set*. Since the intersection of two closed sets of states is again

---

[1] A more formal proof of a somewhat stronger version of this description of irreducible transition matrices is contained in Theorem 3.3.22.

a closed set (or is empty), it follows that two distinct ergodic sets of a Markov chain are disjoint. A state is called an *ergodic state* if it belongs to some ergodic set. A state which is not ergodic is called *transient*. Thus, a Markov chain is irreducible if and only if it contains one ergodic set consisting of all the states.

Since an object can never leave an ergodic set of states, it can never go from an ergodic state to a transient state or from one ergodic set to another ergodic set. It is possible for all the states of a Markov chain to be ergodic, e.g., when the transition matrix is irreducible.[1] However it is not possible for all the states of a Markov chain to be transient.

Theorem 3.3.22 Let A be a transition matrix of order n and let $B = \frac{1}{2}(I + A)$. Then A is irreducible if and only if B is regular. In this case, $B^n > 0$.

PROOF Clearly B is a transition matrix. If A is reducible, then B is also reducible and consequently not regular. Thus, if B is regular, then A is irreducible.

Conversely, let A be irreducible. We wish to show that $B^n > 0$. Let j be fixed and define

$$\mathsf{U}_1 = \{S_i | b_{ij} \neq 0\}$$
$$\mathsf{U}_2 = \{S_i | b_{ik_1} b_{k_1 j} \neq 0 \text{ for some } k_1\}$$
$$\mathsf{U}_3 = \{S_i | b_{ik_1} b_{k_1 k_2} b_{k_2 j} \neq 0 \text{ for some } k_1, k_2\}$$
$$\cdots\cdots\cdots\cdots\cdots\cdots\cdots\cdots\cdots$$

We wish to show that

$$S_j \in \mathsf{U}_1 \subsetneq \mathsf{U}_2 \subsetneq \cdots \subsetneq \mathsf{U}_h = \{S_1, S_2, \ldots, S_n\}$$

for some h; that is, we wish to show that

1 $\mathsf{U}_q \subseteq \mathsf{U}_{q+1}$ for all q.
2 $\mathsf{U}_q = \mathsf{U}_{q+1}$ if and only if U_q consists of all the states.

Let $S_i \in \mathsf{U}_q$. Then

$$b_{ik_1} b_{k_1 k_2} \cdots b_{k_{q-1} j} \neq 0 \quad \text{for some } k_1, k_2, \ldots, k_{q-1}$$

Hence $\qquad b_{ik_1} b_{k_1 k_2} \cdots b_{k_{q-1} j} b_{jj} \neq 0$

since $b_{jj} = \frac{1}{2}(1 + a_{jj}) > 0$. Thus $S_i \in \mathsf{U}_{q+1}$ and

$$\mathsf{U}_q \subseteq \mathsf{U}_{q+1}$$

---

[1] In a Markov chain with infinitely many states it is possible to have closed sets of states but no minimal closed sets of states. Then every state is transient (see Feller [5]).

Now let us assume that $\mathsf{U}_h = \mathsf{U}_{h+1}$ for some h. We will show that U_h is closed. For, suppose $S_r \in \mathsf{U}_h$, $S_s \notin \mathsf{U}_h$. There exists a nonzero product of h terms,

$$b_{rk_1}b_{k_1k_2}\cdots b_{k_{h-1}j} \neq 0$$

But

$$b_{sr}b_{rk_1}b_{k_1k_2}\cdots b_{k_{h-1}j} = 0$$

since $S_s \notin \mathsf{U}_h = \mathsf{U}_{h+1}$. Consequently

$$0 = b_{sr} = \tfrac{1}{2}a_{sr}$$

and an object can never go from a state S_r belonging to U_h to a state S_s not belonging to U_h. Thus U_h is closed. However since A is irreducible, no proper subset of states is closed and it follows that U_h contains all the states.

When U_q does not consist of all the states, then U_{q+1} contains at least one more state than U_q. Thus $h \leq n$, and we have

$$B^n > 0 \qquad ////$$

A Markov chain with transition matrix B can be obtained from the Markov chain whose transition matrix is A in the following way. At the beginning of each step of the new Markov chain some experiment (say, tossing a coin) which has two equally likely outcomes (say, H and T) is performed. If the outcome of the experiment is H, the step consists of the objects remaining in the state it is in. On the other hand, if the outcome of the experiment is T, then, for that step, the object behaves as though the transition matrix were A. Note that the sets of states of the two Markov chains are the same.

In Theorem 3.3.22 and the discussion preceding it we were interested in deciding whether it was possible to go from one state to another. We were not concerned with the numerical probability of this event, only with whether or not it was zero. When this is the case, we can describe the Markov chain by means of a diagram.

EXAMPLE 3.3.23 The movements that are possible for the mouse of Example 3.1.18 to make in one step are described by the arrows of Fig. 3.3.24. These can be summarized in the table

	S_1	S_2	S_3	S_4	S_5	S_6
First step	S_4	S_3 S_5	S_2 S_6	S_1 S_5	S_5	S_6

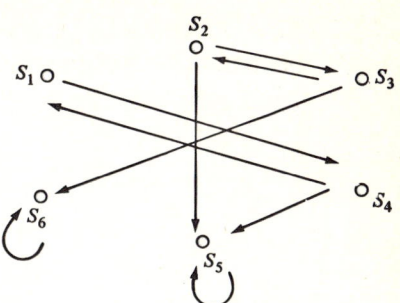

FIGURE 3.3.24

We can follow the arrows of Fig. 3.3.24 to find the additional states that can be reached from each state when more steps are taken.

	S_1	S_2	S_3	S_4	X	X
Second step	S_5	S_6	S_5			
Third step	X	X	X	X	X	X

Thus, no more states can be reached in three (or more) steps than can be reached in two. Clearly $\{S_1, S_4, S_5\}, \{S_2, S_3, S_5, S_6\}, \{S_5\}$, and $\{S_6\}$ are closed sets of states. Consequently $\{S_5\}, \{S_6\}$ are the ergodic sets and S_1, S_2, S_3, S_4 are transient states ////

Theorem 3.3.22 implies that in an irreducible Markov chain it is possible to go from any state to any state in at most $n + 1$ moves. For if

$$B^n = [\tfrac{1}{2}(I + A)]^n = \alpha_0 I + \alpha_1 A + \alpha_2 A^2 + \cdots + \alpha_n A^n > 0$$

where

$$\alpha_i = \frac{n!}{2^n i!(n-i)!} > 0$$

then

$$\alpha_0 A + \alpha_1 A^2 + \alpha_2 A^3 + \cdots + \alpha_n A^{n+1} > 0$$

and for any pair of states S_i and S_j there exists an integer r, $1 \leq r \leq n + 1$, such that the (i, j) entry of A^r is positive. Hence it is possible to go from S_j to S_i in exactly r steps.

Theorem 3.3.18 can be extended (in somewhat weakened form) to irreducible transition matrices by means of Theorem 3.3.22.

Theorem 3.3.25 If A is an irreducible transition matrix, then:
1. The only characteristic vector of A^T corresponding to the root $+1$ is $E^{(n)}$.
2. A has a unique stationary vector.
3. The stationary vector of A is positive.

In fact, if $B = \frac{1}{2}(I + A)$, then $\lim_{k \to \infty} B^k$ exists and is positive and each of its columns is the stationary vector of A.

PROOF Note that X is a characteristic vector of A (or A^T) corresponding to the root λ if and only if X is a characteristic vector of B (or B^T) corresponding to $\frac{1}{2}(1 + \lambda)$. But[1]

$$\tfrac{1}{2}(1 + \lambda) = 1 \quad \text{if and only if} \quad \lambda = 1$$

It now follows from Theorem 3.3.18 that $E^{(n)}$ is the only characteristic vector of A^T corresponding to the characteristic root $+1$.

Since B is a regular transition matrix, Theorem 3.3.18 states that $\lim_{k \to \infty} B^k$ exists and is positive and each of its columns is the unique stationary vector of B (and hence of A). ////

In Sec. 3.5 we will need:

Corollary 3.3.26 If X is the stationary vector of an irreducible transition matrix A and L is the square matrix each of whose columns is X, then there is a nonsingular matrix P such that

$$A = P \begin{bmatrix} 1 & 0 \\ 0 & \tilde{A} \end{bmatrix} P^{-1} \qquad L = P \begin{bmatrix} 1 & 0 \\ 0 & 0 \end{bmatrix} P^{-1}$$

PROOF Let $B = \frac{1}{2}(I + A)$. In the proof of Theorem 3.3.18 we showed that

$$B = P \begin{bmatrix} 1 & 0 \\ 0 & \tilde{B} \end{bmatrix} P^{-1} \qquad L = P \begin{bmatrix} 1 & 0 \\ 0 & 0 \end{bmatrix} P^{-1}$$

for some nonsingular matrix P. Set $\tilde{B} = \frac{1}{2}(I + \tilde{A})$. ////

Let $\mathsf{S}_1, \mathsf{S}_2, \ldots, \mathsf{S}_r$ be the ergodic sets of the transition matrix A and let S_i consist of n_i states. Then A has $n_1 + n_2 + \cdots + n_r$ ergodic states and $l = n - n_1 - n_2 - \cdots - n_r$ transient states. Since the sets $\mathsf{S}_1, \mathsf{S}_2, \ldots, \mathsf{S}_r$ are disjoint, we can permute the states so that A has the form

$$A = \begin{bmatrix} C & 0 & 0 & \cdots & 0 \\ B_1 & D_1 & 0 & \cdots & 0 \\ B_2 & 0 & D_2 & \cdots & 0 \\ \vdots & & & & \\ B_r & 0 & 0 & \cdots & D_r \end{bmatrix} \qquad (3.3.27)$$

---

[1] It does not follow from part 1 of Theorem 3.3.18 that the only characteristic root of an irreducible transition matrix of absolute value 1 is 1. For if λ is a characteristic root of A such that

$$|\lambda| = 1 \qquad \lambda \neq 1$$

then

$$|\tfrac{1}{2}(1 + \lambda)| < 1$$

where D_i is an irreducible $n_i \times n_i$ transition matrix, $i = 1, 2, \ldots, r$, C is a square matrix of order l, and the states corresponding to C are transient. We call (3.3.27) the *standard form* for a transition matrix. Recall that in Example 3.1.11 we found it convenient to put the transition matrix into standard form in order to facilitate the computations.

For any state S_j, the set of states consisting of S_j and all states which can be reached from S_j in a finite number of steps is a closed set. If S_j is a transient state, this set contains a proper closed subset, say $\mathbb{T}_j$, which does not contain S_j. Moreover, some subset of $\mathbb{T}_j$ is an ergodic set, and hence it is always possible to go from a transient state to some ergodic state in a finite number of steps. Let us interpret this result in terms of the matrix of (3.3.27).

Lemma 3.3.28 There exists a positive integer k such that $\gamma(C^k) < 1$.

PROOF When the matrix of (3.3.27) is expressed as

$$A = \begin{bmatrix} C & 0 \\ B & D \end{bmatrix}$$

then
$$A^k = \begin{bmatrix} C^k & 0 \\ B_k & D^k \end{bmatrix} \quad k = 1, 2, \ldots$$

According to the discussion above, for each transient state S_j, there is a positive integer k_j for which at least one of the entries in the jth column of B_{k_j} is positive. Hence the sum of the entries in the jth column of C^{k_j} is less than 1. Moreover, the sum of the entries in the jth column of C^p is less than 1 for every integer $p \geq k_j$. If we set

$$k = \max_{j=1,2,\ldots,l} \{k_j\}$$

then
$$\gamma(C^k) < 1 \qquad ////$$

Corollary 3.3.29 $\lim_{k \to \infty} C^k = 0$, and consequently $1 + C + C^2 + \cdots$ converges to $(I - C)^{-1}$.

PROOF According to Theorem 3.2.10, it is sufficient to show that the absolute value of every characteristic root of C is less than 1. Let λ be a characteristic root of C. Then λ^k is a characteristic root of C^k and, by Lemma 3.3.5, $|\lambda^k| < 1$. Consequently

$$|\lambda| < 1 \qquad ////$$

Once an object has entered one of the states of the ergodic set $\mathbb{S}_i$, it may never leave the states of $\mathbb{S}_i$. The states of the Markov chain other than those of S_i can thereafter be ignored, and the object can be thought of as being in a Markov chain whose transition matrix is D_i.

Theorem 3.3.30 To each ergodic set of states S_i of a finite Markov chain there corresponds a stationary vector X_i having the property that the positions corresponding to the states of S_i contain positive entries and all other entries of X_i are zero. Every stationary vector is a convex combination of these.

PROOF Let A be a transition matrix in standard form and let Y_i be the (unique) stationary vector of the irreducible transition matrix D_i, $i = 1, 2, \ldots, r$. Then the r n-dimensional vectors

$$X_1 = \begin{bmatrix} 0 \\ Y_1 \\ 0 \\ 0 \\ \vdots \\ 0 \end{bmatrix}, \quad X_2 = \begin{bmatrix} 0 \\ 0 \\ Y_2 \\ 0 \\ \vdots \\ 0 \end{bmatrix}, \quad X_3 = \begin{bmatrix} 0 \\ 0 \\ 0 \\ Y_3 \\ \vdots \\ 0 \end{bmatrix}, \quad \ldots, \quad X_r = \begin{bmatrix} 0 \\ 0 \\ 0 \\ 0 \\ \vdots \\ Y_r \end{bmatrix} \begin{matrix} l \\ n_1 \\ n_2 \\ n_3 \\ \\ n_r \end{matrix}$$

(3.3.31)

are linearly independent stationary vectors of A.

Conversely, suppose that

$$X = \begin{bmatrix} Z_0 \\ Z_1 \\ Z_2 \\ \vdots \\ Z_r \end{bmatrix}$$

is a stationary vector of the matrix A of (3.3.27). A direct calculation shows that

$$CZ_0 = Z_0 \quad (3.3.32)$$

However, by Lemma 3.3.5, $+1$ is not a characteristic root of C^k for the value of k found in Corollary 3.3.29. Consequently $+1$ is not a characteristic root of C, and equation (3.3.32) yields

$$Z_0 = 0$$

It is now clear that for each i, $i = 1, 2, \ldots, r$,

$$D_i Z_i = Z_i$$

Hence Z_i is a nonnegative multiple of the unique stationary vector Y_i of D_i, say $Z_i = \alpha_i Y_i$, $\alpha_i \geq 0$. Since the sum of the entries of each of the vectors $X, Y_1, Y_2, \ldots, Y_r$ is $+1$, it follows that

$$\alpha_1 + \alpha_2 + \cdots + \alpha_r = 1$$

and X is a convex combination of the vectors of (3.3.31). ////

EXERCISES

1 Find the ergodic sets and the transient states of the Markov chains in Examples 3.1.15, 3.1.18, and 3.1.21. Find all stationary vectors of these transition matrices by applying Theorem 3.3.30.
2 Find the ergodic sets, the transient states, and the stationary vectors of the Markov chains in Exercises 6 and 9 to 12 of Sec. 3.1.
3 Identify the ergodic sets and the transient states of the Markov chain corresponding to the following mazes:

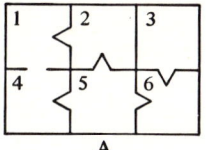
A

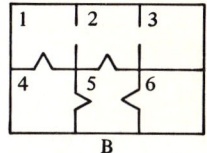
B

Find the standard form of the transition matrix and the stationary vectors.
4 Find the ergodic sets and the transient states of the Markov chain whose transition matrix is:

$$(a) \begin{bmatrix} 0 & 0 & 0 & \frac{1}{2} & 0 \\ 1 & 0 & 0 & \frac{1}{2} & 0 \\ 0 & \frac{1}{2} & 1 & 0 & 0 \\ 0 & \frac{1}{4} & 0 & 0 & 1 \\ 0 & \frac{1}{4} & 0 & 0 & 0 \end{bmatrix} \quad (b) \begin{bmatrix} 0 & 0 & \frac{1}{2} & 0 & 0 \\ 0 & 0 & 0 & \frac{1}{3} & 1 \\ 1 & 0 & \frac{1}{2} & 0 & 0 \\ 0 & 1 & 0 & \frac{1}{3} & 0 \\ 0 & 0 & 0 & \frac{1}{3} & 0 \end{bmatrix} \quad (c) \begin{bmatrix} \frac{1}{2} & \frac{2}{3} & 0 & 1 & 0 \\ \frac{1}{2} & 0 & \frac{1}{2} & 0 & 0 \\ 0 & 0 & \frac{1}{2} & 0 & 1 \\ 0 & \frac{1}{6} & 0 & 0 & 0 \\ 0 & \frac{1}{6} & 0 & 0 & 0 \end{bmatrix}.$$

Find the standard form of each transition matrix and a reordering of the states which yields it.

5 Let A be a regular transition matrix and let X be its stationary vector. Show that
$$\lim_{k \to \infty} P^{(k)} = X$$
for every initial probability vector $P^{(0)}$.

6 Let S_i be a transient state of a Markov chain. Show that
$$\lim_{k \to \infty} p_i^{(k)} = 0$$
for every initial probability vector $P^{(0)}$.

7 Let A_1, A_2 be transition matrices of the same order neither of which has a row consisting only of 0s. Show that $A_1 A_2$ is regular if and only if $A_2 A_1$ is regular. Show that $A_1 A_2$ is irreducible if and only if $A_2 A_1$ is irreducible.

8 Refine the argument of Theorem 3.3.21 to show that $B^{n-1} > 0$.

9 Let A be a regular transition matrix and let
$$L = \lim_{k \to \infty} A^k$$

Show that:
(a) $(A - L)^k = A^k - L$ for $k = 1, 2, \ldots$.
(b) $I + (A - L) + (A^2 - L) + \cdots$ converges to $[I - (A - L)]^{-1}$.
Set $Z = [I - (A - L)]^{-1}$ so that $Z = \lim_{n \to \infty}[(\sum_{k=0}^{n} A^k) - nL]$. Show that
(c) $E_{(n)} Z = E_{(n)}$.
(d) $ZL = LZ = L$.
(e) $I - Z = L - ZA$.

10 Let A be a transition matrix of order n and let $X = [x_i]$ be an n-dimensional row vector. Set $Y = XA = [y_j]$ and

$$M_0 = \max_{i=1,2,\ldots,n} x_i \qquad m_0 = \min_{i=1,2,\ldots,n} x_i$$

$$M_1 = \max_{j=1,2,\ldots,n} y_j \qquad m_1 = \min_{j=1,2,\ldots,n} y_j$$

$$a = \min_{i,j=1,2,\ldots n} a_{ij}$$

Show that $M_1 \leq M_0$, $m_1 \geq m_0$ and

$$M_1 - m_1 \leq (1 - 2a)(M_0 - m_0)$$

Use this result to give another proof that if A is a positive transition matrix, then $\lim_{k \to \infty} A^k$ exists and is positive and all its columns are identical.

11 Complete the following proof that the geometric multiplicity of $+1$ as a characteristic root of a positive transition matrix is 1.

If X is a characteristic vector of A^T corresponding to the characteristic root of $+1$ which is linearly independent with $E^{(n)}$, then there exists a non-negative characteristic vector Y of A^T corresponding to the root $+1$ which has zero components. Contradiction.

Extend this method to prove the corresponding theorem for regular transition matrices.

12 Show that the closed sets of states in the Markov chains with matrices A and $B = \frac{1}{2}(I + A)$ are identical. (Hence the ergodic sets and transient states are identical.)

13 Let A be a transition matrix and define

$$B_m = \frac{1}{m}(I + A + A^2 + \cdots + A^{m-1})$$

Show that A is irreducible if and only if B is regular. Show, moreover, that if A is irreducible, then $\lim_{k \to \infty}(B_m)^k$ exists for $m \geq 2$ and

$$\lim_{k \to \infty}(B_m)^k = \lim_{k \to \infty}(B_q)^k = L \qquad \text{for } m, q \geq 2$$

14 Let A be a transition matrix. Prove that A is singular if and only if there exists distinct *probability vectors* X_1, X_2 such that

$$AX_1 = AX_2$$

[*Hint:* Show that a characteristic vector of A corresponding to a characteristic root other than $+1$ belongs to $\mathbb{N}(E_{(n)})$.]

15 Prove that an irreducible transition matrix A for which $\lim_{k \to \infty} A^k$ exists is regular. (*Hint:* Prove $\lim_{k \to \infty} A^k > 0$.)

16 Show that the first column of the matrix P of Theorem 3.3.18 is a scalar multiple of the stationary vector of A, say $P^{(1)} = \alpha X$. Show that the first row of P^{-1} is a scalar multiple of $E_{(n)}$. In fact, $[P^{-1}]_{(1)} = (1/\alpha)E_{(n)}$.

17 Show that every closed set of states in the Markov chain with transition matrix A is a closed set of states in the Markov chain with transition matrix A^2 but that the converse need not be true. Formulate this statement in terms of the transient states and the ergodic sets of each Markov chain.

18 Prove that if A is a nonsingular transition matrix and $A^{-1} \geq 0$, then A^{-1} is a transition matrix.

19 Show that if A and A^{-1} are both transition matrices, then A is a permutation matrix.

20 Let A be an irreducible transition matrix of order n. Prove that every submatrix of $I - A$ of order $n - 1$ is nonsingular.

21 Let A be an irreducible nonregular transition matrix. Show that A has a characteristic root λ such that $|\lambda| = 1$, $\lambda \neq +1$.

22 Construct an irreducible nonregular transition matrix of order 2. Show that it is unique.

3.4 PERIODIC TRANSITION MATRICES

In this section we will prove that an irreducible nonregular transition matrix has (possibly after a simultaneous permutation of its rows and columns) a particularly simple form. First we will show (in Theorem 3.4.2) one way in which irreducible nonregular transition matrices can be found. Then (in Theorem 3.4.7) we will prove that all irreducible transition matrices are of this type.

We can construct an irreducible nonregular transition matrix as follows. Let $m_1, m_2, \ldots, m_l$, $l \geq 2$, be positive integers and let $B_1, B_2, \ldots, B_l$ be $m_2 \times m_1$, $m_3 \times m_2, \ldots, m_l \times m_{l-1}, m_1 \times m_l$ matrices respectively satisfying the conditions

1 $B_i \geq 0$, $i = 1, 2, \ldots, l$.
2 The sum of the entries in each column of each B_i is 1.

Set

$$A = \begin{bmatrix} 0 & 0 & 0 & \cdots & 0 & B_l \\ B_1 & 0 & 0 & \cdots & 0 & 0 \\ 0 & B_2 & 0 & \cdots & 0 & 0 \\ 0 & 0 & B_3 & \cdots & 0 & 0 \\ \multicolumn{6}{c}{\dotfill} \\ 0 & 0 & 0 & \cdots & B_{l-1} & 0 \end{bmatrix} \quad (3.4.1)$$

Then:

Theorem 3.4.2 The matrix A of (3.4.1) is a nonregular transition matrix. If λ is any lth root of unity[1], then λ is a characteristic root of A (and consequently of A^T) and the vector

$$Y = \begin{bmatrix} E^{(m_1)} \\ \lambda E^{(m_2)} \\ \vdots \\ \lambda^{l-1} E^{(m_l)} \end{bmatrix}$$

is a characteristic vector of A^T corresponding to λ.

Moreover

$$A_1 = B_l B_{l-1} \quad \cdots \quad B_2 B_1$$
$$A_2 = {}_1 B_l \quad \cdots \quad B_3 B_2$$
$$\dotfill$$
$$A_l = B_{l-1} B_{l-2} \quad \cdots \quad B_1 B_l$$

are transition matrices and A is irreducible if and only if $A_1, A_2, \ldots, A_l$ are irreducible.

PROOF A direct computation will verify that if λ is an lth root of unity, then Y is a characteristic vector of A^T corresponding to the characteristic root λ. Also, it is clear that $A_1, A_2, \ldots, A_l$ and A are square nonnegative matrices of orders $m_1, m_2, \ldots, m_l$ and $m_1 + m_2 + \cdots + m_l$, respectively. The sum of the entries in each column of each matrix is 1, and hence $A_1, A_2, \ldots, A_l$ and A are transition matrices.

It remains only to show that A is not regular and that A is irreducible if and only if $A_1, A_2, \ldots, A_l$ are irreducible. First let us compute $A^2, A^3, \ldots, A^l$ from (3.4.1). For each j, $j = 1, 2, \ldots, l$, A^j has exactly

---

[1] Recall that λ is an *l*th *root of unity* if $\lambda^l = 1$. The *l*th roots of unity are

$$\exp \frac{2\pi i p}{l} = \cos \frac{2\pi i p}{l} + i \sin \frac{2\pi i p}{l} \quad p = 0, 1, 2, \ldots, l-1$$

l nonzero blocks, and these are located in the positions j places below the main diagonal and $l-j$ places above the main diagonal. For example,

$$A^2 = \begin{bmatrix} 0 & 0 & \cdots & 0 & B_l B_{l-1} & 0 \\ 0 & 0 & \cdots & 0 & 0 & B_1 B_l \\ B_2 B_1 & 0 & \cdots & 0 & 0 & 0 \\ 0 & B_3 B_2 & \cdots & 0 & 0 & 0 \\ \cdots & \cdots & \cdots & \cdots & \cdots & \cdots \\ 0 & 0 & \cdots & B_{l-1} B_{l-2} & 0 & 0 \end{bmatrix}$$

and

$$A^l = \begin{bmatrix} A_1 & 0 & \cdots & 0 \\ 0 & A_2 & \cdots & 0 \\ \cdots & \cdots & \cdots & \cdots \\ 0 & 0 & \cdots & A_l \end{bmatrix}$$

Since A^{ql} is a diagonal block matrix for $q = 1, 2, \ldots$, it follows that A is not regular.

Finally, suppose that $A_1, A_2, \ldots, A_l$ are irreducible and set

$$m = \max_{i=1, 2, \ldots, l} \{m_i\}$$

According to Theorem 3.3.21,

$$(I + A_i)^m > 0 \qquad i = 1, 2, \ldots, l$$

Hence each of the l diagonal blocks of $(I + A^l)^m$ is positive. More generally, for each j, the l nonzero blocks of $(I + A^l)^m A^j$ are positive. Thus

$$(I + A^l)^m (I + A + A^2 + \cdots + A^{l-1}) > 0$$

If we set $B = \frac{1}{2}(I + A)$, then B is regular and hence A is irreducible.

Conversely, let A be irreducible. Then $(I + A)^n > 0$. By expanding $(I + A)^n$ we see that the ith diagonal block of $(I + A)^n$ is of the form

$$I + \alpha_1 A_i + \alpha_2 A_i^2 + \cdots + \alpha_k A_i^k$$

where $\alpha_1, \alpha_2, \ldots, \alpha_k$ are positive integers and k is the largest integer such that kl is less than or equal to n. Thus

$$(I + A_i)^k > 0$$

and consequently A_i is irreducible. ////

Definition 3.4.3 A transition matrix is *periodic* if it can be brought into the form (3.4.1) by performing the same permutation of its rows and columns.

A Markov chain whose transition matrix is periodic is a *periodic Markov chain*. An object initially in one of the first m_1 states of a Markov chain whose matrix is given by (3.4.1) will be in one of the next m_2 states after the first time period. The object will be in one of the following m_3 states after the second time period, etc. After l time periods have elapsed, the object will be in one of the first m_1 states and the process is repeated. The largest integer l for which the set of states of a Markov chain with matrix A can be partitioned into l disjoint sets such that at each step the object passes from one set of states into the next (and after l steps has returned to the first set of states) is called the *period* of A. That is, A is a periodic transition matrix with period l if it can be brought into the $l \times l$ block form (3.4.1), but it cannot be brought into this form for any integer greater than l by performing the same permutation of its rows and columns.

Theorem 3.4.2 gives a specific example of an irreducible nonregular transition matrix. In Theorem 3.4.7 we will prove that every irreducible nonregular transition matrix can be brought to form (3.4.1) and hence is periodic. The proof of Theorem 3.4.7 is based upon a characteristic root which is not equal to 1 but whose absolute value is 1. In order to establish the existence of this characteristic root we need

Lemma 3.4.4 If A is an irreducible nonregular transition matrix, then $\lim_{k \to \infty} A^k$ does not exist.

PROOF Let $B = \frac{1}{2}(I + A)$. According to Theorem 3.3.25, $\lim_{k \to \infty} B^k$ exists and is positive and each of its columns is the unique stationary vector of B and likewise of A. Now, suppose that $\lim_{k \to \infty} A^k$ exists. Then each column of $\lim_{k \to \infty} A^k$ is a stationary vector of A, and hence

$$\lim_{k \to \infty} A^k = \lim_{k \to \infty} B^k > 0 \qquad (3.4.5)$$

Equation (3.4.5) implies that A is regular, which is a contradiction. ////

Theorem 3.2.9 gives a convenient set of necessary and sufficient conditions that $\lim_{k \to \infty} A^k$ exists for an arbitrary square matrix A with elements from the complex field $\mathbb{C}$. When A is an irreducible nonregular transition matrix, at least one of these conditions must be violated. According to Corollary 3.3.8 the first condition (namely, that the characteristic roots of A have absolute value at most 1) is satisfied for every transition matrix. By Theorem 3.3.25, $E^{(n)}$ is the only characteristic vector of A^T corresponding to the root $+1$. Applying Lemma 3.3.14, we see that the third condition [namely, that $(x - 1)^2$ does not divide the minimal polynomial of A] is also satisfied. Hence the second condition must be violated; that is:

Lemma 3.4.6 An irreducible nonregular transition matrix has a characteristic root λ such that

$$|\lambda| = 1 \qquad \lambda \neq 1$$

It is the existence of this characteristic root that yields form (3.4.1) for an irreducible nonregular transition matrix. The proof is similar to the proof of Lemma 3.3.11 and is applied to A^T since

$$|\lambda| = 1 = \gamma(A) = \rho(A^T)$$

Theorem 3.4.7 An irreducible nonregular transition matrix is periodic.

PROOF Let A be an irreducible nonregular transition matrix. We wish to show that A^T can be brought into a form which is the transpose of (3.4.1) by a simultaneous permutation of its rows and columns. Such a permutation induces the same permutation among the entries of each characteristic vector.

Let $A = [a_{ij}]$, $A^T = [c_{ij}]$ (that is, $c_{ij} = a_{ji}$) and let $Y = [y_j]$ be a characteristic vector of A^T, corresponding to the characteristic root λ of Lemma 3.4.7. Then

$$\sum_{j=1}^{n} c_{ij} y_j = \lambda y_i \qquad i = 1, 2, \ldots, n \qquad (3.4.8)$$

Set

$$M = \max_{j=1,2,\ldots,n} \{|y_j|\} > 0$$

We may assume (possibly after a permutation of the rows and columns of A^T and multiplying Y by a scalar of absolute value $+1$) that

$$M = y_1 = y_2 = \cdots = y_{m_1} \qquad y_j \neq M \quad \text{for} \quad j > m_1 \qquad (3.4.9)$$

Since $\lambda \neq 1$, it follows that $Y \neq E^{(n)}$ and $m_1 < n$. The first m_1 equations of (3.4.8) now become

$$\sum_{j=1}^{n} c_{ij} y_j = \lambda y_i = \lambda M \qquad i = 1, 2, \ldots, m_1 \qquad (3.4.10)$$

and, as in (3.3.7), we have

$$M = |\lambda| M = \left| \sum_{j=1}^{n} c_{ij} y_j \right| \leq \sum_{j=1}^{n} |c_{ij} y_j| = \sum_{j=1}^{n} c_{ij} |y_j|$$

$$\leq \left(\sum_{j=1}^{n} c_{ij} \right) M = \rho_i(A^T) M = M \qquad (3.4.11)$$

Thus there is equality throughout each of the m_1 inequalities of (3.4.11).

Since the first inequality of (3.4.11) is actually an equality, it follows that:

1 For each i, $1 \leq i \leq m_1$, there exists a complex number α_i such that each of the terms $c_{ij} y_j$, $j = 1, 2, \ldots, n$, is a real, nonnegative multiple of α_i. By equation (3.4.10), we may take

$$\alpha_1 = \alpha_2 = \cdots = \alpha_{m_1} = \lambda M$$

and hence, for $1 \leq i \leq m_1$, each of the terms $c_{ij} y_j$ is a real nonnegative multiple of λM.

Since the second inequality of (3.4.11) is actually an equality, it follows that:

2 $|y_j| = M$ whenever $c_{ij} \neq 0$ for some i, $1 \leq i \leq m_1$.

From 1 and 2 we conclude that for $1 \leq i \leq m_1$,

$$c_{ij} = 0 \quad \text{whenever} \quad y_j \neq \lambda M$$

In particular, according to equation (3.4.8), we have

$$c_{ij} = 0 \quad i, j = 1, 2, \ldots, m_1$$

Let us now suppose that

$$y_{m_1+1} = y_{m_1+2} = \cdots = y_{m_1+m_2} = \lambda M$$
$$y_j \neq \lambda M \quad \text{for} \quad j > m_1 + m_2 \quad (3.4.12)$$

Then we also have

$$c_{ij} = 0 \quad \begin{array}{l} i = 1, 2, \ldots, m_1 \\ j > m_1 + m_2 \end{array}$$

and hence the first m_1 rows of A^T can be partitioned into

$$[0 \quad B_1 \quad 0]$$

where the two zero matrices have dimensions $m_1 \times m_1$ and $m_1 \times (n - m_1 - m_2)$, respectively, and B_1 is an $m_1 \times m_2$ real nonnegative matrix for which the sum of the entries in each row is 1. Moreover, the first $m_1 + m_2$ entries of Y have the form

$$\begin{bmatrix} ME^{(m_1)} \\ \lambda M E^{(m_2)} \end{bmatrix}$$

We can continue this process. When we substitute equation (3.4.12) into equation (3.4.8), the next m_2 equations of (3.4.8) (after the first m_1 equations) become

$$\sum_{j=1}^{n} c_{ij} y_j = \lambda y_i = \lambda^2 M \qquad i = m_1 + 1, m_1 + 2, \ldots, m_1 + m_2$$

As before, we conclude that

$$c_{ij} = 0 \qquad \text{whenever} \qquad y_j \neq \lambda^2 M$$

In particular, we know from equation (3.4.12) that

$$y_j = \lambda M \neq \lambda^2 M \qquad j = m_1 + 1, m_1 + 2, \ldots, m_1 + m_2$$

Thus $\quad c_{ij} = 0 \qquad i, j = m_1 + 1, m_1 + 2, \ldots, m_1 + m_2$

At this point two cases arise.

CASE 1 Suppose that $\lambda^2 \neq 1$. When $\lambda^2 \neq 1$, then

$$y_j \neq \lambda^2 M \qquad j = 1, 2, \ldots, m_1$$

and hence
$$c_{ij} = 0 \qquad \begin{array}{l} i = m_1 + 1, m_1 + 2, \ldots, m_1 + m_2 \\ j = 1, 2, \ldots, m_1 \end{array}$$

We may assume

$$y_{m_1+m_2+1} = y_{m_1+m_2+2} = \cdots = y_{m_1+m_2+m_3} = \lambda^2 M$$
$$y_j \neq \lambda^2 M \qquad j > m_1 + m_2 + m_3$$

the first $m_1 + m_2$ rows of A^T can be partitioned into

$$\begin{array}{c} \begin{array}{cccc} m_1 & m_2 & m_3 & n - m_1 - m_2 - m_3 \end{array} \\ \begin{array}{c} m_1 \\ m_2 \end{array}\left[\begin{array}{cccc} 0 & B_1 & 0 & 0 \\ 0 & 0 & B_2 & 0 \end{array}\right] \end{array}$$

and the first $m_1 + m_2 + m_3$ entries of Y have the form

$$\begin{bmatrix} ME^{(m_1)} \\ \lambda ME^{(m_2)} \\ \lambda^2 ME^{(m_3)} \end{bmatrix}$$

We examine the next block row of A^T in exactly the same way and are again faced with two cases, the first of which (namely, $\lambda^3 \neq 1$) can be treated exactly as before. Since the order of A^T is finite, we eventually reach a point at which the first $m_1 + m_2 + \cdots + m_{l-1}$ rows of A^T are given by

$$\begin{array}{c} \\ m_1 \\ m_2 \\ m_3 \\ \\ m_{l-1} \end{array} \overset{\begin{array}{ccccccc} m_1 & m_2 & m_3 & & m_{l-1} & m_l & n-m_1-m_2-\cdots-m_l \end{array}}{\begin{bmatrix} 0 & B_1 & 0 & \cdots & 0 & 0 & 0 \\ 0 & 0 & B_2 & \cdots & 0 & 0 & 0 \\ 0 & 0 & 0 & \cdots & 0 & 0 & 0 \\ \multicolumn{7}{c}{\dotfill} \\ 0 & 0 & 0 & \cdots & 0 & B_{l-1} & 0 \end{bmatrix}}$$

and the first $m_1 + m_2 + \cdots + m_{l-1} + m_l$ entries of Y are given by

$$\begin{bmatrix} ME^{(m_1)} \\ \lambda ME^{(m_2)} \\ \lambda^2 ME^{(m_3)} \\ \vdots \\ \lambda^{l-1} ME^{(m_l)} \end{bmatrix} \quad (3.4.13)$$

CASE 2 $\lambda^l = 1$. For each i, $m_1 + m_2 + \cdots + m_{l-1} \leq i \leq m_1 + m_2 + \cdots + m_{l-1} + m_l$, we have

$$c_{ij} \neq 0 \quad \text{whenever} \quad y_j \neq \lambda^l M = M$$

But we previously required [in equation (3.4.9)] that

$$M = y_1 = y_2 = \cdots = y_{m_1} \quad y_j \neq M \quad \text{for} \quad j > m_1$$

Thus the first $m_1 + m_2 + \cdots + m_l$ rows of A^T have the form

$$\begin{array}{c} \\ m_1 \\ m_2 \\ \\ m_{l-1} \\ m_l \end{array} \overset{\begin{array}{ccccccc} m_1 & m_2 & m_3 & & m_l & n-m_1-\cdots-m_l \end{array}}{\begin{bmatrix} 0 & B_1 & 0 & \cdots & 0 & 0 \\ 0 & 0 & B_2 & \cdots & 0 & 0 \\ \multicolumn{6}{c}{\dotfill} \\ 0 & 0 & 0 & \cdots & B_{l-1} & 0 \\ B_l & 0 & 0 & \cdots & 0 & 0 \end{bmatrix}} \quad (3.4.14)$$

It remains to show that

$$n = m_1 + m_2 + \cdots + m_l$$

[so that A^T has no further rows, and, in fact, the last column of the matrix of (3.4.14) does not actually appear]. If $n > m_1 + m_2 + \cdots + m_l$, we have proved that A^T has the form

$$A^T = \begin{bmatrix} C_{11} & 0 \\ C_{21} & C_{22} \end{bmatrix}$$

where C_{11} is the matrix of (3.4.14) and C_{22} is a square matrix of order $n - (m_1 + m_2 + \cdots + m_l)$. But, A (and consequently A^T) is irreducible, and hence C_{22} does not actually appear, i.e.,

$$n = m_1 + m_2 + \cdots + m_l$$

The matrix A^T of (3.4.14) clearly has the form we required, namely, the transpose of the matrix of (3.4.1). Hence A is periodic. ////

Incidentally Theorem 3.4.7 also shows that a characteristic root λ of an irreducible nonregular (and hence periodic) transition matrix of absolute value 1 is a root of unity. If λ is an hth root of unity, the proof of Theorem 3.4.7 shows that the period l of A is at least h; that is,

$$h \leq l$$

This is the only conclusion we can draw at this time. In Theorem 3.4.21 we will prove that h divides l and consequently λ is an lth root of unity.

Lemma 3.4.15 If λ, μ are characteristic roots of an irreducible periodic matrix A and

$$|\lambda| = 1$$

then μ/λ is also a characteristic root of A.

PROOF According to the proof of Theorem 3.4.7, there exists a characteristic vector

$$Y = [y_j]$$

of A^T corresponding to the characteristic root λ such that

$$|y_j| = 1 \quad j = 1, 2, \ldots, n \quad (3.4.16)$$

[Note that we may set $M = 1$ in (3.4.13).] Clearly

$$Y = DE^{(n)}$$

where D is the diagonal matrix

$$D = \begin{bmatrix} y_1 & 0 & \cdots & 0 \\ 0 & y_2 & \cdots & 0 \\ \cdots & \cdots & \cdots & \cdots \\ 0 & 0 & \cdots & y_n \end{bmatrix}$$

and the equation

$$A^T Y = \lambda Y$$

becomes
$$A^T DE^{(n)} = \lambda DE^{(n)}$$
which we may write as
$$\lambda^{-1} D^{-1} A^T DE^{(n)} = E^{(n)} \quad (3.4.17)$$
Set
$$\lambda^{-1} D^{-1} A^T D = F \quad (3.4.18)$$
so that (3.4.17) becomes
$$FE^{(n)} = E^{(n)} \quad (3.4.19)$$

Since μ is a characteristic root of A, it is a characteristic root of A^T and hence of $D^{-1}A^T D$. Thus μ/λ is a characteristic root of F.

Let $|F|$ denote the matrix whose entries are the absolute values of the entries of F, that is,
$$|F| = [|f_{ij}|] \quad i,j = 1, 2, \ldots, n$$
It follows from (3.4.16) and (3.4.18) that
$$|F| = A^T$$
Combining (3.4.19) and
$$A^T E^{(n)} = E^{(n)}$$
yields
$$1 = \sum_{j=1}^{n} f_{ij} = \sum_{j=1}^{n} a_{ij} = \sum_{j=1}^{n} |f_{ij}| \quad i = 1, 2, \ldots, n$$
However, this is possible only when
$$f_{ij} = |f_{ij}| \quad i,j = 1, 2, \ldots, n$$
that is,
$$F = |F| = A^T$$
Consequently μ/λ is a characteristic root of A. ////

If we set $\mu = 1$ in Lemma 3.4.15, it follows that λ^{-1} is also a characteristic root of A. In the same way, if we set
$$\mu = \lambda \quad \lambda = \lambda^{-1}$$
we see that λ^2 is a characteristic root of A. Continuing this process, we see that $\lambda^3, \lambda^4, \ldots$ are characteristic roots of A. In order to show that $\lambda^{-2}, \lambda^{-3}, \ldots$ are characteristic roots of A we substitute 1 for μ and $\lambda^2, \lambda^3, \ldots$ for λ. Finally we have:

Corollary 3.4.20 If λ, μ are characteristic roots of an irreducible periodic transition matrix A and $|\lambda| = |\mu| = 1$, then

$$\lambda^p \mu^q$$

is a characteristic root of A for all integers p and q.

Now we can prove:

Theorem 3.4.21 Every characteristic root of absolute value 1 of an irreducible periodic transition matrix of period l is an lth root of unity.

PROOF Let v be a characteristic root of absolute value 1 of the irreducible periodic transition matrix A of period l and let $\tilde{l}$ be the smallest positive integer for which

$$v^{\tilde{l}} = 1$$

that is, v is a *primitive $\tilde{l}$th root* of unity. By the discussion above, all the $\tilde{l}$th roots of unity are characteristic roots of A. The $\tilde{l}$th roots of unity can be expressed as

$$\exp \frac{2\pi i \cdot q}{\tilde{l}} = \cos \frac{2\pi \cdot q}{\tilde{l}} + i \sin \frac{2\pi \cdot q}{\tilde{l}} \qquad q = 0, 1, 2, \ldots, \tilde{l} - 1.$$

Since the period of A is l, the lth roots of unity

$$\exp \frac{2\pi i \cdot p}{l} = \cos \frac{2\pi \cdot p}{l} + i \sin \frac{2\pi \cdot p}{l} \qquad p = 0, 1, 2, \ldots, l - 1$$

are also characteristic roots of A. Let d and h be the greatest common divisor and least common multiple of l and $\tilde{l}$, respectively,

$$d = (l, \tilde{l}) \qquad h = [l, \tilde{l}]$$

(See Sec. 1.11.) Recall that

$$h = \frac{l\tilde{l}}{d}$$

There exist integers p and q such that

$$d = p\tilde{l} + ql \qquad (3.4.22)$$

Set

$$\lambda = \exp \frac{2\pi i \cdot q}{\tilde{l}} \qquad \mu = \exp \frac{2\pi i \cdot p}{l}$$

where p and q satisfy (3.4.22). By Corollary 3.4.20,

$$\lambda \mu = \exp \frac{2\pi i \cdot d}{l\tilde{l}} = \exp \frac{2\pi i}{h}$$

is a characteristic root of A. Clearly $\lambda\mu$ is a primitive hth root of unity. However, since A has period l, it follows that

$$h \leq l$$

Consequently $h = l$ and $\bar{l}$ divides l. Thus

$$\lambda^l = 1 \qquad \text{////}$$

The converse of Theorem 3.4.21 is contained in Theorem 3.4.2. Hence:

Corollary 3.4.23 The characteristic roots of absolute value of 1 of an irreducible periodic transition matrix whose period is l are exactly the lth roots of unity.

In Theorem 3.4.2 we showed that A^l is the direct sum of l irreducible transition matrices $A_1, A_2, \ldots, A_l$. According to Corollary 3.4.23, none of the matrices $A_1, A_2, \ldots, A_l$ has a characteristic root of absolute value 1 other than 1. It now follows from Lemma 3.4.6 that:

Corollary 3.4.24 If the matrix A of (3.4.1) is an irreducible transition matrix of period l and $A_1, A_2, \ldots, A_l$ are defined as in Theorem 3.4.2, then $A_1, A_2, \ldots, A_l$ are regular.

EXERCISES

1 Show that the transition matrices of the Markov chains corresponding to the following mazes are periodic. Find the period and the transient states.

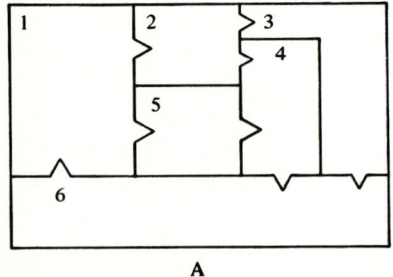
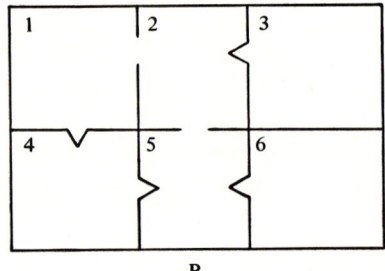

A   B

2 Show that the period of a nonsingular periodic transition matrix divides the order of the matrix.

3 Let A be a periodic transition matrix of period m. Show that if l divides m, then the states of the Markov chain can be numbered so that A has the

$l \times l$ block form (3.4.1). (How has this difficulty been avoided when defining the period of a transition matrix?)

4 Illustrate Exercise 3 using the maze of part (*a*) of Exercise 1.
5 Let A be an irreducible periodic transition matrix of period l. Show that A^{rl+1} is irreducible for $r = 0, 1, 2, \ldots$.
6 Show that if A is an irreducible periodic transition matrix of period l, then A^q is irreducible if and only if q is relatively prime to l.
7 Show that if A is a transition matrix and $|\lambda| = 1$, then $(x - \lambda)^2$ does not divide the minimal polynomial of A.
8 Let A be an irreducible periodic transition matrix in the form of (3.4.1). Let $Z = [Z_j]$ be the stationary vector of A partitioned to be compatible with A; that is, Z_j is an m_j-dimensional column vector. Prove that

$$E_{(m_j)} Z_j = \frac{1}{l}$$

9 Let A be an irreducible period transition matrix of the form (3.4.1) and let $Z_1, Z_2, \ldots, Z_l$ be the stationary vectors of the matrices $A_1, A_2, \ldots, A_l$ of Theorem 3.4.2. Set

$$Z = \frac{1}{l}[Z_j] \qquad j = 1, 2, \ldots, l$$

Show that Z is the stationary vector of A.

3.5 RECURRENCE PROBABILITIES

In Sec. 3.3 we showed that for any transition matrix A, any initial probability vector $P^{(0)}$, and any transient state S_q

$$\lim_{k \to \infty} p_q^{(k)} = 0$$

i.e., as the number of steps of a Markov chain increases, the probability that an object will be in a transient state approaches 0 and the probability that an object will be in some ergodic state approaches 1. In this section we will study several aspects of the behavior of a Markov chain as the number of steps increases. Not surprisingly we will see that this behavior is quite different for transient and ergodic states. For example, we will study f_{pq}, the probability that an object initially in state S_q will be in state S_p at least once if the process is permitted to continue indefinitely. (See Example, 3.1.18; f_{pq} is called the *first-passage probability* from S_q to S_p. In particular, when $p = q$, we call f_{pp} the *recurrence probability* of the state S_p.) In Theorem 3.5.15 we will show an object initially in an ergodic state S_p is *certain* to be in S_p at some subsequent time, that is, $f_{pp} = 1$, while if S_p is transient, the recurrence of S_p is *uncertain*, that is, $f_{pp} < 1$.

Another example of the difference between the long-range behavior of transient and ergodic states is provided by n_{pq}, the number of times an object initially in S_q can be expected to be in S_p. (See Example 3.1.15.) We will show in Theorems 3.5.27 and 3.5.29 that if it is possible to go from S_q to S_p at all, then the object can be expected to be in S_p infinitely many times (in which case we write $n_{pq} = \infty$) if and only if S_p is an ergodic state. In particular, S_p is ergodic if and only if $n_{pp} = \infty$.

For a given transition matrix A, we know that the entry in the (p, q) position of A^s is the probability of being in the state S_p exactly s steps after being in the state S_q, no consideration being made of the intervening steps. By considering these intervening steps we are led to:

Definition 3.5.1 Let $A = [a_{ij}]$ be a transition matrix and let

$$A^s = [a_{ij}{}^{(s)}] \qquad s = 0, 1, 2, \ldots$$

We say that the probability[1] $f_{pq}{}^{(s)}$ that the *first passage* of an object from state S_q to state S_p will take place at the sth step is given recursively by

$$f_{pq}{}^{(1)} = a_{pq}$$

$$f_{pq}{}^{(s)} = a_{pq}{}^{(s)} - a_{pp}{}^{(s-1)}f_{pq}{}^{(1)} - a_{pp}{}^{(s-2)}f_{pq}{}^{(2)} - \cdots - a_{pp}f_{pq}{}^{(s-1)} \qquad (3.5.2)$$

$$= a_{pq}{}^{(s)} - \sum_{h=1}^{s-1} a_{pp}{}^{(s-h)} f_{pq}{}^{(h)} \qquad s = 2, 3, \ldots$$

For each h, $1 \leq h \leq s - 1$, $a_{pp}{}^{(s-h)}f_{pq}{}^{(h)}$ is the probability that the object entered S_p from S_q for the first time at the hth step and was again in S_p $s - h$ steps later. Definition 3.5.1 states that the probability that an object initially in S_q will enter S_p for the first time at the sth step can be found as follows: subtract from $a_{pq}{}^{(s)}$ (that is, the probability that an object initially in S_q will be in S_p at the sth step for the first time or not) all the probabilities that the object entered S_p at some earlier step.

---

[1] In order to justify calling $f_{pq}{}^{(s)}$ a probability it should be shown that

$$0 \leq f_{pq}{}^{(s)} \leq 1$$

In fact, by means of equation (3.5.4) and an induction on s it can be shown that

$$0 \leq f_{pq}{}^{(s)} \leq a_{pq}{}^{(s)} \leq 1$$

An alternate proof of this result follows as an immediate consequence of Exercise 8.

We shall need the identity:

Lemma 3.5.3 For $s = 1, 2, 3, \ldots$

$$f_{pq}^{(s+1)} = \sum_{k \neq p} f_{pk}^{(s)} a_{kq} \qquad (3.5.4)$$

PROOF The proof is by an induction on s. For $s = 1$ we have

$$\sum_{k \neq p} f_{pk}^{(1)} a_{kq} = \sum_{k \neq p} a_{pk} a_{kq} = \sum_{k=1}^{n} a_{pk} a_{kq} - a_{pp} a_{pq}$$

$$= a_{pq}^{(2)} - a_{pp} f_{pq}^{(1)} = f_{pq}^{(2)}$$

as required. Now, suppose the lemma has been proved for all $h < s$. According to equation (3.5.2),

$$\sum_{k \neq p} f_{pk}^{(s)} a_{kq} = \sum_{k \neq p} \left(a_{pk}^{(s)} - \sum_{h=1}^{s-1} a_{pp}^{(s-h)} f_{pk}^{(h)} \right) a_{kq}$$

$$= \sum_{k \neq p} a_{pk}^{(s)} a_{kq} - \sum_{h=1}^{s-1} a_{pp}^{(s-h)} \left(\sum_{k \neq p} f_{pk}^{(h)} a_{kq} \right)$$

Using the induction hypothesis

$$\sum_{k \neq p} f_{pk}^{(s)} a_{kq} = \sum_{k \neq p} a_{pk}^{(s)} a_{kq} - \sum_{h=1}^{s-1} a_{pp}^{(s-h)} f_{pq}^{(h+1)}$$

and adding and subtracting $a_{pp}^{(s)} a_{pq}$, we have

$$\sum_{k \neq p} f_{pk}^{(s)} a_{kq} = \sum_{k=1}^{n} a_{pk}^{(s)} a_{kq} - \sum_{h=1}^{s} a_{pp}^{(s+1-h)} f_{pq}^{(h)}$$

$$= a_{pq}^{(s+1)} - \sum_{h=1}^{s} a_{pp}^{(s+1-h)} f_{pq}^{(h)} = f_{pq}^{(s+1)}$$

since $a_{pp}^{(s)} a_{pq} = a_{pp}^{(s)} f_{pq}^{(1)}$. ////

Clearly $f_{pk}^{(s)} a_{kq}$ is the probability that the first passage from S_q to S_p will take place at the $(s + 1)$th step by means of going from S_q to S_k in one step and then entering S_p from S_k for the first time in exactly s steps.

It will be convenient to write equation (3.5.4) as

$$f_{pq}^{(s+1)} = \sum_{k=1}^{n} f_{pk}^{(s)} a_{kq} - f_{pp}^{(s)} a_{pq} \qquad (3.5.5)$$

For each s, let us define the $n \times n$ matrix $F^{(s)}$ by

$$F^{(s)} = [f_{pq}^{(s)}] \qquad p, q = 1, 2, \ldots, n$$

The term on the left side of equation (3.5.5) is the entry in the (p, q) position of $F^{(s+1)}$. Similarly, the first term on the right side of equation (3.5.5) is the entry in the (p, q) position of $F^{(s)}A$. For any $n \times n$ matrix $X = [x_{pq}]$ let us define dg X (called *diagonal X*) to be the $n \times n$ matrix having the same diagonal entries as X [that is, dg X has x_{ii} in the (i, i) position] and 0s elsewhere. Then the second term on the right side of equation (3.5.5) is the entry in the (p, q) position of $(\text{dg } F^{(s)})A$. Thus equation (3.5.5) is equivalent to

$$F^{(s+1)} = F^{(s)}A - (\text{dg } F^{(s)})A \qquad (3.5.6)$$

When S_q is an ergodic state and S_p is either a transient state or is an ergodic state which is not in the same ergodic set as S_q, then $a_{pq}^{(s)} = 0$ for all s. Consequently $f_{pq}^{(s)} = 0$ for all s, and, for A in standard form (3.3.24) with transient states $S_1, S_2, \ldots, S_l$ and ergodic sets $\mathbb{S}_1, \mathbb{S}_2, \ldots, \mathbb{S}_r$, $F^{(s)}$ has the same form

$$F^{(s)} = \begin{bmatrix} F_0^{(s)} & 0 & 0 & \cdots & 0 \\ G_1^{(s)} & E_1^{(s)} & 0 & \cdots & 0 \\ G_2^{(s)} & 0 & E_2^{(s)} & \cdots & 0 \\ \multicolumn{5}{c}{\dotfill} \\ G_r^{(s)} & 0 & 0 & \cdots & E_r^{(s)} \end{bmatrix}$$

which we call the *standard form* of $F^{(s)}$.

Definition 3.5.7 We say that the *first-passage probability*[1] f_{pq} from the state S_q to the state S_p is given by

$$f_{pq} = \sum_{s=1}^{\infty} f_{pq}^{(s)} \qquad (3.5.8)$$

f_{pp} is called the *recurrence probability* of the state S_p and the $n \times n$ matrix

$$F = [f_{pq}] \qquad p, q = 1, 2, \ldots, n$$

is called the *recurrence matrix* or the *first-passage matrix* of the Markov chain.

Equation (3.5.8) can also be written in matrix form

$$F = \sum_{s=1}^{\infty} F^{(s)}$$

---

[1] It will follow from Corollary 3.5.24 that f_{pq} exists and $0 \leq f_{pq} \leq 1$.

When A is in standard form (3.3.24), F has the same form

$$F = \begin{bmatrix} F_0 & 0 & 0 & \cdots & 0 \\ G_1 & E_1 & 0 & \cdots & 0 \\ G_2 & 0 & E_2 & \cdots & 0 \\ \multicolumn{5}{c}{\dotfill} \\ G_r & 0 & 0 & \cdots & E_r \end{bmatrix} \qquad (3.5.9)$$

which we call the *standard form* of F. In Theorem 3.5.15 we will show that the transient and ergodic states can be characterized by their recurrence probabilities. First we need:

Lemma 3.5.10 For any p and q,

$$f_{pq} = \sum_{k \neq p} f_{pk} a_{kq} + a_{pq} \qquad (3.5.11)$$

PROOF Equation (3.5.11) is a direct consequence of equation (3.5.4) since

$$\sum_{k \neq p} f_{pk} a_{kq} + a_{pq} = \sum_{k \neq p} \left(\sum_{s=1}^{\infty} f_{pq}^{(s)} \right) a_{kq} + a_{pq}$$

$$= \sum_{s=1}^{\infty} \left(\sum_{k \neq p} f_{pk}^{(s)} a_{kq} \right) + a_{pq}$$

$$= \sum_{s=1}^{\infty} f_{pq}^{(s+1)} + f_{pq}^{(1)} = f_{pq} \qquad ////$$

$f_{pk} a_{kq}$ is the probability of going from S_q to S_k in one step and then entering S_p (at least once) from S_k. Equation (3.5.11) can be written as

$$f_{pq} = \sum_{k=1}^{n} f_{pk} a_{kq} + a_{pq} - f_{pp} a_{pq}$$

which is equivalent to the matrix equation

$$F = FA + A - (\mathrm{dg}\, F)A \qquad (3.5.12)$$

Equation (3.5.12) is a special case of a more general relation connecting A and F.

Theorem 3.5.13 For each t, $t = 1, 2, \ldots$,

$$F = FA^t + (A + A^2 + \cdots + A^t) - (\mathrm{dg}\, F)(A + A^2 + \cdots + A^t) \qquad (3.5.14)$$

PROOF Equation (3.5.12) is the case of equation (3.5.14) in which $t = 1$. Equation (3.5.14) can be established by an induction on t and is left as an exercise. ////

As an immediate consequence of Theorem 3.5.13 we have:

Theorem 3.5.15 If S_p is an ergodic state, then $f_{pp} = 1$. In fact, if S_q is any state in the same ergodic set as S_p, then $f_{pq} = f_{qp} = 1$. (For any other state, S_q, $f_{qp} = 0$.)

If S_p is a transient state, then $f_{pp} < 1$.

PROOF First, let S_p be an ergodic state. According to Theorem 3.3.26, there exists a stationary vector

$$X = [x_j]$$

of A for which $x_p \neq 0$. Multiplying equation (3.5.12) on the right by X, we have

$$FX = FAX + AX - (\text{dg } F)AX$$
$$= FX + X - (\text{dg } F)X$$

and consequently

$$0 = x_p - f_{pp} x_p$$

Since $x_p \neq 0$, it follows that

$$f_{pp} = 1$$

Now suppose that S_q is a state in the same ergodic set as S_p. Since it is possible to reach S_q from S_p, there exists a positive integer t such that $a_{qp}^{(t)} \neq 0$. Examining the (p, p) position of the matrices of equation (3.5.14) yields

$$1 = f_{pp} = \sum_{k=1}^{n} f_{pk} a_{kp}^{(t)} + \sum_{m=1}^{t} a_{pp}^{(m)} - f_{pp}\left(\sum_{m=1}^{t} a_{pp}^{(m)}\right) \quad (3.5.16)$$
$$= \sum_{k=1}^{n} f_{pk} a_{kp}^{(t)} \leq \sum_{k=1}^{n} a_{kp}^{(t)} = 1$$

Since the first and last terms of (3.5.16) are equal, there is equality throughout and

$$f_{pk} a_{kp}^{(t)} = a_{kp}^{(t)} \qquad k = 1, 2, \ldots, n$$

But $a_{qp}^{(t)} \neq 0$, and consequently

$$f_{pq} = 1$$

Finally, let S_p be a transient state. There is an (ergodic) state S_q and a positive integer t such that

$$a_{qp}^{(t)} \neq 0 \qquad f_{pq} = 0 \quad (3.5.17)$$

As before, the (p, p) position of the matrices of equation (3.5.14) yields

$$f_{pp} = \sum_{k=1}^{n} f_{pk} a_{kp}^{(t)} + \sum_{m=1}^{t} a_{pp}^{(m)} - f_{pp} \left(\sum_{m=1}^{t} a_{pp}^{(m)} \right)$$

which can be written in the form

$$\left(1 + \sum_{m=1}^{t} a_{pp}^{(m)}\right)(f_{pp} - 1) = \sum_{k=1}^{n} f_{pk} a_{kp}^{(t)} - 1 \quad (3.5.18)$$

It follows from (3.5.17) that

$$\sum_{k=1}^{n} f_{pk} a_{kp}^{(t)} - 1 = \sum_{k \neq q} f_{pk} a_{kp}^{(t)} - 1 \leq \sum_{k \neq q} a_{kp}^{(t)} - 1 = -a_{qp}^{(t)} < 0 \quad (3.5.19)$$

Comparing (3.5.18) and (3.5.19), we have

$$f_{pp} < 1 \qquad ////$$

It is now clear that the choice of notation, $E_1, E_2, \ldots, E_r$ for the diagonal blocks of F in (3.5.9) is consistent with our usual notation since every entry of each of these matrices is 1. Thus far nothing has been said about f_{pq} when S_q is transient (other than the fact that recurrence to S_q is uncertain). It is not difficult to construct examples of Markov chains with passage from a transient state S_q to a state S_p (transient or ergodic) which is certain or uncertain. We can obtain a precise expression for f_{pq} by substituting the standard forms of A and F into equation (3.5.14). Then

$$F_0 = F_0 C^t + (C + C^2 + \cdots + C^t) - (\mathrm{dg}\, F_0)(C + C^2 + \cdots + C^t)$$

Moreover, for $i = 1, 2, \ldots, r$ equation (3.5.12) yields

$$G_i = G_i C + E_i B_i + B_i - (\mathrm{dg}\, E_i) B_i = G_i C + E_i B_i$$

According to Theorem 3.3.26, $\lim_{t \to 0} C^t = 0$ and the infinite series

$$C + C^2 + \cdots$$

converges to $(I - C)^{-1} - I$. Thus

$$F_0 = [I - (\mathrm{dg}\, F_0)][(I - C)^{-1} - I]$$
$$G_i = E_i B_i (I - C)^{-1}$$

For reasons that will become clear shortly we define the $l \times l$ matrix N_0 by

$$N_0 = [n_{ij}] = (I - C)^{-1} = I + C + C^2 + \cdots$$

Corollary 3.5.20 (a) $F_0 = [I - (\mathrm{dg}\, F_0)](N_0 - I)$,

(b) $G_i = E_i B_i N_0, i = 1, 2, \ldots, r$,

from which it follows that:

1. When S_p and S_q are transient,

$$f_{pq} = \frac{n_{pq} - \delta_{pq}}{n_{pp}} \quad \text{where} \quad \delta_{pq} = \begin{cases} 1 & \text{when } p = q \\ 0 & \text{when } p \neq q \end{cases}$$

2. When S_q is transient and S_p and S_s belong to the same ergodic set, then

$$f_{pq} = f_{sq}$$

PROOF The proof is left as an exercise. ////

The second conclusion of the corollary is intuitively clear. For, once an object has entered one state of an ergodic set, it is certain to enter all other states of that set. Thus we can speak of the probability $f_{\mathsf{S}_i, q}$ that an object initially in the transient state S_q will ever enter (and thereafter remain in) the ergodic set S_i. $f_{\mathsf{S}_i, q}$ is the qth entry of the l-dimensional row vector

$$E(n_i) B_i N_0$$

At the end of Example 3.1.18 the special case of equation (3.5.11)

$$f_{11} = \sum_{k \neq 1} f_{1k} a_{k1} + a_{11}$$

was used to evaluate f_{11}. However, a different method was used to obtain f_{1k} when $k \neq 1$, namely, changing S_1 into an absorbing state. We now wish to show that this method and definition 3.5.7 agree. Let A be a transition matrix, S_p a fixed state, and $\tilde{A}$ the matrix formed from A by replacing the pth column of A by $I^{(p)}$, that is, making S_p an absorbing state. Set

$$\tilde{A}^k = [\tilde{a}_{ij}^{(k)}] \qquad i, j = 1, 2, \ldots, n$$

A direct computation shows that

$$\tilde{A}^{s+1} = \tilde{A}^s \tilde{A} = \begin{bmatrix} \tilde{a}_{11}^{(s)} & \tilde{a}_{12}^{(s)} & \cdots & \tilde{a}_{1n}^{(s)} \\ \vdots & & & \vdots \\ \vdots & & & \vdots \\ \tilde{a}_{n1}^{(s)} & \tilde{a}_{n2}^{(s)} & \cdots & \tilde{a}_{nn}^{(s)} \end{bmatrix} \begin{bmatrix} a_{11} & a_{12} & \cdots & 0 & \cdots & a_{1n} \\ \vdots & & & & & \vdots \\ a_{p1} & a_{p2} & \cdots & 1 & \cdots & a_{pn} \\ \vdots & & & & & \vdots \\ a_{n1} & a_{n2} & \cdots & 0 & \cdots & a_{nn} \end{bmatrix}$$

and hence

$$\tilde{a}_{ij}^{(s+1)} = \sum_{k=1}^{n} \tilde{a}_{ik}^{(s)} a_{kj} \qquad j \neq p$$

$$\tilde{a}_{ip}^{(s+1)} = \tilde{a}_{ip}^{(s)} = \begin{cases} 1 & \text{when } i = p \\ 0 & \text{when } i \neq p \end{cases} \qquad (3.5.21)$$

Lemma 3.5.22 Let A be a transition matrix and S_p a fixed state and let $\tilde{A}$ be the matrix defined above. For $q \neq p$ and $s = 1, 2, \ldots$

$$\tilde{a}_{pq}^{(s)} - \tilde{a}_{pq}^{(s-1)} = f_{pq}^{(s)} \qquad (3.5.23)$$

PROOF   The proof is by an induction on s and is immediate for $s = 1$ since

$$\tilde{a}_{pq}^{(1)} - \tilde{a}_{pq}^{(0)} = a_{pq} - 0 = f_{pq}^{(1)}$$

Now suppose equation (3.5.23) holds for some value of s. Substituting (3.5.21) into the left-hand side of equation (3.5.23) and applying Lemma 3.5.3 yields

$$\begin{aligned}
\tilde{a}_{pq}^{(s+1)} - \tilde{a}_{pq}^{(s)} &= \sum_{k=1}^{n} \tilde{a}_{pk}^{(s)} a_{kq} - \sum_{k=1}^{n} \tilde{a}_{pk}^{(s-1)} a_{kq} \\
&= \left[\sum_{k=1}^{n} (\tilde{a}_{pk}^{(s)} - \tilde{a}_{pk}^{(s-1)}) \right] a_{kq} \\
&= \left(\sum_{k \neq p} f_{pk}^{(s)} + \tilde{a}_{pp}^{(s)} - \tilde{a}_{pp}^{(s-1)} \right) a_{kq} \\
&= \sum_{k \neq p} f_{pk}^{(s)} a_{kq} = f_{pq}^{(s+1)} \qquad ////
\end{aligned}$$

We can now justify the procedures used in Example 3.1.18 to find the first-passage probabilities. As an immediate consequence of Lemma 3.5.22 we have:

Corollary 3.5.24   For $q \neq p$

$$f_{pq} = \lim_{m \to \infty} \tilde{a}_{pq}^{(m)}$$

PROOF   It follows from Lemma 3.5.22 that

$$\sum_{s=1}^{m} f_{pq}^{(s)} = \tilde{a}_{pq}^{(m)} - \tilde{a}_{pq}^{(0)} = \tilde{a}_{pq}^{(m)} \qquad m = 1, 2, \ldots \qquad ////$$

In Example 3.1.15 we had a Markov chain in which a mouse was given a piece of cheese every time he was in a certain room of the maze. We were able to calculate the number of pieces of cheese the mouse could be expected to get (i.e., the number of times he could be expected to be in a particular state) if the process were permitted to run indefinitely. We will show that an object will be in a transient state only a finite number of times no matter what its initial state is. On the other hand, we will show that if an object can reach an ergodic state at all, it will be found there infinitely often.

Definition 3.5.25 We say that an object initially in state S_q can be expected to be in state S_p

$$n_{pq} = \sum_{s=0}^{\infty} a_{pq}^{(s)} \quad (3.5.26)$$

times. When the sum on the right-hand side of equation (3.5.26) is unbounded, we say that an object initially in S_q can be expected to be in S_p *infinitely often* and we write

$$n_{pq} = \infty$$

Equation (3.5.25) defined n_{pq} to be the entry in the (p, q) position of the infinite series of matrices

$$I + A + A^2 + \cdots$$

[and $n_{pq} = \infty$ when the sum of the entries in the (p, q) positions of the terms of this series is unbounded]. Clearly, if S_p is an absorbing state, then

$$a_{pp}^{(s)} = 1 \quad s = 1, 2, \ldots$$

and consequently $n_{pp} = \infty$. On the other hand, when S_q is an ergodic state and S_p is either a transient state or an ergodic state not belonging to the same ergodic set as S_q, then

$$a_{pq}^{(s)} = 0 \quad s = 1, 2, \ldots$$

and consequently $n_{pq} = 0$. When S_p and S_q are both transient states (and A is in standard form), then equation (3.5.26) defines n_{pq} to be the entry in the (p, q) position of

$$I + C + C^2 + \cdots$$

that is, n_{pq} is the entry in the (p, q) position of the matrix N_0 which we defined earlier. We now have:

Theorem 3.5.27 If S_p is a transient state, then n_{pq} is finite for every state S_q.

It remains to find n_{pq} when S_q is an ergodic state. For this we need:

Lemma 3.5.28 For any states S_p and S_q we have

$$n_{pq} - \delta_{pq} = f_{pq} n_{pp}$$

where

$$\delta_{pq} = \begin{cases} 1 & \text{when } p = q \\ 0 & \text{when } p \neq q \end{cases}$$

(If $n_{pp} = \infty$, we interpret $f_{pq} n_{pp}$ to be ∞ if $f_{pq} \neq 0$ and 0 if $f_{pq} = 0$.)

PROOF (Note that the special case in which S_p and S_q are both transient has already been proved in Corollary 3.5.20.) Let us write equation (3.5.2) as

$$a_{pq}^{(s)} = f_{pq}^{(s)} + a_{pp}^{(1)} f_{pq}^{(s-1)} + a_{pp}^{(2)} f_{pq}^{(s-2)} + \cdots + a_{pp}^{(s-1)} f_{pq}^{(1)}$$

$$s = 1, 2, \ldots$$

Summing these equations vertically yields

$$n_{pq} - \delta_{pq} = \sum_{s=1}^{\infty} a_{pq}^{(s)} = f_{pq} + a_{pp}^{(1)} \sum_{s=1}^{\infty} f_{pq}^{(s)} + a_{pp}^{(2)} \sum_{s=1}^{\infty} f_{pq}^{(s)} + \cdots$$

$$= f_{pq}(1 + a_{pp}^{(1)} + a_{pp}^{(2)} + \cdots)$$

$$= f_{pq} n_{pp} \qquad \qquad ////$$

Theorem 3.5.29 Let S_p be an ergodic state. If S_q is a state in the same ergodic set as S_p, then

$$n_{pq} = \infty$$

In particular,

$$n_{pp} = \infty$$

PROOF When $p = q$, Lemma 3.5.28 states that

$$n_{pp} - 1 = f_{pp} n_{pp}$$

Since S_p is ergodic, $f_{pp} = 1$ and hence

$$n_{pp} - 1 = n_{pp}$$

Clearly this equation has no finite solutions, and we conclude that

$$n_{pp} = \infty$$

If S_q is a state in the same ergodic set as S_p, then $f_{pq} = 1$ and it follows from Lemma 3.5.28 that $n_{pq} = \infty$. ////

In Theorem 3.5.15 we showed that if S_p and S_q are states belonging to the same ergodic set, then it is certain that an object initially in S_q will enter S_p at some time. However, we did not consider the number of steps that can be expected to elapse before the object enters S_p. (See Example 3.1.21.)

Definition 3.5.30 For states S_p and S_q belonging to the same ergodic set we define the *mean-first-passage time* from S_q to S_p by

$$m_{pq} = \sum_{s=1}^{\infty} sf_{pq}^{(s)}$$

For an irreducible Markov chain, i.e., when all states belong to the same ergodic set, then,

$$M = [m_{pq}] = \sum_{s=1}^{\infty} sF^{(s)}$$

is called the *mean-first-passage matrix*.

The mean-first-passage time from one state to another is not defined unless the passage is certain. It will follow from Exercises 11 to 14 that $m_{pq} < \infty$.

Lemma 3.5.31 Let A be an irreducible transition matrix. Then

$$M - E = MA - (\text{dg } M)A \quad (3.5.32)$$

PROOF Multiplying equation (3.5.6) by s yields

$$sF^{(s+1)} = sF^{(s)}A - [\text{dg}(sF^{(s)})]A$$

which we can write as

$$(s+1)F^{(s+1)} - F^{(s+1)} = sF^{(s)}A - [\text{dg}(sF^{(s)})]A$$

Summing on s, we have

$$\sum_{s=1}^{\infty}(s+1)F^{(s+1)} - \sum_{s=1}^{\infty} F^{(s+1)} = \left(\sum_{s=1}^{\infty} sF^{(s)}\right)A - \left(\text{dg}\sum_{s=1}^{\infty} sF^{(s)}\right)A$$

that is, $\quad M - F = MA - (\text{dg } M)A \quad (3.5.33)$

Since all states belong to the same ergodic set,

$$F = E$$

proving the lemma. ////

From Lemma 3.5.31 we can obtain a simple expression for the diagonal elements of M, which will in turn be used to find the nondiagonal elements of M.

Theorem 3.5.34 Let $X = [x_j]$ be the (unique, positive) stationary vector of an irreducible Markov chain with mean-first-passage matrix M. Then

$$m_{jj} = \frac{1}{x_j} \quad j = 1, 2, \ldots, n$$

PROOF   Multiplying equation (3.5.32) on the right by X,

$$MX - EX = MAX - (\text{dg } M)AX$$
$$MX - E^{(n)} = MX - (\text{dg } M)X$$
$$E^{(n)} = (\text{dg } M)X$$

Thus $\quad\quad\quad\quad 1 = m_{jj}x_j \quad\quad j = 1, 2, \ldots, n \quad\quad\quad ////$

If L is the $n \times n$ matrix each of whose columns is the stationary vector of A, then Theorem 3.5.33 states that

$$\text{dg } M = (\text{dg } L)^{-1} \quad\quad (3.5.35)$$

In order to find the nondiagonal elements of M we will need a lemma.

Lemma 3.5.36[1]   If A is an irreducible matrix, then $I - A + L$ is nonsingular and

$$Z = (I - A + L)^{-1}$$

satisfies

$$Z(I - A) = I - L$$

PROOF   In Sec. 3.3 we proved that there exists a nonsingular matrix P such that

$$A = P \begin{bmatrix} 1 & 0 \\ 0 & \tilde{A} \end{bmatrix} P^{-1} \quad\quad L = P \begin{bmatrix} 1 & 0 \\ 0 & 0 \end{bmatrix} P^{-1}$$

where $\tilde{A}$ is an $(n-1) \times (n-1)$ matrix which does not have 1 as a characteristic root.   Then

$$I - A + L = P \begin{bmatrix} 1 & 0 \\ 0 & I - \tilde{A} \end{bmatrix} P^{-1}$$

which is nonsingular.   A direct computation shows that

$$(I - A + L)(I - L) = I - A + L - L + AL - L^2$$
$$= I - A$$

and the lemma follows. $\quad\quad\quad\quad\quad\quad\quad\quad\quad\quad\quad\quad\quad\quad\quad ////$

---

[1] An alternate proof of Lemma 3.5.36 for the case in which A is regular is described in Exercise 9 of Sec. 3.3.

In preparation for the next theorem let us write equation (3.5.32) as

$$M(I - A) = E - (\text{dg } M)A \qquad (3.5.37)$$

Theorem 3.5.38

$$M = (\text{dg } L)^{-1}[I - Z + (\text{dg } Z)E] \qquad (3.5.39)$$

PROOF The proof consists of two parts. First we verify that the matrix M defined by equation (3.5.39) satisfies equations (3.5.35) and (3.5.37). Using Lemma 3.5.36, we have

$$\begin{aligned}
M(I - A) &= (\text{dg } L)^{-1}[I - Z + (\text{dg } Z)E](I - A) \\
&= (\text{dg } L)^{-1}[(I - A) - Z(I - A) + (\text{dg } Z)E(I - A)] \\
&= (\text{dg } L)^{-1}[(I - A) - (I - L)] = (\text{dg } L)^{-1}(L - A) \\
&= E - (\text{dg } M)A
\end{aligned}$$

Also
$$\begin{aligned}
\text{dg } M &= (\text{dg } L)^{-1}[\text{dg } I - \text{dg } Z + (\text{dg } Z)(\text{dg } E)] \\
&= (\text{dg } L)^{-1}[I - \text{dg } Z + (\text{dg } Z)I] = (\text{dg } L)^{-1}
\end{aligned}$$

Second we show that equations (3.5.35) and (3.5.37) have only one solution. Suppose M and $\tilde{M}$ are both solutions of these equations. Then

$$M(I - A) = \tilde{M}(I - A)$$
$$\text{dg } M = \text{dg } \tilde{M} = (\text{dg } L)^{-1}$$

and consequently
$$(M - \tilde{M})(I - A) = 0$$

By Corollary 3.3.13 each row of $M - \tilde{M}$ is a scalar multiple of $E_{(n)}$. However,

$$\text{dg}(M - \tilde{M}) = \text{dg } M - \text{dg } \tilde{M} = 0$$

and consequently each row of $M - \tilde{M}$ is the zero vector, i.e.,

$$M = \tilde{M} \qquad ////$$

It is not difficult to show that an object initially in a transient state of a Markov chain is certain to enter some ergodic state. (See Exercise 7.) Let A be a transition matrix in standard form

$$A = \begin{bmatrix} C & 0 & 0 & \cdots & 0 \\ B_1 & D_1 & 0 & \cdots & 0 \\ B_2 & 0 & D_2 & \cdots & 0 \\ \multicolumn{5}{c}{\dotfill} \\ B_r & 0 & 0 & \cdots & D_r \end{bmatrix}$$

where C is an $l \times l$ matrix. If we lump together all the ergodic states into a single (absorbing) state (which we will call S'_{l+1}) without disturbing the transient states, the resulting Markov chain has an $(l+1) \times (l+1)$ transition matrix

$$A' = \begin{bmatrix} C & 0 \\ B & 1 \end{bmatrix}$$

where B is a $1 \times l$ matrix. B is uniquely determined by the requirement that the sum of the entries of each column of A' be 1. An object initially in a transient state S_p is certain to enter (and remain in) S'_{l+1}. We define the *mean time before absorption* of an object initially in S_p by

$$m_p = \sum_{s=1}^{\infty} s f_{l+1,p}^{(s)'}$$

where $f_{l+1,p}^{(s)'}$ refers to the Markov chain with transition matrix A'. The $1 \times l$ matrix

$$M'_{21} = [m_1 \quad m_2 \quad \cdots \quad m_l] \qquad (3.5.40)$$

is called the *mean-time-before-absorption matrix*. We can express M'_{21} in terms of C.

Theorem 3.5.41

$$M'_{21} = E_{(l)}(I - C)^{-1} = E_{(l)} N_0$$

PROOF   Let M' be the mean-first-passage matrix of A' and partition M' as

$$M' = \begin{bmatrix} M'_{11} & M'_{12} \\ M'_{21} & m'_{22} \end{bmatrix}$$

where M'_{11} is an $l \times l$ matrix, M'_{21} is a $1 \times l$ matrix, etc. According to Definition 3.5.30, the matrix M'_{21} defined by equation (3.5.40) is identical to the submatrix M'_{21} of M'. In addition it is clear that

$$M'_{12} = 0 \qquad m'_{22} = 1$$

Applying equation (3.5.33) and recalling that an object in a transient state is certain to be absorbed in S'_{l+1}, we have

$$M'_{21} - E_{(l)} = (M'_{21} C + m'_{22} B) - m'_{22} B$$

Thus

$$M'_{21}(I - C) = E_{(l)}$$

and the theorem follows. ////

EXERCISES

1. A man lives at one end of a street 40 yards long. There is a bar at the other end of the street, and streetlight poles are placed at either end of the street and at 10-yard intervals in between. When the man leaves the bar, he heads for home but is only able to get to the next pole before stopping. At each pole except the one in front of his house and the one in front of the bar he is equally likely to start off in either direction, and he keeps going until he reaches the next pole, whereupon the same thing happens. However, when he gets to the pole in front of his house, he goes in and goes to bed. When he finds himself at the pole in front of the bar, he goes in for a drink and starts home. If the man is initially in the bar and has had one drink, how many drinks can he be expected to have before getting home? Is he certain of getting home eventually?

2. A gambler plays the following game. A coin is flipped repeatedly, and a tally is taken of the number of heads and tails which have appeared. The game is over when there are either 3 more heads than tails or 3 more tails than heads. The gambler wins an amount equal to the number of times the coin is flipped. How much can he be expected to win?

3. A cat has fallen down a 3-foot-deep well in which there is a ladder with rungs 1 foot apart. The bottom rung is on the floor of the well, so that if the cat can reach the fourth rung of the ladder, he is out of the well. The cat tries to get out of the well by jumping up the ladder from rung to rung. However, the probability that he can successfully make each jump is $\frac{1}{2}$. If he misses any jump, he falls back to the bottom of the well and has to begin again. Is the cat certain to get out of the well? If so, how many jumps can he be expected to make, and how many times can he be expected to find himself at the bottom of the well before he manages to get out?

4. Let

$$A = \begin{bmatrix} 0 & \frac{1}{2} \\ 1 & \frac{1}{2} \end{bmatrix}$$

Express $F^{(s)}$ as a function of s. Find F by means of Definition 3.5.7 and verify Theorem 3.5.15. Find M by using Definition 3.5.30 and verify equation (3.5.32).

5. Prove Theorem 3.5.13.

6. Prove Corollary 3.5.20.

7. Prove that

$$\sum_{i=1}^{r} f_{\mathbb{S}_i, q} = 1 \qquad (3.5.42)$$

for every transient state S_q. Equation (3.5.42) states that an object initially in a transient state is certain to enter some ergodic set.

8 Prove the following relations:

(a) $f_{ij}^{(s)} = \sum_{k_1, k_2, \ldots, k_{s-1} \neq i} a_{ik_1} a_{k_1 k_2} \cdots a_{k_{s-1} j}.$

(b) $F_0 = (\text{dg } N_0)^{-1}(N_0 - I).$

9 For an irreducible transition matrix A

(a) prove that
$M = MA^k + kE - (\text{dg } M)(A + A^2 + A^3 + \cdots + A^k)$ for $k = 1, 2, \ldots$

(b) Use this result to prove that

$$\lim_{k \to \infty} \frac{A + A^2 + A^3 + \cdots + A^k}{k} = L$$

where L is the matrix each of whose columns is the stationary vector of A.

10 Let A be an irreducible transition matrix and let A_i denote the submatrix of A obtained by deleting the ith row and ith column of A. Let $M_{(i)}$ be the ith row of M and let $\tilde{M}_{(i)}$ be obtained from $M_{(i)}$ by deleting the entry m_{ii}. Show that

$$\tilde{M}_{(i)} = E_{(n-1)}(I - A_i)^{-1}$$

(*Hint:* See Exercise 20 of Sec. 3.3.)

11 For states S_p and S_q of an irreducible Markov chain define

$$m_{pq}^{(k)} = \sum_{s=1}^{k} s f_{pq}^{(s)} \qquad k = 1, 2, 3, \ldots$$

and set

$$M^{(k)} = [m_{pq}^{(k)}]$$

Show that

$$M^{(k+1)} = M^{(k)}A + \sum_{s=1}^{k+1} F^{(s)} - (\text{dg } M^{(k)})A \qquad k = 1, 2, 3, \ldots \qquad (3.5.43)$$

12 Use equation (3.5.43) to show that

$$(\text{dg } M^{(k)})X \leq FX = E$$

and consequently $m_{pp} \leq \dfrac{1}{x_p} \qquad p = 1, 2, \ldots, n$

13 Show that

$$M^{(k+l)} = M^{(k)}A^l + \sum_{s=1}^{k+l} F^{(s)} + \sum_{s=1}^{k+l-1} F^{(s)}A + \cdots + \sum_{s=1}^{k+1} F^{(s)}A^{l-1}$$
$$- \sum_{r=0}^{l-1} (\text{dg } M^{(k+r)}) A^{l-r}$$

for all k and l and consequently

$$(\operatorname{dg} M)(A + A^2 + A^3 + \cdots + A^l) + M^{(k+l)} \geqq M^{(k)} A^l$$

14 For fixed p, q there exists l such that $a_{qp}^{(l)} \neq 0$. Show that

$$m_{pq} \leqq \frac{1}{x_p a_{qp}^{(l)}} (1 + a_{pp} + a_{pp}^{(2)} + \cdots + a_{pp}^{(l)})$$

REFERENCES

There are two quite different ways to approach the subject of finite Markov chains. We have chosen to emphasize the algebraic aspects of the subject in order to make use of the matrix theory developed in Chap. 1. Romanovsky [9] also restricts himself to matrix methods and studies numerous topics related to finite Markov chains in addition to those covered here. He discusses in detail the original papers of Markov and several other Russian mathematicians. The second approach to finite Markov chains treats them as a special type of stochastic process and relies on probability theory rather than matrices. This treatment is used by Chung [1] and Parzen [8]. Karlin [6] and Feller [5] present both approaches to the subject and discuss the relation between them.

It is not a great jump from the Markov chains studied here to those which have a countable number of states [7]. The finite sums of this chapter are then replaced by infinite sums, and many of the results of this chapter can be carried over. Chung [1] and Karlin [6] consider finite Markov chains as a special case of Markov chains with countably many states. In the more general Markov process the state space is continuous [3].

In the Markov chains we have considered, the set of times at which the object may move is discrete. Cox and Miller [3] and Doob [4] treat continuous-parameter Markov chains in which an object may move at any time. Parzen [8] and Feller [5] develop the birth-and-death process as a continuous-parameter Markov chain, and Saaty [10] applies these ideas to the behavior of a queue. Cox [2] studies semi-Markov chains, which consist of a Markov chain together with a probability function describing the amount of time it takes an object to go from one state to another in one step.

1 Chung, K. L.: "Markov Chains with Stationary Transition Probabilities," 2d ed., Springer-Verlag New York Inc., New York, 1967.
2 Cox, D. R. "Renewal Theory," John Wiley & Sons, Inc., New York, 1962.
3 Cox, D. R., and H. D. Miller: "The Theory of Stochastic Processes," John Wiley & Sons, Inc., New York, 1965.
4 Doob, J. L.: "Stochastic Processes," John Wiley & Sons, Inc., New York, 1953.

5 Feller, W.: "An Introduction to Probability Theory and Its Applications," 3d ed., vol. 1, John Wiley & Sons, Inc., New York, 1968.
6 Karlin, S. "A First Course in Stochastic Processes," Academic Press Inc., New York, 1966.
7 Kemeny, J. G., J. L. Snell, and A. W. Knapp: "Denumerable Markov Chains," D. Van Nostrand Company Inc., Princeton, N.J., 1966.
8 Parzen, E.: "Stochastic Processes," Holden-Day, Inc., San Francisco, 1962.
9 Romanovsky, V. I.: "Discrete Markov Chains," trans. from the Russian by E. Seneta, Wolters-Noordhoff Publishing, Groningen, 1960.
10 Saaty, T. L.: "Elements of Queueing Theory," McGraw-Hill Book Company, New York, 1967.

4
THE THEORY OF GRAPHS

4.1 SOME EXAMPLES

In this chapter we introduce the concept of a graph and prove most of the basic theorems of elementary graph theory. We will investigate undirected graphs at some length and then show that an analogous theory can be developed for directed graphs. We will see that there are quite a number of matrices that can be associated with a graph and although the language of graph theory is geometric and we will often find it convenient to employ a diagram, we will strive to obtain as many of the results as we can by matrix methods. In many cases all the entries of these matrices are 0 or 1. For such matrices we may consider 0 and 1 to be real numbers or we may consider them to be elements of the field $\mathbb{Z}_2$ consisting only of the two elements 0 and 1. We will find that unless we are counting things, the latter point of view is usually more fruitful.

This section is devoted to some definitions and a number of examples illustrating various types of problems which can be analyzed and solved using graphs. We will meet two basic types of problems. In one we are primarily concerned with the structure of a graph and, perhaps, of its subgraphs. The other type of problem is a counting problem and asks how many ways some process can be carried out. Unfortunately, many of the most interesting and important applications of graphs require a great deal of "outside" information, e.g., electric networks, or molecular structure of compounds, and these are beyond the scope of this book.

EXAMPLE 4.1.1 The communication network connecting the points (or vertices) $v_1, v_2, \ldots, v_7$ is represented by the *directed graph* of Fig. 4.1.2. Is it

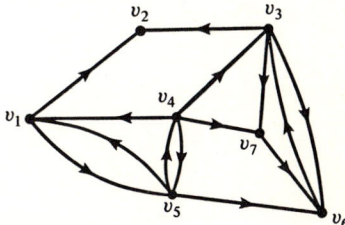

FIGURE 4.1.2

possible to reach v_1 from v_7? If so, which path passes through the smallest number of points; or, equivalently, if each link of the network (or *arc* of the graph) is assumed to have length 1, what is the shortest path from v_1 to v_7? Is it possible to go from v_7 to v_1?

Although these questions can be answered by searching through the graph and testing all possible paths, we are more concerned with finding a systematic way of arriving at an answer. To this end we define the vertex-adjacency matrix of the graph. For each pair of vertices v_i and v_j, let a_{ij} be the number (possibly zero) of arcs from v_i to v_j. Then $A = [a_{ij}]$ is a square matrix called the *vertex-adjacency matrix*[1] of the graph. The order of A is equal to the number of vertices of the graph. Since we will insist that the two vertices at the ends of an arc are distinct, the diagonal elements of the vertex-adjacency matrix

---

[1] If we assign a probability to each arc of the graph in such a way that for each vertex the sum of the assigned probabilities of the arcs directed from that vertex is 1, then we have a Markov chain whose transition matrix has 0s in the same positions as A and which has these probabilities in the other positions. However, in this section we will be primarily interested in counting the number of possible arrangements and not in the probability that each will actually occur. Nevertheless, a number of important concepts from Chap. 3 will reappear in this chapter, though in slightly disguised form.

are 0. For the graph of Fig. 4.1.2, the vertex-adjacency matrix is the 7×7 matrix

$$A = \begin{bmatrix} 0 & 1 & 0 & 0 & 1 & 0 & 0 \\ 0 & 0 & 0 & 0 & 0 & 0 & 0 \\ 0 & 1 & 0 & 0 & 0 & 1 & 1 \\ 1 & 0 & 1 & 0 & 1 & 0 & 1 \\ 1 & 0 & 0 & 1 & 0 & 1 & 0 \\ 0 & 0 & 1 & 0 & 0 & 0 & 0 \\ 0 & 0 & 0 & 0 & 0 & 1 & 0 \end{bmatrix}$$

The matrix A completely describes the network of links connecting the seven points of Fig. 4.1.2.

Let

$$A^k = [a_{ij}^{(k)}] \qquad k = 0, 1, 2, \ldots$$

Then

$$a_{ij}^{(2)} = \sum_{l=1}^{7} a_{il} a_{lj}$$

is equal to the number of paths of length 2 from v_i to v_j, because

$$a_{il} a_{lj} = 1$$

if there are links from v_i to v_l and from v_l to v_j and

$$a_{il} a_{lj} = 0$$

otherwise. Similarly, $a_{ij}^{(k)}$ is the number of paths of length k from v_i to v_j. A computation shows that

$$A^2 = \begin{bmatrix} 1 & 0 & 0 & 1 & 0 & 1 & 0 \\ 0 & 0 & 0 & 0 & 0 & 0 & 0 \\ 0 & 0 & 1 & 0 & 0 & 1 & 0 \\ 1 & 2 & 0 & 1 & 1 & 3 & 1 \\ 1 & 1 & 2 & 0 & 2 & 0 & 1 \\ 0 & 1 & 0 & 0 & 0 & 1 & 1 \\ 0 & 0 & 1 & 0 & 0 & 0 & 0 \end{bmatrix} \quad \text{and} \quad A^3 = \begin{bmatrix} 1 & 1 & 2 & 0 & 2 & 0 & 1 \\ 0 & 0 & 0 & 0 & 0 & 0 & 0 \\ 0 & 1 & 1 & 0 & 0 & 1 & 1 \\ 2 & 1 & 4 & 1 & 2 & 2 & 1 \\ 2 & 3 & 0 & 2 & 1 & 5 & 2 \\ 0 & 0 & 1 & 0 & 0 & 1 & 0 \\ 0 & 1 & 0 & 0 & 0 & 1 & 1 \end{bmatrix}$$

There are no paths of length 1 or 2 from v_1 to v_7 since

$$a_{17}^{(1)} = a_{17}^{(2)} = 0$$

However, there is exactly one path of length 3 from v_1 to v_7 since

$$a_{17}^{(3)} = 1$$

By examining the graph of Fig. 4.1.2 or, preferably, by determining how the 1 in the (1, 7) position of A^3 was formed, we see that

$$v_1 \to v_5 \to v_4 \to v_7$$

is the unique path of length 3 from v_1 to v_7.

Now let us determine whether there is a path from v_7 to v_1. From v_7 it is possible to reach only v_6 in one step. According to the last row of A^2, only v_3 can be reached from v_7 along a path of length 2. The last row of A^3 tells us that v_2, v_6, and v_7 can be reached from v_7 in exactly three steps. Thus, in three or fewer steps we can reach v_2, v_3, v_6, and v_7 from v_7. By examining the last row of A^4 we see that the only points which can be reached from v_7 in four steps are v_3 and v_6. Since each of these points can be reached in fewer than four steps, no further points can be reached by taking further steps. Thus it is not possible to reach v_1 from v_7.

The same argument can be demonstrated more forcefully using the vertex-adjacency matrix. First we must permute the rows and columns of A. Although this would follow naturally from a renumbering of the vertices of the graph, such a step is not really necessary. If we impose the order v_1, v_4, v_5, v_2, v_3, v_6, v_7 on the vertices (rather than the natural order defined by the subscripts), the vertex-adjacency matrix becomes

$$A = \begin{bmatrix} 0 & 0 & 1 & | & 1 & 0 & 0 & 0 \\ 1 & 0 & 1 & | & 0 & 1 & 0 & 1 \\ 1 & 1 & 0 & | & 0 & 0 & 1 & 0 \\ \hline 0 & 0 & 0 & | & 0 & 0 & 0 & 0 \\ 0 & 0 & 0 & | & 1 & 0 & 1 & 1 \\ 0 & 0 & 0 & | & 0 & 1 & 0 & 0 \\ 0 & 0 & 0 & | & 0 & 0 & 1 & 0 \end{bmatrix} \begin{matrix} v_1 \\ v_4 \\ v_5 \\ v_2 \\ v_3 \\ v_6 \\ v_7 \end{matrix}$$

with column labels v_1 v_4 v_5 v_2 v_3 v_6 v_7.

If we express this matrix as

$$A = \begin{bmatrix} A_{11} & A_{12} \\ 0 & A_{22} \end{bmatrix}$$

it is clear that A^k, $k = 1, 2, \ldots,$ has the same form

$$A^k = \begin{bmatrix} A_{11}^{\,k} & A_{12}^{(k)} \\ 0 & A_{22}^{\,k} \end{bmatrix}$$

(It is not necessary to compute the exact value of $A_{12}^{(k)}$.) Thus

$$a_{71}^{(k)} = 0$$

for all k, and consequently it is impossible to reach v_1 from v_7.

One apparent drawback of this approach is that it seems necessary to know the answer to the problem before we begin looking for it. However, the re-ordering of the rows and columns of A is merely a convenience, and the same results could have been obtained without it, though with somewhat more labor.

////

EXAMPLE 4.1.3 The following well-known puzzle can be solved by methods similar to those of Example 4.1.1.

Two missionaries and two cannibals are on a journey. They reach a river which they must cross. There is a rowboat available which can carry two people. All four men can row, but the missionaries feel safer if they are not outnumbered by the cannibals at any time (except when there are no missionaries). Is it possible for the four men to cross the river in such a way that one missionary is never in the company of both cannibals? If so, how many trips across the river are necessary? Suppose it is also required that the two cannibals never be left alone. Can the crossing still be made, and if so, how many trips are necessary?

It will be convenient to think of the missionaries and cannibals as travelling from east to west. We will designate the situation "there are m missionaries and c cannibals and the boat on the east bank of the river" by the triple (m, c, e) and the situation "there are m missionaries and c cannibals and the boat on the west bank of the river" by (m, c, w). Clearly $m = 0, 1, 2$ and $c = 0, 1, 2$, and so there are 18 triples. However, $(1, 2, e)$, $(1, 2, w)$, $(1, 0, e)$, and $(1, 0, w)$ would require one missionary to be with two cannibals, and hence these triples are not permitted. The remaining 14 triples are allowed (at least, until the restrictions added at the end of the problem are considered). We can model this problem as a directed graph whose vertices are the 14 allowable triples and whose arcs describe the possible trips across the river. Set

$$
\begin{array}{ll}
v_1 = (2, 2, e) & v_8 = (2, 2, w) \\
v_2 = (2, 1, e) & v_9 = (2, 1, w) \\
v_3 = (1, 1, e) & v_{10} = (1, 1, w) \\
v_4 = (2, 0, e) & v_{11} = (2, 0, w) \\
v_5 = (0, 2, e) & v_{12} = (0, 2, w) \\
v_6 = (0, 1, e) & v_{13} = (0, 1, w) \\
v_7 = (0, 0, e) & v_{14} = (0, 0, w)
\end{array}
$$

We can now construct the 14×14 vertex adjacency matrix A and use the method of Example 4.1.1 to find the smallest value of k for which

$$a_{18}^{(k)} \neq 0$$

or show that $a_{18}{}^{(k)} = 0$ for all values of k. Note that $a_{18}{}^{(k)}$ can be nonzero only when k is odd. Because the boat crosses the river at each step, A has the form

$$A = \begin{bmatrix} 0 & A_{12} \\ A_{21} & 0 \end{bmatrix} \quad (4.1.4)$$

where A_{12} and A_{21} are 7×7 matrices. Moreover, because the same conditions prevail on the east bank of the river as on the west bank, we have

$$A_{12} = A_{21} \quad (4.1.5)$$

Let us set

$$A_{12} = A_{21} = \tilde{A} = [a_{ij}]$$

It is not difficult to see that for odd values of k

$$A^k = \begin{bmatrix} 0 & \tilde{A}^k \\ \tilde{A}^k & 0 \end{bmatrix}$$

and hence

$$a_{18}{}^{(k)} = \tilde{a}_{11}{}^{(k)}$$

Let us compute $\tilde{A}$. From v_1 it is possible in one step to reach:

1. v_{10} by having one missionary and one cannibal cross the river
2. v_{11} by having both missionaries cross the river
3. v_{12} by having both cannibals cross the river
4. v_{13} by having one cannibal cross the river

Continuing in this way we see that

$$\tilde{A} = \begin{bmatrix} 0 & 0 & 1 & 1 & 1 & 1 & 0 \\ 0 & 1 & 1 & 0 & 1 & 0 & 0 \\ 1 & 1 & 0 & 0 & 0 & 0 & 0 \\ 1 & 0 & 0 & 0 & 0 & 0 & 0 \\ 1 & 1 & 0 & 0 & 0 & 0 & 0 \\ 1 & 0 & 0 & 0 & 0 & 0 & 0 \\ 0 & 0 & 0 & 0 & 0 & 0 & 0 \end{bmatrix}$$

and consequently

$$\tilde{A}^3 = \begin{bmatrix} 0 & 2 & 6 & 4 & 6 & 4 & 0 \\ 2 & 5 & 5 & 2 & 5 & 2 & 0 \\ 6 & 5 & 1 & 0 & 1 & 0 & 0 \\ 4 & 2 & 0 & 0 & 0 & 0 & 0 \\ 6 & 5 & 1 & 0 & 1 & 0 & 0 \\ 4 & 2 & 0 & 0 & 0 & 0 & 0 \\ 0 & 0 & 0 & 0 & 0 & 0 & 0 \end{bmatrix} \quad \text{and} \quad \tilde{A}^5 = \begin{bmatrix} 4 & 18 & 34 & 20 & 34 & 20 & 0 \\ 18 & 29 & 29 & 14 & 29 & 14 & 0 \\ 34 & 29 & 9 & 2 & 9 & 2 & 0 \\ 20 & 14 & 2 & 0 & 2 & 0 & 0 \\ 34 & 29 & 9 & 2 & 9 & 2 & 0 \\ 20 & 14 & 2 & 0 & 2 & 0 & 0 \\ 0 & 0 & 0 & 0 & 0 & 0 & 0 \end{bmatrix}$$

Thus the minimum number of trips in which the missionaries and cannibals can cross the river is 5. However, there are four different ways in which the crossing can be effected in five trips, namely,

$$v_1 \to v_{12} \to v_2 \to v_9 \to v_5 \to v_8$$
$$v_1 \to v_{12} \to v_2 \to v_9 \to v_3 \to v_8$$
$$v_1 \to v_{10} \to v_2 \to v_9 \to v_5 \to v_8$$
$$v_1 \to v_{10} \to v_2 \to v_9 \to v_3 \to v_8$$

Only the last of the four solutions above satisfies the restriction that the two cannibals are never to be alone. Thus this problem can also be solved with a crossing requiring five trips. In this case, the solution is unique. If we had wished to solve only the restricted problem, our computations could have been considerably simplified; for then the situations corresponding to the four points v_4, v_5, v_{11}, and v_{12} would not have been permitted and the same methods could have been used with a 5×5 matrix. The details are left as an exercise. Incidentally, a little thought would have shown that the boat cannot be on a bank of the river with no people, permitting us to eliminate the points v_7 and v_{14}.

Throughout this problem every step that is carried out can be reversed; i.e., for any two vertices of the graph, say v_i and v_j, if v_j can be reached from v_i in one step, then v_i can be reached from v_j in one step. Thus

$$a_{ij} = a_{ji}$$

for all i and j, which means that A is a symmetric matrix. It now follows from equations (4.1.4) and (4.1.5) that $\tilde{A}$ is also symmetric. This property leads to an interesting phenomenon in the graph. Because no arc of the graph can be traversed in only one direction, it is not necessary to specify a direction at all. The 14-vertex graph of the problem without the restriction that the two cannibals

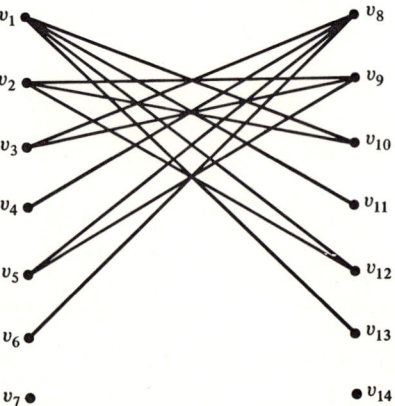

are never alone is illustrated.   (The symmetry about the vertical of this graph is due to the symmetry of $\tilde{A}$.)   A graph in which no directions are assigned to the arcs is called an *undirected graph*.   ////

EXAMPLE 4.1.6   The first serious study of graphs grew out of a puzzle solved by Euler in 1736.   The river Pregel flows through the city of Königsberg (now called Kaliningrad) in East Prussia.   In the river there are two islands, which are connected to each other and to the banks of the river by seven bridges.

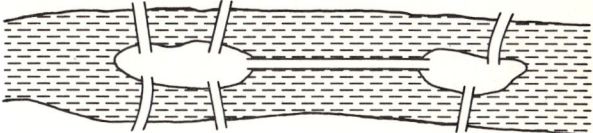

FIGURE 4.1.7

Is it possible to find a path which uses each of the bridges exactly once and which ends at the same place it began?   (Such a path is called an *Euler line*.) Euler proved that such a path does not exist for the Königsberg bridges, and, in fact, he showed that even if the requirement that the path begin and end at the same place is dropped, there still is no solution.

We may represent the four-land area of Fig. 4.1.7 as vertices of a graph and the bridges as arcs of the graph.   Since we have not restricted the direction in which a bridge may be crossed, the graph is undirected.

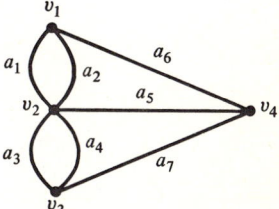

FIGURE 4.1.8

For a path in any graph, beginning at a vertex v_i and ending at a vertex v_j (using each edge of the graph exactly once or not), the number of times each vertex other than v_i and v_j is entered is equal to the number of times it is left. If v_i is the same vertex as v_j, every vertex has the property that the number of times it is entered is equal to the number of times it is left.   On the other hand, if v_i and v_j are distinct, then v_i is left one more time than it is entered and v_j is entered one more time than it is left. When each edge of the graph is used exactly once, the total of the number of times any vertex is entered and is left is equal to the number of arcs incident with that vertex. Thus we have proved that if a graph has a path which uses each arc exactly once, then at every vertex (except possibly for the vertices where the path begins and ends) there are an even number of arcs. Since there are an odd number of arcs at each of the four vertices of the graph of Fig. 4.1.8, no such path exists.

The number of arcs incident with a vertex of a graph is called the *degree* of the vertex. In Sec. 4.2 we will prove that a connected graph in which every vertex has even degree has an Euler line. Also, we will show that a connected graph in which all but two of the vertices have even degree has a path beginning at one of these vertices, ending at the other, and using each arc exactly once.

////

EXAMPLE 4.1.9 Figure 4.1.10 is a map of the oil-pipeline system in South Transylvania. Each pipe is capable of carrying 1 ton of oil per hour. Oil can be transferred from one pipe to another only where pipes meet at one of the pumping stations $v_1, v_2, \ldots, v_{10}$; that is, the pipes between v_1 and v_6 and between v_2 and v_8 do not intersect, and oil cannot be transferred directly from one to the other.

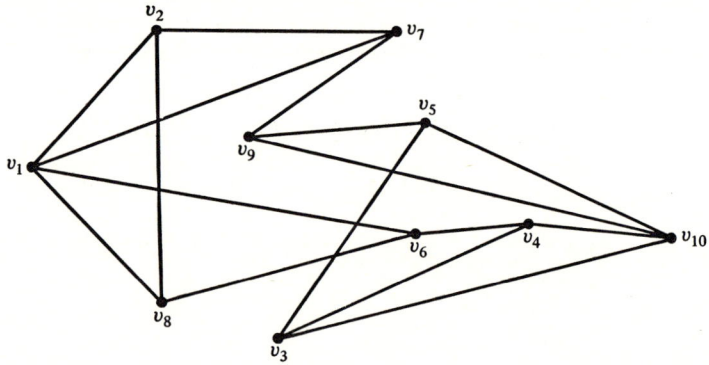

FIGURE 4.1.10

How much oil per hour can the South Transylvanians move from v_1 to v_{10}? What is the smallest number of pipes in Fig. 4.1.10 that the North Transylvanians can block which will prevent the South Transylvanians from shipping any oil from v_1 to v_{10}?

There are numerous ways in which 2 tons of oil per hour can be sent from v_1 to v_{10}. For example, 1 ton per hour can be piped using the route v_1, v_7, v_9, v_{10}, while another ton per hour can be sent using the route $v_1, v_8, v_6, v_4, v_{10}$.

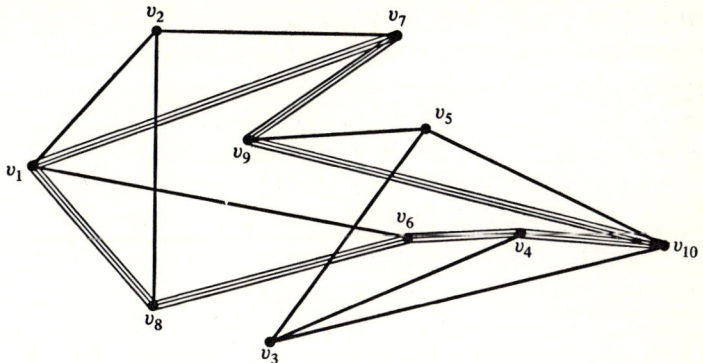

FIGURE 4.1.11

On the other hand, no oil can flow from v_1 to v_{10} if the pipe connecting v_7 and v_9 and the pipe connecting v_6 and v_4 are both blocked, since the remaining network of pipes is no longer connected. The graph of Fig. 4.1.12 consists of two disjoint (connected) components, one of which contains v_1 while the other contains v_{10}.

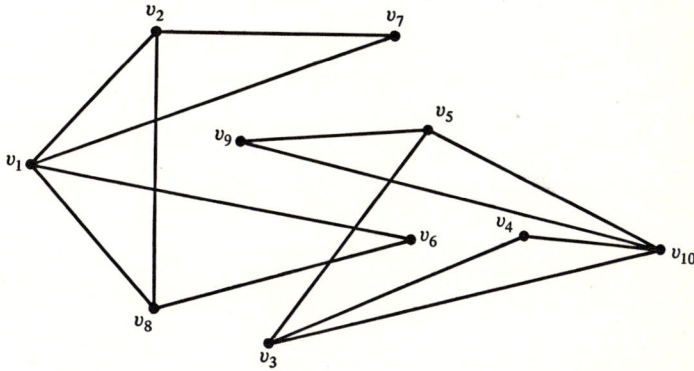

FIGURE 4.1.12

South Transylvania cannot ship more than 2 tons of oil per hour from v_1 to v_{10} since the North Transylvanians are able to stop the flow of oil completely by blocking only two pipes. On the other hand, if the South Transylvanians can ship 2 tons of oil per hour from v_1 to v_{10}, the entire flow of oil cannot be stopped by blocking only one pipe. More generally, it is clear that if in a graph there are k *arc-disjoint paths* (i.e., no arc is used in more than one path) from v_i to v_j and if by the removal of l arcs of the graph there are no paths from v_i to v_j then $k \leq l$. Ford and Fulkerson [4] have shown that if k is the largest number of arc-disjoint paths that exist between v_i and v_j and if l is the smallest number of arcs whose removal from the graph results in there being no paths between v_i and v_j, then $k = l$. ////

EXAMPLE 4.1.13 The South Transylvanians are anxious to defend their oil-pipeline system (Fig. 4.1.10) against attacks by the North Transylvanians but do not wish to commit more of their small army than is absolutely necessary. If one army unit is needed to protect each pipeline segment, and if the South Transylvanians must maintain a system whereby oil can be sent from any city to any other city, how many army units are required?

If any set of pipes form a loop, for example, $v_3 v_4$, $v_4 v_{10}$, $v_{10} v_3$ or $v_1 v_2$, $v_2 v_8$, $v_8 v_6$, $v_6 v_1$, then any one of these pipes is superfluous and need not be defended. For example, if the pipe $v_3 v_4$ is blocked, oil can be shipped from v_3 to v_4 via $v_3 v_{10}$ and $v_{10} v_4$. Thus the system that the South Transylvanians will defend will not contain any loops. Figure 4.1.14 provides an example of a network of pipes connecting all cities in which no pipe may be left undefended. There are many others.

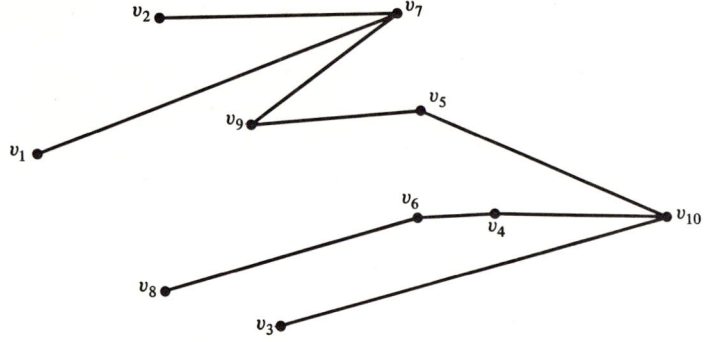

FIGURE 4.1.14

The graph of the pipeline system of Fig. 4.1.14 has nine arcs, and thus nine army units are required. In Sec. 4.4 we will prove that every set of arcs which connects all the vertices of the graph of Fig. 4.1.14 and which has no loops contains precisely nine arcs.[1] ////

The relationship between a graph and the various matrices associated with it is the main topic of this chapter. In Example 4.1.1 we defined the vertex-adjacency matrix A of a directed graph G. If G' is formed from G by reversing the direction of each arc, the vertex-adjacency matrix of G' is A^T, the transpose

---

[1] This problem can also be solved in the more realistic situation where the pipeline segments are not all of the same length; i.e., not all require the same number of army units for protection. See Ore [9].

of A. In Example 4.1.3 we defined the vertex-adjacency matrix of an undirected graph. The two definitions agree if we consider an undirected arc as being directed toward both its vertices. Clearly the vertex-adjacency matrix of an undirected graph is symmetric, i.e.,

$$A^T = A$$

In Example 4.1.6 we defined the degree d_i of the vertex v_i of an undirected graph to be the number of arcs incident with v_i. The *degree matrix* (to be denoted by D) is the diagonal matrix whose diagonal entries are the degree of the vertices. For a directed graph we define the *in degree* δ_i and the *out degree* Δ_i of v_i to be the number of arcs directed to and the number of arcs directed from v_i respectively. As in the undirected case, the *in-degree matrix* δ and the *out-degree matrix* Δ are the diagonal matrices whose diagonal entries are the degrees of the vertices. We leave it as an exercise to show that:

1. For an undirected graph d_i is equal to the entry in the (i, i) position of A^2.
2. For a directed graph δ_i is equal to the entry in the (i, i) position of $A^T A$ whereas Δ_i is equal to the entry in the (i, i) position of AA^T.

Another important matrix associated with a graph is the *vertex-arc incidence matrix* $B = [b_{ij}]$. For a directed graph we define B by

$$b_{ij} = \begin{cases} 1 & \text{if arc } a_j \text{ is directed toward vertex } v_i \\ -1 & \text{if arc } a_j \text{ is directed from vertex } v_i \\ 0 & \text{otherwise} \end{cases}$$

In each column of B there is one 1 and one -1, and the remaining entries are 0. The number of rows of B is equal to the number of vertices of the graph, and the number of columns of B is equal to the number of arcs. For an undirected graph we define B by

$$b_{ij} = \begin{cases} 1 & \text{if } v_i \text{ is one of the vertices of the arc } a_j \\ 0 & \text{otherwise} \end{cases}$$

In each column of B there are two 1s, and the remaining entries are 0.

EXAMPLE 4.1.15 For the graph of Fig. 4.1.8, the vertex-adjacency matrix, the degree matrix, and the vertex-arc incidence matrix are

$$A = \begin{bmatrix} 0 & 2 & 0 & 1 \\ 2 & 0 & 2 & 1 \\ 0 & 2 & 0 & 1 \\ 1 & 1 & 1 & 0 \end{bmatrix} \quad D = \begin{bmatrix} 3 & 0 & 0 & 0 \\ 0 & 5 & 0 & 0 \\ 0 & 0 & 3 & 0 \\ 0 & 0 & 0 & 3 \end{bmatrix}$$

$$B = \begin{bmatrix} 1 & 1 & 0 & 0 & 0 & 1 & 0 \\ 1 & 1 & 1 & 1 & 1 & 0 & 0 \\ 0 & 0 & 1 & 1 & 0 & 0 & 1 \\ 0 & 0 & 0 & 0 & 1 & 1 & 1 \end{bmatrix} \quad ////$$

Since the matrices defined above reflect different properties of the same graph, it is not surprising that the matrices themselves are related. A typical relation is described in:

Theorem 4.1.16 Let G be an undirected graph with m vertices and n arcs. Then
$$BB^T = A + D$$

PROOF   For convenience set
$$BB^T = H = [h_{ij}]$$

Then
$$h_{ij} = \sum_{k=1}^{n} b_{ik} b_{jk} \quad i, j = 1, 2, \ldots, m \quad (4.1.17)$$

CASE 1   Let $i = j$. Clearly
$$b_{ik} b_{ik} = \begin{cases} 1 & \text{if } v_i \text{ is one of the vertices of } a_k \\ 0 & \text{otherwise} \end{cases}$$

It follows from equation (4.1.17) that h_{ij} is equal to the number of arcs which have v_i as one of its vertices, i.e.,
$$h_{ii} = d_{ii}$$

CASE 2   Let $i \neq j$. Then
$$b_{ik} b_{jk} = \begin{cases} 1 & \text{if } v_i \text{ and } v_j \text{ are the vertices of } a_k \\ 0 & \text{otherwise} \end{cases}$$

and consequently h_{ij} is equal to the number of arcs which have v_i and v_j as vertices. Thus
$$h_{ij} = a_{ij} \quad \text{////}$$

EXERCISES

1 Show that the minimum number of trips in which three missionaries and three cannibals using a boat carrying two people can cross a river in such a way that on each of the banks of the river the missionaries are never outnumbered (except when there are no missionaries) is 11. In how many different ways can they cross the river in 11 trips? Show also that if the three cannibals are never permitted to be together, except in the presence of the three missionaries, they cannot get across the river.

2 Use 5×5 matrices to show that there is exactly one way in which two missionaries and two cannibals can cross a river in five trips if the two cannibals are not permitted to be alone together.

3 A farmer has two bulls and three goats and wishes to bring them to town from his farm. The road to town is downhill, and his truck can carry any two animals on each trip. However, on the trips back to the farm his truck can carry only one animal. Unless the farmer is present, there must be more goats than bulls or no goats at all. Show that the farmer can get his livestock to town but once there he cannot get them back to the farm. (*Hint:* Define a matrix to describe the possible trips up the hill and another to describe the possible trips down the hill.)

4 Show that in Example 4.1.1 it would still be possible to reach v_7 from v_1 if the arc from v_4 to v_7 is removed from the graph. What is the shortest path in this case, and how many such paths are there?

5 Show that the analog of Theorem 4.1.16 for directed graphs is

$$BB^T = \delta + \Delta - A - A^T$$

6 For an undirected graph G with m vertices and n arcs, set $\tilde{a}_{ij}$ equal to the number (possibly zero but at most 2) vertices common to the arcs a_i and a_j. The $n \times n$ matrix $\tilde{A} = [a_{ij}]$ is called the *arc-adjacency matrix* of G. Prove that:

(a) $B^T B = \tilde{A}$ (b) $B\tilde{A} - AB = DB$

7 For a directed graph G we define $\tilde{a}_{ij}$ by

$\tilde{a}_{ij} = e_{ij}$ times the number of vertices common to the arcs a_i and a_j,

and when a_i and a_j have vertices in common, e_{ij} is defined by

$$e_{ij} = \begin{cases} 1 & \text{if } a_i \text{ and } a_j \text{ are either both directed toward a common vertex or both directed from a common vertex} \\ -1 & \text{if one of } a_i \text{ and } a_j \text{ is directed toward a common vertex and the other is directed from the common vertex} \end{cases}$$

Prove that $B^T B = \tilde{A}$.

8 Construct the arc-adjacency, vertex-adjacency, and incidence matrices of the graph of Fig. 4.1.8 and verify Theorem 4.1.16 and Exercise 6.

9 Transform the graph of Fig. 4.1.8 into a directed graph by orienting $a_1, a_2, a_3,$ and a_4 toward v_2 and $a_5, a_6,$ and a_7 toward v_4. Construct the arc-adjacency, vertex-adjacency, and incidence matrices and verify Exercises 5 and 7.

4.2 UNDIRECTED GRAPHS

A number of examples involving graphs were presented and analyzed in the last section. A diagram was part of the statement of the problem in several of these examples, and the solution was found by treating the diagram as an abstract graph. In Example 4.1.3 a diagram was not mentioned in the statement of the problem, but we were able to display all the pertinent information by means of a graph and this led to a solution. In order to put the entire subject on a firmer foundation and decrease our reliance on (though not our use of) diagrams we will give some precise definitions and prove some of the "obvious" properties of graphs. In this section we will consider only undirected graphs.

Definition 4.2.1 An $m \times n$ (undirected) *graph* $G = (\mathbb{V}, \mathbb{A})$ consists of a set $\mathbb{V}$ of m objects called *vertices*

$$\mathbb{V} = \{v_1, v_2, \ldots, v_m\}$$

and a set $\mathbb{A}$ of n *arcs*

$$\mathbb{A} = \{a_1, a_2, \ldots, a_n\}$$

where each arc is an unordered pair of distinct vertices. If an arc a consists of the vertices v_1 and v_2, we say that v_1 and v_2 are *incident* with a and that v_1 and v_2 are *adjacent*. Two arcs which have a vertex in common are *adjacent*. When two arcs have both vertices in common, they are *parallel*.

Each arc of a graph is incident with exactly two vertices. A vertex may be incident with any number of arcs, however, and the number of arcs incident with a vertex is called its *degree*.

In order to specify a graph it is sufficient to enumerate the vertices and the adjacency relations among them. It is an easy matter (perhaps too easy) to put all this information into the form of a diagram. As we have seen, this information can also be presented in terms of matrices, and, of course, this is the approach that we will emphasize.

EXAMPLE 4.2.2 The graph with vertices v_1, v_2, v_3, v_4, v_5 and arcs

$$a_1 = (v_1, v_3) \quad a_2 = (v_1, v_4)$$
$$a_3 = (v_2, v_6) \quad a_4 = (v_7, v_1)$$
$$a_5 = (v_1, v_3) \quad a_6 = (v_7, v_4)$$

can be represented by Fig. 4.2.3. ////

The graph of Fig. 4.2.3 is composed of three separate subgraphs:

1. G_1 with vertices v_1, v_4, v_7, v_3 and arcs a_1, a_2, a_4, a_5, a_6
2. G_2 with vertices v_2, v_6 and arc a_3
3. G_3 with vertex v_5 and no arcs

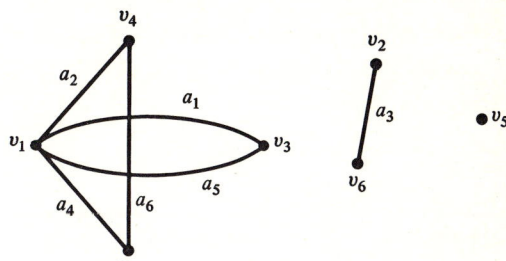

FIGURE 4.2.3

It is much easier to investigate the properties of each of these subgraphs separately than it would be to restrict our attention to the entire graph. In order to characterize these subgraphs we make the following definition.

Definition 4.2.4 A graph is *connected* if it is not possible to partition the set of vertices into two disjoint nonempty subsets in such a way that no arc of the graph is incident with vertices from both subsets.

The graph of Fig. 4.2.3 is clearly not connected. On the other hand, each of the subgraphs G_1, G_2, G_3 described above is connected. An important characterization of connected graphs follows from:

Definition 4.2.5 Let v_1 and v_2 be vertices of the graph G. An ordered set of l distinct arcs of G,

$$(v_1, v_{h_1}), (v_{h_1}, v_{h_2}), (v_{h_2}, v_{h_3}), \ldots, (v_{h_{l-1}}, v_2)$$

is called a *path of length l* between v_1 and v_2. If the vertices $v_{h_1}, v_{h_2}, \ldots, v_{h_{l-1}}$ are distinct from each other and also from v_1 and v_2, the path is said to be *simple*. A simple path between a vertex and itself, that is, $v_1 = v_2$, is called a *loop*.

In addition, we define the empty set of arcs to be a simple path between every vertex and itself. However, the empty set of arcs is not defined to be a loop. An arc is a simple path of length 1 between its vertices. In Fig. 4.2.3 there are two loops, namely,

$$\{a_1, a_5\} \quad \text{and} \quad \{a_2, a_6, a_4\}$$

Also, $\{a_1, a_2, a_6, a_4\}$ is a path between v_3 and v_1, but it is not simple.

If
$$\{(v_1, v_{h_1}), (v_{h_1}, v_{h_2}), (v_{h_2}, v_{h_3}), \ldots, (v_{h_{l-1}}, v_2)\}$$
is a path of length l between v_1 and v_2, then
$$\{(v_2, v_{h_{l-1}}), \ldots, (v_{h_3}, v_{h_2}), (v_{h_2}, v_{h_1}), (v_{h_1}, v_1)\}$$
is a path of length l between v_2 and v_1. It will be convenient not to consider these paths as being distinct (just as we did not consider $\{a_1, a_5\}$ and $\{a_5, a_1\}$ to be distinct loops). Henceforth we will say that a path between v_1 and v_2 is also a path between v_2 and v_1, and we will not specify the direction of the path. However, if another path between v_1 and v_2 can be obtained by rearranging the order of the arcs, this path will be considered to be a different path. For example, in Fig. 4.2.3,

$$\{a_1, a_2, a_6, a_4\} \quad \text{and} \quad \{a_4, a_6, a_2, a_1\}$$

are the same path between v_3 and v_1 (or v_1 and v_3) but

$$\{a_1, a_2, a_6, a_4\} \quad \text{and} \quad (a_1, a_4, a_6, a_2\}$$

are distinct paths between v_1 and v_3.

Theorem 4.2.6 Let v_1, v_2, v_3 be vertices of the graph G.

1. If there is a path between v_1 and v_2, then there is a simple path between v_1 and v_2.
2. If there is a path between v_1 and v_2 and there is a path between v_2 and v_3, then there is a path between v_1 and v_3.

PROOF The proofs of the two parts of the theorem are similar and illustrate a technique frequently used in graph theory.

(1) Let

$$\{a_1 = (v_1, v_{h_1}), a_2 = (v_{h_1}, v_{h_2}), \ldots, a_{l-1} = (v_{h_{l-2}}, v_{h_{l-1}}), a_l = (v_{h_{l-1}}, v_2)\}$$

be a path of length l between v_1 and v_2. Although the arcs of the path are distinct, some of the vertices may occur more than once. For convenience of notation, set

$$v_1 = v_{h_0} \quad v_2 = v_{h_l}$$

If the l vertices $v_{h_0}, v_{h_1}, v_{h_2}, \ldots, v_{h_{l-1}}$ are distinct, the path is simple. If the l vertices are not distinct, let i be the smallest integer such that v_{h_i} occurs more than once. Let j be the largest integer such that

$$v_{h_i} = v_{h_j}$$

(Note that the special case $j = l$ is not excluded.) Then

$$\{a_1, a_2, \ldots, a_i, a_{j+1}, \ldots, a_l\} \quad (4.2.7)$$

is a path between v_1 and v_2 containing fewer than l arcs. When $i = 0$ and $j \neq l$, (4.2.7) is interpreted to mean $\{a_{j+1}, \ldots, a_l\}$. When $i \neq 0$ and $j = l$, (4.2.7) is interpreted to mean $\{a_1, a_2, \ldots, a_i\}$. Finally, when $i=0$ and $j=l$, (4.2.7) is interpreted to be the empty set of arcs. If the path of (4.2.7) is simple we have finished. If not, the process is repeated. In a finite number of steps we reach a single path between v_1 and v_2.

(2) Let

$$\{a_1 = (v_1, v_{h_1}), a_2 = (v_{h_1}, v_{h_2}), \ldots, a_k = (v_{h_{k-1}}, v_2)\}$$

and $\{a_{k+1} = (v_2, v_{h_{k+1}}), a_{k+2} = (v_{h_{k+1}}, v_{h_{k+2}}), \ldots, a_l = (v_{h_{l-1}}, v_3)\}$

be paths between v_1 and v_2 and between v_2 and v_3 respectively. For convenience of notation set

$$v_1 = v_{h_0} \qquad v_2 = v_{h_k} \qquad v_3 = v_{h_l}$$

According to Definition 4.2.5, $a_1, a_2, \ldots, a_k$ are distinct. Likewise, $a_{k+1}, a_{k+2}, \ldots, a_l$ are distinct. If these l arcs are distinct, they form a path between v_1 and v_3. If these arcs are not distinct, let i be the smallest integer such that the arc a_i occurs more than once. Consequently a_i occurs twice, once in the path between v_1 and v_2 and once in the path between v_2 and v_3. Let a_i appear in the form

$$a_i = (v_{h_p}, v_{h_{p+1}}) \qquad 0 \leq p \leq k-1$$

in the path between v_1 and v_2 and as

$$a_i = (v_{h_q}, v_{h_{q+1}}) \qquad k \leq q \leq l-1$$

in the path between v_2 and v_3. Now we must consider two cases.

CASE 1 Suppose

$$v_{h_p} = v_{h_q}$$

and consequently

$$v_{h_{p+1}} = v_{h_{q+1}}$$

Then $\{a_1, a_2, \ldots, a_i, a_{j+1}, \ldots, a_l\}$ is the required path.

CASE 2 Suppose

$$v_{h_p} = v_{h_{q+1}}$$

and consequently

$$v_{h_{p+1}} = v_{h_q}$$

Then $\{a_1, a_2, \ldots, a_{i-1}, a_{j+1}, \ldots, a_l\}$ is the required path.

The interpretation of the cases $i = 0$ and $j = l$ follows the same pattern as in part 1 of the theorem. ////

Let us define the relation $\sim$ on the vertices of a graph G by

$$v_i \sim v_j \Leftrightarrow \text{there is a path between } v_i \text{ and } v_j$$

Clearly $\sim$ is reflexive. Because we have defined the empty set of arcs to be a path between each vertex of a graph and itself, $\sim$ is symmetric. Also, part 2 of Theorem 4.2.6 states that $\sim$ is transitive. Thus $\sim$ is an equivalence relation and partitions the vertices of G into disjoint equivalence classes. We can extend this equivalence to the arcs of G since an arc is a path of length 1 between its two vertices and consequently both vertices of an arc belong to the same equivalence class. Let us now add each arc to the equivalence class of its vertices and denote the resulting sets (of vertices and arcs) by $G_1, G_2, \ldots, G_k$. Then:

1. Each G_i is a graph, $i = 1, 2, \ldots, k$.
2. For $i \neq j$, the graphs G_i and G_j are both arc- and vertex-distinct.
3. G is the union of $G_1, G_2, \ldots, G_k$.

We call $G_1, G_2, \ldots, G_k$ the *components* of G.

Theorem 4.2.8 A graph G is connected if and only if there is a path between any two vertices of G; that is, G consists of only one component.

PROOF Let v be a vertex of the connected graph G. Denote by $\mathbb{C}(v)$ the set of all vertices of G which are equivalent to v and by $\mathbb{D}(v)$ the remaining vertices. We wish to show that $\mathbb{D}(v)$ is empty. Let us suppose that $\mathbb{D}(v)$ is not empty. Since G is connected, there is at least one arc, say (v_s, v_t), of G which is incident with both a vertex belonging to $\mathbb{C}(v)$, say v_s, and a vertex belonging to $\mathbb{D}(v)$, say v_t. Since $v_s \in \mathbb{C}(v)$, there is a path between v and v_s. Moreover the single arc (v_s, v_t) is a path between v_s and v_t. It follows from part 2 of Theorem 4.2.6 that there is a path between v and v_t and hence $v_t \notin \mathbb{D}(v)$. This is a contradiction, and consequently $\mathbb{D}(v)$ is empty.

Conversely, let G be a graph such that between any two vertices there is a path. Let us suppose that G is not connected. Then the vertices of G can be partitioned into two disjoint nonempty sets, say $\mathbb{V}_1$ and $\mathbb{V}_2$, in such a way that no arc of G is incident with vertices from both $\mathbb{V}_1$ and $\mathbb{V}_2$. Let v_1 be any vertex belonging to $\mathbb{V}_1$ and v_2 be any vertex belonging to $\mathbb{V}_2$. By hypothesis there is a path

$$\{(v_1, v_{h_1}), (v_{h_1}, v_{h_2}), \ldots, (v_{h_{l-1}}, v_2)\}$$

between v_1 and v_2. Set
$$v_1 = v_{h_0} \qquad v_2 = v_{h_l}$$
Let i be the smallest integer for which v_{h_i} belongs to $\mathbb{V}_2$. Since $v_{h_l} \in \mathbb{V}_2$, such an integer exists. Also, because $v_{h_0} \notin \mathbb{V}_2$, it follows that $1 \leq i \leq l$. We now have found an arc $(v_{h_{i-r}}, v_{h_i})$ which is incident with a vertex of $\mathbb{V}_1$ and with a vertex of $\mathbb{V}_2$. This is a contradiction, and hence the supposition that G is not connected is false. ////

Corollary 4.2.9 Each component of a graph is connected. Consequently every graph is a union of disjoint connected subgraphs.

EXAMPLE 4.2.10 Applying Corollary 4.2.9 to the graph of Fig. 4.2.3 shows that there are three components:
$$G_1 = \{\mathbb{V}_1 = \{v_1, v_3, v_4, v_7\}, \mathbb{A}_1 = \{a_1, a_2, a_4, a_5\}\}$$
$$G_2 = \{\mathbb{V}_2 = \{v_2, v_6\}, \mathbb{A}_2 = \{a_3\}\}$$
$$G_3 = \{\mathbb{V}_3 = \{v_5\}, \mathbb{A}_3 = \varnothing\} \qquad ////$$

Henceforth, unless we specify otherwise, we will assume that all graphs under consideration are connected. In Sec. 4.4 we will present an alternate way of describing connected graphs.

We conclude this brief introduction to undirected graphs with a proof of the theorem we used (without proof) in Example 4.1.6 to show that the Königsberg bridge problem has no solution.

Theorem 4.2.11 If the degree of every vertex of a graph G is even, then G has an Euler line, i.e., a path between a vertex and itself which contains all the arcs of G.

PROOF Let v be any vertex of G. Among all the paths between v and itself there is one which is of maximal length, say
$$P = \{a_1 = (v, v_{h_1}), a_2 = (v_{h_1}, v_{h_2}), \ldots, a_l = (v_{h_{l-1}}, v)\}$$
It will be convenient to set
$$v = v_{h_0} = v_{h_l}$$
We wish to prove that P is an Euler line.

Let us suppose that P is not an Euler line. First we need to show that at least one vertex of P is incident with fewer arcs than its degree. Let us suppose that this is false. Then both vertices of every arc of G which is not in P do not belong to P. Consequently G is not connected. This is a contradiction, and we have now shown that at least one vertex of P is incident with fewer arcs than its degree. Let us call this vertex v_{k_0} (so that $v_{k_0} = v_{h_i}$ for some i, $0 \leq i \leq l$). There is an arc $\bar{a}_1 = (v_{k_0}, v_{k_1})$ which does not belong to P. We now see that v_{k_1} is incident with an even number of arcs of P and with $\bar{a}_1$. (Note that we have not specified whether or not v_{k_1} belongs to P.) Since the degree of v_{k_1} is even, there is an arc other than $\bar{a}_1$ which is incident with v_{k_1} and which does not belong to P. Denote this arc by $\bar{a}_2 = (v_{k_1}, v_{k_2})$. Unless $v_{k_2} = v_{k_0}$, the same argument can be applied to v_{k_2} to prove that there is an arc $\bar{a}_3 = (v_{k_2}, v_{k_3})$ which does not belong to P. Continuing in this way, we eventually return to v_{k_0}, and then we have a path $\bar{P}$,

$$\bar{P} = \{\bar{a}_1 = (v_{k_0}, v_{k_1}), \bar{a}_2 = (v_{k_1}, v_{k_2}), \ldots, \bar{a}_m = (v_{k_{m-1}}, v_{k_0})\}$$

between v_{k_0} and itself consisting of arcs which do not belong to P. If we now form

$$P' = \{a_1, a_2, \ldots, a_i, \bar{a}_1, \bar{a}_2, \ldots, \bar{a}_m, a_{i+1}, \ldots, a_l\}$$

P' is a path between v and itself and, moreover, the length of P' is equal to the sum of the lengths of P and $\bar{P}$. This contradicts the maximality of the length of P. Hence P is an Euler line. ////

Corollary 4.2.12 If G has exactly two vertices whose degree is odd, then there is a path between them which contains all the arcs of G.

PROOF Let v_1 and v_2 be the two vertices of odd degree. Form the graph G' from G by adding the arc $a_0 = (v_1, v_2)$. Every vertex of G' has even degree, and hence G' has an Euler line P'. We may consider a_0 to be the last arc of P'. If a_0 is removed from P', the remaining path, P, satisfies the conditions of the corollary. ////

EXAMPLE 4.2.13 The graph in Fig. 4.2.14 contains an Euler line since

$$d_3 = 4 \qquad d_1 = d_2 = d_4 = d_5 = 2$$

Although there are two Euler lines between v_1 and itself

$$P_1 = \{a_1, a_4, a_5, a_6, a_2, a_3\} \qquad P_2 = \{a_1, a_6, a_5, a_4, a_2, a_3\}$$

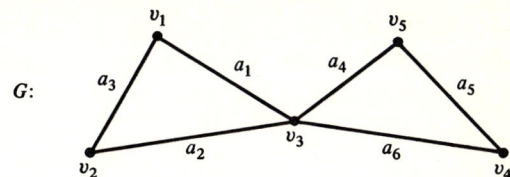

FIGURE 4.2.14

there are four Euler lines between v_3 and itself

$Q_1 = \{a_1, a_3, a_2, a_6, a_5, a_4\}$ $Q_2 = \{a_2, a_3, a_1, a_6, a_5, a_4\}$
$Q_3 = \{a_1, a_3, a_2, a_4, a_5, a_6\}$ $Q_4 = \{a_2, a_3, a_1, a_4, a_5, a_6\}$ ////

An Euler line describes the order of the arcs in a path between a vertex and itself. If

$$P = \{a_1 = (v_{h_1}, v_{h_2}), a_2, \ldots, a_{k-1}, a_k = (v_{l_1}, v_{l_2}), \ldots, a_n\}$$

is an Euler line between v_{h_1} and itself, then

$$P' = \{a_k = (v_{l_1}, v_{l_2}), \ldots, a_n, a_1 = (v_{h_1}, v_{h_2}), a_2, \ldots, a_{k-1}\}$$

is an Euler line between v_{l_1} and itself. We do not exclude the case $v_{h_1} = v_{l_1}$. We will say that two Euler lines of a graph belong to the same *circuit* if one of them can be obtained from the other by the process described above. Thus, a circuit in an $m \times n$ graph consists of n Euler lines. Each Euler line in a circuit is uniquely determined by specifying its first arc. In Example 4.2.13, P_1, Q_1, and Q_4 belong to the same circuit. Similarly, P_2, Q_2, and Q_3 belong to the same circuit. These are the only circuits of the graph. Find the other Euler lines in each circuit.

Lemma 4.2.15 Let G be a graph which has c circuits. If the degree of the vertex v_i is d_i, there are $\frac{1}{2}d_i c$ Euler lines between v_i and itself.

PROOF Let

$$C = \{a_1 = (v_{h_1}, v_{h_2}), a_2 = (v_{h_2}, v_{h_3}), \ldots, a_n = (v_{h_n}, v_{h_1})\} \quad (4.2.16)$$

be a circuit of G. The number of Euler lines between a fixed vertex, say v_i, and itself is equal to the number of arcs in (4.2.16) whose first vertex is v_i. But this number is $\frac{1}{2}d_i c$.   ////

EXERCISES

1. Find the circuits of the component G_1 of the graph of Fig. 4.2.3. Find the Euler lines of each circuit.

2. Let v_1 and v_2 be vertices of the graph G. We define the *distance* between v_1 and v_2 by

$$d(v_1, v_2) = \begin{cases} \text{length of shortest path between } v_1 \text{ and } v_2 & \text{if } v_1 \text{ and } v_2 \text{ belong to same component of } G \\ \infty & \text{If } v_1 \text{ and } v_2 \text{ do not belong to same component of } G \end{cases}$$

   Show that d satisfies the properties of a *metric*, i.e.,
   (a) $d(v_1, v_2) \geq 0$ for all $v_1, v_2 \in \mathbb{G}$ and $d(v_1, v_2) = 0$ if and only if $v_1 = v_2$.
   (b) $d(v_1, v_2) = d(v_2, v_1)$ for all v_1, v_2.
   (c) $d(v_1, v_2) + d(v_2, v_3) \geq d(v_1, v_3)$ for all v_1, v_2, v_3.
   For the graph of Fig. 4.1.10, find $d(v_1, v_3)$ and $d(v_7, v_{10})$.

3. Let G be a graph which has an Euler line and let v_1 and v_2 be vertices of G. Show that if P is a path between v_1 and v_2, then there is a path P' between v_1 and v_2 which is arc-disjoint from P. [*Hint:* Consider the case $v_1 = v_2$ separately. For $v_1 \neq v_2$ there is an arc $a_1 = (v_1, v_{h_1})$ which does not belong to P.]

4. Let G be a connected graph. We define the *diameter* of G by

$$\text{diam } G = \max_{v_1, v_2} d(v_1, v_2)$$

   Find the diameter of the graphs of Figs. 4.1.8, 4.1.10, and 4.1.14.

5. Let G be a connected graph with m vertices and n arcs. Show that diam $G \leq m - 1$. Also show that if G has an Euler line, then diam $G \leq n/2$.

6. Let A be the vertex-adjacency matrix of the connected graph G. Prove that diam G is the smallest positive integer d such that every entry of $(I + A)^d$ is positive.

7. Show that a graph contains exactly

$$\frac{1}{2} \sum_{i=1}^{m} d_i(d_i - 1)$$

   paths of length 2, where m is the number of vertices and the d_i's are the degrees of the vertices.

8. A *complete graph* is defined to be a graph such that between every pair of distinct vertices there is exactly one arc. Show that a complete graph with m vertices has $\frac{1}{2}m(m-1)$ arcs and $\frac{1}{2}m!$ simple paths of length $m - 1$. For which values of m does a complete graph have an Euler line?

9 Let G be a connected graph with m vertices. Prove that the following conditions are equivalent:
(*a*) G has no loops.
(*b*) G has $m - 1$ arcs.
(*c*) Between every pair of vertices of G there is exactly one path.
A graph satisfying one of these conditions (and hence all) is called a *tree*

4.3 THE VECTOR SPACE OF SETS

In Exercise 6 of Chap. 0 a method was outlined for constructing an infinite number of fields, one for each positive prime integer. The field $\mathbb{Z}_p$ corresponding to the prime p contains p elements. In this chapter we will make extensive use of the field $\mathbb{Z}_2$ consisting of the two elements 0 and 1. Addition and multiplication within $\mathbb{Z}_2$ are given by the following tables:

ADDITION			MULTIPLICATION		
	0	1		0	1
0	0	1	0	0	0
1	1	0	1	0	1

(4.3.1)

Let S be a set consisting of n elements

$$S = \{s_1, s_2, \ldots, s_n\}$$

and let $\mathbb{S}$ be the collection of all subsets of S. Among the elements of $\mathbb{S}$ are

1 The empty set $\emptyset$
2 n elements $\{s_1\}, \{s_2\}, \ldots, \{s_n\}$, each consisting of a single element of S
3 $\frac{1}{2}n(n-1)$ elements $\{s_i, s_j\}$ each consisting of a pair of distinct elements of S, etc.

In total, $\mathbb{S}$ contains 2^n elements. We will show that with the proper definitions of addition and scalar multiplication (the scalars coming from the field $\mathbb{Z}_2$) $\mathbb{S}$ has the properties of a vector space given in Sec. 1.3. We will also show that there is a natural way of representing each element of $\mathbb{S}$ as an n-dimensional (row or column) vector with entries from $\mathbb{Z}_2$. Once this representation has been made, $\mathbb{S}$ and the vector space of n-dimensional vectors are (mathematically) identical.

Let x be an element of $\mathbb{S}$. Then x is a subset of S. The set of elements of S which do not belong to x is a subset of S and hence belongs to $\mathbb{S}$. We call this set the *complement* of x and denote it by x'. Clearly

$$(x')' = x$$

Thus if y is the complement of x, then x is the complement of y. Note that $\emptyset$ and S are complements of each other.

We can define an operation on the elements of $\mathbb{S}$ which, as we will see, behaves very much like addition.

Definition 4.3.2 Let x and y be subsets of S. The set of all elements of S which belong to exactly one of the subsets x and y is a subset of S called the *symmetric difference* of x and y and is denoted by $x + y$.

All the following properties are easy to verify.

Corollary 4.3.3. For all $x, y \in \mathbb{S}$,

1. $x + y = y + x$.
2. $x + (y + z) = (x + y) + z$.
3. $x + \emptyset = \emptyset + x = x$.
4. $x + S = S + x = x'$.
5. $x + x' = x' + x = S$.
6. $x + x = \emptyset$.

Thus $\emptyset$ acts as an additive identity, and every element of $\mathbb{S}$ is its own additive inverse.

The equations

$$1 + 1 = 0 \quad \text{and} \quad x + x = \emptyset$$

in (4.3.1) and part 6 of Corollary 4.3.3 respectively suggest that addition in $\mathbb{Z}_2$ and the symmetric difference behave in a similar manner. Let us define a *scalar multiplication* between elements of $\mathbb{Z}_2$ and elements of $\mathbb{S}$ by

$$1 \cdot x = x \qquad 0 \cdot x = \emptyset$$

for all $x \in \mathbb{S}$. Then

Lemma 4.3.4 For all $x, y \in \mathbb{S}$ and $\alpha, \beta \in \mathbb{Z}_2$,

1. $(\alpha + \beta)x = \alpha x + \beta x$.
2. $\alpha(x + y) = \alpha x + \alpha y$.
3. $(\alpha \beta)x = \alpha(\beta x)$.

With these operations we call $\mathbb{S}$ *the vector space of subsets of S over the field* $\mathbb{Z}_2$. All the properties of a vector space of $m \times n$ matrices as defined in Sec. 1.3 are shared by $\mathbb{S}$. In fact, there is a natural way whereby we can identify $\mathbb{S}$ with $\mathbb{F}^n$ or $\mathbb{F}_n$, where $\mathbb{F}$ is $\mathbb{Z}_2$. For $x \in \mathbb{S}$, let $X = [x_i]$ be the n-dimensional column vector defined by

$$x_i = \begin{cases} 1 & \text{if } s_i \in x \\ 0 & \text{if } s_i \notin x \end{cases} \quad (4.3.5)$$

Equation (4.3.5) gives rise to a mapping

$$\Pi : x \in \mathbb{S} \to X \in \mathbb{F}^n$$

of the 2^n elements of $\mathbb{S}$ to the 2^n elements of $\mathbb{F}^n$. This mapping is a one-to-one correspondence between the elements of $\mathbb{S}$ and $\mathbb{F}^n$. Moreover the arithmetic operations on $\mathbb{S}$ and $\mathbb{F}^n$ are preserved, i.e.,

$$\Pi(x + y) = \Pi(x) + \Pi(y)$$
$$\Pi(\alpha x) = \alpha \Pi(x)$$

for all $x, y \in \mathbb{S}$ and $\alpha \in \mathbb{Z}_2$. We call Π the *natural correspondence* between the elements of $\mathbb{S}$ and $\mathbb{F}^n$. In the same way there is a natural correspondence between the elements of $\mathbb{S}$ and $\mathbb{F}_n$. We will phrase our discussion in terms of the natural correspondence between $\mathbb{S}$ and $\mathbb{F}^n$. Clearly the same results are valid for $\mathbb{S}$ and $\mathbb{F}_n$.

Under the natural correspondence the zero vector of $\mathbb{F}^n$ corresponds to the empty subset of S, and E, the vector all of whose entries are equal to 1, corresponds to S.

We can use the natural correspondence to transfer properties of sets from $\mathbb{S}$ to $\mathbb{F}^n$ and to transfer properties of vectors and vector spaces from $\mathbb{F}^n$ to $\mathbb{S}$. For example, let $X, Y \in \mathbb{F}^n$ correspond to $x, y \in \mathbb{S}$ respectively. We say that X is *contained* in Y and write $X \subseteq Y$ if $x \subseteq y$, considering x and y as subsets of S. Similarly we define X and Y to be *disjoint* if x and y are disjoint, and we say that X is the *complement* of Y and write $X = Y'$ if $x = y'$. Thus, for every vector X of $\mathbb{F}^n$, X and X' are disjoint and $X + X' = E$. Every vector of $\mathbb{F}^n$ is contained in E. Clearly X is contained in Y if whenever $x_i = 1$, then $y_i = 1$. Similarly X and Y are disjoint if whenever $x_i = 1$, then $y_i = 0$.

Let $\mathbb{T}$ be a subspace of $\mathbb{S}$. An element $x \in \mathbb{T}$ is *minimal* if $x \neq \emptyset$ and there does not exist $y \in \mathbb{S}$, $y \neq \emptyset$ such that $y \subseteq x$. Thus the minimal elements of $\mathbb{S}$ are those subsets of S which consist of a single element. Let $\mathbb{V}$ be the set of vectors of $\mathbb{F}^n$ which correspond to the elements of $\mathbb{T}$ under the natural correspondence. Then $\mathbb{V}$ is a subspace of $\mathbb{F}^n$, and we say that X is *minimal* in $\mathbb{V}$ if the element of $\mathbb{T}$ corresponding to X is minimal in $\mathbb{T}$.

As an example of a property of vectors being carried across the natural correspondence from $\mathbb{F}^n$ to $\mathbb{S}$ we say that the elements $x, y, \ldots, z$ of $\mathbb{S}$ are *linearly independent* if the corresponding elements $\Pi(x), \Pi(y), \ldots, \Pi(z)$ are linearly independent vectors of $\mathbb{F}^n$.

Lemma 4.3.6 Let $\mathbb{V}$ be a subspace of $\mathbb{F}^n$ and let $X \in \mathbb{V}$. Then X can be expressed as a sum of disjoint minimal vectors of $\mathbb{V}$ which are contained in X.

PROOF If X is not itself minimal, then it contains a minimal vector, say Y. Set

$$Z = X + Y \qquad (4.3.7)$$

Since $Y \subseteq X$, it follows that $Z \subseteq X$ and that Y and Z are disjoint. Moreover, equation (4.3.7) is equivalent to

$$X = Y + Z \qquad (4.3.8)$$

If Z is a minimal vector, we have expressed X as a sum of disjoint minimal vectors of $\mathbb{V}$ contained in X. If Z is not minimal, then we use the same argument to express Z as

$$Z = U + V \qquad (4.3.9)$$

where U and V are vectors contained in Z (and hence are contained in X) and U is minimal. Substituting equation (4.3.9) into equation (4.3.8) yields

$$X = Y + U + V$$

If V is a minimal vector, we are finished. If V is not minimal, the process is repeated once again. Since the number of entries which are equal to 1 is decreasing at each step (that is, X has more 1s than Z and Z has more 1s than V), this process must eventually stop. ////

Note that we do not claim that the minimal elements are either unique for a given vector or are the same for different vectors. In Example 4.3.11 we will show that neither of these properties need be true.

It is an immediate consequence of Lemma 4.3.6 that the minimal vectors of a vector space form a spanning set, and hence:

Theorem 4.3.10 Every subspace of $\mathbb{F}^n$ has a basis of minimal vectors.

EXAMPLE 4.3.11 Let $\mathbb{V}$ be the set of all vectors of $\mathbb{F}^4$ (where $\mathbb{F} = \mathbb{Z}_2$) which contain an even number of 1s. Then $\mathbb{F}^4$ contains 16 elements, and it is not difficult to verify that $\mathbb{V}$ is a subspace of $\mathbb{F}^4$ and has 8 elements. There are 6 vectors belonging to $\mathbb{V}$ which have exactly two 1s, and these are the minimal vectors of $\mathbb{V}$. It can easily be shown that $\mathbb{V}$ does not have a basis of disjoint vectors. However, one basis for $\mathbb{V}$ consisting of minimal vectors is

$$\left\{ \begin{bmatrix} 1 \\ 1 \\ 0 \\ 0 \end{bmatrix}, \begin{bmatrix} 0 \\ 0 \\ 1 \\ 1 \end{bmatrix}, \begin{bmatrix} 0 \\ 1 \\ 1 \\ 0 \end{bmatrix} \right\}$$

////

In the next section we will make extensive use of the natural correspondence. The set S will be chosen in several different ways according to our needs at the time. Among the choices we will make are: (1) the set of vertices of a graph, (2) the set of arcs of a graph, (3) the set of loops of a graph.

EXERCISES

1 For the vector space $\mathbb{V}$ of Example 4.3.11, express the following vectors in terms of the given basis:

$$(a) \begin{bmatrix} 1 \\ 0 \\ 0 \\ 1 \end{bmatrix} \quad (b) \begin{bmatrix} 1 \\ 0 \\ 1 \\ 0 \end{bmatrix} \quad (c) \begin{bmatrix} 1 \\ 1 \\ 1 \\ 1 \end{bmatrix}$$

2 Let x_1, x_2, and x_3 be subsets of S. Show that $x_1 + x_2 + x_3$ consists of all elements of S which belong to exactly one of x_1, x_2, and x_3 or which belong to all three. Formulate and prove a corresponding result for the sum of any finite number of subsets of S.

3 Show that for every choice of n the set $\mathbb{E}$ of all vectors of $\mathbb{F}^n$ (where $\mathbb{F} = \mathbb{Z}_2$) containing an even number of 1s is a vector space. What are the minimal vectors of $\mathbb{E}$? Find a basis for $\mathbb{E}$ consisting of minimal vectors. What is the dimension of $\mathbb{E}$, and how many vectors does $\mathbb{E}$ contain?

4 Show that for $n \geq 5$ the subspace of $\mathbb{F}^n$ spanned by all vectors containing four 1s is $\mathbb{E}$.

5 Show that for $n \geq 4$ the subspace of $\mathbb{F}^n$ spanned by all vectors containing three 1s is $\mathbb{F}^n$.

6 Prove that every k-dimensional subspace of $\mathbb{F}^n$ contains 2^k vectors.

4.4 MATRICES AND UNDIRECTED GRAPHS

Our prime purpose in this chapter is to illustrate how the theory of matrices can be applied to graph theory. As we have already shown, a matrix provides a systematic and efficient means for displaying and manipulating the properties of a graph. This can be a particularly useful method when we are interested in counting the number of times, if any, that a specific type of configuration appears in a given graph (see, for example, Examples 4.1.1 and 4.1.3). However, there is a second and more basic way in which the theory of matrices can be used to obtain information about graphs. We will be able to formulate many of the important properties of a graph (e.g., connectedness, loops, etc.) in terms of properties of one or more of the matrices associated with it. We will then be able to prove theorems about graphs by demonstrating the corresponding theorems about their matrices. Often the theorem concerning graphs is considerably more complex than the corresponding theorem about matrices, e.g., Theorem 4.4.14. In Sec. 4.1 we defined the vertex-adjacancy matrix A, the arc-adjacancy matrix $\tilde{A}$, the vertex-arc incidence matrix B, and the degree matrix D of an undirected graph. For a graph with m vertices and n arcs these matrices have dimensions $m \times m$, $n \times n$, $m \times n$, and $m \times m$ respectively. It will be convenient to call such a graph *an $m \times n$ graph*.

In Sec. 4.1 the field from which the entries of the matrices A, $\tilde{A}$, B, and D were taken was not specified, though certainly for A, $\tilde{A}$, and D the field of real (or rational) numbers was implied. However, since all the entries of B are either 0 or 1 and every field contains both a 0 and a 1, we are at liberty to choose whichever field we wish to contain the entries of B. If fact, we will occasionally decide to change fields. B is a typical example of an incidence matrix. Its entries, 1 and 0, denote the answer yes or no to a question of incidence. The entries of A, $\tilde{A}$, and D answer the question: How many?. Later in this section we will define several other incidence matrices associated with a graph. For this reason we will call B the *incidence matrix* of its graph.

In order for Theorem 4.1.15 to have meaning, the entries of B must be taken from the same field as those of A and D, namely, the real field. Although we will occasionally revert to the real field again, we will find that it is more profitable to consider the entries of an incidence matrix as coming from the field $\mathbb{Z}_2$, which we discussed at length in the last section. Thus, unless we specify otherwise, we will henceforth assume that the field $\mathbb{F}$ from which the entries of B, as well as the other incidence matrices to be introduced later, are taken is $\mathbb{Z}_2$.

The incidence matrix of an $m \times n$ graph is an $m \times n$ matrix. Since every column of B contains exactly two 1s, we have

$$E_{(m)} B = 0 \qquad (4.4.1)$$

where $E_{(m)}$ is the m-dimensional row vector all of whose entries are 1. The rows of B are linearly dependent, and it follows that the rank of the incidence matrix of an $m \times n$ graph is at most $m - 1$. In Theorem 4.4.4 we will prove that the rank of an $m \times n$ incidence matrix is uniquely determined by the number of components of the graph.

Let $G_1, G_2, \ldots, G_k$ be the components of the graph G and let G_i consist of m_i vertices and n_i arcs. If the vertices and arcs of G are numbered so that

$$G_1 = \{\mathbb{V}_1 = \{v_1, v_2, \ldots, v_{m_1}\}, \mathbb{A}_1 = \{a_1, a_2, \ldots, a_{n_1}\}\}$$
$$G_2 = \{\mathbb{V}_2 = \{v_{m_1+1}, v_{m_1+2}, \ldots, v_{m_1+m_2}\}, \mathbb{A}_2 = \{a_{n_1+1}, a_{n_1+2}, \ldots, a_{n_1+n_2}\}\}$$
$$\cdots\cdots\cdots\cdots\cdots\cdots\cdots\cdots\cdots\cdots\cdots\cdots\cdots\cdots\cdots\cdots$$

then the incidence matrix of G has the form

$$B = \begin{bmatrix} B_1 & 0 & \cdots & 0 \\ 0 & B_2 & \cdots & 0 \\ \cdots\cdots\cdots\cdots\cdots \\ 0 & 0 & \cdots & B_k \end{bmatrix} \quad (4.4.2)$$

where B_i is the $m_i \times n_i$ incidence matrix of G_i.[1] Conversely, a graph whose incidence matrix can be written in the form of (4.4.2) can be expressed as the union of k disjoint subgraphs, one for each of the matrices $B_1, B_2, \ldots, B_k$. Since a connected graph is one which consists of a single component, we can prove:

Lemma 4.4.3 If B is the incidence matrix of a graph G with m vertices, then G is connected if and only if the only row vectors X such that

$$XB = 0$$

are $X = 0$ and $X = E_{(m)}$.

PROOF If G is not connected, it has more than one component and we may assume that B has the form (4.4.2). Clearly

$$X = [E_{(m_1)} \ 0 \ \cdots \ 0]$$

satisfies the equation

$$XB = 0$$

and $X \neq 0$, $X \neq E_{(m_i)}$.

---

[1] A vertex is isolated if there are no arcs incident with it. We are then faced with the paradox that the incidence matrix of an isolated point has one row and no columns. This need cause no trouble, however. If G contains at least one arc but the component G_i does not, then there is a row of B corresponding to G_i and every entry of that row is 0.

Conversely, let G be a connected graph and suppose that there exists a vector X such that

$$XB = 0 \quad X \neq 0 \quad X \neq E_{(m)}$$

Since each entry of X is either 0 or 1, we may assume that the vertices of G are numbered so that

$$X = [E_{(q)} \quad 0] \quad 1 \leq q \leq m - 1$$

We now partition B as

$$B = \begin{bmatrix} B_1 \\ B_2 \end{bmatrix}$$

where B_1 and B_2 contain q rows and $m - q$ rows respectively. Then

$$E_{(q)} B_1 = 0$$

and consequently every column of B_1 contains either two 1s or no 1s. We may assume that the arcs of G are numbered so that each of the first p columns of B_1 contains two 1s and each of the remaining columns of B_1 contains no 1s. We can now partition B as

$$B = \begin{bmatrix} B_{11} & 0 \\ 0 & B_{22} \end{bmatrix}$$

where B_{11} and B_{22} are $q \times p$ and $(m - q) \times (n - p)$ matrices respectively. (If $p = 0$, then

$$B = \begin{bmatrix} 0 \\ B_{22} \end{bmatrix}$$

whereas if $p = n$, then

$$B = \begin{bmatrix} B_{11} \\ 0 \end{bmatrix}$$

Either way, G contains isolated vertices.) Since the graph is connected, we have reached a contradiction. Thus, no such vector X exists. ////

Lemma 4.4.3 states that a graph is connected if and only if the rows of its incidence matrix are a *minimal linearly dependent set*; i.e., the m rows of B are linearly dependent, but every subset consisting of fewer than m rows is linearly independent. Thus we have an algebraic characterization of a connected graph. The same idea can be used to identify the components of a nonconnected graph. Thus Lemma 4.4.3 is a special case of the following theorem.

Theorem 4.4.4 If B is the incidence matrix of an $m \times n$ graph with k components the rank of B is given by

$$r(B) = m - k$$

PROOF We may assume that the vertices and arcs of the graph are numbered so that B has the form (4.4.2). The rank of B can be obtained by means of (the row version of) Theorem 1.6.1, i.e.,

$$r(B) = m - v_R(B)$$

where $v_R(B)$ is the row nullity of B. Recall that the row nullity of B is the dimension of $\mathbb{N}_R(B)$, the row null space of B,

$$\mathbb{N}_R(B) = \{X \mid XB = 0, X \in \mathbb{F}_m\}$$

Thus, in order to prove the theorem it is sufficient to show that

$$v_R(B) = k \qquad (4.4.5)$$

It follows from equation (4.4.1) that the k vectors

$$\begin{aligned} Y_1 &= [E_{(m_1)} \quad 0 \quad 0 \quad \cdots \quad 0 \;] \\ Y_2 &= [\; 0 \quad E_{(m_2)} \quad 0 \quad \cdots \quad 0 \;] \\ &\cdots \cdots \cdots \cdots \cdots \cdots \cdots \cdots \\ Y_k &= [\; 0 \quad 0 \quad 0 \quad \cdots \quad E_{(m_k)}] \end{aligned}$$

belong to $\mathbb{N}_R(B)$. Clearly $Y_1, Y_2, \ldots, Y_k$ are linearly independent. In order to establish equation (4.4.5) it remains only to prove that $Y_1, Y_2, \ldots, Y_k$ span $\mathbb{N}_R(B)$.

Let $X \in \mathbb{F}_m$. Partition X as

$$X = [X_1 \quad X_2 \quad \cdots \quad X_k]$$

where X_i is an m_i-dimensional row vector. Clearly $XB = 0$ if and only if

$$X_i B_i = 0 \qquad i = 1, 2, \ldots, k$$

and, by Lemma 4.4.3, we have

$$X_i = 0 \quad \text{or} \quad X_i = E_{(m_i)}$$

Consequently X can be expressed as a linear combination of $Y_1, Y_2, \ldots, Y_k$. ////

Theorem 4.4.4 illustrates an important advantage gained by choosing the field $\mathbb{Z}_2$ for $\mathbb{F}$. In proving the theorem we incidentally showed that in the natural correspondence between the vector space of subsets of the set of vertices of an

$m \times n$ graph G and $\mathbb{F}_m$, the components of G correspond to the minimal elements of $\mathbb{N}_R(B)$.[1] In this natural correspondence the elements of the vector space of sets of components of G correspond to the elements of $\mathbb{N}_R(B)$. The various natural correspondences associated with a graph form one of the strongest tools in graph theory.

Because of Theorem 4.4.4 and the simple form of (4.4.2) we can once again limit our discussion to graphs which are connected. In all cases the extension to graphs with more than one component is a trivial matter. It will be convenient for us to look at the incidence matrices of a simple path and a loop before investigating the properties of incidence matrices generally.

EXAMPLE 4.4.6 For a pair of vertices v_1 and v_l, $v_1 \neq v_l$, a path P of length $l - 1$ between v_1 and v_l is a graph which consists of l vertices and $l - 1$ arcs. When these vertices and arcs can be numbered so that

$$P = \{a_1 = (v_1, v_2), a_2 = (v_2, v_3), \ldots, a_{l-1} = (v_{l-1}, v_l)\}$$

then the incidence matrix of P is the $l \times (l - 1)$ matrix

$$B_P = \begin{bmatrix} 1 & & & & \\ 1 & 1 & & & \\ & 1 & 1 & & \\ & & \cdots & \cdots & \\ & & & 1 & 1 \\ & & & & 1 \end{bmatrix}$$

A loop L of length l is a graph which consists of l vertices and l arcs. When the vertices and arcs of L are numbered so that

$$L = \{a_1 = (v_1, v_2), a_2 = (v_2, v_3), \ldots, a_{l-1} = (v_{l-1}, v_l), a_l = (v_l, v_1)\}$$

then the incidence matrix of L is the $l \times l$ matrix

$$B_L = \begin{bmatrix} 1 & & & & & 1 \\ 1 & 1 & & & & \\ & 1 & 1 & & & \\ & & \cdots & \cdots & & \\ & & & 1 & 1 & \\ & & & & 1 & 1 \end{bmatrix} \quad (4.4.7)$$

[1] More accurately, the natural correspondence is between the minimal elements of $\mathbb{N}_R(B)$ and those sets of vertices of G which are the vertices of a component of G. We have tacitly identified a component with the set of vertices of that component.

Clearly L can be formed from P by adding the arc $a_l = (v_l, v_1)$, and therefore B_L can be formed from B_P by adding the column corresponding to a_l. Since P and L are connected graphs with l vertices, it follows from Theorem 4.4.4 that

$$r(B_P) = r(B_L) = l - 1$$

(The rank can also be obtained directly from the matrix.) Since B_P has $l - 1$ columns, they are linearly independent. However, each row of B_L has exactly two 1s, and consequently

$$B_L E^{(l)} = 0$$

According to Theorem 1.6.1, the (column) nullity of B_L is 1, and hence $X = E^{(l)}$ is the only nonzero vector such that

$$B_L X = 0$$

Thus the columns of B_L form a minimal linearly dependent set. We will see that this property characterizes a loop. ////

Lemma 4.4.8 Let B be the incidence matrix of the graph G. A necessary and sufficient condition that a set L of arcs of G (and the vertices incident with them) form a loop is that the columns of B corresponding to the arcs of L form a minimal linearly dependent set.

PROOF In the discussion above we showed that a set of columns of B which correspond to the arcs of a loop form a minimal linearly dependent set. Conversely, let L be any subgraph of G which has the property that the columns of B corresponding to the arcs contained in L form a minimal linearly dependent set. We wish to prove that L is a loop. Let L contain l arcs and assume that the arcs of G are numbered so that the arcs of L are $a_1, a_2, \ldots, a_l$. Then we can partition B as

$$B = [B_1 \quad B_2]$$

where B_1 consists of the l columns corresponding to the arcs of L. Since the columns of B_1 form a minimal linearly dependent set, we have

$$B_1 E^{(l)} = 0 \qquad (4.4.9)$$

and $X = E^{(l)}$ is the only nonzero vector such that

$$B_1 X = 0$$

It follows from equation (4.4.9) that every row of B_1 contains an even number of 1s.

We wish to show that there is some numbering of the vertices and arcs of G (or L) for which B_1 has the form (4.4.7), possibly augmented by rows consisting entirely of 0s. We may assume that the vertices of G are numbered so that

$$a_1 = (v_1, v_2)$$

Then the first column of B_1 is

$$\begin{bmatrix} 1 \\ 1 \\ 0 \\ 0 \\ \vdots \\ 0 \end{bmatrix}$$

Since the second row of B_1 possesses an even number of 1s, there is another column of B_1 which contains a 1 in the second row. We may assume that the second column of B_1 has this property. However, every column of B_1 (in particular, the second column) has two 1s, and we must consider two cases. If the other 1 of the second column is in the first row, then the first two columns of B_1 are identical. Hence they form a minimal linearly dependent set, and their arcs, being parallel, form a loop of length 2. In this case we are finished. On the other hand, if the other 1 of the second column of B_1 is not in the first row, then we may assume that it is in the third row, i.e.,

$$a_2 = (v_2, v_3)$$

and the first two columns of B_1 are

$$\begin{bmatrix} 1 & 0 \\ 1 & 1 \\ 0 & 1 \\ 0 & 0 \\ \vdots & \vdots \\ 0 & 0 \end{bmatrix}$$

Suppose now that we have continued this process until we have

$$a_1 = (v_1, v_2), \quad a_2 = (v_2, v_3), \quad \ldots, \quad a_k = (v_k, v_{k+1}) \qquad (4.4.10)$$

and the first k columns of B_1 are

$$\begin{array}{c} \\ v_1 \\ v_2 \\ v_3 \\ v_4 \\ \\ v_k \\ v_{k+1} \\ \\ \end{array} \begin{array}{c} \begin{array}{cccccc} a_1 & a_2 & a_3 & & a_k \end{array} \\ \left[\begin{array}{ccccc} 1 & 0 & 0 & \cdots & 0 \\ 1 & 1 & 0 & \cdots & 0 \\ 0 & 1 & 1 & \cdots & 0 \\ 0 & 0 & 1 & \cdots & 0 \\ \cdot & \cdot & \cdot & \cdots & \cdot \\ 0 & 0 & 0 & \cdots & 1 \\ 0 & 0 & 0 & \cdots & 1 \\ \cdot & \cdot & \cdot & \cdots & \cdot \\ 0 & 0 & 0 & \cdots & 0 \end{array} \right] \end{array}$$

As before, the $(k + 1)$st row of B_1 contains at least two 1s, and we may assume that the $(k + 1)$st column of B_1 has a 1 in the $(k + 1)$st row. Now several cases arise.

CASE 1 If the second 1 of the $(k + 1)$st column is in the first row, we may partition the first $k + 1$ columns of B_1 into

$$B_1 = \begin{bmatrix} B_{11} \\ 0 \end{bmatrix}$$

where B_{11} is a $(k + 1) \times (k + 1)$ matrix of the form of (4.4.7). The first $k + 1$ columns of B_1 are linearly dependent and hence constitute all of B_1, that is, $k + 1 = l$.

In this case we also have

$$a_{k+1} = (v_{k+1}, v_1)$$

and a_{k+1} together with the arcs of (4.4.10) form a loop of length $k + 1 = l$. Note that the rows of B_1 which do not belong to B_{11} correspond to the vertices of G which are not incident with the arcs of the loop.

CASE 2 Suppose that the second 1 of the $(k + 1)$st column of B_1 is in the rth row, where $2 \leq r \leq k$. Then the rth, $(r + 1)$st, ..., kth, and $(k + 1)$st columns of B_1 are linearly dependent since their sum is 0. This contradicts the fact that the columns of B_1 form a minimal linearly dependent set. Thus case 2 can never occur.

CASE 3 If the second 1 of the $(k + 1)$st column of B_1 is not in any of the first $k + 1$ rows, we may assume that it is in the $(k + 2)$nd row and this process continues. Since there are only a finite number of columns in B_1, we must eventually reach case 1. ////

Theorem 4.4.11 Under the natural correspondence between the vector space of sets of arcs of G and $\mathbb{F}^n$ the loops of G correspond to the minimal vectors of the vector space

$$\mathbb{N}(B) = \{X \mid BX = 0, \, X \in \mathbb{F}^n\}$$

[Recall that $\mathbb{N}(B)$ is the (column) null space of B.]

Lemma 4.4.8 and Theorem 4.4.11 apply equally to connected and nonconnected graphs. Since the rank of an $m \times n$ graph with k components is $m - k$, it follows that

$$\dim \mathbb{N}(B) = \nu(B) = n - m + k$$

For a connected $m \times n$ graph we have

$$\dim \mathbb{N}(B) = \nu(B) = n - m + 1$$

Unlike the situation we encountered with the minimal vectors of $\mathbb{N}_R(B)$, the minimal vectors of $\mathbb{N}(B)$ need not be disjoint.

EXAMPLE 4.4.12 The incidence matrix of the 5×7 graph G

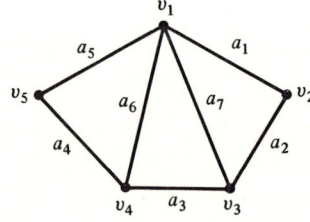

is the 5×7 matrix

$$B = \begin{array}{c} \\ \\ \end{array} \begin{array}{c} a_1 \ a_2 \ a_3 \ a_4 \ a_5 \ a_6 \ a_7 \\ \begin{bmatrix} 1 & 0 & 0 & 0 & 1 & 1 & 1 \\ 1 & 1 & 0 & 0 & 0 & 0 & 0 \\ 0 & 1 & 1 & 0 & 0 & 0 & 1 \\ 0 & 0 & 1 & 1 & 0 & 1 & 0 \\ 0 & 0 & 0 & 1 & 1 & 0 & 0 \end{bmatrix} \end{array} \begin{array}{c} v_1 \\ v_2 \\ v_3 \\ v_4 \\ v_5 \end{array}$$

Since G is a connected graph, the rank of B is 4. According to Theorem 1.6.1, the dimension of $\mathsf{N}(B)$ is 3. It is not difficult to verify that the three vectors

$$X_1 = \begin{bmatrix} 1 \\ 1 \\ 0 \\ 0 \\ 0 \\ 0 \\ 1 \end{bmatrix} \quad X_2 = \begin{bmatrix} 0 \\ 0 \\ 1 \\ 0 \\ 0 \\ 1 \\ 1 \end{bmatrix} \quad X_3 = \begin{bmatrix} 0 \\ 0 \\ 0 \\ 1 \\ 1 \\ 1 \\ 0 \end{bmatrix}$$

corresponding to the three loops

$$L_1 = \{a_1, a_2, a_7\} \quad L_2 = \{a_3, a_6, a_7\} \quad L_3 = \{a_4, a_5, a_6\}$$

form a basis for $\mathsf{N}(B)$. The remaining vectors belonging to $\mathsf{N}(B)$ are 0 and

$$X_4 = X_1 + X_2 = \begin{bmatrix} 1 \\ 1 \\ 1 \\ 0 \\ 0 \\ 1 \\ 0 \end{bmatrix} \quad X_5 = X_1 + X_3 = \begin{bmatrix} 1 \\ 1 \\ 0 \\ 1 \\ 1 \\ 1 \\ 1 \end{bmatrix} \quad X_6 = X_2 + X_3 = \begin{bmatrix} 0 \\ 0 \\ 1 \\ 1 \\ 1 \\ 0 \\ 1 \end{bmatrix} \quad X_7 = X_1 + X_2 + X_3 = \begin{bmatrix} 1 \\ 1 \\ 1 \\ 1 \\ 1 \\ 0 \\ 0 \end{bmatrix}$$

All the nonzero vectors of $\mathsf{N}(B)$ are minimal except X_5. X_5 is not minimal since it contains both X_1 and X_3. The loops corresponding to X_4, X_6, and X_7 are

$$L_4 = \{a_1, a_2, a_3, a_6\} \quad L_6 = \{a_3, a_4, a_5, a_7\} \quad L_7 = \{a_1, a_2, a_3, a_4, a_5\}$$

whereas X_5 corresponds to the set of arcs belonging to L_1 and L_3. ////

The natural correspondence described in Theorem 4.4.11 between the minimal elements of $\mathsf{N}(B)$ and the loops of G can be extended to include all the elements of $\mathsf{N}(B)$. For this we need:

Definition 4.4.13 A subgraph of a graph is called a *loop set* if its set of arcs can be expressed as the sum of the arcs of arc-disjoint loops.

Definition 4.4.13 is somewhat clumsily worded because we do not require the loops of a loop set to be vertex-disjoint. In Example 4.4.12, the subgraph L_5 consisting of the set of arcs corresponding to the vector X_5, and the vertices incident with them, is a loop set. In order to simplify our discussion we will say that

$$L_5 = L_1 + L_3$$

(ignoring the problem caused by the fact that L_1 and L_3 are not vertex-disjoint). Note that a loop is a loop set consisting of a single loop.

Theorem 4.4.14 Under the natural correspondence between the vector space of sets of arcs of G and $\mathbb{F}^n$ the loop sets of G correspond to the elements of $\mathbb{N}(B)$.

PROOF Let L be a loop set of G. Since a loop of G corresponds to a minimal vector of $\mathbb{N}(B)$, the vector corresponding to L is a sum of minimal vectors of $\mathbb{N}(B)$ and consequently belongs to $\mathbb{N}(B)$.

Conversely, let $X \in \mathbb{N}(B)$. According to Lemma 4.3.6, X can be expressed as a sum of disjoint minimal vectors of $\mathbb{N}(B)$. Therefore the graph whose set of arcs corresponds to X can be expressed as a sum of arc-disjoint loops and hence is a loop set. ////

Since $\mathbb{N}(B)$ is closed under the operation of addition, it follows from Theorem 4.4.14 that the collection of loop sets of G has the same property. Thus the sum of (possibly not arc-disjoint) loops of G can be expressed as the sum of arc-disjoint loops. In Example 4.4.12, the sum $L_2 + L_7$ is not arc-disjoint, but it can be expressed as the arc-disjoint sum $L_1 + L_3$.

According to Theorem 4.4.14, the number of loop sets of a graph is equal to the number of vectors contained in $\mathbb{N}(B)$. We have previously shown that for a connected $m \times n$ graph G the dimension of $\mathbb{N}(B)$ is $n - m + 1$ and $\mathbb{N}(B)$ contains 2^{n-m+1} vectors. Therefore G contains 2^{n-m+1} loop sets, including the empty set. It is generally not possible to specify the number of loops of G. The 5×7 graph of Example 4.4.12 has eight loop sets, of which six are loops. We leave it as an exercise to show that if the arc $a_7 = (v_1, v_3)$ of the graph of Example 4.4.12 is replaced by $a_7 = (v_2, v_5)$, the resulting 5×7 graph also has eight loop-sets but seven of them are loops.

It is an immediate consequence of Lemma 4.4.8 that a necessary and sufficient condition for a set of arcs of a graph not to contain a loop is that the columns of the incidence matrix corresponding to these arcs be linearly independent. This result allows us to characterize those subgraphs of a graph which do not contain loops.

Definition 4.4.15 A subgraph T of a connected graph G is called a *tree* if T is a maximal subgraph of G containing no loops; that is, T contains no loops, but every subgraph of G which contains T properly also contains a loop.

It does not follow from Definition 4.4.15 that all trees of a graph have the same number of arcs. However, in Theorem 4.4.16 we shall show that this is, in fact, the case. Lemma 4.4.8 states that a set of arcs of G form a tree if and only if the corresponding columns of B form a basis for $\mathbb{C}(B)$, the column space of B. We can now apply some of the theorems concerning bases of a vector space which we proved in Chap. 1. Then we have:

Theorem 4.4.16 For an $m \times n$ graph G:

1. Every tree of G has exactly $m - 1$ arcs.
2. Every subgraph of G which contains $m - 1$ arcs and has no loops is a tree.
3. If S is any set of arcs of G which contains no loops, then G has a tree which contains S.

PROOF By Theorem 4.4.4,

$$r(B) = m - 1$$

Parts 1, 2, and 3 of the theorem are restatements of Corollary 1.5.3, Corollary 1.5.4 and Theorem 1.5.1 respectively. ////

There are numerous trees in the graph of Example 4.4.12. In Sec. 4.6 we shall calculate the exact number. Every tree of the graph has four arcs; e.g.,

$$T_1 = \{a_1, a_2, a_3, a_4\} \qquad T_2 = \{a_1, a_7, a_5, a_6\}$$
$$T_3 = \{a_5, a_6, a_3, a_2\} \qquad T_4 = \{a_5, a_6, a_7, a_2\}$$

are trees. Since the incidence matrix of a tree with m vertices has rank $m - 1$, a tree is connected.

Definition 4.4.17 Let T be a tree of the graph G. An arc of G which does not belong to T is called a *chord* of G (with respect to T).

A graph with n arcs containing a tree with $m - 1$ arcs has $n - m + 1$ chords. Recall that

$$\dim \mathbb{N}(B) = v(B) = n - m + 1$$

In Theorem 4.4.21 we will show that the chords of a graph with respect to a given tree determine a basis for $\mathbb{N}(B)$.

Lemma 4.4.18 Each chord of a tree is contained in a unique loop which contains no other chords of the tree.

PROOF Let T be a tree of G and let $c = (v_1, v_2)$ be a chord of G with respect to T. Since T is connected, there is a path P between v_1 and v_2

consisting of arcs of T. Let L be the union of P and $\{c\}$. Then L is a loop consisting of c and some of the arcs of T. It is left as an exercise to show that L is the only loop with this property. ////

The set of $n - m + 1$ loops of a graph corresponding to the $n - m + 1$ chords of a tree, as described by Lemma 4.4.18, is called a *fundamental set of loops*. Under the natural correspondence between sets of arcs of G and $\mathbb{F}^n$, the vectors corresponding to the loops of a fundamental set constitute a set of $n - m + 1$ vectors of $\mathbb{N}(B)$. We will show that they form a basis for $\mathbb{N}(B)$.

Definition 4.4.19 Let $L_1, L_2, \ldots, L_p$ be the loops of G (in some order). The $n \times p$ matrix $L = [l_{jk}]$, where

$$l_{jk} = \begin{cases} 1 & \text{when arc } a_j \text{ belongs to loop } L_k \\ 0 & \text{when arc } a_j \text{ does not belong to loop } L_k \end{cases}$$

is called the *loop matrix* of G.

The loop matrix of the graph of Example 4.4.12 is the 7×6 matrix

$$L = \begin{matrix} & \begin{matrix} L_1 & L_2 & L_3 & L_4 & L_6 & L_7 \end{matrix} & \\ & \begin{bmatrix} 1 & 0 & 0 & 1 & 0 & 1 \\ 1 & 0 & 0 & 1 & 0 & 1 \\ 0 & 1 & 0 & 1 & 1 & 1 \\ 0 & 0 & 1 & 0 & 1 & 1 \\ 0 & 0 & 1 & 0 & 1 & 1 \\ 0 & 1 & 1 & 1 & 0 & 0 \\ 1 & 1 & 0 & 0 & 1 & 0 \end{bmatrix} & \begin{matrix} a_1 \\ a_2 \\ a_3 \\ a_4 \\ a_5 \\ a_6 \\ a_7 \end{matrix} \end{matrix}$$

There are several relations between the incidence matrix and the loop matrix of a graph.

Theorem 4.4.20

1. $BL = 0$.
2. $\mathbb{C}(L) = \mathbb{N}(B)$.
3. $r(L) = n - m + 1$.
4. $\mathbb{N}_R(L) = \mathbb{R}(B)$.

PROOF The kth column of L is the vector of $\mathbb{F}^n$ corresponding to the set of arcs of L_k under the natural correspondence. Applying Theorem 4.4.11, we see that the columns of L are the minimal elements of $\mathbb{N}(B)$. Thus

$BL = 0$. Moreover, by Theorem 4.3.10, the columns of L span $\mathbb{N}(B)$, and consequently

$$\mathbb{C}(L) = \mathbb{N}(B)$$

Thus
$$r(L) = v(B) = n - m + 1$$

Finally, it is clear from part 2 of the theorem that

$$\mathbb{R}(B) \subseteq \mathbb{N}_R(L)$$

Since
$$r(B) = m - 1 = v_R(L)$$

it follows that $\mathbb{N}_R(L) = \mathbb{R}(B)$. ////

In Definition 4.4.19 we did not specify any particular ordering of the arcs and loops of G. Because of Lemma 4.4.18 there is an ordering which produces an interesting form for L. Let T be a tree of G and let us number the arcs of G so that $a_1, a_2, \ldots, a_{n-m+1}$ are the chords of T. Also, let us number the loops of G so that L_i is the unique loop contained in $T \cup \{a_i\}$, $i = 1, 2, \ldots, n - m + 1$. Then it is easily seen that L has the form

$$L = \begin{bmatrix} I & L_{12} \\ L_{21} & L_{22} \end{bmatrix}$$

where I is an identity matrix of order $n - m + 1$. $L_1, L_2, \ldots, L_{n-m+1}$ form a fundamental set of loops of T, and the vectors corresponding to them are the first $n - m + 1$ columns of B. Clearly these columns are linearly independent. Since the rank of L is $n - m + 1$, we have proved:

Theorem 4.4.21 *The vectors of a fundamental set of loops of a graph form a basis for $\mathbb{C}(L) = \mathbb{N}(B)$.*

In Example 4.1.9 we were concerned with finding the smallest set of arcs of a connected graph such that the removal of these arcs leaves a graph which is not connected (and also has the property that two specified vertices are in different components). Such a set of arcs is called a *cut*. Let us give a formal definition.

Definition 4.4.22 Let G be a connected graph. A set C of arcs of G is a *cut* if:

1. The removal of the arcs of C (but not the vertices of these arcs) from G leaves a graph which is not connected.
2. C is minimal; i.e., the removal of any proper subset of the arcs of C from G leaves a connected graph.

Note that a cut is defined only for a connected graph. Since a cut is defined to be a minimal set which disconnects a graph (rather than the smallest set with this property), different cuts of a graph may have different numbers of arcs. There are numerous cuts in the graph of Example 4.4.12; e.g.,

$$C_1 = \{a_2, a_5, a_6, a_7\} \qquad C_2 = \{a_3, a_5, a_6\}$$
$$C_3 = \{a_1, a_2\} \qquad C_4 = \{a_1, a_5, a_6, a_7\}$$

The cuts of a graph can be described in terms of its incidence matrix. Let C be a cut of the $m \times n$ graph G and let $\tilde{G}$ be the (not connected) graph which remains when the arcs of C are removed from G. It will be convenient to number the arcs of G so that $\tilde{G}$ contains $a_1, a_2, \ldots, a_r$ and C consists of $a_{r+1}, a_{r+2}, \ldots, a_n$. Then B can be partitioned as

$$B = [B_1 \quad B_2] \qquad (4.4.23)$$

where B_1 and B_2 are $m \times r$ and $m \times (n - r)$ matrices respectively. Moreover, by Definition 4.4.22, B_1 and B_2 satisfy the following two conditions:

1 $r(B_1) \leq m - 2$ (since $\tilde{G}$ is not connected).
2 An $m \times (r + 1)$ matrix formed from B_1 and any column of B_2 has rank $m - 1$ (since the graph consisting of $\tilde{G}$ and any arc of C is connected). (4.4.24)

Consequently $r(B_1) = m - 2$, and $\tilde{G}$ contains exactly two components.

Conversely, if B is an incidence matrix which can be partitioned as in (4.4.23) so that the two conditions of (4.4.24) are satisfied, then the set of arcs of G corresponding to the columns of B_2 form a cut.

Since the rank of B_1 is $m - 2$, there exists an m-dimensional row vector X, $X \neq 0$, $X \neq E_{(m)}$, such that

$$XB_1 = 0 \qquad (4.4.25)$$

Moreover the only other vector satisfying equation (4.4.25) is

$$\bar{X} = E_{(m)} - X$$

According to condition 2 of (4.4.24), if $\tilde{B}$ is an $m \times (r + 1)$ matrix formed from B_1 and any column of B_2, then

$$X\tilde{B} \neq 0$$

Thus, if we denote the columns of B_2 by $B^{(r+1)}, B^{(r+2)}, \ldots, B^{(n)}$, then

$$XB^{(r+1)} = XB^{(r+2)} = \cdots = XB^{(n)} = 1$$

that is,
$$XB_2 = E_{(n-r)} \qquad (4.4.26)$$

We can use this result to prove:

Theorem 4.4.27 For every loop L and every cut C of a graph the intersection of L and C contains an even number of arcs.

PROOF We may assume that B has the form (4.4.23). We may further assume that $a_1, a_2, \ldots, a_s$ are the arcs of L which do not belong to C and that $a_{r+1}, a_{r+2}, \ldots, a_{r+t}$ are the arcs of L which do belong to C. Thus

$$L = \{a_1, a_2, \ldots, a_s, a_{r+1}, a_{r+2}, \ldots, a_{r+t}\}$$

and we wish to show that t is even. It follows from Lemma 4.4.8 that

$$0 = B^{(1)} + B^{(2)} + \cdots + B^{(s)} + B^{(r+1)} + B^{(r+2)} + \cdots + B^{(r+t)} \quad (4.4.28)$$

Let X be a vector satisfying equations (4.4.25) and (4.4.26). Then

$$XB^{(1)} = XB^{(2)} = \cdots = XB^{(s)} = 0$$
$$XB^{(r+1)} = XB^{(r+2)} = \cdots = XB^{(r+t)} = 1$$

Multiplying equation (4.4.28) on the left by X yields

$$0 = XB^{(1)} + XB^{(2)} + \cdots + XB^{(s)} + XB^{(t+1)} + XB^{(r+2)} + \cdots + XB^{(r+t)}$$
$$= 0 + 0 + \cdots + 0 + 1 + 1 + \cdots + 1$$

Thus t is even. ////

At this point we can introduce another important matrix associated with a graph.

Definition 4.4.29 Let $C_1, C_2, \ldots, C_q$ be the cuts of the connected $m \times n$ graph G. The $q \times n$ matrix $C = [c_{hj}]$, where

$$c_{hj} = \begin{cases} 1 & \text{when arc } a_j \text{ belongs to cut } C_h \\ 0 & \text{when arc } a_j \text{ does not belong to cut } C_h \end{cases}$$

is called the *cut matrix* of G.

The hth row of C is the vector of $\mathbb{F}_n$ which corresponds to the set of arcs of the cut C_i under the natural correspondence. Because of Theorem 4.4.27 we have:

Corollary 4.4.30 $CL = 0$.

As we will see, there are surprising similarities between the properties of the set of loops and the set of cuts of a graph. In Theorem 4.4.21 we showed that the chords of a tree determine a fundamental set of loops and that these loops form a basis for the vector space of loop sets. Similarly, let T be a tree of the connected graph G. Let a_i be an arc of T and denote by T_i the graph which remains when a_i is removed from T. Since T_i contains m vertices and $m - 2$ arcs, T_i is not connected. Moreover, because the addition of the single arc a_i to T_i yields a connected graph, T_i consists of exactly two components. Let us call these components $\tilde{T}_i$ and $\tilde{\tilde{T}}_i$. Since T_i is not connected, the set of arcs of G which do not belong to T_i contains a cut, say C_i. Moreover C_i contains a_i. The two components of the graph which remains when C_i is removed from G have the same sets of vertices as $\tilde{T}_i$ and $\tilde{\tilde{T}}_i$, though possibly they contain more arcs. We have now proved:

Lemma 4.4.31 Each arc of a tree of a connected graph is contained in a unique cut which contains no other arcs of the tree.

A set of $m - 1$ cuts corresponding to the $m - 1$ arcs of a tree, as described by Lemma 4.4.31, is called a *fundamental set of cuts*. If we number the arcs and cuts of a graph so that the arcs of a given tree are $a_1, a_2, \ldots, a_{m-1}$ and the corresponding cuts are $C_1, C_2, \ldots, C_{m-1}$, the cut matrix C has the form

$$C = \begin{bmatrix} I & C_{12} \\ C_{21} & C_{22} \end{bmatrix} \quad (4.4.32)$$

where I is an identity matrix of order $m - 1$. The first $m - 1$ rows of C are the vectors corresponding to the cuts of the fundamental set under the natural correspondence. We will show that they form a basis for $\mathbb{R}(C)$.

Lemma 4.4.33 $\mathbb{R}(C) = \mathbb{R}(B)$

PROOF Combining Corollary 4.4.30 and part 4 of Theorem 4.4.20, we have

$$\mathbb{R}(C) \subseteq \mathbb{N}_R(L) = \mathbb{R}(B) \quad (4.4.34)$$

In order to show that we have equality throughout equation (4.4.34) it is sufficient to show that

$$r(C) = r(B)$$

From equation (4.4.34) we have

$$r(C) \leq r(B)$$

However, when C is written in the form of (4.4.32), then the identity submatrix of order $m - 1$ is a submatrix of C and consequently

$$r(C) \geq m - 1 = r(B)$$

completing the proof. ////

Theorem 4.4.35 Under the natural correspondence between the vector space of sets of arcs of G and $\mathbb{F}_n$ the cuts of G correspond to the minimal vectors of $\mathbb{R}(B)$.

PROOF It follows immediately from Corollary 4.4.30 that if X is the n-dimensional row vector corresponding to a cut of G, then

$$X \in \mathbb{R}(C) = \mathbb{N}_R(L) = \mathbb{R}(B)$$

It remains to show that X is a minimal element of $\mathbb{R}(B)$.

However, we will prove the converse first. That is, we will show that the set of arcs of a minimal element of $\mathbb{R}(B)$ is a cut of G. If Y is a minimal vector of $\mathbb{R}(B)$, then Y can be expressed as

$$Y = WB \quad (4.4.36)$$

for some vector $W \in \mathbb{F}_m$. Further, $W \neq 0, E$ since $Y \neq 0$. Let us now assume that the arcs of G are numbered so that

$$Y = [0 \ E] \quad B = [B_1 \ B_2] \quad (4.4.37)$$

i.e., the columns forming B_2 are those which correspond to the arcs of Y whereas the columns forming B_1 are those which correspond to the arcs of G that do not belong to Y. B_1 is the incidence matrix of the graph, say G_1, which remains when the arcs of Y are removed from G. Substituting equations (4.4.37) into equation (4.4.36), we have

$$0 = WB_1 \quad E = WB_2$$

Since $W \in \mathbb{N}_R(B_1)$ and $W \neq 0, E$ it follows that

$$r(B_1) \leq m - 2$$

and consequently G_1 is not connected. We have now shown that Y contains a cut. We wish to show that, in fact, Y is a cut. Let Z be any cut contained in Y. By the first paragraph of the proof,

$$Z \in \mathbb{R}(C) = \mathbb{R}(B)$$

However, because Y is a minimal element of $R(B)$, we now have

$$Y = Z$$

and thus Y is a cut.

We now return to the first part of the proof. It remains to show that a vector corresponding to a cut of G is a minimal element of $\mathbb{R}(B)$. Let X be a cut. If X is not minimal, then X contains a minimal element Y of $\mathbb{R}(B)$. By the second part of the proof given above, Y is a cut of G. But one cut can not properly contain another. Thus $X = Y$ and X is a minimal element of $\mathbb{R}(B)$. ////

The analogy between loops and cuts can be carried further. It follows from Theorem 4.4.35 that the vector corresponding to the sum of two cuts of G belongs to $\mathbb{R}(B)$ and, by Lemma 4.3.6, can be expressed as the disjoint union of minimal elements of $\mathbb{R}(B)$. In other words, the sum of two cuts of a graph can be expressed as the arc-disjoint union of cuts. For example, in the graph of Example 4.4.12,

$$C_1 + C_4 = \{a_1, a_2\} = C_3$$

Also, let us set

$$C_5 = \{a_1, a_4, a_6, a_7\} \qquad C_6 = \{a_4, a_5\}$$

then
$$C_1 + C_5 = \{a_1, a_2, a_4, a_5\} = C_3 + C_6$$

i.e., the nondisjoint sum of cuts $C_1 + C_5$ can be expressed as the disjoint sum of cuts $C_3 + C_6$. We define a *cut set* of a graph to be any set of arcs which can be expressed as a disjoint union of cuts. Then we have proved:

Theorem 4.4.38 Under the natural correspondence between the vector space of sets of arcs of G and $\mathbb{F}_n$, the cut sets of G correspond to the elements of $\mathbb{R}(B)$.

For a large class of graphs the similarity which we have observed in our study of the loops and the cuts of a graph can be illustrated more dramatically. At the same time we can present an interesting application of the dimensions of the vector spaces associated with a graph.

Definition 4.4.39 Let G and $\tilde{G}$ be connected graphs with the same number of arcs. We say that $\tilde{G}$ is a *dual* of G if the arcs of G and $\tilde{G}$ can be numbered so that

$$L = \{a_{h_1}, a_{h_2}, \ldots, a_{h_k}\}$$

is a loop of G if and only if

$$\tilde{C} = \{\tilde{a}_{h_1}, \tilde{a}_{h_2}, \ldots, \tilde{a}_{h_k}\}$$

is a cut of $\tilde{G}$.

We call $\tilde{G}$ a dual of G (rather than *the* dual of G) because a graph may have more than one dual (see Exercise 5). Although dual graphs have the same number of arcs, they generally do not have the same number of vertices. By combining Theorems 4.4.11 and 4.4.35 we can express the duality of graphs in terms of their incidence matrices. Specifically:

Theorem 4.4.40 Let G and $\tilde{G}$ be graphs with the same number of arcs. Then $\tilde{G}$ is a dual of G if and only if the arcs of G and $\tilde{G}$ can be numbered so that

$$\mathbb{N}(B) = \mathbb{R}(\tilde{B})^T$$

where B and $\tilde{B}$ are the incidence matrices of G and $\tilde{G}$ respectively and where $\mathbb{R}(\tilde{B})^T$ is the vector space consisting of the transposes of the vectors of $\mathbb{R}(\tilde{B})$; that is, $\mathbb{R}(\tilde{B})^T = \mathbb{C}(\tilde{B}^T)$.

EXAMPLE 4.4.41 It is not difficult to verify that a dual of the 6×9 graph

G:
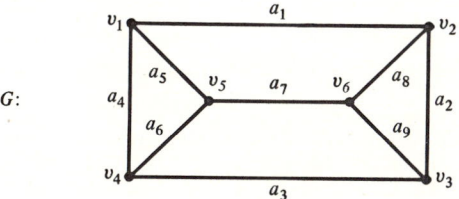

is the 5×9 graph

$\tilde{G}$:
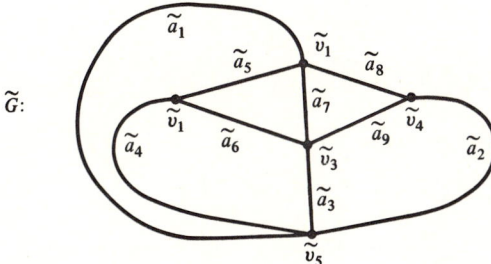

The relation of duality between G and $\tilde{G}$ can be seen more easily when the graphs are positioned as in Fig. 4.4.42.

Note that when the diagrams of G and $\tilde{G}$ are superimposed, as in Fig. 4.4.42, each arc of each graph crosses the corresponding arc of the other graph

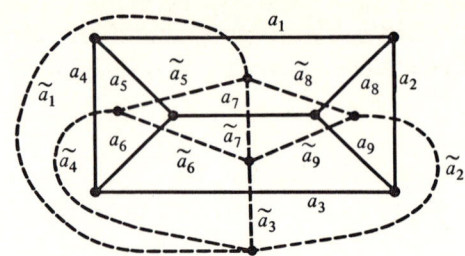

FIGURE 4.4.42

and intersects no other arc of either graph (except at the vertices). In other words, the arcs of each graph meet only at the vertices of the graph, and a_i and $\tilde{a}_i$ intersect for $i = 1, 2, \ldots, 9$, but a_i and $\tilde{a}_j$ do not intersect when $i \neq j$. Each loop L of G partitions the vertices of $\tilde{G}$ into two disjoint sets, namely, the set of vertices of $\tilde{G}$ which are "inside" the loop and the set of vertices of $\tilde{G}$ which are "outside" the loop. When all the arcs of $\tilde{G}$ which intersect L are removed from $\tilde{G}$, the resulting graph consists of two components, one inside L and the other outside L. The arcs that have been removed from $\tilde{G}$ are those which correspond to the arcs of L, and they form a cut of $\tilde{G}$.   ////

The description we have just given of the properties of the two graphs of Example 4.4.41 is symmetric; that is, G bears the same relation to $\tilde{G}$ as $\tilde{G}$ does to G. We next show that the same thing can be said of all pairs of dual graphs.

Theorem 4.4.43 $\tilde{G}$ is a dual of G if and only if G is a dual of $\tilde{G}$.

PROOF According to Exercise 27 of Sec. 1.7,

$$\mathbb{N}(B) = \mathbb{C}(\tilde{B}^T) \quad \text{if and only if} \quad \mathbb{N}_R(\tilde{B}^T) = \mathbb{R}(B) \quad (4.4.44)$$

But the right-hand equation of (4.4.44) is equivalent to

$$\mathbb{N}(\tilde{B}) = \mathbb{C}(B^T)$$

The theorem now follows from Theorem 4.4.40.   ////

The material to be presented from this point to the end of this section leans heavily on the geometric picture of a graph and is developed here in a more informal and less rigorous manner than we have previously used. A rigorous treatment of this subject requires large doses of algebraic topology and is beyond the scope of this book.

Definition 4.4.45 A graph G is *planar* if it can be drawn in the plane in such a way that the arcs of G intersect only at the vertices of G.

Clearly the graphs G and $\tilde{G}$ of Example 4.4.41 are planar. It is not always apparent when a graph is planar. The 4×6 graph

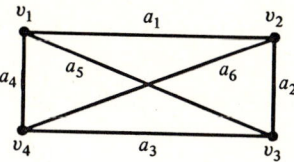

is planar since it can be drawn as

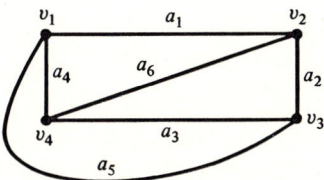

In Examples 4.4.50 and 4.4.51 we will construct two graphs which we will prove are not planar. A planar graph often can be drawn in several different ways. For example, the graph G of Example 4.4.41 can also be drawn as

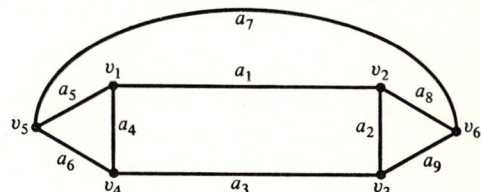

FIGURE 4.4.46

For reasons which will become clear shortly we will restrict our attention to connected planar graphs in which the degree of every vertex is at least 2. Since there are no vertices of degree 1, every arc of the graph belongs to at least one loop. Such a graph divides the plane into a number of regions. A *region* can be defined as an area of the plane which is bounded by arcs of the graph and which does not contain any arcs in its interior. The regions of a graph depend on the way the graph is drawn. The regions of the graph G of Example 4.4.41 and the graph of Fig. 4.4.46 are not the same although the graphs are the same graph (in the sense that the incidence relations among the vertices and arcs are identical in the two cases). However, we will see that the number of regions of a planar graph does not depend on how the graph is drawn. The regions of the graph of Figure 4.4.46 are

R_1, bounded by the loop $\{a_4, a_5, a_6\}$
R_2, bounded by the loop $\{a_1, a_2, a_3, a_4\}$
R_3, bounded by the loop $\{a_2, a_8, a_9\}$
R_4, bounded by the loop $\{a_1, a_8, a_7, a_5\}$
R_5, outside the loop $\{a_3, a_9, a_7, a_6\}$

One of the regions of a planar graph will have infinite extent, for example, R_5 above, and is called the *infinite region*. All other regions are said to be *finite*. Every arc of a planar graph is a portion of the boundary of exactly two regions.

Theorem 4.4.47 The loops bounding the finite regions of a planar graph form a basis for the vector space of loop sets.

PROOF Let $R_1, R_2, \ldots, R_k$ be distinct finite regions of a planar graph and let $L_1, L_2, \ldots, L_k$ be the loops which enclose them. The area

$$R_1 \cup R_2 \cup \cdots \cup R_k$$

is bounded by the loop set

$$L_1 + L_2 + \cdots + L_k$$

Since $R_1 \cup R_2 \cup \cdots \cup R_k$ is neither the empty set nor the entire plane, it has a boundary consisting of a nonempty set of arcs, i.e.,

$$L_1 + L_2 + \cdots + L_k \neq 0$$

Thus $L_1, L_2, \ldots, L_k$ are linearly independent elements of $\mathbb{N}(B)$.

It remains only to show that every loop of the graph can be expressed as a sum of loops bounding finite regions. However, any loop of the graph encloses an area which is the union of a set of finite regions and therefore can be expressed as the sum of the loops of these regions.   ////

Since the dimension of the vector space of loop sets of an $m \times n$ graph is $n - m + 1$, we have established:

Theorem 4.4.48 Euler's formula. A planar $m \times n$ graph contains $n - m + 2$ regions of which $n - m + 1$ are finite.

Thus, even though a planar graph may be represented by a diagram in several different ways, the number of regions is the same for all such representations. Euler's formula can be used to demonstrate that some graphs are not planar. First we need:

Lemma 4.4.49 If every loop of a planar $m \times n$ graph contains at least k arcs, then

$$n \geq \tfrac{1}{2}k(n - m + 2)$$

PROOF We have shown that there are $n - m + 2$ regions. Each region is bounded by a loop containing at least k arcs. Thus the total number of arcs, counting each arc as many times as it belongs to loops which bound regions, is at least

$$k(n - m + 2)$$

However, each arc belongs to exactly two such loops. Thus the number of arcs is at least $\tfrac{1}{2}k(n - m + 2)$. ////

EXAMPLE 4.4.50 Let us apply Lemma 4.4.49 to the complete graph with five vertices.

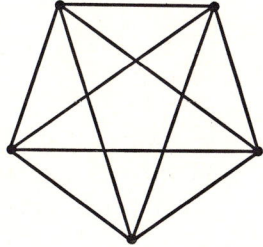

In this case

$$m = 5 \quad n = 10 \quad k = 3$$

and consequently $\quad \tfrac{1}{2}k(n - m + 2) = \tfrac{1}{2} \cdot 3 \cdot 7 = 10\tfrac{1}{2} > 10 = n$

Thus the complete graph with five (or more) vertices is not planar. ////

EXAMPLE 4.4.51 For the graph[1]

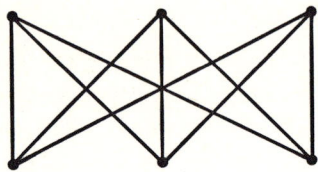

$$m = 6 \quad n = 9 \quad k = 4$$

This graph is not planar since

$$9 = n < \tfrac{1}{2}k(n - m + 2) = \tfrac{1}{2} \cdot 4 \cdot 5 = 10 \qquad ////$$

Clearly every graph which contains either the graph of Example 4.4.50 or the graph of Example 4.4.51 as a subgraph is not planar. In 1930 the Polish mathematician Kazimierz Kuratowski [7] proved the converse, namely, that every nonplanar graph contains as a subgraph one of the two nonplanar graphs we have described.[2] The proof of this theorem is difficult and will not be given here. The graphs of Example 4.4.50 and 4.4.51 are called the *Kuratowski graphs*.

The dual pair of graphs G and $\tilde{G}$ of Example 4.4.41 are both planar. The principle of superimposing a graph and its dual, as in Fig. 4.4.42, can be used to construct a dual of any planar graph. Moreover, we will see that the dual graph obtained in this way is also planar.

Theorem 4.4.52 Every planar graph has a dual which is planar.

PROOF Let G be a planar graph. Form $\tilde{G}$ as follows. Place a vertex of $\tilde{G}$ in each region (finite or infinite) of G. For any two distinct regions of G, count the number of arcs common to the loops bounding them. We will construct an equal number of arcs of $\tilde{G}$ between the vertices of $\tilde{G}$ which we have just placed inside these regions. For each common arc between

---

[1] The nonplanarity of the graph of Example 4.4.51 is the basis for a well-known puzzle. "There are three houses on one side of the street and three utilities on the other side of the street. Can lines be drawn from each utility to each house in such a way that no two lines cross?"

[2] If additional vertices are added along the arc of either of the nonplanar graphs, the resulting graph has a different set of incidence relations and is also nonplanar. For Kuratowski's theorem all vertices of degree 2 are eliminated. For a proof of the theorem see Berge [1].

two regions of G draw an arc of $\tilde{G}$ between the vertices of $\tilde{G}$ in these regions in such a way that it cuts the corresponding arc of G. (This will result in parallel arcs when two regions of G have more than one arc common to their bounding loops. What would happen in this construction if we permitted G to have arcs which do not belong to any loops?) It is easily seen that the resulting graph $\tilde{G}$ is a dual of G and is planar.

////

The converse of Theorem 4.4.52, namely, that any graph which has a dual is planar, was proved in 1932 by Hassler Whitney [11], an American mathematician. In addition he gave an alternate proof of Kuratowski's theorem by showing directly that a graph has a dual if and only if it does not contain either of the Kuratowski graphs as a subgraph. The proof of Whitney's theorem (in either version) is difficult and will not be given here.

EXERCISES

1 Find the cut matrix of the graph of Example 4.4.12. Using the incidence matrix and the loop matrix given in the text, verify Theorem 4.4.20, Corollary 4.4.30, and Lemma 4.4.33.

2 Find the incidence matrix, the loop matrix, and the cut matrix of the graphs G and $\tilde{G}$ of Example 4.4.41. Show that G and $\tilde{G}$ are duals of each other by verifying that their incidence matrices satisfy Theorem 4.4.40.

3 Use the construction of Theorem 4.4.52 to obtain a dual for the graph of Example 4.4.12.

4 Show that when the construction of the dual of a graph, as described in Theorem 4.4.52, is applied to the graph of Figure 4.4.46, the resulting graph is the graph $\tilde{G}$ of Example 4.4.41.

5 Let G be the graph of Example 4.4.12 and let G' be the graph

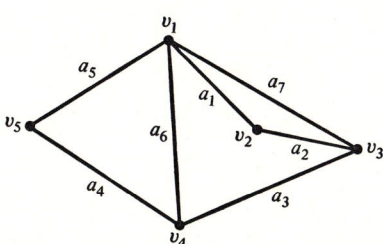

Show that G and G' are the same graph (in the sense that they have the same incidence relations). Use Theorem 4.4.52 to construct duals $\tilde{G}$ and $\tilde{G}'$ of G and G', respectively. Show that $\tilde{G}$ is a dual of G' and that $\tilde{G}'$ is a dual of G. Also, show that $\tilde{G}$ and $\tilde{G}'$ are not the same graph.

6 Construct a graph whose loop space has the vectors

$$\begin{bmatrix}1\\1\\1\\1\\1\\1\\0\end{bmatrix} \begin{bmatrix}1\\0\\1\\0\\1\\1\\1\end{bmatrix} \begin{bmatrix}1\\0\\1\\1\\0\\0\\1\end{bmatrix}$$

as a basis. Assuming that the graph has no isolated vertices, what are its dimensions? (*Hint:* In order to construct the graph, find the minimal vectors of the vector space spanned by the given vectors.)

7 Show that there does not exist a graph whose loop space has the vectors

$$\begin{bmatrix}1\\1\\1\\1\\0\\0\\0\end{bmatrix} \begin{bmatrix}0\\1\\1\\0\\1\\0\\0\end{bmatrix} \begin{bmatrix}0\\0\\1\\1\\0\\1\\0\end{bmatrix} \begin{bmatrix}0\\0\\1\\1\\1\\0\\1\end{bmatrix}$$

as a basis. (*Hint:* Find the minimal vectors of the vector space spanned by the given vectors and try to construct the graph.)

8 Find a connected graph whose cut space has the vectors

$$[1\ 1\ 1\ 1\ 0\ 1\ 0]$$
$$[1\ 0\ 0\ 1\ 1\ 1\ 0]$$
$$[1\ 0\ 1\ 1\ 1\ 0\ 1]$$
$$[0\ 1\ 1\ 1\ 1\ 1\ 1]$$

as a basis. (*Hint:* Find the loop space of the graph.)

9 Prove that every arc of a graph belongs to at least one cut.

10 Prove that two arcs of a graph belong to exactly the same cut sets if and only if they are parallel.

11 Show that each arc of an $m \times n$ graph belongs to either no loop-sets or to 2^{n-m} loop sets. (*Hint:* Consider the graph which remains when the arc is removed. Find a necessary and sufficient condition that an arc belong to no loop sets.)

12 Show that each arc of an $m \times n$ graph belongs to exactly 2^{m-2} cut sets.
13 Show that a maximal subgraph with no loops of an $m \times n$ graph with k components has $n - m + k$ arcs and k components. (Such a subgraph is sometimes called a *forest*.)
14 Let G be an $m \times n$ graph and let $\bar{B}$ be the $(m-1) \times n$ submatrix of B formed by deleting any one row of B. Show that there is a one-to-one correspondence between the trees of G and the nonsingular submatrices of $\bar{B}$ of order $m - 1$.
15 Prove that the subgraph H of G is a tree if and only if H contains all the vertices of G and between any two of these vertices there is exactly one path.
16 A vertex of a connected graph is called a *cut vertex* if when that vertex and all the arcs incident to it are removed from the graph, the remaining graph is not connected. Prove that a vertex is not a cut vertex if and only if the row of B corresponding to it is a minimal element of $\mathbb{R}(B)$.
17 Use Theorem 4.4.52 to give an alternate proof of Euler's formula. [*Hint:* If $\tilde{G}$ is the dual of G constructed using Theorem 4.4.52, how many vertices does $\tilde{G}$ have? What is $r(\tilde{B})$?]
18 Let $a = (v_1, v_2)$ be an arc of the connected graph G and let G have at least three vertices. Show that there is a cut in G such that v_1 and v_2 belong to the same component of the graph which results when the arcs of the cut are removed from G.
19 Show that every cut set of a complete graph is a cut.
20 Show that a connected graph with m vertices which has the property that every cut set is a cut is (or contains) the complete graph with m vertices.

4.5 DIRECTED GRAPHS

The purpose of this section is to do for directed graphs what we did for undirected graphs in the last section. There are, of course, many similarities between the two, but there are enough differences to warrant treating the two topics separately. One immediate difference is that we can no longer take the entries of our matrices from the field $\mathbb{F}_2$. Instead, throughout this section we will let $\mathbb{F}$ denote the field of real numbers unless we specify otherwise.

In a directed graph each arc is oriented, i.e., the arc is directed *from* one of its vertices *toward* the other. This orientation is most conveniently shown in a diagram for the graph by attaching an arrowhead to the arc. We understand then that the orientation of the arc is in the direction that the arrowhead is pointing. When we wish to show there is orientation both from the vertex v_1 to the vertex v_2 and from v_2 to v_1, we will use parallel arcs, each with a single direction (see Fig. 4.5.1).

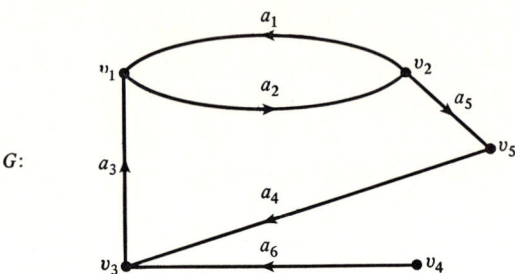

FIGURE 4.5.1

Let us give a precise definition of a directed graph.

Definition 4.5.2 An $m \times n$ *directed graph* $G = (\mathbb{V}, \mathbb{A})$ consists of a set $\mathbb{V}$ of m objects called *vertices*
$$\mathbb{V} = \{v_1, V_2, \ldots, v_m\}$$
and a set $\mathbb{A}$ of n arcs
$$\mathbb{A} = \{a_1, a_2, \ldots, a_n\}$$
where each arc is an ordered pair of distinct vertices. We write
$$a_j = (v_i, v_k)$$
to denote that the arc a_j is incident with the vertices v_i and v_k and is directed from v_i to v_k.

As with an undirected graph, the incidence relations between the vertices and arcs of a directed graph can be conveniently expressed in matrix form. In addition, the matrix contains the directional relations between the vertices and arcs.

Definition 4.5.3 Let G be a directed graph with vertices $v_1, v_2, \ldots, v_m$ and arcs $a_1, a_2, \ldots, a_n$. The $m \times n$ matrix
$$B = [b_{ij}] \quad \begin{array}{l} i = 1, 2, \ldots, m \\ j = 1, 2, \ldots, n \end{array}$$
defined by
$$b_{ij} = \begin{cases} +1 & \text{if } a_j \text{ is directed toward } v_i \\ -1 & \text{if } a_j \text{ is directed from } v_i \\ 0 & \text{if } a_j \text{ is not incident with } v_i \end{cases}$$
is called the *(vertex-arc) incidence matrix* of G.[1]

---

[1] Since we will often be working with the real field and with $\mathbb{Z}_2$ simultaneously in this section, we will denote the real number one by $+1$ and the nonzero element of $\mathbb{Z}_2$ by 1.

Associated with any directed graph G there is an undirected graph (which we will denote by $|G|$) formed from G by ignoring the orientations of the arcs. $|G|$ is called the *underlying (undirected) graph* of G. Since $|G|$ is an undirected graph, the entries of its incidence matrix (denoted by $|B|$) are still elements of $\mathbb{Z}_2$. In fact, $|B|$ can be obtained from B by changing all -1s and $+1$s to 1s.

EXAMPLE 4.5.4 For the graph of G of Figure 4.5.1 we have

$$B = \begin{bmatrix} +1 & -1 & +1 & 0 & 0 & 0 \\ -1 & +1 & 0 & 0 & -1 & 0 \\ 0 & 0 & -1 & +1 & 0 & +1 \\ 0 & 0 & 0 & 0 & 0 & -1 \\ 0 & 0 & 0 & -1 & +1 & 0 \end{bmatrix}$$

The underlying graph of G is

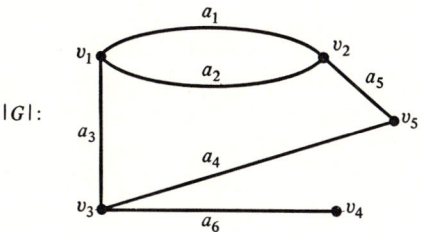

$|G|$:

and

$$|B| = \begin{bmatrix} 1 & 1 & 1 & 0 & 0 & 0 \\ 1 & 1 & 0 & 0 & 1 & 0 \\ 0 & 0 & 1 & 1 & 0 & 1 \\ 0 & 0 & 0 & 0 & 0 & 1 \\ 0 & 0 & 0 & 1 & 1 & 0 \end{bmatrix}$$

////

Many of the properties of a directed graph can be obtained from the incidence matrix. This includes properties which are, in reality, functions of the underlying undirected graph. The methods we will use for directed graphs often are precisely the same as those which we used in the last section. For example, we define a directed graph to be *connected* if its underlying graph is connected. More generally, the subgraph H of G is called a *component* of G if $|H|$ is a component of $|G|$. Then the analog of Lemma 4.4.3 is:

Lemma 4.5.5 If B is the incidence matrix of a directed graph G with m vertices, then G is connected if and only if the only row vectors X such that

$$XB = 0$$

are the scalar multiples of $E_{(m)} = [+1 \quad +1 \quad \cdots \quad +1]$.

PROOF Since every column of B has exactly one $+1$, one -1, and $m-2$ 0s, we have

$$E_{(m)}B = 0$$

If G is not connected, then the vertices and arcs of G can be ordered so that B has the form

$$B = \begin{bmatrix} B_1 & 0 & \cdots & 0 \\ 0 & B_2 & \cdots & 0 \\ \cdots & \cdots & \cdots & \cdots \\ 0 & 0 & \cdots & B_k \end{bmatrix}$$

where $B_1, B_2, \ldots, B_k$ are the incidence matrices of the components of G. If B_i is an $m_i \times n_i$ matrix for $i = 1, 2, \ldots, k$, then each of the k vectors

$$Y_1 = [E_{(m_1)} \quad 0 \quad 0 \quad \cdots \quad 0]$$
$$Y_2 = [0 \quad E_{(m_2)} \quad 0 \quad \cdots \quad 0]$$
$$\cdots \cdots \cdots \cdots \cdots \cdots \cdots \cdots \cdots$$
$$Y_k = [0 \quad 0 \quad 0 \quad \cdots \quad E_{(m_k)}]$$

satisfies

$$XB = 0$$

and is not a scalar multiple of $E_{(m)}$.

The proof of the converse requires one additional step. Let G be a connected directed graph and suppose there exists a vector X such that

$$XB = 0 \quad X \neq \alpha E_{(m)} \text{ for any real number } \alpha \quad (4.5.6)$$

Set $X = [x_i]$ and note that since X satisfies (4.5.6), so does

$$Y = X - x_n E_{(m)}$$

Let the number of entries of Y which are equal to 0 be s. Since the last entry of Y is 0, we have

$$s \geq 1$$

On the other hand, $Y \neq 0$, and hence

$$m - 1 \geq s$$

We may assume that the vertices of G are ordered so that

$$Y = [Z \ 0]$$

where Z is an $(m - s)$-dimensional row vector none of whose entries is 0. Partition B as

$$B = \begin{bmatrix} B_1 \\ B_2 \end{bmatrix}$$

where B_1 and B_2 contain $m - s$ and s rows, respectively. Then

$$ZB_1 = 0$$

Since none of the entries of Z is 0, each column of B_1 contains either two nonzero terms (and these must be $+1$ and -1) or contains no nonzero terms. We may assume that the arcs of G are numbered so that the first p columns of B_1 have two nonzero terms and the remaining $n - p$ columns contain only 0's. We can now partition B as

$$B = \begin{bmatrix} B_{11} & 0 \\ 0 & B_{22} \end{bmatrix}$$

where B_{11} and B_{22} are $(m - s) \times p$ and $s \times (n - p)$ matrices, respectively. Since G is a connected graph, we have reached a contradiction. Consequently, no such vector X exists. ////

The next theorem follows from Lemma 4.5.5 in the same way that Theorem 4.4.4 follows from Lemma 4.4.3.

Theorem 4.5.7 If B is the incidence matrix of an $m \times n$ directed graph with k components, then the rank of B is given by

$$r(B) = m - k$$

It is interesting to see that although we have changed the field from which the entries of our matrices are chosen, both the statement and the proof of Lemma 4.5.5 and Theorem 4.5.7 differ only slightly from their counterparts in the last section. However, connectedness is actually a property of the underlying undirected graph rather than of the graph itself. The field which is used will play a more prominent role when the direction of the arcs in a graph becomes a more significant factor. Even then, however, many of the results of the last section can be extended to directed graphs with little or no change.

Definition 4.5.8 A set of l distinct arcs of a directed graph G,

$$\{(v_{h_0}, v_{h_1}), (v_{h_1}, v_{h_2}), \ldots, (v_{h_{l-1}}, v_{h_l})\}$$

is called a *directed path* of length l from v_{h_0} to v_{h_l}. If the vertices $v_{h_1}, v_{h_2}, \ldots, v_{h_{l-1}}$ are distinct from each other and from v_{h_0} and v_{h_l}, then the path is said to be *simple*. A simple path from a vertex to itself (that is, $v_{h_0} = v_{h_l}$) is called a *directed loop*. A subgraph H of G is a loop of *length* l if $|H|$ is a loop of length l of $|G|$.

If the directions of all the arcs of a directed loop are reversed, the resulting graph is again a directed loop. Every directed loop is a loop, and every loop differs from a directed loop only in the orientation of some of its arcs. In the graph of Example 4.5.4, there are three loops,

$$L_1 = \{a_1, a_2\} \quad L_2 = \{a_3, a_2, a_5, a_4\} \quad L_3 = \{a_1, a_5, a_4, a_3\}$$

L_1 and L_2 are directed loops L_3 is not a directed loop.

EXAMPLE 4.5.9 When the vertices and arcs of a directed loop are numbered so that

$$L = \{a_1 = (v_1, v_2), a_2 = (v_2, v_3), \ldots, a_{l-1} = (v_{l-1}, v_l), a_l = (v_l, v_1)\}$$

then the incidence matrix of L is the $l \times l$ matrix

$$B_L = \begin{bmatrix} -1 & & & & & +1 \\ +1 & -1 & & & & \\ & +1 & -1 & & & \\ & & \cdots & \cdots & & \\ & & & +1 & -1 & \\ & & & & +1 & -1 \end{bmatrix} \quad (4.5.10)$$

The incidence matrix of a loop has the same form as (4.5.10) except that within each column the $+1$ and the -1 may be interchanged, reflecting the change in the orientation of the corresponding arc.   ////

For the incidence matrix B of a directed loop we have

$$BE = 0$$

and the only vectors X such that

$$BX = 0$$

are the scalar multiples of E.

On the other hand, suppose B is the incidence matrix of a loop L. We can form a directed loop $\tilde{L}$ from L by changing the direction of some of the arcs of L. (This may be done in two different ways.) If $F = [f_j]$ is the column vector defined by

$$f_j = \begin{cases} +1 & \text{when orientation of } a_j \text{ is the same in } L \\ & \text{as in } \tilde{L} \\ -1 & \text{when orientation of } a_j \text{ in } L \text{ is opposite} \\ & \text{to its orientation in } \tilde{L} \\ 0 & \text{when } a_j \text{ is not an arc of } L \end{cases} \quad (4.5.11)$$

then
$$BF = 0$$

As before, the only vector X such that

$$BX = 0$$

are the scalar multiples of F.

Lemma 4.5.12 Let B be the incidence matrix of the directed graph G. A necessary and sufficient condition that a set L of arcs of G form a loop is that the columns of B corresponding to the arcs of L be a minimal linearly dependent set.

PROOF The proof is similar to the proof of Lemma 4.4.8 and is left as an exercise. ////

The natural correspondence between sets of arcs of G and n-dimensional vectors which was introduced in Sec. 4.3 and used extensively in Sec. 4.4 can be carried over to the case where the entries of our matrices come from the field $\mathbb{F}$ of real numbers. More generally:

Definition 4.5.13 Let $S = \{s_1, s_2, \ldots, s_n\}$ be any set of n objects. An expression of the form

$$x = \alpha_1 s_1 + \alpha_2 s_2 + \cdots + \alpha_n s_n \quad \alpha_1, \alpha_2, \ldots, \alpha_n \in \mathbb{F}$$

is called a *formal sum* of the elements of S with coefficients from $\mathbb{F}$. Denote the set of all formal sums of elements of S be $\mathbb{S}$. We define addition and scalar multiplication in $\mathbb{S}$ as follows. Let

$$y = \beta_1 s_1 + \beta_2 s_2 + \cdots + \beta_n s_n$$

Set
$$x + y = (\alpha_1 + \beta_1)s_1 + (\alpha_2 + \beta_2)s_2 + \cdots + (\alpha_n + \beta_n)s_n$$
$$\alpha x = \alpha\alpha_1 s_1 + \alpha\alpha_2 s_2 + \cdots + \alpha\alpha_n s_n$$

With these operations we call $\mathbb{S}$ the *vector space of formal sums* of elements of S.

There is a natural way whereby we can identify $\mathbb{S}$ with $\mathbb{F}^n$ or $\mathbb{F}_n$. Let $x \in \mathbb{S}$,
$$x = \alpha_1 s_1 + \alpha_2 s_2 + \cdots + \alpha_n s_n$$
and let X be the n-dimensional column vector

$$X = \begin{bmatrix} \alpha_1 \\ \alpha_2 \\ \vdots \\ \alpha_n \end{bmatrix}$$

The mapping
$$\Pi : x \in \mathbb{S} \to X \in \mathbb{F}^n$$
is a one-to-one correspondence between the elements of $\mathbb{S}$ and $\mathbb{F}^n$. Moreover,
$$\Pi(x + y) = \Pi(x) + \Pi(y)$$
$$\Pi(\alpha x) = \alpha \Pi(x)$$

for all $x, y \in \mathbb{S}$ and $\alpha \in \mathbb{F}$. We call Π the *natural correspondence* between elements of $\mathbb{S}$ and $\mathbb{F}^n$. In the same way there is a natural correspondence between the elements of $\mathbb{S}$ and $\mathbb{F}_n$.

One more step is necessary before we can make use of the natural correspondence. Let L be a loop of a directed graph G and let $\tilde{L}$ be (either) one of the two directed loops that can be formed from L by changing the orientation of some of its arcs. We will identify L with the formal sum of arcs of G,
$$L = \varepsilon_1 a_1 + \varepsilon_2 a_2 + \cdots + \varepsilon_n a_n \quad (4.5.14)$$

where $\varepsilon_j = \begin{cases} +1 & \text{when orientation of } a_j \text{ is the same in } L \text{ as in } \tilde{L} \\ -1 & \text{when orientation of } a_j \text{ in } L \text{ is opposite to its orientation in } \tilde{L} \\ 0 & \text{when } a_j \text{ is not an arc of } L \end{cases}$

These coefficients are precisely the ones described in (4.5.11). Thus both the

n-dimensional column vector $\mathbb{F}$ defined in (4.5.11) and the formal sum (4.5.14) correspond to the loop L, and these correspond to each other by means of the natural correspondence. Later we will have reason to identify other subgraphs of a graph with formal sums of arcs.

In order to continue to preserve the language of Sec. 4.4 we will define a vector X of a subspace $\mathbb{V}$ of $\mathbb{F}^n$ or $\mathbb{F}_n$ to be *minimal* if the only vectors Y of $\mathbb{V}$ which have 0s in all the positions in which X has a zero are the scalar multiples of X. Two minimal vectors of $\mathbb{V}$ which have 0s in the same positions are scalar multiples of each other. For, suppose that $U = [u_j]$ and $V = [v_j]$ are two minimal vectors of $\mathbb{V}$ which have 0s in the same positions. Let u_k be any nonzero entry of U and set

$$Z = u_k V - v_k U$$

Z has 0s wherever U and V have 0s and, moreover, in the kth component. Since U and V are minimal vectors, it follows that $Z = 0$ and therefore

$$u_k V = v_k U$$

We will say that two minimal vectors are *distinct* if they are not scalar multiples of each other.

Combining Lemma 4.5.12 with the discussion above, we have:

Theorem 4.5.15 Under the natural correspondence between the vector space of formal sums of arcs of G and $\mathbb{F}^n$, the loops of G correspond to the distinct minimal vectors of $\mathbb{N}(B)$. Each minimal element of $\mathbb{N}(B)$ is (a scalar multiple of) a vector containing only 0s, +1s, and −1s among its entries. The directed loops of G are those which correspond to minimal elements of $\mathbb{N}(B)$ all of whose nonzero entries have the same sign.

EXAMPLE 4.5.16 The incidence matrix B of the graph of Fig. 4.5.1 is a 5×6 matrix whose rank is 4. Hence $v(B) = 2$. The distinct minimal vectors of $\mathbb{N}(B)$ are

$$L_1 = \begin{bmatrix} +1 \\ +1 \\ 0 \\ 0 \\ 0 \\ 0 \end{bmatrix} \quad L_2 = \begin{bmatrix} 0 \\ +1 \\ +1 \\ +1 \\ +1 \\ 0 \end{bmatrix} \quad L_3 = \begin{bmatrix} -1 \\ 0 \\ +1 \\ +1 \\ +1 \\ 0 \end{bmatrix}$$

corresponding to the formal sums

$$L_1 = a_1 + a_2$$
$$L_2 = a_2 + a_3 + a_4 + a_5$$
$$L_3 = -a_1 + a_3 + a_4 + a_5$$

which, in turn, correspond to the loops

$$L_1 = \{a_1, a_2\} \quad L_2 = \{a_2, a_3, a_4, a_5\} \quad L_3 = \{a_1, a_3, a_4, a_5\}$$

Note that

$$L_3 = L_2 - L_1 \qquad ////$$

We say that a vector is *unimodular* if all its entries are $+1$, -1, or 0. A vector space is *unimodular* if all its minimal vectors are scalar multiples of unimodular vectors. Clearly $\mathbb{F}^n$ and $\mathbb{F}_n$ are unimodular vector spaces. According to Theorem 4.5.15, $\mathbb{N}(B)$ is unimodular, and it follows from Lemma 4.5.5 and Theorem 4.5.7 that $\mathbb{N}_R(B)$ is unimodular.

The concept of a loop set can be extended to directed graphs. However, because the field of real numbers contains infinitely many elements, we will be able to form infinitely many loop sets from a set of loops.

Definition 4.5.17 A formal sum of arcs of a directed graph is a *loop set* if it can be expressed as a linear combination of loops.

Corollary 4.5.18 Under the natural correspondence between the vector space of formal sums of arcs of G and $\mathbb{F}^n$ the loop sets correspond to the elements of $\mathbb{N}(B)$.

Another important concept which can be carried over from undirected graphs to directed graphs is the loop matrix.

Definition 4.5.19 Let $L_1, L_2, \ldots, L_p$ be the loops of G. For each loop L_k, let $\tilde{L}_k$ be one of the two directed loops that can be formed from L_k by changing the direction of some of its arcs. The $n \times p$ matrix $L = [l_{jk}]$, where

$$l_{jk} = \begin{cases} +1 & \text{when orientation of } a_j \text{ is the same in } L_k \\ & \text{as in } \tilde{L}_k \\ -1 & \text{when orientation of } a_j \text{ in } L_k \text{ is opposite} \\ & \text{to its orientation in } \tilde{L}_k \\ 0 & \text{when } a_j \text{ is not an arc of } L_k \end{cases}$$

is called the *loop matrix* of G.

By comparing Definition 4.5.19 with (4.5.11), we see that the kth column of L is the vector corresponding to the loop L_k. The ambiguity in the definition of the vector corresponding to a loop is carried over to the loop matrix since either of the two directed loops associated with L_k may be chosen for $\tilde{L}_k$. When the choice of $\tilde{L}_k$ is changed, the sign of the entries of the kth column of L are changed. We will see that this causes no difficulty. The loop matrix of the graph of Fig. 4.5.1 is the 6×3 matrix

$$L = \begin{array}{c} \phantom{\begin{bmatrix}}L_1 L_2 L_3 \phantom{\end{bmatrix}} \\ \begin{bmatrix} +1 & 0 & -1 \\ +1 & +1 & 0 \\ 0 & +1 & +1 \\ 0 & +1 & +1 \\ 0 & +1 & +1 \\ 0 & 0 & 0 \end{bmatrix} \begin{array}{c} a_1 \\ a_2 \\ a_3 \\ a_4 \\ a_5 \\ a_6 \end{array} \end{array}$$

At this point in Sec. 4.4 we defined a tree and showed how the chords of a tree determine a fundamental set of loops. We then proved that the loops of a fundamental set form a basis for $\mathbb{C}(L)$. We leave it as an exercise to show that the same treatment carries over to directed graphs. On the basis of these remarks we conclude that:

Theorem 4.5.20 For an $m \times n$ connected directed graph,

1. $BL = 0$.
2. $\mathbb{C}(L) = \mathbb{N}(B)$.
3. $r(L) = n - m + 1$.
4. $\mathbb{N}_R(L) = \mathbb{R}(B)$.

Just as we have defined loops and directed loops for a directed graph, we can define cuts and directed cuts. We will call a set C of arcs of a connected directed graph G a *cut* if $|C|$ is a cut of $|G|$. When the arcs of C are removed from G, the resulting graph, say G', consists of two components, say G_1 and G_2. Every arc of C has one vertex which belongs to G_1 and one vertex which belongs to G_2. We say that C is a *directed cut from G_1 to G_2* if every arc of C is directed from its vertex in G_1 to its vertex in G_2. If the orientation of every arc of a directed cut from G_1 to G_2 is reversed, then the resulting set of arcs form a directed cut from G_2 to G_1. Thus, every cut can be transformed into a directed cut by changing the direction of some of its arcs. In fact, this can be done in two different ways. We will see that this ambiguity causes no difficulty.

Let C be a cut and let $\tilde{C}$ be one of the two directed cuts associated with C. We will identify C with both the formal sum of arcs of G,

$$C = \delta_1 a_1 + \delta_2 a_2 + \cdots + \delta_n a_n$$

where $\delta_j = \begin{cases} +1 & \text{when orientation of } a_j \text{ is the same in } C \text{ as in } \tilde{C} \\ -1 & \text{when orientation of } a_j \text{ in } C \text{ is opposite to its orientation in } \tilde{C} \\ 0 & \text{when } a_j \text{ is not an arc of } C \end{cases}$

and with the n-dimensional row vector

$$D = [\delta_1 \quad \delta_2 \quad \cdots \quad \delta_n]$$

As before, C and D correspond to each other under the natural correspondence between the vector space of formal sums of arcs of G and $\mathbb{F}_n$.

Definition 4.5.21 A formal sum of arcs of a connected directed graph is a *cut set* if it can be expressed as a linear combination of cuts.

The collection of cut sets of a graph is a subspace of $\mathbb{F}_n$ or, equivalently, of the vector space of formal sums of arcs. We will show that the cut sets of a directed graph play the same role as the cut sets of undirected graphs.

EXAMPLE 4.5.22 Let us denote the cuts of the graph G of Fig. 4.5.1 by

$$C_1 = \{a_1, a_2, a_3\} \quad C_2 = \{a_3, a_4\}$$
$$C_3 = \{a_4, a_5\} \quad C_4 = \{a_1, a_2, a_5\}$$
$$C_5 = \{a_1, a_2, a_4\} \quad C_6 = \{a_3, a_5\}$$
$$C_7 = \{a_6\}$$

Let $\tilde{C}_1, \tilde{C}_2, \ldots, \tilde{C}_7$ be directed cuts corresponding to $C_1, C_2, \ldots, C_7$ respectively. We must specify an orientation for each of the directed cuts. In each case, let us select $\tilde{C}$ so that it is directed toward the component of G' which contains v_1. Then the cuts $C_1, C_2, \ldots, C_7$ correspond to the vectors

$$C_1 = [+1 \quad -1 \quad +1 \quad 0 \quad 0 \quad 0]$$
$$C_2 = [0 \quad 0 \quad +1 \quad -1 \quad 0 \quad 0]$$
$$C_3 = [0 \quad 0 \quad 0 \quad +1 \quad -1 \quad 0]$$
$$C_4 = [+1 \quad -1 \quad 0 \quad 0 \quad +1 \quad 0]$$
$$C_5 = [+1 \quad -1 \quad 0 \quad +1 \quad 0 \quad 0]$$
$$C_6 = [0 \quad 0 \quad +1 \quad 0 \quad -1 \quad 0]$$
$$C_7 = [0 \quad 0 \quad 0 \quad 0 \quad 0 \quad +1]$$

respectively. ////

Definition 4.5.23 Let $C_1, C_2, \ldots, C_q$ be the cuts of the connected directed graph G. For each cut C_h let $\tilde{C}_h$ be either one of the two directed cuts that can be formed from C_h by changing the direction of some of its arcs. The $q \times n$ matrix $C = [c_{hj}]$, where

$$c_{hj} = \begin{cases} +1 & \text{when orientation of } a_j \text{ is the same in } C_h \text{ as in } \tilde{C}_h \\ -1 & \text{when orientation of } a_j \text{ in } C_h \text{ is opposite to its orientation in } \tilde{C}_h \\ 0 & \text{when } a_j \text{ is not an arc of } C_h \end{cases}$$

is called the *cut matrix* of G.

The 7×6 matrix

$$C = \begin{bmatrix} +1 & -1 & +1 & 0 & 0 & 0 \\ 0 & 0 & +1 & -1 & 0 & 0 \\ 0 & 0 & 0 & +1 & -1 & 0 \\ +1 & -1 & 0 & 0 & +1 & 0 \\ +1 & -1 & 0 & +1 & 0 & 0 \\ 0 & 0 & +1 & 0 & -1 & 0 \\ 0 & 0 & 0 & 0 & 0 & +1 \end{bmatrix} \begin{matrix} C_1 \\ C_2 \\ C_3 \\ C_4 \\ C_5 \\ C_6 \\ C_7 \end{matrix}$$

with columns labeled $a_1, a_2, a_3, a_4, a_5, a_6$, is the cut matrix of the graph of Fig. 4.5.1.

Lemma 4.5.24 Let C be a cut of G and let the graph G' which results when the arcs of C are removed from G have the two components G_1 and G_2. Let $\tilde{C}$ be the directed cut from G_2 to G_1 associated with C. If $X = [x_i]$ is the m-dimensional row vector of the vertices of G_1, that is,

$$x_i = \begin{cases} +1 & \text{when } v_i \in G_1 \\ 0 & \text{when } v_i \notin G_1 \end{cases}$$

then the vector $C \in \mathbb{F}_n$ corresponding to the cut C (with respect to $\tilde{C}$) is given by

$$C = XB$$

where B is the incidence matrix of G.

PROOF We may assume that the vertices and arcs of G are numbered so that

$$G_1 = \{v_1, v_2, \ldots, v_s; a_1, a_2, \ldots, a_t\}$$
$$G_2 = \{v_{s+1}, v_{s+2}, \ldots, v_m; a_{t+1}, a_{t+2}, \ldots, a_r\}$$
$$C = \{a_{r+1}, a_{r+2}, \ldots, a_n\}$$

Then B is partitioned as

$$B = [B_1 \quad B_2]$$

where B_1 is the $m \times r$ incidence matrix of G'. Moreover, since G_1 and G_2 are both arc- and vertex-disjoint,

$$B_1 = \begin{bmatrix} B_{11} & 0 \\ 0 & B_{22} \end{bmatrix}$$

where B_{11} and B_{22} are the $s \times t$ and $(m-s) \times (r-t)$ incidence matrices of G_1 and G_2 respectively. By definition, the vector of the vertices of G_1 is

$$X = [\overset{s}{E} \quad \overset{m-s}{0}]$$

According to Lemma 4.5.5, we have

$$XB_1 = 0$$

Since X is not a scalar multiple of E and the rank of any matrix formed with B_1 and a column of B_2 is $m - 1$, it follows that

$$XB^{(i)} \neq 0 \qquad i = r+1, r+2, \ldots, n$$

where $B^{(i)}$ is the column of B corresponding to the arc a_i of C. If a_i is directed toward G_1, then $B^{(i)}$ has a $+1$ in one of its first s rows, a -1 in one of its last $m - s$ rows, and 0s elsewhere. Consequently $XB^{(i)} = +1$. On the other hand, if a_i is directed away from G_1, then $B^{(i)}$ has a -1 in one of its first s rows, a $+1$ in one of its last $m - s$ rows, and 0s elsewhere. In this case $XB^{(i)} = -1$. We have now proved that the entries of XB are identical to the entries of the vector C corresponding to the cut C.
////

Theorem 4.5.25 For a connected directed graph:

1. $\mathbb{R}(C) \subseteq \mathbb{R}(B)$.
2. $CL = 0$.

PROOF By Lemma 4.5.24, each row of C is contained in $\mathbb{R}(B)$. Hence $\mathbb{R}(C) \subseteq \mathbb{R}(B)$. The second part of the theorem follows immediately from part 4 of Theorem 4.5.20.
////

Part 1 of Theorem 4.5.25 can be strengthened considerably. We can prove that $\mathbb{R}(C) = \mathbb{R}(B)$ in exactly the same way it was proved for undirected graphs. The arcs of any tree determine a fundamental set of cuts, consisting of $m - 1$ cuts. By numbering the arcs and cuts of the graph properly we can show that $r(C) \geq m - 1$. Combining this relation with Theorem 4.5.7 and the first part of Theorem 4.5.25, we have

$$\mathbb{R}(C) = \mathbb{R}(B) = \mathbb{N}_R(L)$$
$$\mathbb{N}(C) = \mathbb{N}(B) = \mathbb{C}(L)$$

We can complete the analogy between directed and undirected graphs with the following theorem.

Theorem 4.5.26 Under the natural correspondence between the vector space of sets of arcs of G and $\mathbb{F}_n$, the cuts of G correspond to the minimal vectors of $\mathbb{R}(B)$, and the cut sets of G correspond to the elements of $\mathbb{R}(B)$.

PROOF The proof of part 1 of the theorem is similar to the proof of Theorem 4.4.35 and is left as an exercise. Part 2 follows immediately from part 1. ////

As a consequence of Theorem 4.5.26 we have:

Corollary 4.5.27 $\mathbb{R}(B)$ is unimodular.

In Sec. 4.4 we used Theorem 4.4.27 to establish the relation

$$CL = 0$$

We obtained the same result for directed graphs (in the second part of Theorem 4.5.25) by means of Lemma 4.5.24. However, the analog of Theorem 4.4.27 for directed graphs has some interesting ramifications. Let L be a loop and C be a cut of a connected directed graph G and let $\tilde{L}$ and $\tilde{C}$ be a directed loop and a directed cut respectively associated with L and C. If $L = [l_j]$ and $C = [c_j]$ are the n-dimensional column and row vectors corresponding to L and C (with respect to $\tilde{L}$ and $\tilde{C}$), then according to the second part of Theorem 4.5.25,

$$\sum_{j=1}^{n} c_j l_j = CL = 0 \qquad (4.5.28)$$

Set

$$S = C \cap L = \{a_1, a_2, \ldots, a_s\}$$

Theorem 4.4.27 states that s is even. For $a_j \in S$ the following situations are possible:

1 The orientation of a_j in L is the same as in $\tilde{L}$ (in which case $l_j = +1$).
2 The orientation of a_j in L is opposite to its orientation in $\tilde{L}$ (in which case $l_j = -1$).
1' The orientation of a_j in C is the same as in $\tilde{C}$, (in which case $c_j = +1$).
2' The orientation of a_j in C is opposite to its orientation in $\tilde{C}$ (in which case $c_j = -1$).

Then the term $c_j l_j$ in (4.5.28) is given by

$$c_j l_j = \begin{cases} +1 & \text{when either 1 and } 1' \text{ or } 2 \text{ and } 2' \text{ occur} \\ -1 & \text{when either 1 and } 2 \text{ or } 2 \text{ and } 1' \text{ occur} \\ 0 & \text{when } a_j \notin \mathbb{S} \end{cases}$$

t follows from equation (2.5.28) that the number of arcs of S for which $c_j l_j = +1$ is equal to the number of arcs of S for which $c_j l_j = -1$, that is, $\tfrac{1}{2}s$. Thus we have proved:

Theorem 4.5.29 Let L be a loop and C be a cut of a connected directed graph and let $\tilde{L}$ and $\tilde{C}$ be a directed loop and a directed cut associated with L and C respectively. Set $S = C \cap L$. The number of arcs of S whose orientation is either (a) the same in both L and $\tilde{L}$ and in C and $\tilde{C}$ or (b) opposite in both L and $\tilde{L}$ and in C and $\tilde{C}$ is equal to the number of arcs of S for which the orientation is the same in one case and opposite in the other.

The most interesting situation occurs when L is a directed loop (so that we may set $L = \tilde{L}$). Then condition 1 above always occurs, and condition 2 never takes place. In this case we have:

Theorem 4.5.30 If L is a directed loop and C is a cut of the connected directed graph G, and if G_1 and G_2 are the components of the graph G' which results when the arcs of C are removed from G, then exactly half the arcs of $S = C \cap L$ are directed from G_1 to G_2 and exactly half are directed from G_2 to G_1.

PROOF Let $\tilde{C}$ be a directed cut, say from G_1 to G_2, associated with C. Since we have set $L = \tilde{L}$, case (b) of Theorem 4.5.29 cannot occur. Similarly, an arc of S for which the orientation is the same in one case and opposite in the other must have opposite orientations in C and $\tilde{C}$. Thus the number of arcs of S which have the same orientation in C and $\tilde{C}$ is the same as the number of arcs of S which have opposite orientation in C and $\tilde{C}$. ////

There is a simple geometric proof of Theorem 4.5.30. The components G_1 and G_2 can be thought of as being contained in the two regions K_1 and K_2 of Fig. 4.5.31. Each arc of C has one vertex in G_1 and one vertex in G_2. By following a path along the directed loop L (in the direction of the arrowheads) we alternately pass from G_1 to G_2 and from G_2 to G_1. The arcs of L which do not belong to C are indicated in G_1 and G_2 by wavy lines. The arcs of C which do not belong to L are not included in the diagram.

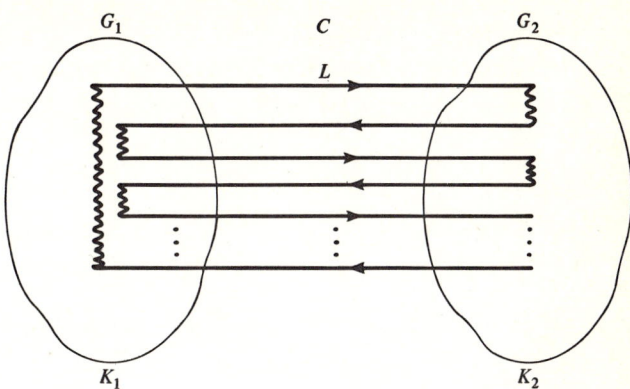

FIGURE 4.5.31

We have emphasized the close relation between directed and undirected graphs. The development of the theory of directed graphs in this section parallels the development of Sec. 4.3. However, we close this section with a theorem pertaining specifically to directed graphs. This theorem depends on one of the fundamental properties of the real field, namely, *a sum of squares of real numbers is zero only when the numbers themselves are all zero*.

Lemma 4.5.32 Let G be a connected directed $m \times n$ graph and let $\mathbb{S}$ be the vector space of formal sums of arcs of G. If $\mathbb{L}$ and $\mathbb{C}$ are the subspaces of $\mathbb{S}$ which consist of all loop sets and all cut sets respectively, then $\mathbb{L} \cap \mathbb{C} = 0$.

PROOF Let $X \in \mathbb{L} \cap \mathbb{C}$. Since X is a loop set, X corresponds to an n-dimensional column vector belonging to $\mathbb{N}(B)$, say $Y = [y_j]$, under the natural correspondence between loop sets and elements of $\mathbb{N}(B)$. Similarly, since X is a cut set, X corresponds to an n-dimensional row vector belonging to $\mathbb{R}(B)$, say Z, under the natural correspondence between cut sets and elements of $\mathbb{R}(B)$. Thus we have

$$ZY = 0 \quad (4.5.33)$$

However, because Y and Z correspond to the same formal sum of arcs under the two natural correspondences, we have

$$Z = Y^T$$

Then equation (4.5.33) becomes

$$ZY = Y^T Y = y_1^2 + y_2^2 + \cdots + y_n^2 = 0$$

and consequently
$$y_1 = y_2 = \cdots = y_n = 0$$
Thus $X = 0$. ////

Theorem 4.5.34 Every formal sum of arcs of a connected directed graph C can be expressed in exactly one way as the sum of a loop set and a cut set.

PROOF Let G be an $m \times n$ graph. It follows from Corollary 4.5.18 and Theorem 4.5.26 that

$$\dim \mathbb{L} = n - m + 1 \qquad \dim \mathbb{C} = m - 1 \qquad (4.5.35)$$

Combining equations (4.5.35) with Lemma 4.5.32, we have

$$\mathbb{S} = \mathbb{L} \oplus \mathbb{C}$$

where $\mathbb{S}$ is the vector space of all formal sums of arcs of G. ////

EXERCISES

1 Denote by G the directed graph formed from the undirected graph of Example 4.4.12 by orienting each arc toward the vertex which has the larger subscript. Form the incidence matrix, the loop matrix, and the cut matrix of G. Verify Theorems 4.5.20 and 4.5.25.
2 Repeat Exercise 1 with the graph G of Example 4.4.41.
3 Express each of the arcs of the graph of Fig. 4.5.1 as a sum of loop sets and cut sets.
4 Show that Lemma 4.5.32 and Theorem 4.5.34 are not generally true for undirected graphs (and $\mathbb{F} = \mathbb{Z}_2$). Specifically, show that Lemma 4.5.32 and Theorem 4.5.34 are false when G is a loop of even length. Show, however, that the lemma and theorem remain true when G is a loop of odd length.
5 Give an example of two unimodular vector spaces whose sum is not unimodular.
6 Show that Theorem 4.5.29 can be obtained from Theorem 4.5.30 by changing the orientation of some of the arcs of the loop. (This provides a proof of Theorem 4.5.29 based on the ideas suggested by Fig. 4.5.31.)

4.6 MATRICES AND TREES

We have seen in Secs. 4.4 and 4.5 that for a directed graph those properties which are not directly related to the orientations of the arcs, e.g., connectedness, undirected loops, and undirected cuts, can be studied equally well in terms of the matrices of the directed graph G or in terms of the matrices of the underlying undirected graph $|G|$. Recall that the entries of the matrices associated with G are from the real field whereas the entries of the matrices associated with $|G|$ are from $\mathbb{Z}_2$. In this section we will work with both fields simultaneously.

A subgraph T of a directed graph G is called a *tree* if $|T|$ is a tree of $|G|$. It follows from Lemmas 4.4.8 and 4.5.12 that a set T of $m - 1$ arcs of an $m \times n$ graph form a tree:

1. If and only if the columns of B corresponding to the arcs of T are linearly independent.
2. If and only if the columns of $|B|$ corresponding to the arcs of $|T|$ are linearly independent.

We can conclude from the equivalence of these two conditions that a set of columns of an incidence matrix B are linearly independent if and only if the corresponding columns of $|B|$ are linearly independent. The same result can be obtained as a corollary to:

Theorem 4.6.1 The determinant of every square submatrix of an incidence matrix B is equal to 0, $+1$, or -1.

PROOF The proof is by an induction on the order of the submatrix and is immediate for square submatrices of order 1. Now suppose that the theorem is true for all $k \times k$ submatrices of B and let M be a square submatrix of B of order $k + 1$. Each column of M contains at most two nonzero entries, at most one $+1$, and at most one -1. If every column of M contains either two nonzero entries or no nonzero entries, then M is singular since

$$EM = 0$$

and consequently det $M = 0$. On the other hand, suppose that some column of M, say the qth column, contains exactly one nonzero entry. Then, expressing det M by means of the qth column-expansion of the determinant [see equation (1.9.31)], we see that det M is equal (up to a possible change in sign) to the determinant of one of its submatrices, say N, of order k. By the induction hypothesis, det $N = 0$, $+1$, or -1, and therefore det $M = 0$, $+1$, or -1. ////

Corollary 4.6.2 Let M be a square submatrix of B and let $|M|$ be the corresponding submatrix of $|B|$. Then

1. det $|M| = 0$ if and only if det $M = 0$.
2. det $|M| = 1$ if and only if det $M = +1$ or -1.

PROOF Let the order of M be k. The determinant of M (or $|M|$) is a sum of $k!$ terms, each of which is $+1$, -1, or 0 (or is 1 or 0). It follows from Theorem 4.6.1 that $\det M = 0$ if and only if an even number of the terms of the determinant are nonzero. Similarly, $\det |M| = 0$ if and only if an even number of the $k!$ terms of the determinant are nonzero. Finally, a term of $\det M$ is nonzero if and only if the corresponding term of $\det |M|$ is nonzero. ////

Theorem 4.6.1 leads directly to a simple and beautiful formula for the number of trees contained in a connected graph. First we need a lemma.

Lemma 4.6.3 Let B be the incidence matrix of an $m \times n$ connected graph G and let $\bar{B}$ be formed from B by deleting any one row of B. (It does not matter which row of B is deleted.) There is a one-to-one correspondence between the trees of G and the nonsingular submatrices of $\bar{B}$ of order $m - 1$.

PROOF Since $\bar{B}$ has $m - 1$ rows, a submatrix, say M, of $\bar{B}$ of order $m - 1$ consists of $m - 1$ columns of $\bar{B}$. By Corollary 4.6.2, M is nonsingular if and only if $|M|$ is nonsingular. Hence, to complete the proof of the lemma it is sufficient to show that there is a one-to-one correspondence between the trees of G and those sets of $m - 1$ columns of $|\bar{B}|$ which are linearly independent.

According to part 2 of Theorem 4.4.16 and Lemma 4.4.8, a necessary and sufficient condition for a set of $m - 1$ arcs of G to form a tree is that the corresponding columns of $|B|$ be linearly independent. We wish to prove that the same result holds true when $|B|$ is replaced by $|\bar{B}|$. Specifically, we wish to show that if a set of columns of $|B|$ are linearly independent, then the corresponding columns of $|\bar{B}|$ are also linearly independent. The converse of this statement is clear.

Let $|H|$ be an $m \times k$ submatrix of $|B|$ whose columns are linearly independent and let $|\bar{H}|$ be the corresponding submatrix of $|\bar{B}|$. Also, let $H_{(0)}$ denote the row of H which is deleted from H to form $\bar{H}$. According to equation (4.4.1),

$$E_{(m)}|H| = 0 \qquad (4.6.4)$$

Equation (4.6.4) can be expressed as

$$|H_{(0)}| + E_{(m-1)}|\bar{H}| = 0$$

and consequently

$$|H_{(0)}| = E_{(m-1)}|\bar{H}| \qquad (4.6.5)$$

Now, let us suppose that the columns of $|\overline{H}|$ are linearly dependent. Then there exists a vector X, $X \neq 0$, such that

$$|\overline{H}| X = 0$$

It follows from equation (4.6.5) that

$$|H_{(0)}| X = 0$$

and hence

$$|H| X = 0$$

This is a contradiction since the columns of $|H|$ are linearly independent.

////

Now we can prove:

Theorem 4.6.6 Let B be the incidence matrix of a connected graph G and let $\overline{B}$ be formed from B by deleting any one row of B. Then the number of trees contained in G is equal to det $\overline{B}\overline{B}^T$ [1].

PROOF The proof consists of evaluating det $\overline{B}\overline{B}^T$ by means of the Cauchy-Binet theorem (see Exercise 24 of Sec. 1.9). For each subset Θ of $\mathbb{N} = \{1, 2, \ldots, n\}$ consisting of $m - 1$ elements of $\mathbb{N}$, define $\overline{B}_\Theta$ to be the square matrix of order $m - 1$ formed from $\overline{B}$ by deleting all columns $\overline{B}^{(j)}$ for $j \notin \Theta$. Similarly, define $\overline{B}_\Theta{}^T$ to be the square matrix of order $m - 1$ formed from $\overline{B}^T$ by deleting all rows $\overline{B}^T_{(j)}$ for $j \notin \Theta$. Clearly $\overline{B}_\Theta{}^T = (\overline{B}_\Theta)^T$. Then

$$\det \overline{B}\overline{B}^T = \sum_\Theta (\det \overline{B}_\Theta)(\det \overline{B}_\Theta{}^T)$$

$$\sum_\Theta (\det \overline{B}_\Theta)[\det (\overline{B}_\Theta)^T]$$

$$\sum_\Theta (\det \overline{B}_\Theta)^2$$

where the sum is taken over all $n!/[(m-1)!(n-m+1)!]$ subsets of $\mathbb{N}$ which contain $m - 1$ elements. However, combining Theorem 4.6.1 and Lemma 4.6.3, we have

$$\det \overline{B}_\Theta = \begin{cases} \pm 1 & \text{when arcs of } G \text{ corresponding to elements of } \Theta \text{ form a tree} \\ 0 & \text{otherwise} \end{cases}$$

Hence det BB^T is equal to a sum of 0s and $+1$s, and the number of $+1$s is the same as the number of trees of G.

////

---

[1] This relation was known to Maxwell in 1892 and probably even to Kirchhoff in 1847. However, the first formal proof was given by Brooks, Smith, Stone and Tutte [2] in 1940.

Theorem 4.6.6 remains true when G is not connected. In this case

$$r(B) \leq m - 2$$

and consequently

$$r(\bar{B}) \leq m - 2$$

Thus $\bar{B}\bar{B}^T$ is singular, and det $\bar{B}\bar{B}^T = 0$.

The formula for the number of trees in a graph can be expressed in a somewhat different way.

Corollary 4.6.7 If B is the incidence matrix of a graph, then all the entries of adj BB^T are equal and each entry is equal to the number of trees in the graph.

PROOF For convenience we will assume that in Theorem 4.6.6, $\bar{B}$ was formed from B by deleting the last row of B. Then

$$B = \begin{bmatrix} \bar{B} \\ B_{(m)} \end{bmatrix} \quad \text{and} \quad BB^T = \begin{bmatrix} \bar{B}\bar{B}^T & \bar{B}B_{(m)}^T \\ B_{(m)}\bar{B}^T & B_{(m)}B_{(m)}^T \end{bmatrix}$$

Theorem 4.6.6 states that det $\bar{B}\bar{B}^T$ is equal to the number of trees contained in G. However, det $\bar{B}\bar{B}^T$ is the cofactor of the entry in the last row and last column of BB^T and thus is equal to the entry in the last row and last column of adj BB^T. In the same way, we are able to show that all the diagonal entries of adj BB^T are equal to the number of trees contained in G. It remains only to prove that all the entries of adj BB^T are equal.

According to part (a) of Exercise 25 in Sec. 1.7,

$$r(BB^T) \leq r(B) \quad (4.6.8)$$

We will consider two cases. When G is not a connected graph, the rank of B is at most $m - 2$ and every submatrix of BB^T of order $m - 1$ is singular. Thus adj $BB^T = 0$, as required. Now, let us suppose that G is connected. G contains some trees, and, according to the preceding paragraph, the diagonal entries of adj BB^T are positive. By part (c) of Exercise 9 in Sec. 1.9,

$$r(BB^T) \geq m - 1 \quad (4.6.9)$$

Combining (4.6.8) and (4.6.9), we see that

$$r(BB^T) = m - 1 \quad v_R(BB^T) = 1$$

But

$$E_{(m)}BB^T = 0$$

and consequently $E_{(m)}$ is a basis for $\mathbb{N}_R(BB^T)$. Applying Theorem 1.9.32, we have

$$[\text{adj } BB^T](BB^T) = (\det BB^T)I = 0$$

Every row of adj BB^T is contained in $\mathbb{N}_R(BB^T)$, and thus is a scalar multiple of $E_{(m)}$; that is all its entries are equal. ////

EXAMPLE 4.6.10 For the graph of Example 4.4.12,

$$BB^T = \begin{bmatrix} 4 & -1 & -1 & -1 & -1 \\ -1 & 2 & -1 & 0 & 0 \\ -1 & -1 & 3 & -1 & 0 \\ -1 & 0 & -1 & 3 & -1 \\ -1 & 0 & 0 & -1 & 2 \end{bmatrix}$$

no matter what orientation is given to the arcs. The entry in the first row and first column of adj BB^T is the cofactor of 4 in the matrix above.

$$\det BB^T = \det \begin{bmatrix} 2 & -1 & 0 & 0 \\ -1 & 3 & -1 & 0 \\ 0 & -1 & 3 & -1 \\ 0 & 0 & -1 & 2 \end{bmatrix} = 21$$

Verify that every entry of adj BB^T is 21. List the 21 trees of G. How many subgraphs does G contain consisting of four arcs? List the subgraphs of G containing four arcs which are not trees. ////

The concept of a tree can be extended to a directed graph. A subgraph T of a directed graph G is a *tree* if $|T|$ is a tree of $|G|$. Moreover:

Definition 4.6.11 A tree T of a directed graph is a *directed tree, rooted at the vertex v*, if every vertex of T, other than v, has exactly one arc of T directed from it.

A directed tree has at least one arc directed to its root, and the (unique) path in the tree from any vertex to the root of the tree is a directed path, directed toward the root.

Let v be a vertex of the tree T of a directed graph. T can be transformed into a directed tree rooted at v by changing the orientations of some of the arcs of T. Moreover, this can be done in exactly one way.

EXAMPLE 4.6.12 Let G be the directed graph

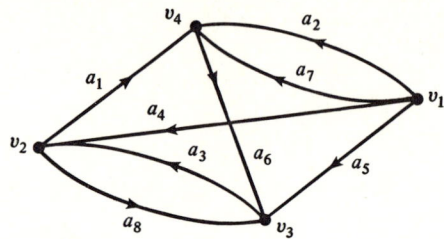

There are no directed trees rooted at v_1 since there are no arcs of G directed toward v_1. On the other hand,

$$T_1 = (a_1, a_2, a_3) \quad T_2 = (a_1, a_3, a_4)$$
$$T_3 = (a_1, a_3, a_5) \quad T_4 = (a_1, a_3, a_7)$$

are the directed trees of G rooted at v_4. Find the directed trees, if any, of G rooted at v_2 and at v_3. ////

We will see that the number of trees of a directed graph rooted at a particular vertex can be expressed by means of determinants. Recall that in Example 4.1.1 we defined the vertex-adjacency matrix $A = [a_{ij}]$ of a directed graph G by setting a_{ij} equal to the number of arcs of G directed from v_i to v_j. Also, in the discussion preceding Example 4.1.15, we defined the out-degree matrix Δ of a directed graph. For the graph of Example 4.6.12,

$$A = \begin{bmatrix} 0 & +1 & +1 & +2 \\ 0 & 0 & +1 & +1 \\ 0 & +1 & 0 & 0 \\ 0 & 0 & +1 & 0 \end{bmatrix} \quad \Delta = \begin{bmatrix} +4 & 0 & 0 & 0 \\ 0 & +2 & 0 & 0 \\ 0 & 0 & +1 & 0 \\ 0 & 0 & 0 & +1 \end{bmatrix}$$

We set

$$M = \Delta - A$$

M is called the *tree matrix* of G. The sum of the entries in each row of M is 0 and consequently

$$ME^{(m)} = 0$$

The tree matrix of the graph of Example 4.6.12. is

$$M = \begin{bmatrix} +4 & -1 & -1 & -2 \\ 0 & +2 & -1 & -1 \\ 0 & -1 & +1 & 0 \\ 0 & 0 & -1 & +1 \end{bmatrix}$$

In Theorem 4.6.14 we will show that the number of directed trees of a graph rooted at any vertex can be obtained from its tree matrix. First we will prove a special case of the general theorem which will be used later as a lemma.

Lemma 4.6.13 Let G be a directed $m \times n$ graph, let $\bar{G}$ be an $m \times (m-1)$ subgraph of G, and let $M = [m_{ij}]$ be the tree matrix of $\bar{G}$. For each vertex v_h of G, $1 \leq h \leq m$, the cofactor of m_{hh} in M is

$$\begin{cases} +1 & \text{if } \bar{G} \text{ is a directed tree rooted at } v_h \\ 0 & \text{otherwise} \end{cases}$$

PROOF Partition M as

$$M = \begin{bmatrix} M_{11} & M_{12} \\ M_{21} & M_{22} \end{bmatrix}$$

where M_{11}, M_{12}, M_{21}, and $M_{22} = [m_{m\,m}]$ are $(m-1) \times (m-1)$, $(m-1) \times 1$, $1 \times (m-1)$, and 1×1 matrices, respectively. It is sufficient to prove that

$$\det M_{11} = \begin{cases} +1 & \text{if } \bar{G} \text{ is a directed tree rooted at } v_m \\ 0 & \text{otherwise} \end{cases}$$

If any vertex of G other than v_m does not have an arc of $\bar{G}$ directed from it, then the corresponding row of M (and of M_{11}) consists solely of 0s and we have $\det M_{11} = 0$, as required.

Henceforth we will assume that every vertex of G other than v_m does have an arc of $\bar{G}$ directed from it. Then $\bar{G}$ is a directed tree of G rooted at v_m if and only if $|\bar{G}|$ is a tree of $|G|$. All the diagonal entries of M_{11} are equal to $+1$. Moreover, each row of M_{11} contains one -1 unless the arc of $\bar{G}$ from the vertex corresponding to that row is directed toward v_m.

Let us suppose that $\bar{G}$ is not a directed tree of G rooted at v_m. Since $|\bar{G}|$ is not a tree of $|G|$, it follows that $|\bar{G}|$ (and therefore $\bar{G}$) is not connected. Let $\bar{G}_1$ be the component of $\bar{G}$ which does not contain v_m. M_{11} can be partitioned as

$$M_{11} = \begin{bmatrix} M_1 & 0 \\ 0 & M_2 \end{bmatrix}$$

where the rows and columns of M_1 correspond to the vertices of $\bar{G}_1$. Since

$$M_1 E = 0$$

M_1 is singular and consequently M_{11} is singular. Thus $\det M_{11} = 0$.

It remains to prove that if $\bar{G}$ is a directed tree of G rooted at v_m, then det $M_{11} = +1$. Since all the diagonal entries of M_{11} are equal to $+1$, the term of det M_{11} corresponding to the identity permutation I, is $+1$. We will show that all other terms of det M_{11} are equal to 0. Suppose there exists a permutation σ, $\sigma \neq I$, for which the corresponding term of det M_{11}, namely,

$$\text{sgn } \sigma \, m_{1\sigma(1)} \, m_{2\sigma(2)} \cdots m_{m-1 \, \sigma(m-1)}$$

is not 0. Then there is an integer k_1 such that

$$\sigma(k_1) = k_2 \neq k_1$$

Hence $m_{k_1 k_2} = -1$, and $\bar{G}$ contains the arc $a_1 = (v_{k_1}, v_{k_2})$.

Let $\sigma(k_2) = k_3$. Clearly $k_3 \neq k_2$, and two cases arise.

CASE 1 Suppose $k_3 = k_1$. Then $m_{k_2 k_1} = -1$, and $\bar{G}$ contains the arc $a_2 = (v_{k_2}, v_{k_1})$. However (a_1, a_2) is a loop. Since $\bar{G}$ was assumed to be a tree, we have reached a contradiction.

CASE 2 Suppose $k_3 \neq k_1$. Then $m_{k_2 k_3} = -1$, and $a_2 = (v_{k_2}, v_{k_3})$ is an arc of $\bar{G}$. Let $\sigma(k_3) = k_4$. Clearly $k_4 \neq k_2, k_3$. If $k_4 = k_1$, we are again in case 1 and it follows that $\bar{G}$ contains a loop of length 3. If $k_4 \neq k_1$, then this process is continued. Eventually we must return to case 1 and reach the desired contradiction. ////

Theorem 4.6.14 (Tutte, 1948) Let M be the tree matrix of the directed $m \times n$ graph G. For each h, $1 \leq h \leq m$, all the entries in the hth column of adj M are equal and are equal to the number of directed trees of G rooted at v_h.

PROOF Since

$$ME^{(m)} = 0$$

M is singular and

$$M(\text{adj } M) = (\det M)I = 0$$

It follows that the entries in each column of adj M are equal. For, if $r(M) \leq m - 2$, then adj $M = 0$, and if $r(M) = m - 1$, then $E^{(m)}$ forms a basis for $\mathbb{N}(M)$.

Let us partition M as

$$M = \begin{bmatrix} M_{11} & M_{12} \\ M_{21} & M_{22} \end{bmatrix}$$

where M_{11}, M_{12}, M_{21}, and M_{22} are $(m-1) \times (m-1)$, $(m-1) \times 1$, $1 \times (m-1)$, and 1×1 matrices, respectively. In order to prove the theorem it is sufficient to show that the number of directed trees of G rooted at v_m is equal to det M_{11}.

The number of directed trees of G rooted at v_m does not change when all arcs of the graph directed from v_m are removed. Moreover, when these arcs are removed, only the last row of M is affected. Thus we will henceforth assume that these arcs have been deleted, and M has the form

$$M = \begin{bmatrix} M_{11} & M_{12} \\ 0 & 0 \end{bmatrix}$$

Since there are Δ_i arcs of G directed from v_i [where Δ_i is the entry in the (i, i) position of Δ], there are $\Delta_1 \Delta_2 \cdots \Delta_{m-1}$ $m \times (m-1)$ subgraphs of G each of which contains one arc directed from each vertex of G other than v_m. For convenience, set

$$\Delta_1 \Delta_2 \cdots \Delta_{m-1} = d$$

Let us denote these graphs by $\overline{G}_1, \overline{G}_2, \ldots, \overline{G}_d$ and the corresponding tree matrices by $M^{(1)}, M^{(2)}, \ldots, M^{(d)}$. All the entries of the last row of each $M^{(q)}$, $q = 1, 2, \ldots, d$, are equal to 0. Partition $M^{(q)}$ as

$$M^{(q)} = \begin{bmatrix} M_{11}^{(q)} & M_{12}^{(q)} \\ 0 & 0 \end{bmatrix}$$

where $M_{11}^{(q)}$ and $M_{12}^{(q)}$ are $(m-1) \times (m-1)$ and $(m-1) \times 1$ matrices, respectively. In Lemma 4.6.13 we proved that

$$\det M_{11}^{(q)} = \begin{cases} +1 & \text{if } \overline{G}^{(q)} \text{ is a directed tree rooted at } v_m \\ 0 & \text{otherwise} \end{cases}$$

In order to complete the proof of the theorem it remains to show that

$$\det M_{11} = \sum_{q=1}^{d} \det M_{11}^{(q)} \quad (4.6.15)$$

We will prove that equation (4.6.15) follows directly from the repeated use of Lemma 1.9.19. First let us apply Lemma 1.9.19 to the first row of M_{11}. The entry in the (1, 1) position of M_{11} is Δ_1, and the remaining entries of the first row of M_{11} are nonpositive integers whose sum is at most $-\Delta_1$. The first row of M_{11} can be expressed (in exactly one way) as a sum of Δ_1 $(m-1)$-dimensional row vectors, say $X_1, X_2, \ldots, X_{\Delta_1}$, each of which has a $+1$ in the first position and the remaining entries are either all 0s or consist of $m-2$ 0s and a single -1. There is a one-to-one correspondence between arcs of G directed from v_1 and the vectors $X_1, X_2, \ldots, X_{\Delta_1}$ such that:

1. $a_i = (v_1, v_m)$ corresponds to
$$X_i = [+1 \quad 0 \quad 0 \quad \cdots \quad 0]$$

2. $a_i = (v_1, v_k)$, $k \neq m$, corresponds to a vector X_i in which the first entry is $+1$, the kth entry is -1, and all other entries are 0.

Let

$$\tilde{M}^{(1)}, \tilde{M}^{(2)}, \ldots, \tilde{M}^{(\Delta_1)} \qquad (4.6.16)$$

be the square matrices of order $m - 1$ such that:

1. The first row of $\tilde{M}^{(q)}$ is X_q, $q = 1, 2, \ldots, \Delta_1$.
2. Each row of $\tilde{M}^{(q)}$, other than the first, is equal to the corresponding row of M_{11}.

According to Lemma 1.9.19,

$$\det M_{11} = \sum_{q=1}^{\Delta_1} \det \tilde{M}^{(q)}$$

Now let us apply the same procedure to the second row of each of the matrices of (4.6.16). Let $\tilde{M}^{(q)}$ be one of these matrices. For convenience we will set $q = 1$. From $\tilde{M}^{(1)}$ we construct Δ_2 square matrices of order $m - 1$ by expressing the second row of $\tilde{M}^{(1)}$ as a sum of Δ_2 vectors, each with a $+1$ in the second position, and either all the remaining entries are 0s or one of the remaining entries is a -1 and the rest are 0s. Denote these matrices by $\tilde{\tilde{M}}^{(1)}, \tilde{\tilde{M}}^{(2)}, \ldots, \tilde{\tilde{M}}^{(\Delta_2)}$. Then, by Lemma 1.9.19,

$$\det \tilde{M}^{(1)} = \sum_{s=1}^{\Delta_2} \det \tilde{\tilde{M}}^{(s)}$$

In this way we construct a total of $\Delta_1 \Delta_2$ matrices [Δ_2 matrices from each of the Δ_1 matrices of (4.6.16)], say

$$\tilde{\tilde{M}}^{(1)}, \tilde{\tilde{M}}^{(2)}, \ldots, \tilde{\tilde{M}}^{(\Delta_2)}, \ldots, \tilde{\tilde{M}}^{(\Delta_1 \Delta_2)} \qquad (4.6.17)$$

Again, by Lemma 1.9.19

$$\det M_{11} = \sum_{s=1}^{\Delta_1 \Delta_2} \det \tilde{\tilde{M}}^{(s)}$$

Moreover, there is a one-to-one correspondence between the matrices of (4.6.17) and all pairs of arcs of G, (a_i, a_j), such that a_i is directed from v_1 and a_j is directed from v_2.

Continuing this process, we express det M_{11} as the sum of the determinants of d square matrices of order $m - 1$. These are, in fact, the matrices $M_{11}^{(1)}, M_{11}^{(2)}, \ldots, M_{11}^{(d)}$ of equation (4.6.15). ////

EXAMPLE 4.6.18   Let us illustrate the successive steps of the proof of Theorem 4.6.14 using the graph of Example 4.6.12. When those arcs of G directed from v_4, (that is, a_6) are removed from G, the resulting graph is

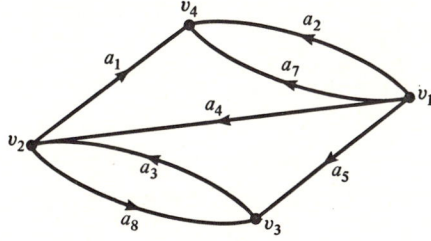

and the tree matrix is

$$\begin{bmatrix} +4 & -1 & -1 & -2 \\ 0 & +2 & -1 & -1 \\ 0 & -1 & +1 & 0 \\ 0 & 0 & 0 & 0 \end{bmatrix}$$

Then, maintaining the notation used in the proof of Theorem 4.6.14,

$$M_{11} = \begin{bmatrix} +4 & -1 & -1 \\ 0 & +2 & -1 \\ 0 & -1 & +1 \end{bmatrix}$$

and the matrices of (4.6.16) are

$$\tilde{M}_{11}^{(1)} = \begin{bmatrix} +1 & -1 & 0 \\ 0 & +2 & -1 \\ 0 & -1 & +1 \end{bmatrix} \quad \tilde{M}_{11}^{(2)} = \begin{bmatrix} +1 & 0 & -1 \\ 0 & +2 & -1 \\ 0 & -1 & +1 \end{bmatrix}$$

$$\tilde{M}_{11}^{(3)} = \begin{bmatrix} +1 & 0 & 0 \\ 0 & +2 & -1 \\ 0 & -1 & +1 \end{bmatrix} = \tilde{M}_{11}^{(4)}$$

At the next stage we have the matrices of (4.6.17)

$$\tilde{M}_{11}^{(1)} = \begin{bmatrix} +1 & -1 & 0 \\ 0 & +1 & -1 \\ 0 & -1 & +1 \end{bmatrix} \quad \tilde{M}_{11}^{(2)} = \begin{bmatrix} +1 & -1 & 0 \\ 0 & +1 & 0 \\ 0 & -1 & +1 \end{bmatrix}$$

$$\tilde{M}_{11}^{(3)} = \begin{bmatrix} +1 & 0 & -1 \\ 0 & +1 & -1 \\ 0 & -1 & +1 \end{bmatrix} \quad \tilde{M}_{11}^{(4)} = \begin{bmatrix} +1 & 0 & -1 \\ 0 & +1 & 0 \\ 0 & -1 & +1 \end{bmatrix}$$

$$\tilde{M}_{11}^{(5)} = \begin{bmatrix} +1 & 0 & 0 \\ 0 & +1 & -1 \\ 0 & -1 & +1 \end{bmatrix} = M_{11}^{(7)} \quad M_{11}^{(6)} = \begin{bmatrix} +1 & 0 & 0 \\ 0 & +1 & 0 \\ 0 & -1 & +1 \end{bmatrix} = M_{11}^{(8)}$$

Since $\Delta_3 = +1$, another step is not needed. It is easily seen that

$$\det \tilde{M}_{11}^{(1)} = \det \tilde{M}_{11}^{(3)} = \det \tilde{M}_{11}^{(5)} = \det \tilde{M}_{11}^{(7)} = 0$$

and
$$\det \tilde{M}_{11}^{(2)} = \det \tilde{M}_{11}^{(4)} = \det \tilde{M}_{11}^{(6)} = \det \tilde{M}_{11}^{(8)} = +1$$

G has four directed trees rooted at v_4. These are given in Example 4.6.12. However, we have also shown that G contains eight subgraphs which consist of one arc from each vertex of G other than v_4 (including the four trees). Find them (a) directly from the graph and (b) by using the matrices $\tilde{M}_{11}^{(q)}$, $q = 1, 2, \ldots, 8$. ////

There is an interesting relation between the directed trees and the Euler lines of a graph. In Sec 4.2 we discussed undirected graphs which contain an Euler line. It is natural to extend the same idea to a directed graph. A directed path from a vertex to itself in a directed graph is called a *directed Euler line* if it contains all the arcs of the graph. For directed Euler lines the analog of Theorem 4.2.10 is:

Theorem 4.6.19 A connected directed graph contains a directed Euler line if and only if for each vertex v_k, the in degree δ_k is equal to the out degree Δ_k.

PROOF It is clear that a directed Euler line leaves each vertex of a graph as many times as it enters it. Thus, if a directed Euler line exists, the in degree and the out degree of every vertex are equal.

The proof that a connected directed graph in which

$$\delta_k = \Delta_k$$

for each vertex v_k has a directed Euler line is similar to the proof of Theorem 4.2.10 and is left as an exercise. ////

Corollary 4.6.20 In a connected directed graph for which:

1. $\Delta_1 - \delta_1 = +1$
2. $\Delta_2 - \delta_2 = -1$
3. $\Delta_i = \delta_i$ for $i \neq 1, 2$

there is a directed path from v_1 to v_2 which contains all the arcs of the graph.

Let v_k be a vertex of a directed graph G and let P be a directed Euler line from v_k to itself. P induces an ordering on the arcs of G. We say that a_i *precedes* a_j with respect to P if it occurs before a_j in P. Similarly, for each vertex v_l of G, P induces an ordering on the Δ_l arcs of G directed from v_l; that is, $a_i = (v_l, v_{h_1})$ *precedes* $a_j = (v_l, v_{h_2})$ with respect to P if a_i occurs before a_j in P. For each vertex v of G, $v \neq v_k$, the arc by which P leaves v for the last time is called the *last exit* of v with respect to P. Note that v_k does not have a last exit with respect to P. The relation between directed trees and directed Euler lines of a graph is described in the next two lemmas.

Lemma 4.6.21 Let P be a directed Euler line from v_k to itself in the directed $m \times n$ graph G. The set of last exits with respect to P is a directed tree of G rooted at v_k.

PROOF Let the set of last exits of P be denoted by T. T contains $m - 1$ arcs, one directed from each vertex of G except v_k. In order to prove that T is a directed tree rooted at v_k it remains only to show that T does not contain a loop.

Let us suppose that L is a loop of G contained in T. Since every vertex in T has one arc directed from it, L is a directed loop. Furthermore, v_k does not belong to L because T does not contain an arc directed from v_k. Let us denote the arc of L which occurs last in P by a_i and set $a_i = (v_{h_1}, v_{h_2})$. Since L is a directed loop, it also contains an arc directed from v_{h_2}, say $a_j = (v_{h_2}, v_{h_3})$. According to the way in which T was formed, a_j is the last exit of v_{h_2} with respect to P. However, a_i occurs in P after a_j, and the arc in P immediately after a_i also leaves v_{h_2}. We have now reached a contradiction, and consequently we must conclude that T does not contain a loop. ////

Lemma 4.6.21 describes a method for obtaining a directed tree rooted at v_k from a directed Euler line from v_k to itself. However, different Euler lines will yield the same directed tree if they have the same set of last exits.

Lemma 4.6.22 Let G be a connected directed $m \times n$ graph such that

$$\Delta_i = \delta_i \quad i = 1, 2, \ldots, m$$

and let T be a directed tree of G rooted at v_k. The number of directed Euler lines from v_k to itself for which T is the set of last exits is

$$\Delta_k \prod_{i=1}^{m} [(\Delta_i - 1)!] \quad (4.6.23)$$

PROOF At each vertex of G we order the arcs directed from that vertex. The only requirement imposed on the ordering is that each of the arcs of T receive the highest number at the vertex from which it is directed. Thus, if a_j belongs to T and is directed from v_i, then a_j must be the Δ_ith arc at v_i. Since T does not contain an arc directed from v_k, the number of orderings which satisfy this requirement is given by (4.6.23). If P is a directed Euler line from v_k to itself which has T as its set of last exits, then the ordering of the arcs at each vertex induced by P satisfies the requirement above. We wish to show that, conversely, every ordering with this property determines a unique directed Euler line whose set of last exits is T.

We begin by selecting the first arc at v_k, say (v_k, v_{h_1}). Next we choose the first arc at v_{h_1}, say (v_{h_1}, v_{h_2}). We then take the first arc at v_{h_2} which has not been used. (If $v_{h_2} \neq v_k$, the desired arc is the first arc at v_{h_2}. If $v_{h_2} = v_k$, we take the second arc at v_k since the first one has already been used.) Continuing in this way, we choose the arc at each vertex which is first among the arcs that have not been previously used. Clearly the ordering of the arcs at each vertex uniquely determines the path from v_k which we construct. This path must end at v_k since, for every other vertex, whenever it is entered, there is an unused arc directed from it. Also, when the path ends, there remain no unused arcs directed from v_k, or else we would continue the process. Since there are no unused arcs directed from v_k when the process ends, there are also no unused arcs directed toward v_k. Finally, it is clear that if all the arcs of G are used, then the arcs of T is the set of last exits of the path. It now remains to show that all arcs of G are used.

Suppose that there is an arc of G which is never selected in the process described in the last paragraph. Let this arc be directed from v_{l_1}. We have already shown that $v_{l_1} \neq v_k$. Because of the method of selecting arcs at each vertex, the arc of T directed from v_{l_1}, say (v_{l_1}, v_{l_2}), is also never selected. Since (v_{l_1}, v_{l_2}) is directed toward v_{l_2}, it follows that there is an arc of G directed from v_{l_2} which is never used. As before, the arc of T directed from v_{l_2}, say (v_{l_2}, v_{l_3}), is never used. There is a directed path in

T from v_{l_1} to v_k, and hence this argument leads eventually to an arc directed toward v_k which is never selected. This is a contradiction. Therefore every arc of G is used in the construction described in the last paragraph, and the path we have formed in this way is a directed Euler line from v_k to itself. ////

If
$$P = \{a_1 = (v_{h_1}, v_{h_2}), a_2, \ldots, a_{k-1}, a_k = (v_{l_1}, v_{l_2}), \ldots, a_n\}$$
is a directed Euler line from v_{h_1} to itself, then
$$P' = \{a_k = (v_{l_1}, v_{l_2}), \ldots, a_n, a_1 = (v_{h_1}, v_{h_2}), a_2, \ldots, a_{k-1}\}$$
is a directed Euler line from v_{l_1} to itself. As in the case of undirected graphs, we say that two directed Euler lines of a graph belong to the same *circuit* if one of them can be obtained from the other in this way. A circuit in an $m \times n$ directed graph consists of n directed Euler lines, and each directed Euler line in a circuit is uniquely determined by specifying its first arc.

Theorem 4.6.24 (van Aardenne-Ehrenfest and de Brujin, 1950)
If G is a connected directed $m \times n$ graph in which
$$\Delta_i = \delta_i \qquad i = 1, 2, \ldots, m$$
then the number of directed trees of G rooted at any vertex is equal to the number of directed trees rooted at any other vertex. Denote this number by t. Moreover, the number of circuits in G is
$$t \prod_{i=1}^{m} [(\Delta_i - 1)!]$$

PROOF Let v_k be any vertex of G and let T be a directed tree rooted at v_k. By Lemma 4.6.22 there are
$$\Delta_k \prod_{i=1}^{m} [(\Delta_i - 1)!]$$
directed Euler lines from v_k to itself whose set of last exits is T. If G contains t_k directed trees rooted at v_k, then there are a total of
$$t_k \Delta_k \prod_{i=1}^{m} [(\Delta_i - 1)!]$$
directed Euler lines in G from v_k to itself. Since v_k has Δ_k arcs directed from it and each of these can be the first arc of a directed Euler line, it

follows that each circuit contains Δ_k directed Euler lines from v_k to itself. Hence there are

$$t_k \prod_{i=1}^{m} [(\Delta_i - 1)!]$$

circuits in G. The number of circuits in G is independent of the vertex v_k and therefore

$$t_1 = t_2 = \cdots = t_m = t \qquad ////$$

See Exercise 8 for a different proof of Theorem 4.6.24.

EXERCISES

1 Let $G=$

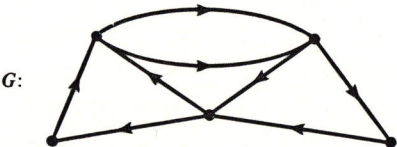

Find the tree matrix of G, the number of directed trees rooted at each vertex, the number of Euler lines from each vertex to itself, and the number of circuits. Find the number of trees in the underlying undirected graph of G.

2 Let m be an odd positive integer and let G be the complete graph with m vertices. For $1 \leq i < j \leq m$, define the orientation of the arc between v_i and v_j by

$$a = \begin{cases} (v_i, v_j) & \text{if } i - j \text{ is even} \\ (v_j, v_i) & \text{if } i - j \text{ is odd} \end{cases}$$

Prove that G has a directed Euler line.

3 Let $m = 2k + 1$ and let G be the complete graph with m vertices. For $1 \leq i < j \leq m$, define the orientation of the arc between v_i and v_j by

$$a = \begin{cases} (v_i, v_j) & \text{if } |i - j| \leq k \\ (v_j, v_i) & \text{if } |i - j| > k \end{cases}$$

Prove that G has a directed Euler line. Show that for $m = 5$ the graph described in Exercise 2 can be transformed into the graph described above by a renumbering of the vertices.

4 Show that a complete undirected graph with m vertices, $m \geq 2$, contains $m^{(m-2)}$ trees. [*Hint:* Let $f(n, a)$ denote the determinant of a square matrix of order n which has a's in the diagonal positions and -1s elsewhere. Prove that

$$f(n, a) = af(n - 1, a) - (n - 1)(a + 1)^{n-2}$$

and $f(n, n - 1) = 0$.

5 Show that if one edge is removed from a complete undirected graph with m vertices, $m \geq 2$, the resulting graph has $(m - 2)m^{(m-3)}$ trees. [*Hint:* Let $h(n, a)$ denote the determinant of a square matrix of order n which has $a - 1$ in the (1, 1) position, a's in the remaining diagonal positions, and -1s elsewhere. Prove that

$$h(n, a) = (a - 1)f(n - 1, a) - (n - 1)(a + 1)^{n-2}$$

where $f(n, a)$ is defined in Exercise 4.]

6 Present an alternate proof of Exercise 5 by finding the number of trees in a complete undirected graph that contain a given arc.

7 Find the number of trees in a complete graph with m vertices which contain the arc $a = (v_1, v_2)$ but no other arc incident with v_1. Find the number of trees which contain both a and $a' = (v_1, v_3)$ but no other arcs incident with v_1. (*Hint:* Use Exercises 4 and 5.)

8 Let G be a connected graph. Prove that the number of directed trees rooted at a vertex is the same for all vertices if and only if

$$EM = 0$$

where M is the tree matrix of G, and that this equation is satisfied if and only if G contains a directed Euler line.

9 Prove that a directed complete graph with m vertices in which each arc is directed at the vertex with a lower subscript has $(m - 1)!$ directed trees rooted at v_1 and no directed trees rooted at any other vertex.

REFERENCES

Euler solved the Königsberg bridge problem in 1736, and since then there has been a vast amount written about graphs. However, until recently graph theory as an independent branch of mathematics and as a part of the mathematical curriculum has been largely ignored. Undoubtedly much of the impetus for the renewed interest in graphs which has taken place in recent years comes from the many varied applications to which graph theory has been put. In addition to the classical applications to the physical sciences, (e.g., electric network theory), graph theory now plays a prominent role in the social sciences, (e.g., communication and transportation problems), and in computer technology. Moreover, it is only the advent of large-scale computers which has made many of the algorithms associated with solving graphical problems feasible.

Several texts present mathematical treatments of graph theory. Harary [5] is primarily concerned with undirected graphs, whereas Harary, Norman, and Cartwright [6] is devoted to directed graphs. Busacker and Saaty [3] treat both undirected and directed graphs and present a great variety of applications. Seshu and Reed [10] develop the fundamentals of electrical analysis and network synthesis in terms of graphs. To a large extent they follow Kirchhoff's development of Kirchhoff's laws. In Example 4.1.9 we discussed the maximal flow in a network and the minimum cut necessary to prevent any flow. Ford and Fulkerson [4] combine graph theory with linear programming to solve these and related problems dealing with minimal cost and optimal assignment of flows. Massey [8] applies the methods of algebraic topology to graphs and illustrates the relation between a graph and its fundamental group.

1 Berge, C.: "The Theory of Graphs and its Applications," trans. from the French by A. Doig, Methuen & Co., Ltd., London, 1962.
2 Brooks, R., C. Smith, A. Stone, and W. Tutte: The Dissection of Rectangles into Squares, *Duke Math. J.*, **7**: 312–340 (1940).
3 Busacker, R., and T. Saaty: "Finite Graphs and Networks: An Introduction with Applications," McGraw-Hill Book Company, New York, 1965.
4 Ford, L., and D. Fulkerson: "Flows in Networks," Princeton University Press, Princeton, N.J., 1962.
5 Harary, F.: "Graph Theory," Addison-Wesley Publishing Company, Inc., Reading, Mass., 1969.
6 Harary, F., R. Norman, and D. Cartwright: "Structural Models: An Introduction to the Theory of Directed Graphs," John Wiley & Sons, Inc., New York, 1965.
7 Kuratowski, K.: Sur le problème des courbes gauches en topologie, *Fund. Math.*, **15**: 271–283 (1930).
8 Massey, W. "Algebraic Topology: An Introduction," Harcourt, Brace & World, Inc., New York, 1967.
9 Ore, O., "Graphs and Their Uses," Random House, New York, 1963.
10 Seshu, S., and M. Reed: "Linear Graphs and Electrical Networks," Addison-Wesley Publishing Company, Inc., Reading, Mass., 1961.
11 Whitney, H.: Non-separable and Planar Graphs, *Trans. Am. Math. Soc.*, **34**: 339–362 (1932).

ANSWERS TO SELECTED EXERCISES

CHAPTER 0 FIELDS

0.1 Fields

1 (a), (b), (e)

CHAPTER 1 MATRICES

1.2 The Arithmetic of Matrices

1 (a) $\begin{bmatrix} x & y \\ 0 & x \end{bmatrix}$ x, y arbitrary (b) $\begin{bmatrix} x & y \\ 0 & x+y \end{bmatrix}$ x, y arbitrary

3 (a) 0, (b) 0, (c) 0, (d) 0, (e) $\dfrac{1}{12}\begin{bmatrix} 6 & -2 \\ 3 & 1 \end{bmatrix}$

4 (a) $\begin{bmatrix} 2x & 2y \\ -x & -y \end{bmatrix}$ x, y arbitrary

5 $x\begin{bmatrix} yz & y^2 \\ -z^2 & -yz \end{bmatrix}$ x, y, z arbitrary

1.3 Vector Spaces

1 $\begin{bmatrix} x \\ y \\ z \end{bmatrix} = \tfrac{1}{2}(3x+y-z)\begin{bmatrix} 1 \\ 1 \\ 0 \end{bmatrix} + \tfrac{1}{2}(-x+y+z)\begin{bmatrix} 0 \\ 1 \\ 1 \end{bmatrix} + \tfrac{1}{2}(-x-y+z)\begin{bmatrix} 1 \\ 2 \\ -1 \end{bmatrix}$

2 Since $2A_1 + 3A_2 - A_3 = 0$, $\mathbb{S}$ is not a basis for $\mathbb{V}$. Any two of these three vectors form a basis for $\mathbb{V}$.

3 $\left\{\begin{bmatrix} 0 & 0 \\ 1 & 1 \end{bmatrix}, \begin{bmatrix} 0 & 1 \\ -1 & 1 \end{bmatrix}, \begin{bmatrix} 1 & 1 \\ 1 & 0 \end{bmatrix}\right\}$ is a basis for $\mathbb{V} + \mathbb{W}$; $\left\{\begin{bmatrix} 1 & 2 \\ 0 & 1 \end{bmatrix}\right\}$ is a basis for $\mathbb{V} \cap \mathbb{W}$.

10 (a), (b), (d), (f)

1.4 The Row Space and the Column Space of a Matrix

1 Any two rows of A form a basis for $\mathbb{R}(A)$. Any two columns of A form a basis for $\mathbb{C}(A)$.

2 $A = \begin{bmatrix} -\frac{1}{2} & -2 & \frac{3}{2} \\ 0 & -1 & 2 \end{bmatrix}$ is the only matrix which satisfies $AB = C$.

3 Yes.

5 $\left\{ \begin{bmatrix} 4 \\ -1 \\ 5 \end{bmatrix} \right\}$ **6** $x \neq 9$ **7** $x = 1, y = \frac{1}{2}$

1.5 The Steinitz Replacement Theorem

1 (a) Either of the first two row of A forms a basis for $\mathbb{R}(A)$. Any column of A forms a basis for $\mathbb{C}(A)$. One basis for $\mathbb{N}(A)$ is

$$\left\{ \begin{bmatrix} 3 \\ -1 \\ 0 \end{bmatrix}, \begin{bmatrix} 2 \\ 0 \\ 1 \end{bmatrix} \right\}$$

2 (a) The row rank and the column rank are both 2, the nullity is 5.

3 $\left\{ \begin{bmatrix} 0 \\ 2 \\ -1 \\ 0 \end{bmatrix}, \begin{bmatrix} 0 \\ 1 \\ 1 \\ 0 \end{bmatrix}, \begin{bmatrix} 1 \\ 0 \\ 0 \\ 0 \end{bmatrix}, \begin{bmatrix} 0 \\ 0 \\ 0 \\ 1 \end{bmatrix} \right\}$

5 (a) $\left\{ \begin{bmatrix} 1 & 0 \\ 0 & 1 \end{bmatrix}, \begin{bmatrix} 0 & 1 \\ 4 & 0 \end{bmatrix} \right\}$

(b) Since $\{X | XA = BX\} = 0$, the empty set is the only basis.

6 (a) $\{[-1 \quad 2 \quad 0]\}$

(b) $\{[1 \quad 1 \quad 0], [0 \quad 1 \quad 0], [1 \quad 0 \quad 1]\}$. No.

(c) The empty set.   (d) $\left\{ \begin{bmatrix} 0 \\ 0 \\ 1 \end{bmatrix}, \begin{bmatrix} 2 \\ 1 \\ -2 \end{bmatrix} \right\}$. Yes.

7 $\left\{ \begin{bmatrix} 5 \\ 7 \\ 0 \end{bmatrix} \right\}$

1.6 Nonsingular Matrices

1 (a) $[x \quad y \quad z] = (x - y)[1 \quad 0 \quad 1] + (x - z)[1 \quad 1 \quad 0]$
$ + (-x + y - z)[1 \quad 1 \quad 1]$

2 $A^{-1} = \frac{1}{3}\begin{bmatrix} 0 & -3 \\ 1 & 1 \end{bmatrix}$ 3 $A^{-1} = \begin{bmatrix} 1 & 1 & -1 \\ -1 & 0 & 1 \\ 0 & -1 & 1 \end{bmatrix}$

4 (a) The column rank is 2, the nullity is 5.
   (b) The column rank is 2, the nullity is 0.
5 No.   6 $x \neq 9$.

1.7 The Elementary Operations

1 (a) $\begin{bmatrix} 0 & 0 & 1 & 1 \\ 2 & -4 & 2 & 0 \\ -1 & 2 & 0 & 1 \end{bmatrix} \xrightarrow{C_{13}} \begin{bmatrix} 1 & 0 & 0 & 1 \\ 2 & -4 & 2 & 0 \\ 0 & 2 & -1 & 1 \end{bmatrix} \xrightarrow{R_{21}(-2)}$

$\begin{bmatrix} 1 & 0 & 0 & 1 \\ 0 & -4 & 2 & -2 \\ 0 & 2 & -1 & 1 \end{bmatrix} \xrightarrow{C_{41}(-1)} \begin{bmatrix} 1 & 0 & 0 & 0 \\ 0 & -4 & 2 & -2 \\ 0 & 2 & -1 & 1 \end{bmatrix} \xrightarrow{R_{32}(\frac{1}{2})}$

$\begin{bmatrix} 1 & 0 & 0 & 0 \\ 0 & -4 & 2 & -2 \\ 0 & 0 & 0 & 0 \end{bmatrix} \xrightarrow{C_2(-\frac{1}{4})} \begin{bmatrix} 1 & 0 & 0 & 0 \\ 0 & 1 & 2 & -2 \\ 0 & 0 & 0 & 0 \end{bmatrix} \xrightarrow{C_{32}(-2)}$

$\begin{bmatrix} 1 & 0 & 0 & 0 \\ 0 & 1 & 0 & -2 \\ 0 & 0 & 0 & 0 \end{bmatrix} \xrightarrow{C_{42}(2)} \begin{bmatrix} 1 & 0 & 0 & 0 \\ 0 & 1 & 0 & 0 \\ 0 & 0 & 0 & 0 \end{bmatrix}$

3 (a) The first two rows of A form a basis for $\mathbb{R}(A)$.
   (b) The first two columns of A form a basis for $\mathbb{C}(A)$.
   (c) $\left\{ \begin{bmatrix} 0 \\ 1 \\ 2 \\ 0 \\ 0 \end{bmatrix}, \begin{bmatrix} 1 \\ 0 \\ 0 \\ -1 \\ -1 \end{bmatrix}, \begin{bmatrix} 0 \\ 0 \\ 1 \\ -1 \\ 1 \end{bmatrix} \right\}$ is a basis for $\mathbb{N}(A)$.

5 No, because the last row of any such matrix would consist only of zeros.
6 $C_{rs} = R_{rs}$, $C_r(\lambda) = R_r(\lambda)$, $C_{rs}(\alpha) = R_{sr}(\alpha)$.

1.8 Permutations

1 (a) $\begin{pmatrix} 1 & 2 & 3 & 4 & 5 \\ 5 & 4 & 1 & 3 & 2 \end{pmatrix}$ (b) $\begin{pmatrix} 1 & 2 & 3 & 4 & 5 \\ 1 & 2 & 4 & 5 & 3 \end{pmatrix}$

   (f) $\begin{pmatrix} 1 & 2 & 3 & 4 & 5 \\ 2 & 1 & 5 & 4 & 3 \end{pmatrix} = (1 \ \ 2)(3 \ \ 5)$

2 $(1 \ \ 3)(1 \ \ 7)(2 \ \ 6)(2 \ \ 4)(2 \ \ 8)(2 \ \ 9)$ is one expression for σ as a product of transpositions. σ is even.

3 If $\sigma(a) = b$, then let $\tilde{\sigma}(b) = a$.

4 (a) $\begin{pmatrix} 1 & 2 & 3 & 4 & 5 \\ 2 & 1 & 4 & 5 & 3 \end{pmatrix}$ (b) $\begin{pmatrix} 1 & 2 & 3 & 4 & 5 \\ 5 & 3 & 1 & 4 & 2 \end{pmatrix}$

5 $\sigma^{-1} = (2\ 9)(2\ 8)(2\ 4)(2\ 6)(1\ 7)(1\ 3)$

16 (a) $\begin{bmatrix} 0 & 1 & 0 & 0 & 0 \\ 1 & 0 & 0 & 0 & 0 \\ 0 & 0 & 0 & 1 & 0 \\ 0 & 0 & 0 & 0 & 1 \\ 0 & 0 & 1 & 0 & 0 \end{bmatrix}$ (b) $\begin{bmatrix} 0 & 0 & 0 & 0 & 1 \\ 0 & 0 & 1 & 0 & 0 \\ 1 & 0 & 0 & 0 & 0 \\ 0 & 0 & 0 & 1 & 0 \\ 0 & 1 & 0 & 0 & 0 \end{bmatrix}$ (c) $\begin{bmatrix} 0 & 0 & 1 & 0 & 0 \\ 0 & 0 & 0 & 0 & 1 \\ 0 & 0 & 0 & 1 & 0 \\ 0 & 1 & 0 & 0 & 0 \\ 1 & 0 & 0 & 0 & 0 \end{bmatrix}$

17 (a) $\begin{bmatrix} 0 & 1 & 0 & 0 & 0 \\ 1 & 0 & 0 & 0 & 0 \\ 0 & 0 & 0 & 0 & 1 \\ 0 & 0 & 1 & 0 & 0 \\ 0 & 0 & 0 & 1 & 0 \end{bmatrix}$ (b) $\begin{bmatrix} 0 & 0 & 1 & 0 & 0 \\ 0 & 0 & 0 & 0 & 1 \\ 0 & 1 & 0 & 0 & 0 \\ 0 & 0 & 0 & 1 & 0 \\ 1 & 0 & 0 & 0 & 0 \end{bmatrix}$ (c) $\begin{bmatrix} 0 & 0 & 0 & 0 & 1 \\ 0 & 0 & 0 & 1 & 0 \\ 1 & 0 & 0 & 0 & 0 \\ 0 & 0 & 1 & 0 & 0 \\ 0 & 1 & 0 & 0 & 0 \end{bmatrix}$

1.9 Determinants

1 (a) -3,   (b) -9,   (c) 4,   (e) 3

3 $2, -3$

4 $\dfrac{1}{4}\begin{bmatrix} -5 & -7 & 10 \\ 1 & -1 & 2 \\ -7 & -9 & 14 \end{bmatrix}$ 5 $\begin{bmatrix} -7 & -7 & 7 \\ 5 & 5 & -5 \\ 3 & 3 & -3 \end{bmatrix}$

6 $\det A = (-1)^{n+1} a_{1n} a_{21} a_{32} \cdots a_{n,n-1}$

8 (a) $\det \alpha A = \alpha^n \det A$ (b) $\operatorname{adj} \alpha A = \alpha^{n-1} \operatorname{adj} A$

1.10 Linear Transformations and Invariant Spaces

1 $\begin{bmatrix} & & & & 1 \\ & & & 1 & \\ & & 1 & & \\ & \cdots & & & \\ 1 & & & & \end{bmatrix}$ 2 $\begin{bmatrix} 0 & 1 & 0 & 0 & \cdots & 0 \\ 0 & 0 & 2 & 0 & \cdots & 0 \\ 0 & 0 & 0 & 3 & \cdots & 0 \\ \multicolumn{6}{c}{\dotfill} \\ 0 & 0 & 0 & 0 & \cdots & n-1 \end{bmatrix}$

3 $\mathscr{B}: \begin{bmatrix} 1 \\ 1 \\ 1 \end{bmatrix} \to \begin{bmatrix} 5 \\ 7 \\ -3 \end{bmatrix}$ $\mathscr{B}: \begin{bmatrix} 7 \\ -19 \\ -5 \end{bmatrix} \to \begin{bmatrix} 1 \\ 1 \\ 1 \end{bmatrix}$ $B = \begin{bmatrix} 1 & -1 & 5 \\ 3 & 0 & 4 \\ 0 & 1 & -4 \end{bmatrix}$

4 $\begin{bmatrix} 0 \\ 4 \\ 1 \end{bmatrix}$ 5 $P = \begin{bmatrix} x & y \\ 0 & y \end{bmatrix}$ $x, y \neq 0$ $Q = \begin{bmatrix} x & y \\ y & y \end{bmatrix}$ $x \neq y, y \neq 0$

8 $\lambda_1 = 0, \lambda_2 = 1, \lambda_3 = 2$

$$X_1 = \begin{bmatrix} -3 \\ 1 \\ 1 \end{bmatrix} \quad X_2 = \begin{bmatrix} 2 \\ -1 \\ 0 \end{bmatrix} \quad X_3 = \begin{bmatrix} 1 \\ 1 \\ -1 \end{bmatrix} \quad P = \begin{bmatrix} -3 & 2 & 1 \\ 1 & -1 & 1 \\ 1 & 0 & -1 \end{bmatrix}$$

$Y_1 = [1 \ \ 2 \ \ 3] \quad Y_2 = [1 \ \ 1 \ \ 2] \quad Y_3 = [1 \ \ 2 \ \ 1]$

$$Q = \begin{bmatrix} 1 & 2 & 3 \\ 1 & 1 & 2 \\ 1 & 2 & 1 \end{bmatrix}$$

9 $\lambda_1 = \lambda_2 = -1, \lambda_3 = -2$

$$X_1 = \begin{bmatrix} 1 \\ -1 \\ 0 \end{bmatrix} \quad X_2 = \begin{bmatrix} 0 \\ 0 \\ 1 \end{bmatrix} \quad X_3 = \begin{bmatrix} 3 \\ -4 \\ 3 \end{bmatrix} \quad P = \begin{bmatrix} 1 & 0 & 3 \\ -1 & 0 & -4 \\ 0 & 1 & 3 \end{bmatrix}$$

10 $\lambda_1 = -1, \lambda_2 = \lambda_3 = 2 \quad X_1 = \begin{bmatrix} 0 \\ 1 \\ 2 \end{bmatrix} \quad X_2 = \begin{bmatrix} 1 \\ 0 \\ 1 \end{bmatrix} \quad$ No.

1.11 Polynomials

1 (a) $x^3 + x$, (b) $x^3 + x$, (c) 1
2 (a) $x^3 + x = (-\frac{2}{3}x - 1)(x^5 - x) + (\frac{1}{3}x - \frac{2}{3})(2x^5 - x^4 + 2x^3 - x^2)$
3 (a) $x^3 + x$, (b) $1, 1 = 2x(x^3 + x) - 2x(x^3 - x) - (4x^2 - 1)$
5 (1a) $x^2(x - 1)(x + 1)(x^2 + 1)(2x - 1)$,
   (1b) $x^2(x - 1)(x + 1)(x^2 + 1)(x - 2)$,
   (1c) $(x^5 - x)(x^4 - 4x^2 + 4)$,
   (3a) $x^2(x - 1)(x + 1)(x^2 + 1)(2x - 1)(x - 2)$
8 (a) $x(x^2 - 1)^3 + 3x(x^2 - 1)^2 + (5x - 1)(x^2 - 1) + (4x + 2)$
   (b) $2(x - 2)^4 + 13(x - 2)^3 + 28(x - 2)^2 + 22(x - 2) - 3$
10 (a) $x^2 - 6x + 5$, (b) $(x - 2)^2$, (d) $x^2 + x$
11 (a) $x^3 - x^2$

1.12 The Decomposition Theorems

1 (a) $m(x) = x - 1$; the elementary divisors are $x - 1, x - 1$. Any two linearly independent vectors can be used to generate two one-dimensional cyclic spaces whose direct sum is $\mathbb{F}^2$.
   (b) $m(x) = (x - 1)(x - 2)$; the elementary divisors are $x - 1, x - 2$. $\mathbb{F}^2$ is the direct sum of the two one-dimensional cyclic spaces generated by $\begin{bmatrix} 1 \\ 0 \end{bmatrix}$ and $\begin{bmatrix} 0 \\ 1 \end{bmatrix}$.

(c) $m(x) = (x-1)^2$; the elementary divisor is $(x-1)^2$. $\mathbb{F}^2$ is an irreducible cyclic space. Any vector which is not a scalar multiple of $\begin{bmatrix} 0 \\ 1 \end{bmatrix}$ generates $\mathbb{F}^2$.

(f) $m(x) = x(x-2)$; the elementary divisors are x, $x-2$. $\mathbb{F}^2$ is the direct sum of the two one-dimensional cyclic spaces generated by $\begin{bmatrix} 1 \\ -1 \end{bmatrix}$ and $\begin{bmatrix} 1 \\ 1 \end{bmatrix}$.

2 (a) $m(x) = x^2 + 1$ When $\mathbb{F}$ is the complex field, the elementary divisors are $x - i$ and $x + i$. $\mathbb{F}^2$ is the direct sum of the two one-dimensional cyclic spaces generated by $\begin{bmatrix} 1 \\ 1-i \end{bmatrix}$ and $\begin{bmatrix} 1 \\ 1+i \end{bmatrix}$.

3 $m(x) = x(x-2)(x+1)$. The elementary divisors are x, $x-2$ and $x+1$. $\mathbb{F}^3$ is the direct sum of the three one-dimensional cyclic spaces generated by
$\begin{bmatrix} 1 \\ 0 \\ 1 \end{bmatrix} \begin{bmatrix} 1 \\ -1 \\ 2 \end{bmatrix}$ and $\begin{bmatrix} 1 \\ -1 \\ 1 \end{bmatrix}$.

4 $m() = (x+1)^2$. The elementary divisors are $(x+1)^2$ and $x+1$. $\mathbb{F}^3$ is the direct sum of the two-dimensional cyclic space generated by $\begin{bmatrix} 1 \\ 0 \\ 0 \end{bmatrix}$ and the one-dimensional cyclic space generated by $\begin{bmatrix} 1 \\ 0 \\ -1 \end{bmatrix}$.

5 $m(x) = (x-3)^3$. The canonical basis for $\mathbb{F}^3$ with generator $Y = \begin{bmatrix} 1 \\ 0 \\ 0 \end{bmatrix}$ is

$$Y = \begin{bmatrix} 1 \\ 0 \\ 0 \end{bmatrix} \quad (A-3I)Y = \begin{bmatrix} -1 \\ 1 \\ 2 \end{bmatrix} \quad (A-3I)^2 Y = \begin{bmatrix} 1 \\ -1 \\ -1 \end{bmatrix}$$

The cyclic basis with generator Y is

$$Y = \begin{bmatrix} 1 \\ 0 \\ 0 \end{bmatrix} \quad AY = \begin{bmatrix} 2 \\ 1 \\ 2 \end{bmatrix} \quad A^2 Y = \begin{bmatrix} 4 \\ 5 \\ 11 \end{bmatrix}$$

1.13 The Canonical Forms

1 (a) A is already in both its rational canonical form and its classical canonical form.

[1.13] ANSWERS TO SELECTED EXERCISES 429

(b) A is already in both its rational canonical form and its classical canonical form.

(c) To obtain the rational canonical form of A set $P = \begin{bmatrix} 0 & 1 \\ 1 & 1 \end{bmatrix}$. Then $P^{-1}AP = \begin{bmatrix} 0 & -1 \\ 1 & 2 \end{bmatrix}$. To obtain the classical canonical form set $P = \begin{bmatrix} 0 & 1 \\ 1 & 0 \end{bmatrix}$. Then $P^{-1}AP = \begin{bmatrix} 1 & 0 \\ 1 & 1 \end{bmatrix}$.

(d) A is already in its classical canonical form. The rational canonical form of A is the same as the rational canonical form of the matrix of (c). To obtain the rational canonical form of A set $P = \begin{bmatrix} 1 & 1 \\ 0 & 1 \end{bmatrix}$. Then $P^{-1}AP = \begin{bmatrix} 0 & -1 \\ 1 & 2 \end{bmatrix}$.

2 (3) Both the rational canonical form and the classical canonical form of

$$A = \begin{bmatrix} -3 & 1 & 3 \\ 3 & -1 & -3 \\ -5 & 1 & 5 \end{bmatrix}$$

are obtained by setting

$$P = \begin{bmatrix} 1 & 1 & 1 \\ 0 & -1 & -1 \\ 1 & 2 & 1 \end{bmatrix}$$

Then

$$P^{-1}AP = \begin{bmatrix} 0 & 0 & 0 \\ 0 & 2 & 0 \\ 0 & 0 & -1 \end{bmatrix}$$

(4) The rational canonical form of A is obtained by setting

$$P = \begin{bmatrix} 1 & 0 & 1 \\ 0 & 1 & 0 \\ 0 & -1 & -1 \end{bmatrix}$$

Then

$$P^{-1}AP = \begin{bmatrix} 0 & -1 & 0 \\ 1 & -2 & 0 \\ 0 & 0 & -1 \end{bmatrix}$$

The classical canonical form of A is obtained by setting

$$P = \begin{bmatrix} 1 & 1 & 1 \\ 0 & 1 & 0 \\ 0 & -1 & -1 \end{bmatrix}$$

Then

$$P^{-1}AP = \begin{bmatrix} -1 & 0 & 0 \\ 1 & -1 & 0 \\ 0 & 0 & -1 \end{bmatrix}$$

3 $m(x) = x^2(x^2 - 7x - 2)(x - 1)$. The characteristic polynomial of A is $-x^4(x^2 - 7x - 2)(x - 1)$. The rational canonical form and the classical canonical form of A are both equal to

$$\begin{bmatrix} 0 & 0 & 0 & 0 & 0 & 0 & 0 \\ 0 & 0 & 0 & 0 & 0 & 0 & 0 \\ 0 & 0 & 0 & 0 & 0 & 0 & 0 \\ 0 & 0 & 1 & 0 & 0 & 0 & 0 \\ 0 & 0 & 0 & 0 & 0 & 2 & 0 \\ 0 & 0 & 0 & 0 & 1 & 7 & 0 \\ 0 & 0 & 0 & 0 & 0 & 0 & 1 \end{bmatrix}$$

4 $m(x) = (x - 2)^3$. The characteristic polynomial of A is $-(x - 2)^7$. The rational canonical form and the classical canonical form are, respecitively

$$\begin{bmatrix} 0 & -4 & 0 & 0 & 0 & 0 & 0 \\ 1 & 4 & 0 & 0 & 0 & 0 & 0 \\ 0 & 0 & 0 & -4 & 0 & 0 & 0 \\ 0 & 0 & 1 & 4 & 0 & 0 & 0 \\ 0 & 0 & 0 & 0 & 0 & 0 & 8 \\ 0 & 0 & 0 & 0 & 1 & 0 & -12 \\ 0 & 0 & 0 & 0 & 0 & 1 & 6 \end{bmatrix} \quad \text{and} \quad \begin{bmatrix} 2 & 0 & 0 & 0 & 0 & 0 & 0 \\ 1 & 2 & 0 & 0 & 0 & 0 & 0 \\ 0 & 0 & 2 & 0 & 0 & 0 & 0 \\ 0 & 0 & 1 & 2 & 0 & 0 & 0 \\ 0 & 0 & 0 & 0 & 2 & 0 & 0 \\ 0 & 0 & 0 & 0 & 1 & 2 & 0 \\ 0 & 0 & 0 & 0 & 0 & 1 & 2 \end{bmatrix}$$

CHAPTER 2 THE THEORY OF GAMES

2.1 Examples of Games

1 (a) $\left(\bar{P} = [\tfrac{2}{3} \ \tfrac{1}{3}], v = \tfrac{1}{3}, \bar{Q} = \begin{bmatrix} \tfrac{2}{3} \\ \tfrac{1}{3} \end{bmatrix}\right)$

(b) $\left(\bar{P} = [1 \ 0], v = 1, \bar{Q} = \begin{bmatrix} q \\ 1-q \end{bmatrix} \text{ for any } q, 0 \leq q \leq \tfrac{3}{5}\right)$

(c) $\left(\bar{P} = [1 \ 0], v = 0, \bar{Q} = \begin{bmatrix} 1 \\ 0 \end{bmatrix}\right)$

(d) $\left(\bar{P} = [\tfrac{1}{2} \ \tfrac{1}{2}], v = \tfrac{1}{2}, \bar{Q} = \begin{bmatrix} \tfrac{5}{8} \\ \tfrac{3}{8} \end{bmatrix}\right)$

2 (a) $\left(\bar{P} = [\tfrac{16}{23} \ \tfrac{7}{23}], v = \tfrac{22}{23}, \bar{Q} = \begin{bmatrix} 0 \\ \tfrac{10}{23} \\ \tfrac{13}{23} \end{bmatrix}\right)$

(b) $\left(\bar{P} = [p \ 1-p] \text{ for any } p, \tfrac{1}{4} \leq p \leq \tfrac{3}{4}, v = -1, \bar{Q} = \begin{bmatrix} 0 \\ 1 \\ 0 \end{bmatrix}\right)$

[2.1]

(c) $\left(\bar{P} = [\tfrac{1}{2} \; \tfrac{1}{2} \; 0], v = 1, \bar{Q} = \begin{bmatrix} \tfrac{1}{6} \\ \tfrac{5}{6} \\ 0 \end{bmatrix}\right)$

3 Yes.

4 $A = \begin{bmatrix} 0 & 2 & -3 & 0 \\ -2 & 0 & 0 & 3 \\ 3 & 0 & 0 & -4 \\ 0 & -3 & 4 & 0 \end{bmatrix}$ Yes. Find them.

5 $\left(\bar{P} = [\tfrac{1}{3} \; \tfrac{2}{3}], v = \tfrac{1}{3}, \bar{Q} = \begin{bmatrix} \tfrac{2}{3} \\ \tfrac{1}{3} \end{bmatrix}\right)$

6 Let
$$S_3 = \begin{cases} \text{if the card drawn is H, then fold} \\ \text{if the card drawn is L, then bet} \end{cases}$$
$$S_4 = \begin{cases} \text{if the card drawn is H, then fold} \\ \text{if the card drawn is L, then fold} \end{cases}$$

$$A = \begin{bmatrix} 0 & 1 \\ \tfrac{1}{2} & 0 \\ -\tfrac{3}{2} & 0 \\ -1 & -1 \end{bmatrix}$$

7 $A = \begin{bmatrix} -\tfrac{2}{5} & 1 \\ \tfrac{1}{5} & -\tfrac{1}{5} \end{bmatrix}$ $\left(\bar{P} = [\tfrac{2}{9} \; \tfrac{7}{9}], v = \tfrac{1}{15}, \bar{Q} = \begin{bmatrix} \tfrac{2}{3} \\ \tfrac{1}{3} \end{bmatrix}\right)$

9 For $n = 2k - 1$, the kth row of A is

$$\begin{bmatrix} 1 & 1 - \tfrac{2}{n} & 1 - \tfrac{4}{n} & \cdots & \tfrac{1}{n} & \overset{k}{0} & \overset{k+1}{\tfrac{1}{n}} & \tfrac{1}{n} & \cdots & \tfrac{1}{n} & \tfrac{1}{n} \end{bmatrix}$$

For $n = 2k$, the kth and $(k+1)$st rows of A are

$$\begin{bmatrix} 1 & 1 - \tfrac{2}{n} & 1 - \tfrac{4}{n} & \cdots & \tfrac{2}{n} & 0 & 0 & 0 & \cdots & 0 & 0 \end{bmatrix}$$

and

$$\begin{bmatrix} 1 & 1 - \tfrac{2}{n} & 1 - \tfrac{4}{n} & \cdots & \tfrac{2}{n} & 0 & 0 & \tfrac{2}{n} & \cdots & \tfrac{2}{n} & \tfrac{2}{n} \end{bmatrix}$$

respectively.

10 For $n = 4$ the matrix of the duel is

$$A = \begin{bmatrix} 0 & -\tfrac{1}{4} & -\tfrac{1}{2} & -\tfrac{3}{4} & -1 \\ \tfrac{1}{4} & 0 & -\tfrac{1}{8} & -\tfrac{5}{16} & -\tfrac{3}{4} \\ \tfrac{1}{2} & \tfrac{1}{8} & 0 & \tfrac{1}{8} & 0 \\ \tfrac{3}{4} & \tfrac{5}{16} & -\tfrac{1}{8} & 0 & \tfrac{1}{2} \\ 1 & \tfrac{3}{4} & 0 & -\tfrac{1}{2} & 0 \end{bmatrix}$$

For n = 5 the matrix of the duel is

$$A = \begin{bmatrix} 0 & -\frac{1}{5} & -\frac{2}{5} & -\frac{3}{5} & -\frac{4}{5} & -1 \\ \frac{1}{5} & 0 & -\frac{3}{25} & -\frac{7}{25} & -\frac{11}{25} & -\frac{3}{5} \\ \frac{2}{5} & \frac{3}{25} & 0 & -\frac{1}{25} & -\frac{2}{25} & -\frac{1}{5} \\ \frac{3}{5} & \frac{7}{25} & -\frac{1}{25} & 0 & \frac{7}{25} & \frac{1}{5} \\ \frac{4}{5} & \frac{11}{25} & \frac{2}{25} & -\frac{7}{25} & 0 & \frac{3}{5} \\ 1 & \frac{3}{5} & \frac{1}{5} & -\frac{1}{5} & -\frac{3}{5} & 0 \end{bmatrix}$$

11 The matrix of the duel is

$$A = \begin{bmatrix} \frac{1}{20} & -\frac{1}{20} & -\frac{7}{20} & -\frac{10}{20} \\ \frac{4}{20} & \frac{2}{20} & \frac{2}{20} & 0 \\ \frac{8}{20} & \frac{1}{20} & -\frac{1}{20} & \frac{10}{20} \\ \frac{12}{20} & \frac{4}{20} & -\frac{12}{20} & 0 \end{bmatrix}$$

The solution of the duel is

$$\left(\bar{P} = [0 \ \tfrac{11}{13} \ \tfrac{2}{13} \ 0], v = \tfrac{1}{13}, \bar{Q} = \begin{bmatrix} 0 \\ 0 \\ \tfrac{10}{13} \\ \tfrac{3}{13} \end{bmatrix} \right)$$

16 $\tilde{A} = \begin{bmatrix} 0 & A & -E \\ -A^T & 0 & E \\ E & -E & 0 \end{bmatrix}$

where A is the matrix of G and E is the appropriate row or column vector all of whose entries are 1.

..2 Maxmin and Minmax

(Example 2.1.1) maxmin = -1, minmax = 1
(Example 2.1.2) maxmin = 0, minmax = 1
(Example 2.1.4) maxmin = -1, minmax = 2
(Example 2.1.10) maxmin = -1, minmax = 1

2 (a) minmax = 2, maxmin = 1
3 $\min_{y \in \mathbb{R}} F(x, y) = -4x - x^2$ for all x. Thus $\max_{x \in \mathbb{R}} \left(\min_{y \in \mathbb{R}} F(x, y) \right) = 4$ at $x = y = -2$. However, $\max_{x \in \mathbb{R}} F(x, y)$ does not exist for any value of y.
4 (b), (d), (g).
10 (b) Every point (x, y) for which $y = x^2$.
  (d) There are no extreme points.
  (g) Every point (x, y) for which $x^2 + y^2 = 1$.

2.3 A MATHEMATICAL INTRODUCTION TO THE THEORY OF GAMES

6 (a) $a_{13} = 1$ is a saddle point.
   (b) $a_{33} = 3$ is a saddle point. (Note that $a_{31} = 3$ is not a saddle point.)
   (c) $a_{42} = a_{43} = -1$ are both saddle points.

13 The pure strategies

$$S = \begin{cases} \text{if I have } V, \text{ call } X \\ \text{if I have } X, \text{ call } X \end{cases} \qquad T = \begin{cases} \text{do not believe } V \\ \text{believe } X \end{cases}$$

are optimal strategies. The value of the game is 0.

2.4 Linear Inequalities

1 (b), (f), (h)
2 $\mathbb{K}(A) = \mathbb{K}(C)$ if and only if

$$C = A \begin{bmatrix} \beta_1 & 0 \\ 0 & \beta_2 \end{bmatrix} \quad \text{or} \quad C = A \begin{bmatrix} 0 & \beta_1 \\ \beta_2 & 0 \end{bmatrix}, \quad \beta_1, \beta_2 > 0$$

4 $Z \geq AX$, $X \geq 0$ has a solution if and only if Z is contained in the cone spanned by $\begin{bmatrix} 0 \\ 1 \end{bmatrix}$ and $\begin{bmatrix} 2 \\ -1 \end{bmatrix}$.

6 $Z = AX$, $X \geq 0$ does not have a solution for any positive vector Z (as well as for other vectors).

2.5 The Fundamental Theorem

2 $P = [\frac{2}{22} \ \frac{2}{22} \ \frac{3}{22} \ \frac{1}{22} \ \frac{14}{22}]$ is an optimal strategy for both players. The value of the game is 0.

6 The game with matrix B is the symmetrization of the game with matrix

$$A = \begin{bmatrix} 6 & 0 \\ 2 & 6 \end{bmatrix}$$

which has solution ($\bar{P} = [\frac{2}{5} \ \frac{3}{5}]$, $\bar{Q} = \begin{bmatrix} \frac{3}{5} \\ \frac{2}{5} \end{bmatrix}$, $v = \frac{18}{5}$). Thus

$$X = [\tfrac{6}{25} \ \tfrac{4}{25} \ \tfrac{9}{25} \ \tfrac{6}{25}]$$

is an optimal strategy for both players in the game with matrix B.

2.6 Relations among Games

1. Example 2.1.1 $A^* = A$, Example 2.1.2 $A^* = A$, Example 2.1.4
$$A^* = \begin{bmatrix} 1 & 2 \\ 3 & 1 \end{bmatrix}$$

3. The second and third strategies for each player are the essential strategies. Thus $A^* = \begin{bmatrix} 0 & 0 \\ 0 & 0 \end{bmatrix}$.

4. (a) $\bar{P} = [p \ \ 1-p]$, where $\frac{3}{5} \leq p \leq 1$, $\bar{Q} = \begin{bmatrix} 1 \\ 0 \end{bmatrix}$, $A^* = \begin{bmatrix} 2 \\ 2 \end{bmatrix}$

   (b) $\bar{P} = [p \ \ 1-p]$, where $\frac{2}{3} \leq p \leq \frac{2}{3}$, $\bar{Q} = \begin{bmatrix} 0 \\ 1 \\ 0 \end{bmatrix}$, $A^* = \begin{bmatrix} -1 \\ -1 \end{bmatrix}$

5. $A^* = \begin{bmatrix} 0 & 1 & 2 & 3 & 4 \\ 4 & 3 & 2 & 1 & 0 \end{bmatrix}$, $\bar{P} = [\frac{1}{2} \ \ 0 \ \ 0 \ \ 0 \ \ \frac{1}{2}]$ is the only optimal strategy for player I.

$$\left\{ \begin{bmatrix} \frac{1}{2} \\ 0 \\ 0 \\ 0 \\ \frac{1}{2} \end{bmatrix}, \begin{bmatrix} 0 \\ \frac{1}{2} \\ 0 \\ \frac{1}{2} \\ 0 \end{bmatrix}, \begin{bmatrix} 0 \\ 0 \\ 1 \\ 0 \\ 0 \end{bmatrix}, \begin{bmatrix} \frac{1}{3} \\ 0 \\ 0 \\ \frac{2}{3} \\ 0 \end{bmatrix} \right\}$$

is a maximal set of linearly independent optimal strategies for player II.

12. (a) $A_{(2)}$ dominates $A_{(1)}$, and $A^{(2)}$ dominates $A^{(3)}$. Hence $A_{(1)}$ and $A^{(3)}$ can be eliminated. In the remaining matrix the second row is dominated by the first row and can be eliminated, leaving $[1 \ \ 1]$.

    (b) $A_{(1)}$ is dominated by $\frac{1}{4}A_{(3)} + \frac{3}{4}A_{(4)}$. $A_{(2)}$ is dominated by $\frac{1}{2}A_{(3)} + \frac{1}{2}A_{(4)}$. Thus $A_{(1)}$ and $A_{(2)}$ can be eliminated.

2.7 Extreme Optimal Strategies

1. (a) $A' = \begin{bmatrix} 2 & 4 \\ 2 & -1 \end{bmatrix}$ yields $v = 2$, $\bar{P} = [\frac{3}{5} \ \ \frac{2}{5}]$, $\bar{Q} = \begin{bmatrix} 1 \\ 0 \end{bmatrix}$

   $A' = [2] = [a_{11}]$ yields $v = 2$, $\bar{P} = [1 \ \ 0]$, $\bar{Q} = \begin{bmatrix} 1 \\ 0 \end{bmatrix}$

   (b) $A' = \begin{bmatrix} 2 & -1 \\ -3 & -1 \end{bmatrix}$ yields $v = -1$, $\bar{P} = [\frac{2}{5} \ \ \frac{3}{5}]$, $\bar{Q} = \begin{bmatrix} 0 \\ 1 \\ 0 \end{bmatrix}$

   $A' = \begin{bmatrix} -1 & -4 \\ -1 & 5 \end{bmatrix}$ yields $v = -1$, $\bar{P} = [\frac{2}{3} \ \ \frac{1}{3}]$, $\bar{Q} = \begin{bmatrix} 0 \\ 1 \\ 0 \end{bmatrix}$

[2.7]

(c) $A' = \begin{bmatrix} 3 & -3 \\ -2 & 5 \end{bmatrix}$ yields $v = \frac{9}{13}$, $\bar{P} = [\frac{7}{13} \quad \frac{6}{13}]$, $\bar{Q} = \begin{bmatrix} \frac{8}{13} \\ 0 \\ \frac{5}{13} \end{bmatrix}$

(d) $A' = \begin{bmatrix} 2 & 2 \\ 4 & -6 \end{bmatrix}$ yields $v = 2$, $\bar{P} = [1 \quad 0]$, $\bar{Q} = \begin{bmatrix} \frac{4}{5} \\ 0 \\ \frac{1}{5} \end{bmatrix}$

$A' = \begin{bmatrix} 2 & 2 \\ 5 & -6 \end{bmatrix}$ yields $v = 2$, $\bar{P} = [1 \quad 0]$, $\bar{Q} = \begin{bmatrix} 0 \\ \frac{8}{11} \\ \frac{3}{11} \end{bmatrix}$

$A' = [2] = [a_{13}]$ yields $v = 2$, $\bar{P} = [1 \quad 0]$, $\bar{Q} = \begin{bmatrix} 0 \\ 0 \\ 1 \end{bmatrix}$

2 $A' = \begin{bmatrix} 0 & 4 \\ 4 & 0 \end{bmatrix}$ yields $v = 2$, $\bar{P} = [\frac{1}{2} \quad 0 \quad 0 \quad 0 \quad \frac{1}{2}]$, $\bar{Q} = \begin{bmatrix} \frac{1}{2} \\ 0 \\ 0 \\ 0 \\ \frac{1}{2} \end{bmatrix}$

$A' = \begin{bmatrix} 1 & 3 \\ 3 & 1 \end{bmatrix}$ yields $v = 2$, $\bar{P} = [\frac{1}{2} \quad 0 \quad 0 \quad 0 \quad \frac{1}{2}]$, $\bar{Q} = \begin{bmatrix} 0 \\ \frac{1}{2} \\ 0 \\ \frac{1}{2} \\ 0 \end{bmatrix}$

$A' = \begin{bmatrix} 0 & 2 \\ 4 & 2 \end{bmatrix}$ yields $v = 2$, $\bar{P} = [\frac{1}{2} \quad 0 \quad 0 \quad 0 \quad \frac{1}{2}]$, $\bar{Q} = \begin{bmatrix} 0 \\ 0 \\ 1 \\ 0 \\ 0 \end{bmatrix}$

$A' = \begin{bmatrix} 0 & 3 \\ 4 & 1 \end{bmatrix}$ yields $v = 2$, $\bar{P} = [\frac{1}{2} \quad 0 \quad 0 \quad 0 \quad \frac{1}{2}]$, $\bar{Q} = \begin{bmatrix} \frac{1}{3} \\ 0 \\ 0 \\ \frac{2}{3} \\ 0 \end{bmatrix}$

$A' = \begin{bmatrix} 1 & 4 \\ 3 & 0 \end{bmatrix}$ yields $v = 2$, $\bar{P} = [\frac{1}{2} \quad 0 \quad 0 \quad 0 \quad \frac{1}{2}]$, $\bar{Q} = \begin{bmatrix} 0 \\ \frac{2}{3} \\ 0 \\ 0 \\ \frac{1}{3} \end{bmatrix}$

2.8 Linear Programming

2 The maximization problem is unbounded.

3 (a) The maximization problem is unbounded.

(b) $X = \begin{bmatrix} \frac{7}{5} \\ \frac{1}{5} \end{bmatrix}$, $Y = [\frac{8}{5} \quad \frac{1}{5}]$, $v = \frac{19}{5}$

(c) $X = \alpha \begin{bmatrix} 0 \\ 2 \\ 2 \end{bmatrix} + \beta \begin{bmatrix} \frac{10}{4} \\ \frac{3}{4} \\ \frac{3}{4} \end{bmatrix}$, where $\alpha + \beta = 1$ $\begin{matrix} \alpha \geq 0 \\ \beta \geq 0 \end{matrix}$

$Y = [\frac{1}{3} \quad \frac{8}{3} \quad 0]$, $v = 12$.

4 If the farmer must grow the vegetables to feed his animals, then he raises 60 animals and 10 tons of vegetables for a profit of $2,700. If he can buy vegetables at $30 per ton, then he raises 70 animals, buys $\frac{70}{6}$ tons of vegetables, and increases his profit to $2,800.

CHAPTER 3 FINITE MARKOV CHAINS

3.1 Examples of Finite Markov Chains

4 The probability that player I will lose is $\frac{3}{4}$.

5 The probability that a mouse initially in room 2 will enter room 3 is 1. The probability that a mouse initially in room 3 will enter room 2 is $\frac{1}{3}$.

6 If the rooms of the maze are ordered, 2, 3, 1, 4, then

$$I + A + A^2 + \cdots = \begin{bmatrix} \frac{4}{3} & \frac{2}{3} & 0 & 0 \\ \frac{2}{3} & \frac{4}{3} & 0 & 0 \\ \infty & \infty & \infty & \infty \\ \infty & \infty & \infty & \infty \end{bmatrix}$$

(Note that if the mouse is initially in room 2 and the cheese is in room 4, then the mouse may get no cheese. However, his expectation is still infinite.)

7 The probability that a mouse initially in room 2 will ever enter room 4 is $\frac{1}{3}$. The probability that he will enter room 1 is $\frac{2}{3}$.

8 2 minutes

9 $f_{51} = \frac{1}{7}$ $f_{52} = \frac{2}{7}$ $f_{53} = \frac{2}{7}$ $f_{54} = \frac{3}{7}$
$f_{61} = \frac{6}{7}$ $f_{62} = \frac{5}{7}$ $f_{63} = \frac{5}{7}$ $f_{64} = \frac{4}{7}$
$f_{12} = \frac{3}{5}$ $f_{13} = \frac{3}{5}$ $f_{14} = \frac{1}{5}$
$f_{21} = \frac{1}{2}$ $f_{23} = 1$ $f_{24} = \frac{1}{3}$
$f_{31} = \frac{3}{13}$ $f_{32} = \frac{6}{13}$ $f_{34} = \frac{2}{13}$
$f_{41} = \frac{1}{3}$ $f_{42} = \frac{2}{3}$ $f_{43} = \frac{2}{3}$
$f_{11} = \frac{3}{10}$ $f_{22} = \frac{11}{18}$ $f_{33} = \frac{6}{13}$ $f_{44} = \frac{2}{9}$

10 The probability that the biased coin will be used for the third toss is $\frac{43}{72}$. The probability that the third toss will be heads is $\frac{259}{432}$.

3.2 Infinite Processes

1 (a) $\begin{bmatrix} 1 & -1 & 1 \\ 0 & 0 & 0 \\ 0 & 0 & 0 \end{bmatrix}$ (b) $\begin{bmatrix} 0 & 1 & -1 \\ 0 & 1 & -1 \\ 0 & 0 & 0 \end{bmatrix}$ (c) 0 (d) 0

2 $\begin{bmatrix} \frac{1}{2} & -\frac{1}{2} & \frac{3}{2} \\ 1 & -1 & 3 \\ \frac{1}{2} & -\frac{1}{2} & \frac{3}{2} \end{bmatrix}$ 3 $\begin{bmatrix} 1 & -2 & 0 \\ 0 & 2 & 0 \\ 2 & -4 & 1 \end{bmatrix}$

4 $I + A + A^2 + \cdots = \begin{bmatrix} 5 & -2 & 4 \\ 2 & 0 & 2 \\ -1 & 0 & 0 \end{bmatrix}$

10 One pair is $A = \begin{bmatrix} 0 & 2 \\ 0 & 0 \end{bmatrix}$, $B = \begin{bmatrix} 0 & 0 \\ 1 & 0 \end{bmatrix}$.

12 (a) $\begin{bmatrix} e & 0 \\ 0 & e \end{bmatrix}$ (b) $\begin{bmatrix} e^2 & 0 \\ e^2 & e^2 \end{bmatrix}$ (c) $\begin{bmatrix} 2e^2 & e^2 \\ -e^2 & 0 \end{bmatrix}$

3.3 A Mathematical Introduction to the Theory of Finite Markov Chains

(Example 3.1.15) 1, 2, and 3 are transient; 4 forms an ergodic set.
(Example 3.1.18) 1, 2, 3, and 4 are transient; 5 and 6 each form ergodic sets.
(Example 3.1.20) 1, 2, and 3 form an ergodic set.

2 (Exercise 3.1.6) 1 and 4 each form an ergodic set. The stationary vectors are

$$\begin{bmatrix} p \\ 0 \\ q \end{bmatrix} \quad \begin{array}{l} p \geq 0 \\ q \geq 0 \\ p + q = 1 \end{array}$$

(Exercise 3.1.9) 5 and 6 each form an ergodic set. The other states are transient.
(Exercises 3.1.10 to 3.1.12) The only ergodic set is the set of all states.

3 (a) There are two ergodic sets, {1, 4} and {6}; 2, 3, and 5 are transient.
4 (a) 3 forms an ergodic set. All other states are transient.
   (b) There are two ergodic sets, {1, 3} and {2, 4, 5}. There are no transient states.

3.4 Periodic Transition Matrices

1 (a) The period is 4. If the states are ordered 1, 2, 5, 3, 4, 6, then

$$A = \begin{bmatrix} 0 & 0 & 0 & 0 & 0 & 1 \\ \frac{1}{2} & 0 & 0 & 0 & 0 & 0 \\ \frac{1}{2} & 0 & 0 & 0 & 0 & 0 \\ 0 & \frac{1}{2} & 0 & 0 & 0 & 0 \\ 0 & \frac{1}{2} & 1 & 0 & 0 & 0 \\ 0 & 0 & 0 & 1 & 1 & 0 \end{bmatrix}$$

There are no transient states.

(b) The period is 2. If the states are ordered 1, 3, 5, 2, 4, 6, then

$$A = \begin{bmatrix} 0 & 0 & 0 & \frac{1}{2} & 0 & 0 \\ 0 & 0 & 0 & 0 & 0 & 0 \\ 0 & 0 & 0 & \frac{1}{2} & 1 & 1 \\ \frac{1}{2} & 1 & 1 & 0 & 0 & 0 \\ \frac{1}{2} & 0 & 0 & 0 & 0 & 0 \\ 0 & 0 & 0 & 0 & 0 & 0 \end{bmatrix}$$

3 and 6 are transient states.

4 If the states are ordered 1, 3, 4, 2, 5, 6, then

$$A = \begin{bmatrix} 0 & 0 & 0 & 0 & 0 & 1 \\ 0 & 0 & 0 & \frac{1}{2} & 0 & 0 \\ 0 & 0 & 0 & \frac{1}{2} & 1 & 0 \\ \frac{1}{2} & 0 & 0 & 0 & 0 & 0 \\ \frac{1}{2} & 0 & 0 & 0 & 0 & 0 \\ 0 & 1 & 1 & 0 & 0 & 0 \end{bmatrix}$$

3.5 Recurrence Probabilities

1 4 drinks. Yes.
2 He can expect to win 9.
3 14 jumps. He can be expected to be at the bottom of the well 4 times
4 $F^{(1)} = A$

$$F^{(2)} = \begin{bmatrix} \frac{1}{2} & \frac{1}{4} \\ 0 & \frac{1}{2} \end{bmatrix} \quad F^{(s)} = \begin{bmatrix} \frac{s}{2^{s-1}} & \frac{s}{2^s} \\ 0 & 0 \end{bmatrix} \quad s \geq 3 \quad M = \begin{bmatrix} \frac{3}{2} & 2 \\ 1 & \frac{3}{2} \end{bmatrix}$$

CHAPTER 4 THE THEORY OF GRAPHS

4.1 Some Examples

1 The missionaries and cannibals can cross the river in 11 trips in exactly four ways:

$(3, 3, e)$ $\begin{array}{c}(1, 1, w)\\(0, 2, w)\end{array}$ $(3, 2, e) \to (0, 3, w) \to (3, 1, e) \to (2, 2, w) \to (2, 2, e)$

$\to (3, 1, w) \to (0, 3, e) \to (3, 2, w)$ $\begin{array}{c}(0, 2, e)\\(1, 1, e)\end{array}$ $(3, 3, w)$

2 There are two shortest paths of length 4:

$$(v_1, v_5) \to (v_5, v_4) \to (v_4, v_3) \to (v_3, v_7)$$
$$(v_1, v_5) \to (v_5, v_6) \to (v_6, v_3) \to (v_3, v_7)$$

4.2 Undirected Graphs

1

$$\{a_1, a_2, a_6, a_4, a_5\} \quad \{a_1, a_4, a_6, a_2, a_5\}$$
$$\{a_5, a_4, a_6, a_2, a_1\} \quad \{a_5, a_2, a_6, a_4, a_1\}$$

are the circuits of G_1 and also are the Euler lines from v_3 to itself. There are eight Euler lines from v_1 to itself.

4 (Fig. 4.1.8) 2 (Fig. 4.1.10) 4 (v_2 to v_3)

4.3 The Vector Space of Sets

1 (a) $\begin{bmatrix}1\\0\\0\\1\end{bmatrix} = \begin{bmatrix}1\\1\\0\\0\end{bmatrix} + \begin{bmatrix}0\\1\\1\\0\end{bmatrix} + \begin{bmatrix}0\\0\\1\\1\end{bmatrix}$ (b) $\begin{bmatrix}1\\0\\1\\0\end{bmatrix} = \begin{bmatrix}1\\1\\0\\0\end{bmatrix} + \begin{bmatrix}0\\1\\1\\0\end{bmatrix}$

3 The minimal vectors of $\mathbb{E}$ are those vectors with two 1s.

$$\begin{bmatrix}1\\1\\0\\0\\\vdots\\0\\0\end{bmatrix}, \begin{bmatrix}0\\1\\1\\0\\\vdots\\0\\0\end{bmatrix}, \ldots, \begin{bmatrix}0\\0\\0\\0\\\vdots\\1\\1\end{bmatrix}$$

is a basis for $\mathbb{E}$. Thus dim $\mathbb{E} = n - 1$.

4.4 Matrices and Undirected Graphs

1

$$C = \begin{bmatrix} 0 & 0 & 0 & 1 & 1 & 0 & 0 \\ 0 & 0 & 1 & 1 & 0 & 1 & 0 \\ 0 & 1 & 1 & 0 & 0 & 0 & 1 \\ 1 & 1 & 0 & 0 & 0 & 0 & 0 \\ 1 & 0 & 0 & 0 & 1 & 1 & 1 \\ 0 & 1 & 0 & 0 & 1 & 1 & 1 \\ 0 & 0 & 1 & 0 & 1 & 1 & 0 \\ 1 & 0 & 0 & 1 & 0 & 1 & 1 \\ 1 & 0 & 1 & 0 & 0 & 0 & 1 \\ 0 & 1 & 0 & 1 & 0 & 1 & 1 \end{bmatrix}$$

2 The incidence and loop matrices of *G* are

$$B = \begin{bmatrix} 1 & 0 & 0 & 1 & 1 & 0 & 0 & 0 & 0 \\ 1 & 1 & 0 & 0 & 0 & 0 & 0 & 1 & 0 \\ 0 & 1 & 1 & 0 & 0 & 0 & 0 & 0 & 1 \\ 0 & 0 & 1 & 1 & 0 & 1 & 0 & 0 & 0 \\ 0 & 0 & 0 & 0 & 1 & 1 & 1 & 0 & 0 \\ 0 & 0 & 0 & 0 & 0 & 0 & 1 & 1 & 1 \end{bmatrix}$$

$$L = \begin{bmatrix} 1 & 0 & 0 & 0 & 1 & 1 & 0 & 0 & 1 & 1 & 0 & 1 & 1 & 1 \\ 0 & 1 & 0 & 0 & 1 & 0 & 1 & 1 & 1 & 0 & 0 & 0 & 1 & 1 \\ 0 & 0 & 1 & 0 & 1 & 1 & 1 & 1 & 0 & 0 & 1 & 1 & 0 & 1 \\ 0 & 0 & 0 & 1 & 1 & 0 & 0 & 1 & 0 & 1 & 1 & 1 & 1 & 0 \\ 1 & 0 & 0 & 1 & 0 & 1 & 0 & 1 & 1 & 0 & 1 & 0 & 0 & 1 \\ 0 & 0 & 1 & 1 & 0 & 1 & 1 & 0 & 0 & 1 & 0 & 0 & 1 & 1 \\ 1 & 0 & 1 & 0 & 0 & 0 & 1 & 1 & 1 & 1 & 0 & 1 & 0 \\ 1 & 1 & 0 & 0 & 0 & 1 & 1 & 0 & 1 & 0 & 1 & 0 & 0 \\ 0 & 1 & 1 & 0 & 0 & 1 & 0 & 0 & 1 & 0 & 1 & 1 & 1 & 0 \end{bmatrix}$$

6

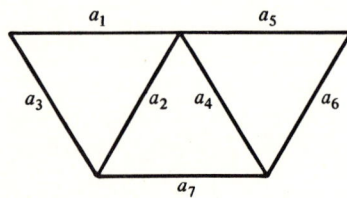

4.5 Directed Graphs

1
$$B = \begin{bmatrix} -1 & 0 & 0 & 0 & -1 & -1 & -1 \\ +1 & -1 & 0 & 0 & 0 & 0 & 0 \\ 0 & +1 & -1 & 0 & 0 & 0 & +1 \\ 0 & 0 & +1 & -1 & 0 & +1 & 0 \\ 0 & 0 & 0 & +1 & +1 & 0 & 0 \end{bmatrix}$$

$$L = \begin{bmatrix} -1 & 0 & 0 & -1 & 0 & -1 \\ -1 & 0 & 0 & -1 & 0 & -1 \\ 0 & -1 & 0 & -1 & +1 & -1 \\ 0 & 0 & -1 & 0 & +1 & -1 \\ 0 & 0 & +1 & 0 & -1 & +1 \\ 0 & +1 & -1 & +1 & 0 & 0 \\ +1 & -1 & 0 & 0 & +1 & 0 \end{bmatrix}$$

$$C = \begin{bmatrix} 0 & 0 & 0 & +1 & +1 & 0 & 0 \\ 0 & 0 & +1 & -1 & 0 & +1 & 0 \\ 0 & +1 & -1 & 0 & 0 & 0 & +1 \\ +1 & -1 & 0 & 0 & 0 & 0 & 0 \\ +1 & 0 & 0 & 0 & +1 & +1 & +1 \\ 0 & +1 & 0 & 0 & +1 & +1 & +1 \\ 0 & 0 & +1 & 0 & +1 & +1 & 0 \\ +1 & 0 & 0 & -1 & 0 & +1 & +1 \\ +1 & 0 & -1 & 0 & 0 & 0 & +1 \\ 0 & +1 & 0 & -1 & 0 & +1 & +1 \end{bmatrix}$$

The signs in any row of C and in any column of L may be changed.

2 $a_1 = \frac{1}{7}(3L_1 - L_2 + C_1 + C_2 + C_3)$ $a_3 = \frac{1}{7}(L_1 + 2L_2 + 5C_1 - 2C_2 - 2C_3)$
$a_6 = C_4$
where

$C_1 = a_1 - a_2 + a_3$ $L_1 = a_1 + a_2$
$C_2 = a_1 - a_2 + a_4$ $L_2 = -a_1 + a_3 + a_4 + a_5$
$C_3 = a_1 - a_2 + a_5$
$C_4 = a_6$

4.6 Matrices and Trees

$$M = \begin{bmatrix} +2 & -2 & 0 & 0 & 0 \\ 0 & +2 & -1 & -1 & 0 \\ 0 & 0 & +1 & -1 & 0 \\ -1 & 0 & 0 & +2 & -1 \\ -1 & 0 & 0 & 0 & +1 \end{bmatrix}$$

There are 4 directed trees rooted at each vertex, 4 circuits, and 33 trees in the underlying undirected graph.

3 If the vertices of G are renumbered according to

$$v_1 \to v_1 \qquad v_2 \to v_4 \qquad v_3 \to v_2 \qquad v_4 \to v_5 \qquad v_5 \to v_3$$

then the resulting graph is identical to the graph of Exercise 2.

Subject Index

Absorbing state, 264
Adjoint of a matrix, 87
Arc(s) of a graph, 333, 346, 388
    adjacent, 346
    parallel, 346
Arc-adjacency matrix, 345
Arc-disjoint paths, 341
Associate polynomials, 116

Basis of a vector space, 24
    canonical, 147
    cyclic, 137
Bluffing, 179
    symmetrization of, 185

Canonical basis, 147
Cauchy-Binet theorem, 94
Central optimal strategy, 224
Characteristic equation, 110
Characteristic polynomial, 110
Characteristic root, 99
    algebraic multiplicity of, 110
    geometric multiplicity of, 110

Characteristic vector, 99
Chord, 371
Closed set of states, 273, 292
Cofactor, 85, 87
Column:
    cone of a matrix, 203
    dimension of a matrix, 6
    expansion of a determinant, 87
    null space, 17
    nullity, 40
    rank, 40
    space of a matrix, 30
    sum of a matrix, 284
    vector, 8
Companion matrix, 157
Complement, 204, 356, 357
Complete graph, 354
Component of a graph, 350, 389
Cone, 203
Connected graph, 347, 389
Convex combination, 189
Convex set, 189
    extreme point of, 196
Cramer's rule, 93
Cut, 373, 397

444 SUBJECT INDEX

Cut:
   directed, 397
   fundamental set of, 376
   matrix, 375, 399
   set, 378, 398
   vertex, 387
Cyclic basis, 137
Cyclic space, 137

Degree of a vertex, 340, 346
Degree matrix, 343
Derivative, 96
Determinant of a matrix, 75, 83
   column expansion of, 87
   Laplace expansion of, 94
   row expansion of, 85
Diagonal X, 316
Dimension:
   of a matrix, 6
   of a vector, 8
   of a vector space, 40
Direct sum(s):
   fundamental theorem of, 107
   of matrices, 89
   of vector spaces, 42
Directed cut, 397
Directed Euler line, 416
Directed graph, 333, 388
Directed loop, 392
Directed path, 392
Distributive law for matrices, 14
Division algorithm, 114
Divisor, 116
Dominance, 228
   strict, 228
Double subscript notation, 6
Dual graphs, 378
Dual linear-programming problems, 245
Duelling, 184, 187

Elementary divisors, 151
Elementary matrices, 54–59
Elementary operations, 54–59
Entry of a matrix, 6, 7
Ergodic set, 292
Ergodic state, 293
Essential strategy, 219
Essential subgame, 221
Essential submatrix, 221

Euclidean algorithm, 117
Euler line, 339, 351
   directed, 416
Euler's formula, 383
Expectation, 172, 197
Extreme optimal strategy, 229
Extreme optimal vector, 252
Extreme point of a set, 196

Farkas' theorem, 205
Feasible vector, 245
Field, 1
Finite Markov chain, 254, 283
First passage, 314
   matrix, 316
   mean, 324
   probability, 313, 316
Formal sums, 393
Fundamental set of loops, 372

Game, 168, 197
   saddle point of, 200
   solution of, 169, 197
   symmetric, 185, 213
   value of, 177, 197
Generator of a vector space, 137
Graph, 333, 339
   arc of, 333, 346
   complete, 354
   component of, 350
   connected, 347
   diameter of, 354
   directed, 333, 388
   dual, 378
   planar, 380
      region of, 381
   underlying, 389
   undirected, 339, 346
   vertices of, 333, 346
Greatest common divisor, 116

Hamilton-Cayley theorem, 162
Hypercompanion matrix, 163

Identity element of a field, 2
Identity matrix, 15
Identity permutation, 67

Incidence matrix, 346, 360
   cut, 375
   loop, 372
   vertex-arc, 343, 388
Inconsistent linear-programming problem, 246
In-degree of a vertex, 343
Inessential strategy, 219
Initial population vector, 258
Initial probability vector, 255, 283
Invariant subspace, 99
Inverse:
   of a matrix, 50
   of a permutation, 72
Irreducible Markov chain, 292
Irreducible polynomial, 123
Irreducible vector space, 140

Jordan canonical form, 164

Königsberg bridge problem, 339
Kuratowski graphs, 384

Laplace expansion of a determinant, 94
Least common multiple, 125
Length of a loop, 392
Length of a path, 392
Linear combination, 19
Linear dependence, 23
   minimal, 362
Linear independence, 23, 358
Linear-programming problem, 245
   dual of, 245
   feasible vector of, 245
   inconsistent, 246
   unbounded, 246
Linear transformation, 96
   image of, 112
   kernel of, 113
   matrix of, 98
Loop, 347, 364
   directed, 392
   fundamental set of, 372
   matrix, 372, 396
   set, 369, 396

Markov chain, 254, 283
   irreducible, 292

Markov chain:
   periodic, 304
   reducible, 292
   regular, 289
   stationary vector of, 258, 284
Markov process, 254
Matching pennies, 169, 171, 173
Matrix(ices), 6
   addition of, 6
   adjoint of, 87
   arc-adjacency, 345
   classical canonical form of, 163
   column rank of, 40
   column space of, 30
   column sum of, 284
   companion, 157
   cut, 375
   degree, 343
   determinant of, 75, 83
   diagonal, 15
   elementary, 54–59
   entries of, 6, 7
   equal, 7
   equivalent, 64
   first-passage, 316
   hypercompanion, 163
   identity, 15
   in-degree, 343
   irreducible, 292
   Jordan canonical form of, 164
   linear transformation of, 98
   loop, 372, 396
   lower triangular, 92
   mean-first-passage, 342
   mean-time-before-absorption, 327
   minimal polynomial of, 130
   multiplication: by a matrix, 9
      by a scalar, 9
   nonnegative, 202
   nonsingular, 50
   norm of, 274
   order of, 8
   out-degree, 343
   partitioned, 7
   permutation, 72
   polynomial, 114
   positive, 202
   rank of, 61
   rational canonical form of, 157
   recurrence, 316
   reducible, 292

Matrix(ices):
  row rank of, 40
  row space of, 30
  row sum of, 284
  scalar, 15
  semipositive, 202
  similar, 51, 105
  singular, 50
  skew-symmetric, 92
  square, 8
  symmetric, 92
  transition, 255, 283
    periodic, 303
    regular, 289
  transpose of, 65
  tree, 410
  vertex-adjacency, 333
  vertex-arc incidence, 343, 388
  zero, 8
Mean-first-passage matrix, 324
Mean-time-before-absorption, 327
Minimal polynomial:
  of a matrix, 130
  of a vector, 138
Minimal vector, 357, 395
  distinct, 395
Minor, 87
Multiple, 116

n-linear function, 83
Natural correspondence, 357, 394
Nonnegative row complement, 204
Nonsingular matrix, 50
Norm of a matrix, 274
Null space of a matrix, 17
  column, 17
  row, 28
Nullity of a matrix, 40
  row, 363

Optimal strategy, 169, 193, 197
  central, 224
  extreme, 229
Optimal vector, 245
  extreme, 252
Order of a matrix, 8
Out-degree of a vertex, 343

Path(s), 334, 347, 364
  arc-disjoint, 341

Path(s):
  directed, 393
  simple, 347
Periodic Markov chain, 304
Periodic transition matrix, 303
Permutation, 67
  even, 71
  identity, 67
  matrix, 72
  odd, 71
  signature of, 71
Polynomial(s):
  associate, 116
  companion matrix of, 157
  greatest common divisor of, 116
  hypercompanion matrix of, 163
  irreducible, 123
  least common multiple of, 125
  minimal, 130, 138
  monic, 116
  relatively prime, 122
Primary decomposition theorem, 132
Probability:
  first-passage, 313, 316
  recurrence, 313, 316
    certain, 314
    uncertain, 314
  vector, 256, 283
$p(x)$-elementary divisors, 151

Rank of a matrix, 61
  column, 40
  row, 40
Rational canonical form, 157
Recurrence matrix, 316
Recurrence probability, 313, 316
Reducible Markov chain, 292
Region of a graph, 381
  finite, 382
  infinite, 382
Regular Markov chain, 289
Regular transition matrix, 289
Ring, $13n$.
Root:
  of a directed tree, 409
  of unity, $302n$.
    primitive, 311
Row:
  characteristic equation, 101
  characteristic polynomial, 101

Row:
   characteristic root, 101
   characteristic vector, 101
   cone of a matrix, 203
   dimension of a matrix, 6
   expansion of a determinant, 85
   null space, 28, 363
   nullity of a matrix, 363
   rank, 40
   space of a matrix, 30
   sum of a matrix, 284
   vector, 8

Saddle point of a game, 200
Scalar, 6
   matrix, 15
   multiplication, 9, 356
Secondary decomposition theorem, 140
Semipositive matrix, 202
Signature of a permutation, 71
Similar matrices, 51, 105
Singular matrix, 50
Slack vector, 208
Solution:
   of a game, 169, 197
   of a linear-programming problem, 245
Span, 19
Spanning set, 19
State(s) of a Markov chain, 254
   absorbing, 264
   closed set of, 273, 292
   ergodic, 293
   transient, 293
Stationary vector, 258, 284
Steinitz replacement theorem, 35
Strategy, 169
   mixed, 171
   optimal, 169, 193, 197
      central, 224
      extreme, 229
   pure, 171, 197
      essential, 219
      inessential, 219
   sensible, 180
   space, 190
Submatrix, 7
   essential, 221
Subspace of a vector space, 16
   invariant, 99

Symmetric difference, 355
Symmetric game, 185, 213
Symmetrization of a game, 185, 215

Transient state of a Markov chain, 293
Transition matrix, 255, 283
   periodic, 303
   regular, 289
   standard form of, 297
Transition probability, 255
Transpose of a matrix, 64
Transposition, 68
Tree, 355, 370, 405
   chord of, 371
   directed, 409
   matrix, 410

Unbounded linear-programming problem, 246
Undirected graph, 339
Unique factorization theorem, 123
Unit matrices, 39
Unit vectors, 20

Value:
   of a game, 177, 197
   of a linear-programming problem, 245
Vector(s), 8
   column, 8
   disjoint, 357
   feasible, 245
   initial probability, 255, 283
   linearly independent, 23, 258
   minimal, 357, 395
      distinct, 395
   minimal polynomial of, 138
   optimal, 245
      extreme, 254
   population, 258
   probability, 256, 283
      initial, 255, 283
      k-step, 255
   row, 8
   slack, 208
   stationary, 258, 284
   unimodular, 396
   zero, 8

Vector space(s), 16
  dimension of, 40
  direct sum of, 41
  of formal sums, 394
  generator of, 137
  intersection of, 20
  irreducible, 140
  of sets, 357
  subspace of, 16
  sum of, 21
  unimodular, 396
  zero, 17

Vertex(ices), 333, 346, 388
  adjacent, 346
  cut, 387
  degree of, 333, 346
  isolated, 361n.
Vertex-adjacency matrix, 333
Vertex-arc incidence matrix, 343

Zero, 2
  matrix, 8
  space, 17
  vector, 8

Index of Symbols

$\mathscr{A}$, 96
$\|A\|$, 274
A^*, 221
$A^{\cdot}$, 230
A^{-1}, 50
$A_1 + A_2$, 39
$(a\ b)$, 68
$A > B$, 203
$A \geqq B$, 203
$A \geqq B$, 203
$A^{(i)}$ 14
$A_{(i)}$, 14
a_{ij}, 7
$a_{ij}{}^{(s)}$, 314
$(A, P^{(0)})$, 283
A_{pq}, 85
A^T, 65
(A, Z_1, Z_2), 245
adj A, 87

$B\mathbb{V}$, 28

$\mathbb{C}(A)$, 30
C_{rs}, 59

$\mathscr{C}_{rs}$, 59
$C_r(\lambda)$, 59
$\mathscr{C}_r(\lambda)$, 59
$C_{rs}(\alpha)$, 59
$\mathscr{C}_{rs}(\alpha)$, 59
$\mathbb{C}_n{}^m$, 273
C_σ, 73
$\mathscr{C}_\sigma$, 73

D, 83
$\mathscr{D}$, 96
det A, 75
diam G, 354
dim $\mathbb{V}$, 40
dg X, 316

E, 201
$E^{(m)}$, 215
$E_{(n)}$, 209
$E(P, Q)$, 172

$\mathbb{F}$, 1
$f(A)$, 113

450 INDEX OF SYMBOLS

$\mathbb{F}^m$ 17
$\mathbb{F}_n$, 17
$\mathbb{F}_n^m$, 17
$F^{(s)}$, 315
f_{pp}, 313, 316
f_{pq}, 265, 313, 316
$f_{pq}^{(s)}$, 314
$[f_1(x), f_2(x), \ldots, f_k(x)]$, 116
$[f_1(x), f_2(x), \ldots, f_k(x)]$, 125

$|G|$, 389
G^*, 221

I, 15, 67
$I^{(i)}$, 20
$I^{(i,j)}$, 39
I_n, 15
im $\mathscr{A}$, 112

$\mathbb{K}$, 203
$\mathbb{K}(A)$, 203
$\mathbb{K}_R(A)$, 203
ker $\mathscr{A}$, 113

$\lim_{k \to \infty} A^k$, 261, 275

m^*, 221
$m^{\cdot}$, 230
$m \times n$, 6, 346, 360

$\mathbb{N}$, 67
n^*, 221
$n^{\cdot}$, 230
$\mathbb{N}(A)$, 17
$\mathbb{N}_R(A)$, 28
n_{pq}, 322

$\mathbb{O}(\mathbb{S})$, 196, 204

P^*, 221
$P^{(k)}$, 284
$\mathbb{P}_m$, 190

$\mathbb{P}^n$, 256
(P, Q, ν), 197

Q^*, 221
$\mathbb{Q}^n$, 190

$\mathbb{R}$, 189n.
$r(A)$, 61
$\mathbb{R}(A)$, 30
$r_C(A)$, 40
$\mathbb{R}m$, 189
$r_R(A)$, 40
R_{rs}, 55
$\mathscr{R}_{rs}$, 55
$R_r(\lambda)$, 56
$\mathscr{R}_r(\lambda)$, 56
$R_{rs}(\alpha)$, 58
$\mathscr{R}_{rs}(\alpha)$, 58
R_σ, 71
$\mathscr{R}_\sigma$, 71

$\mathbb{S}_n$, 67
sgn σ, 71

$\mathbb{U}_1 + \mathbb{U}_2$, 41

$\mathbb{V}$, 16
$\mathbb{V}C$, 28
$(\mathbb{V}, \mathbb{A})$, 346, 388
$\mathbb{V}_1 + \mathbb{V}_2$, 20
$\mathbb{V}_1 \cap \mathbb{V}_2$, 20

$\mathbb{Z}$, 4
$\mathbb{Z}_p$, 4

$\gamma(A)$, 284
δ_{pq}, 322
$\nu(A)$, 40
$\nu_R(A)$, 363
$\rho(A)$, 284
σ^{-1}, 72

O, 2, 8, 17
ϕ, 355

64010 QA 263 P34

PEARL, MARTIN
MATRIX THEORY AND FINITE
MATHEMATICS.

64010
PEARL, MARTIN
MATRIX THEORY AND FINITE
MATHEMATICS.

QA 263 P34

NOV 21

1362

FERNALD LIBRARY
COLBY-SAWYER COLLEGE
NEW LONDON, N.H. 03257